Seed Physiology

Editors
K. Vanangamudi
Malika Vanangamudi
M. Djanaguiraman
K. Natarajan
N. Natarajan
T. Saravanan
K. Raja

2020
Associated Publishing Company
A Division of
Astral International (P) Ltd
New Delhi 110 002

First Published, 2006
Reprinted, 2020

ISBN: 978-81-85211-65-7 (Hardbound)

Published by : **Associated Publishing Company**
A Division of
Astral International Pvt. Ltd.
– ISO 9001:2015 Certified Company –
4736/23, Ansari Road, Darya Ganj
New Delhi-110 002
Ph. 011-43549197, 23278134
E-mail: info@astralint.com
Website: www.astralint.com

Printed at : **Replika Press Pvt. Ltd.**

PRINT IN INDIA

Contents

S. No		Page.No
	The Contributors	v
	Foreword	vii
1.	Seed and man – An introduction *T. Saravanan, K. Natarajan and K. Vanangamudi*	1
2.	Seed development and maturation *R. Jerlin, K. Raja, S. Kavitha, M. Kathiravan and P. Murugesan*	27
3.	Seed – The structural function and biochemical variations. *P.Srimathi, K.Natarajan and K.Vanangamudi*	69
4.	Physiology of seed germination *Mallika Vanangamudi, J. Annie Sheeba, M. Djanaguiraman and P. Saravanane*	102
5.	Seed germination under various abiotic stresses *J. Annie Sheeba, M. Djanaguiraman, Mallika Vanangamudi, K. Vanangamudi and G. Kayalvizhi*	143
6.	Physiology and biochemistry of seed germination and vigour due to seed management technologies *S. Kavitha, M. Bhaskaran and K. Vanangamudi*	176
7.	Physiological seed conditioning – An overview *M. Djanaguiraman, J. Annie Sheeba, K.R.V. Sathyasheela and Mallika Vanangamudi*	200
8.	Seed dormancy *S. Kavitha, K. Natarajan and R. Renuka*	223
9.	Ecological aspects of germination *K. Sivasubramaniam and E. Sujatha*	258
10.	Physiological mechanism of seed deterioration *M. Djanaguiraman, J. Annie Sheeba, Mallika Vanangamudi and G. Thilagavathi*	280
11.	Seed Vigour *K. Natarajan, T. Saravanan and A.P. Sivamurugan*	308
12.	Tree seed physiology *M. Djanaguiraman, Mallika Vanangamudi and R. Arunkumar*	355
13.	Seed source on tree seed quality *K. Raja, A. Bharathi, S. Marimuthu, Mallika Vanangamudi and S. Kavitha*	381
14.	Seed testing for tree crops *K. Raja, A. Bharathi, K. Vanangamudi, Mallika Vanangamudi and S. Kavitha*	396

The Contributors

ANNIE SHEEBA, J. Department of Crop Physiology, Tamil Nadu Agricultural University, Coimbatore - 641 003, Tamil Nadu, India.

ARUNKUMAR, R. Department of Horticulture, Adhiparasakthi Agricultural College, G.B.Nagar, Kalavai – 632 506, Tamil Nadu, India

BHARATHI, A. Department of Seed Science and Technology, Tamil Nadu Agricultural University, Coimbatore - 641 003, Tamil Nadu, India.

DJANAGUIRAMAN, M. Agricultural Research Station, Bhavanisagar – 638 451, Tamil Nadu, India

JERLIN, R. Agricultural Research Station, Bhavanisagar – 638 451, Tamil Nadu, India.

KATHIRAVAN, M. Agricultural Research Station, Bhavanisagar – 638 451, Tamil Nadu, India.

KAVITHA, S. Central Institute for Cotton Research, Coimbatore – 641 003, Tamil Nadu, India.

KAYALVIZHI, G. Department of Plant Breeding and Genetics, Adhiparasakthi Agricultural College, G.B.Nagar, Kalavai – 632 506, Tamil Nadu, India

MALLIKA VANANGAMUDI, Department of Crop Physiology, Tamil Nadu Agricultural University, Coimbatore - 641 003, Tamil Nadu, India.

MARIMUTHU, S. Department of Soil and Crops, Adhiparasakthi Agricultural College, G.B.Nagar, Kalavai – 632 506, Tamil Nadu, India

MURUGESAN, P. National Research Centre for Oil palm, Pedavegi – 534 450, Andhra Pradesh, India.

NATARAJAN, K. Department of Plant Breeding and Genetics, Adhiparasakthi Agricultural College, G.B.Nagar, Kalavai – 632 506, Tamil Nadu, India.

RAJA, K. Central Institute for Cotton Research, Coimbatore – 641 003, Tamil Nadu, India.

RENUKA, R. Department of Plant Protection, Adhiparasakthi Agricultural College, G.B.Nagar, Kalavai – 632 506, Tamil Nadu, India.

SARAVANAN, T. Department of Plant Breeding & Genetics, Adhiparasakthi Agricultural College, G.B.Nagar, Kalavai - 632 506, Tamil Nadu, India.

SARAVANANE, P. Department of Soil and Crops, Adhiparasakthi Agricultural College, G.B.Nagar, Kalavai – 632 506, Tamil Nadu, India.

SATHYASHEELA, K.R.V. Department of Plant Breeding and Genetics, Adhiparasakthi Agricultural College, G.B.Nagar, Kalavai – 632 506, Tamil Nadu, India.

SIVAMURUGAN, A.P. Department of Soil and Crops, Adhiparasakthi Agricultural College, G.B.Nagar, Kalavai – 632 506, Tamil Nadu, India.

SIVASUBRAMANIAM, K. Department of Seed Science and Technology, Agricultural College and Research Institute,Tamil Nadu Agricultural University, Madurai - 625104, Tamil Nadu, India.

SRIMATHI, P. Department of Seed Science and Technology, Forest College and Research Institute, Mettupalayam – 641 301, Tamil Nadu, India.

SUJATHA, E. Department of Soil and Crops, Adhiparasakthi Agricultural College, G.B.Nagar, Kalavai – 632 506, Tamil Nadu, India.

THILAGAVATHI, G. Department of Plant Protection, Adhiparasakthi Agricultural College, G.B.Nagar, Kalavai – 632 506, Tamil Nadu, India.

VANANGAMUDI, K. Department of Plant Breeding and Genetics, Adhiparasakthi Agricultural College, G.B. Nagar, Kalavai-632 506, Tamil Nadu, India.

TAMILNADU AGRICULTURAL UNIVERSITY
COIMBATORE - 641 003, INDIA

Dr. S. RAMANATHAN, Ph.D.,
Director of Research

Phone : 5511254, 247, 447
FAX : 0422-5511454
Grams : Farmvar
Email : dr_aris2000@yahoo.com

FOREWORD

The first evergreen revolution bloomed as a result of development of high yielding varieties and timely availability of good quality seeds. Modern agriculture is unsustainable, as it cannot continue to produce enough food for the ever-growing population. To achieve this meticulous task, exploitation of good quality seed is a must. A good quality seed should be genetically pure, physiologically sound with good health and vigour. Among the above qualities, physiological sound seed will have better vigour, which could increase the yield of the plant. Hence, understanding the basic physiological process and metabolism in seed during seed germination, growth and development is indispensable for scientists in plant sciences. Improvement of the quality of seed can be brought by physiological manipulation of vigour or breaking dormancy. In this issue, this book may be highly useful. In this book, the authors have dealt with the modern concept of seed physiology, mechanism of dormancy, germination, deterioration, vigour, seed coatings, development and maturation and stress tolerance. It is noteworthy that the authors have also dealt with tree seed physiology.

In recent years, significant advancements have been achieved in research technology, such as applications of biotechnology and analyses of the microstructure of genes at a DNA level. Nevertheless, it is my belief that the research achievements reviewed in this book are extremely important with regard to the development of seed. Much of the information presented in the book is technical and scientific in nature, while some chapters synthesise what little is known on a particular subject.

The information contained in the book will be useful for the student, researchers and teachers of various universities. I congratulate the authors Dr. K. Vanangamudi, Dr. Mallika Vanangamudi, Dr. M. Djanaguiraman, Dr. K. Natarajan, Dr. N. Natarajan, T. Saravanan and Dr. K. Raja for their efforts in publishing this book.

S. RAMANATHAN
Director of Research

1

Seed and Man—An Introduction

T. Saravanan, K. Natarajan and K. Vanangamudi
Department of Plant Breeding and Genetics
Adhiparasakthi Agricultural College
G. B. Nagar, Kalavai - 632 506

1. INTRODUCTION

Initially, man collected leaves, roots, tubers, fruits and seed, to be consumed with hunted meat and fish. The question why man started deliberately to plant seeds for cultivation will probably always remain open to speculation. Agriculture was probably first practiced in the Middle East approximately 10,000 years ago. The general assumption is that it spread from there to Europe, Africa and Asia. The first 'cultivated plants' may have grown from spontaneously germinating seeds and sprouting roots or tubers in waste heaps around homesteads. At a certain point in history, grains or tubers were saved, with the purpose of sowing them at a determined place near the homestead. Saving the best seeds, roots or tubers from consumption, storing and planting them became associated with 'cultivation' (Almekinders and Louwaars, 1999).

From the beginning of crop husbandry to our time, seeds have been the key ingredient in the establishment, expansion, diversification and improvement of crop production. Seed is the most vital input in crop production and quality seed is indispensable for a bumper crop.

This chapter analyses various important facts about the seeds and its relationship with man, obviously with farmers, plant breeders and seedsmen.

2. FIFTEEN BASIC FACTS ABOUT SEEDS

The word basic is defined as constituting or serving as the basis or starting point. Most persons, who have a concept of nature, think of a seed as the starting point, the beginning of new life. Those of us more closely associated with crop culture recognize that the true role of a seed is to continue or propagate life rather than start it.

Regardless of whether you are just beginning to develop a serious interest in seeds or have been involved in one or more aspects of the seed industry for many years, there are some basic facts and truths about seeds which must be continually recalled to be successful "husbands" of nature's most vital resource, the seed. In the daily routine, many of us tend to forget or occasionally ignore the basic facts about seeds, but our forgetfulness or ignorance does not change or negate their importance.

In this chapter, a brief description has been made about 15 seeds facts which, like the law of gravity, are always in effect. Each person should learn these seed facts; place them in the subconscious memory to be recalled before, not after, we experience the consequence.

2.1. Seed is the only crop production input that increases in quantity when used

The first fact about seeds has its primary impact of seed marketing activities. When a farmer uses a fertilizer, insecticide or herbicide it is gone. When he needs more the next season, he must purchase it from his supplier. However, when a farmer plants a quintal of soybean seed, in a few months he can harvest 30 or 40 times the about he "used". Every salesman has competition, but only seedsmen intentionally sell their strongest competitors – the farmer – the materials necessary for them to become an even stronger competitor, and sell it at the lowest reasonable price! The fertilizer or pesticide salesman knows that when the farmer uses his products there will be an opportunity for another sale next season. Seedsmen have little such assurance except with hybrid varieties.

Table 1: Germination percentages of high quality seed lots of twelve species during storage under ambient conditions

Kind	Storage Period (Months)		
	0	12	24
Snap bean	98	96	92
Red clover	94	88	60
Field corn	98	96	90
Tall fesue,	95	85	37
Onion	96	42	0
Peanut, Shelled*	96	60	0
Radish	98	98	95
Sorghum	96	93	82
Soybean*	96	85	42
Wheat	98	97	92

*Hand-shelled

2.2. Seeds of different species vary in their natural life spans

A major factor involved in repeat sales, essential for a profitable business operation, leads to a second basic fact about seeds. The data presented in Table 1 clearly demonstrate this basic fact. You will note that at the start of the study from which this data was obtained the germination level of the seed of each species was well above 90%. After 12 months, under identical storage conditions, germination of the onion and peanut seed had decreased dramatically, and all seeds were dead after 24 months storage. In contrast, look at the germination of the snapbean, radish and wheat seed.

As indicated in Fig. 1, the natural life span of seed of the same species also varies among different lines or varieties. How long will a seed remain alive? Many factors are involved; however, let's consider basic seed fact number three.

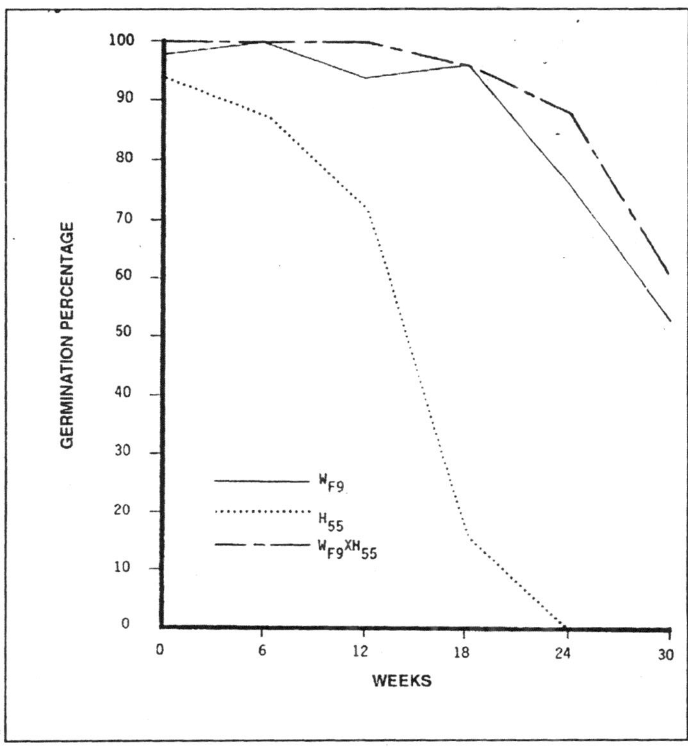

Fig. 1: Differences in longevity of seed of two inbred lines of corn and the single cross hybrid under conditions of 86° F and 75 % relative humidity. The seed were produced at the same time and location

2.3. Moisture content and temperature are the two most important factors which influence the rate of deterioration in seeds

The role and importance of moisture content in the life of seeds of our economically important field, horticultural and flower seeds are illustrated in Fig. 2. Note that after seed reach physiological maturity, the higher the moisture content the more rapid the rate of the deteriorative processes, whether from respiration, disease or insects. When seed moisture content is above that necessary for safe bulk storage, about 15%, biological activity in the seed mass can produce sufficient heat to injure the living seed within a few hours, unless they are well-aerated.

Seed high in moisture content are more susceptible to heat damage than the same seed at lower moisture content. This is especially important to remember when drying seed with artificially heated air. In the normal temperature range, the biological activity of seeds, insects and most molds increase as temperature increases.

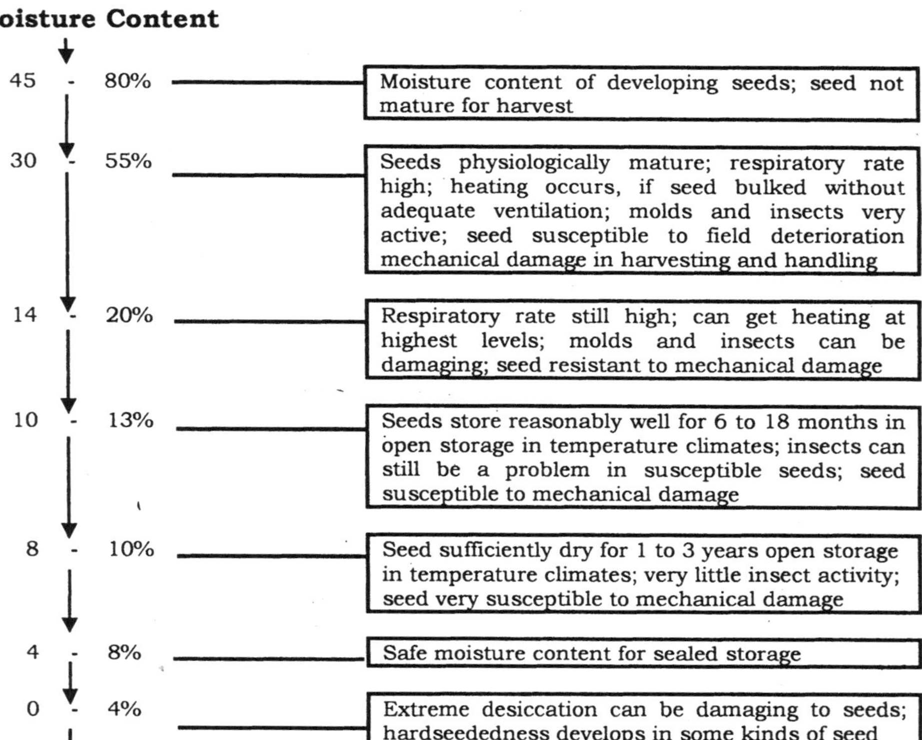

Fig. 2: Role and importance of moisture content in the life of seeds

The temperature and moisture effects may compensate or magnify each other in some ways. For example, the higher the seed moisture content, the more adverse the effects of increased temperature, as the seed dry, however, their thermal death point increases. Seeds very high in moisture can be damaged by sub-freezing temperatures, but seed with 10% moisture are unaffected by the lowest temperatures that occur in nature.

Not only the moisture content of the seed, but also the moisture content of the air which surrounds the seed is important.

2.4. Seeds are hygroscopic

This means that at a given temperature a seed loses moisture to the atmosphere or absorbs moisture from it until the vapour pressures of seed moisture and atmospheric moisture (relative humidity) reach equilibrium. When seed moisture and the relative humidity of the air are in equilibrium, there is no net moisture movement. However, any change in (a) the seed moisture, (b) the relative humidity of the air surrounding the seed, or (c) the temperature, upsets this equilibrium and a net movement of water occurs. This process is virtually continuous under ambient conditions, but is not instantaneous, often requiring several days and sometimes several months before equilibrium is established.

The action of hygroscopicity and seed moisture equilibration is paralleled by the action of a playground see-saw. The seed, which hold water, is on one end of the see-saw and the air, which also holds water, is on the other end. As indicated in Fig. 3; 3-A, when seed moisture and relative humidity of the air around the seed are in equilibrium, there is no net exchange of moisture. If the moisture content of the air is increased (Fig. 3-B) there will be a net movement of water from the air to the seed until once again equilibrium is reached (Fig. 3-C). If the amount of water in the air is reduced (Fig.3-D); the seed will lose water to the air until an equilibrium is again attained (Fig. 3-E). If we increase the temperature of the air, the relative humidity goes down (Fig. 3-E) and the seed will lose water to the air until again equilibrium is reached (Fig. 3-F).

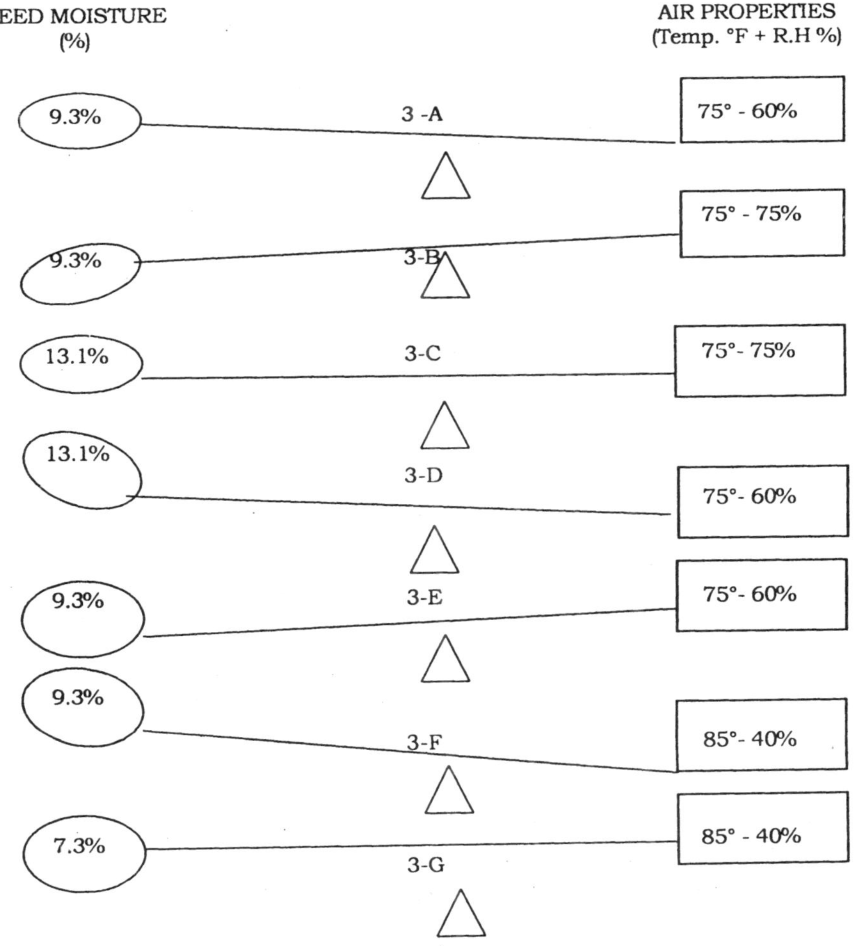

Fig. 3: Seed moisture migration as influenced by moisture content of the seed in relation to the temperature and relative humidity of the air

2.5. Seeds dry from the surface to the interior

The fifth basic fact about seeds relates to drying and wetting of seed during the hygroscopic equilibrium process. This fact can be demonstrated by looking at a schematic cross-section of a corn seed which permits us to visualize the water molecules in both the seed the surrounding air (Fig. 4). When the relative humidity of the air around the seed is first reduced, the seed still has its original moisture content (Fig. 4-A; 4-B), but the amount of water at the seed's surface has been reduced. Now, recall basis seed fact IV; seeds are hygroscopic. Because the relative humidity of the air at the seed's surface has been reduced, the vapour pressure of the water in the interior of the seed is relatively higher. To re-establish the moisture equilibrium, the water molecules just under the seed's surface move to the surface to replace the water on the surface that has evaporated (Fig. 4-C). Do not develop the impression that the water in the seed simply rushes to the surface. Because of the laws of physics, a moisture gradient is established from the surface to the center of the seed. If the lower relative humidity is maintained long enough, the seed will once again come into equilibrium (Fig. 4-D) at lower moisture content.

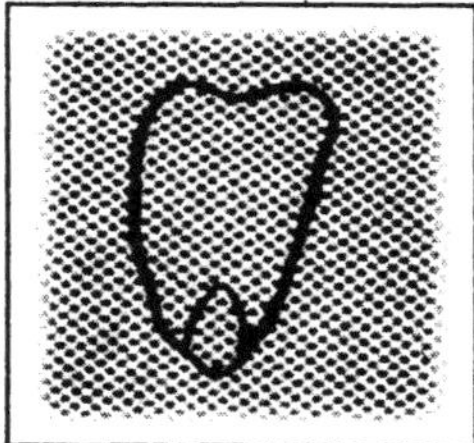

Fig.4A: Seed moisture 15%
Relative humidity 75%
Equilibrium

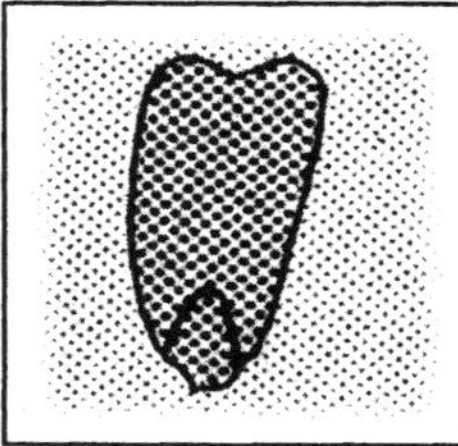

Fig. 4B: Seed moisture 15%
Relative humidity 75%

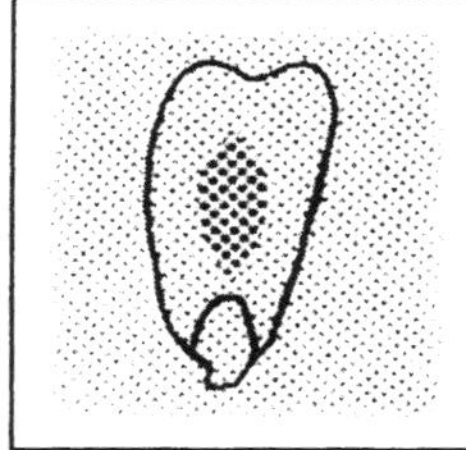

Fig. 4C: Seed moisture 14%
Relative humidity 55%

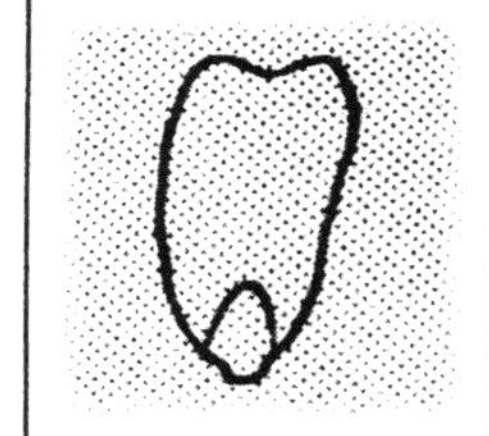

Fig.4D: Seed moisture 13%
Relative humidity Equilibrium

Fig.4: How a seed dries

2.6. The lower the moisture content of a seed, the greater the amount of energy required to further reduce its moisture content

The sixth seed fact relates to one of the most serious economic problems that we face today is energy. In application, this seed fact means that more energy are required to reduce the moisture content of seed from 15% to 13% than from 17% to 15%. This fact is true regardless of the energy source: gas, diesel, propane, wood, coal, etc.

The moisture in seeds is present in two forms, free and bound. For our purposes, we can think of free water molecules in the interspaces of the seed. This water reacts to heat energy in much the same manner as water on the surface of a lake. On the other hand, the bound water in a seed is part of the physical structure of the molecular and colloidal material in seed. A seed high in moisture has a greater percentage of free than bound water. As a seed dries, it gives up both free and bound water; however, because of the stronger attachment of the bound water, a greater percentage of free water decreases, greater amounts of energy are required to break the physical bonds of the increasing percentage of the bound water which must be removed to reduce moisture content.

2.7. Seed deterioration is not reversible

All living things degenerate over time and die. The apparent, sequence of seed deterioration is given in Figure 5.

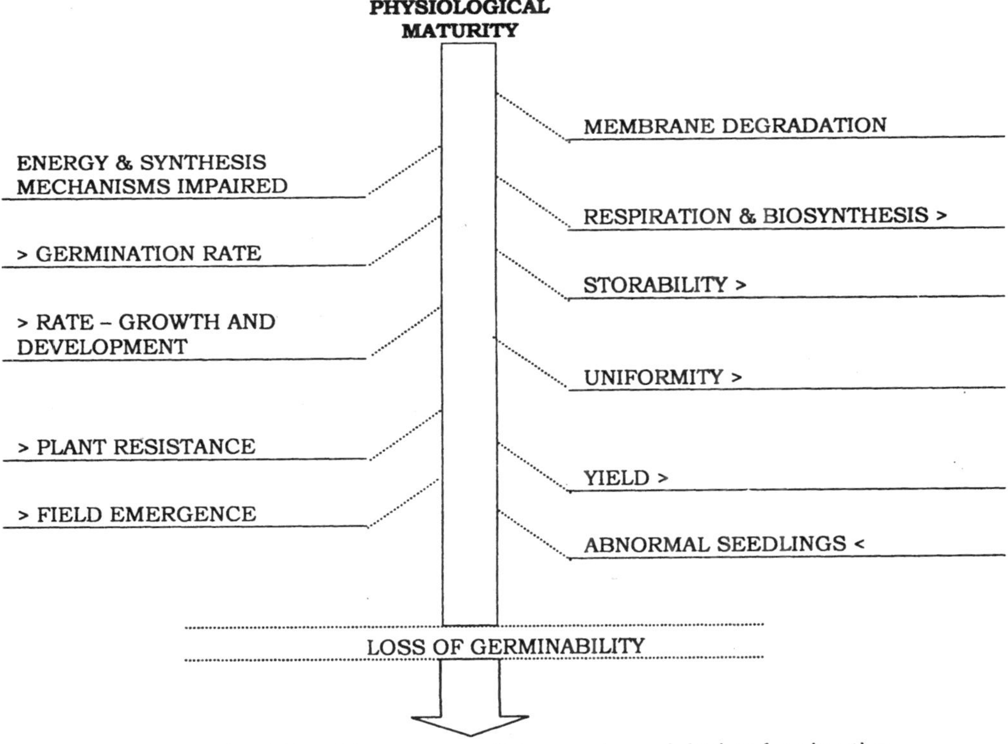

Fig. 5: Possible sequence of changes in seed during deterioration

Air dry seed, contrary to most other living organisms, have neither the capability to replenish their energy supply nor to redistribute the stored foods which are located inside the seed. As a consequence, when any of the life sustaining processes in a seed are impaired or lost, they can not be repaired. The application of seed treatment chemicals or other treatments will not reverse the process of deterioration, although their proper application will permit the seed in a lot to express their maximum potential.

2.8. Mechanical separation of seeds from their contaminants is possible only when there is a mechanically detectable difference in at least one of their physical characteristics

The physical characteristics used to separate good seeds from contaminating material are listed in Table 2. Some of the machines used to clean and up-grade seed utilize differences in two or more of these physical characteristics to make the desired separation.

Most of the problem encountered in making the desired mechanical separations arise from the fact that, although there is a visual and often a physically measurable difference between the good seed and the undesired material, these differences are not sufficiently large to be detected (measured) by the machine's separating action. This is true even when the operator is very proficient in equipment selection and adjustment and knowledgeable concerning differences in physical characteristics of each component of the seed lot. Every processor knows that the good seeds in each lot vary in their physical characteristics. However, we seem prone to forget that the contaminating materials like-wise differ in their physical characteristics.

Table 2: Physical properties of seeds and their contaminants and the machines which utilizes differences in these properties to make separations

	Physical Property	Machine
1.	Gross Size	Air-screen cleaner (screen section)
2.	Length	Indented cylinder, disc separator
3.	Width	Width & thickness separator
4.	Thickness	Width & thickness separator
5.	Weight	Gravity table, aspirator, stone, air-screen cleaner (fan section)
6.	Surface Texture	Roll mill, Magnetic separator
7.	Shape	Spiral separator, Roll mill
8.	Colour	Colour sorter

2.9. Microorganisms are present on all seeds

Not all microorganisms will attack a live seed; however, we must remember that even the air surrounding the seed contains microorganisms some of which, under the proper conditions, are capable of destroying the seed's value for planting. Most microorganisms found on seeds are not of economic importance, but they are present. On the other hand, most of the microorganisms which are carried inside a seed can be very destructive; loose smuts, pod and stem blight, etc. The application of fungicides is not always the solution for this.

It seems probable that in the not too distant future the results of a "seed health" test will be a labelling requirement, just as the germination test is today. But, this should not occur until the seed pathologists accurately determine the association between seed infection levels and crop performance.

2.10. Seeds must be exposed to the proper temperature, moisture and oxygen conditions to initiate and maintain germination

Earlier, it has been discussed that moisture and temperature could be seed's worst enemies. That is true, until we want the seed to germinate and grow. For any seed to germinate there must be an adequate supply of moisture and a favourable temperature. The oxygen requirement, while small in terms of volume, must be met because certain physiological processes of a germinating seed cannot continue in the absence of oxygen. Some seeds even require light to trigger germination, especially when they are fresh.

2.11. The embryo of every true seed is formed from the union of a sperm cell from the male parent and an egg cell from female parent

With the advent of the sexual revolution and more recently the birth of the so-called test tube baby, everyone is aware of this basic fact of life. Of importance to seedsmen is that this same union, between the male and female gametes, is the fundamental basis of the seed industry. Whether producing genetically pure seed, hybrid seed or eliminating unwanted outcrosses, seedsmen are directly affected by the results of this union since it determines the "genetics" of the seed produced. These gametes contain the "life" and life pattern that are passed from generation to generation.

In plants, unlike in animal reproduction, a second union, between a second male sperm cell and two other cells (polar nuclei) in the embryo is also required for the seed to be formed. It is the endosperm formed by the second union that provides food for the embryo's growth.

2.12. A seed must be alive to be of value for planting

In reality, this fact is the total basis for the entire seed industry. If a seed is not alive, it is not really a seed, but a storage unit for carbohydrates, proteins and fats – a grain. The technology is available to keep seeds alive for a hundred years. The technology is also available to kill any seed in a fraction of a second.

2.13. Seeds reach their maximum potential at physiological maturity

As depicted in Figure 6, which portrays the maturation and deterioration cycle of a seed as a reproductive unit, physiological maturity of seeds occurs when they attain their maximum dry weight, maximum level of viability and their highest physiological capacity. While it is true that some seeds enter a stage of dormancy shortly after attaining physiological maturity, the dormancy period is not a period of improvement, rather one of preservation of the status attained at physiological maturity.

Since the moisture content of seeds at physiological maturity ranges from 30 to 55% among the species the seed must further dry down before they can be mechanically harvested. It is during this post maturation pre-harvest period, referred to as the field weathering period that the deteriorative phase of seed life sets in. Keep in mind seeds produced on any one plant and certainly in any field, do not all reach physiological maturity at the same time, nor are they exposed to exactly the same environments or levels of insect and disease infestation. Thus, there is some variation in the physiological quality among the seeds of every lot.

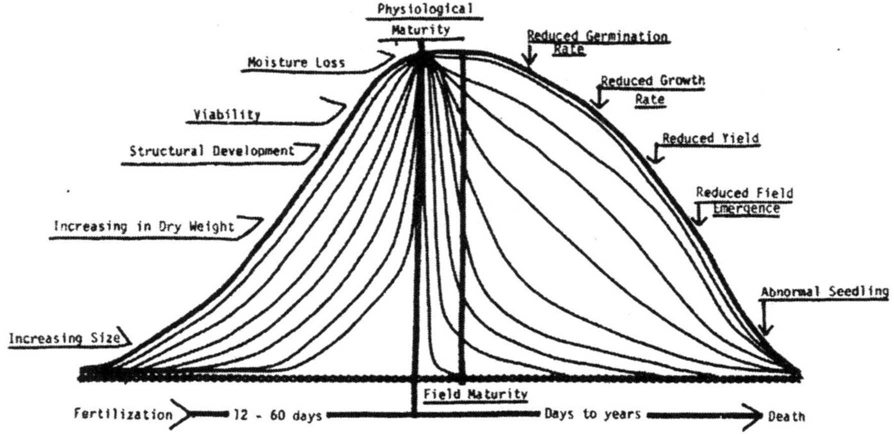

Fig.6: Life cycle of a seed

2.14. The seeds that are alive in each lot can differ in physiological quality

The term "Vigour", as applied to seeds, is very controversial in some segments of the seed industry. There is no question that within every seed lot and among seed lots, there is variation in the physiological capacity of the seeds to reach certain levels of growth in a given time period and set of environmental conditions. However to date, there is a wide variety of opinions concerning how to accurately measure these differences, and especially the "degree" of difference required to affect plant performance. Regardless of your own opinion concerning "vigour" do not ignore the basic fact that differences do exist in the physiological quality of seeds within the among seed lots.

2.15. Seeds are basic to man's survival

3. SEEDS AND CROP PROPAGATION

Seeds are, first of all, a most efficient and effective means of crop propagation. They are the principal means by which plant populations have been and are distributed over both time and space through man and other factors. Compared to the vegetative propagules used for some crops, seeds are smaller, hence very convenient to store and transport, hardier and longer-lived, easier to sow, relatively free of diseases, and a much smaller portion of the production is required for propagation (Delouche and Potts, 1983).

4. SEEDS, FARMERS AND CROP IMPROVEMENT

Sexual processes in plants continuously combine the heritable variability in the inter-breeding population in different ways in the progeny produced. Seeds are the product of the sexual process in higher plants and the repository of the heritable traits of the progeny. Some of the plants propagated from seed exhibit traits or combinations of traits that make them 'different' from other members of the population. For thousands of years man has slowly improved the crops he cultivates by saving seed for propagation from plants exhibiting traits he perceived to be most useful. In this century, the process of crop improvement has been tremendously accelerated through systematic collection and preservation of the variability in crop species and exploitation of the science of genetics. The improvements genetically engineered in small populations of plants by modern plant breeders are most efficiently maintained, transmitted and multiplied by seeds. Seeds, therefore, are the principal means for extending improvements effected in crops, i.e., improved cultivars, form the breeding nursery to farmer's fields throughout the area of adaptation (Delouche and Potts, 1983).

Farmers can modify the natural selection process through the way they select the seeds or cultivate the crop. Early agriculturists must have selected and saved seeds, tubers or roots that they liked most: the ones which were largest tasted best and cooked fastest, seeds that did not easily shatter, from plants without thorns, which yielded highest, and showed the least insect and disease injury, etc. As a result, characteristics that determine the success of plants in surviving in a natural environment, like early shattering of seeds, disappeared from the cultivated plant populations. Weeding eliminated the slow and late emerging crop seeds resulting in more rapid and even seed germination. Seeds with long dormancy would not germinate and did not contribute to the harvest in the next season (Almekinders and Louwaars, 1999). Simply speaking, seeds play a 'key role' in crop improvement programmes and without which millions and millions of money invested in breeding programmes would not have become fruitful!

5. SEEDS AND GREEN REVOLUTION

The world's worst recorded food disaster occurred in 1943 in British-ruled India, known as the Bengal famine, an estimated four million people died of hunger that year in eastern India (which included today's Bangladesh) (Fig. 7).

Fig. 7: Picture depicting a man in hunger due to Bengal famine

Initially, this catastrophe was attributed to an acute shortfall in food production in the area. However, Indian economist Amartya Sen (recipient of the Nobel Prize for Economics, 1998) has established that while food shortage was a contributor to the problem, a more potent factor was the result of hysteria related to World War II, which made food supply a low priority for the British rulers.

When the British left India in 1947, India continued to be haunted by memories of the Bengal Famine. It was therefore natural that food security was one of the main items on free India's agenda. This awareness led, on one hand, to the Green Revolution in India and, on the other, legislative measures to ensure that businessmen would never again be able to hoard food for reasons of profit. The Green Revolution, spreading over the period from1967/68 to 1977/78, changed India's status from a food-deficient country to one of the world's leading agricultural nations. Until 1967 the government largely concentrated on expanding the farming areas. But the population was growing at a much faster rate than food production. This called for an immediate and drastic action to increase yield. The action came in the form of the Green Revolution. The term 'Green Revolution' is a general one that is applied to successful agricultural experiments in many developing countries. India is one of the countries where it was most successful.

There were three basic elements in the method of the Green Revolution

i) Continuing expansion of farming areas

ii) Double-cropping in the existing farmland

iii) Using seeds with improved genetics.

Using seeds with superior genetics was the scientific aspect of the Green Revolution. The Indian Council for Agricultural Research (which was established by the British in 1929) was reorganized in 1965 and then again in 1973. It developed new strains of high yield variety seeds, mainly wheat and rice and also millet and corn.

Since the beginning of agriculture, people have been working to improving seed quality and variety. But the term 'Green Revolution' was coined in the 1960s after improved varieties of wheat dramatically increased yields in test plots in northwest Mexico. The reason why these 'modern varieties' produced more than traditional varieties were that they were more responsive to controlled irrigation and to petrochemical fertilizers. With a big boost from the international agricultural research centres created by the Rockefeller and Ford Foundations, the 'miracle' seeds quickly spread to Asia, and soon new strains of rice and corn were developed as well.

By the 1970s the new seeds, accompanied by chemical fertilizers, pesticides, and, for the most part, irrigation, had replaced the traditional farming practices of millions of farmers in developing countries. By the 1990s, almost 75% of the area under rice cultivation in Asia was growing these new varieties. The same was true for almost half of the wheat planted in Africa and more than half of that in Latin America and Asia, and more than 50% of the world's corn as well. Overall, a very large percentage of farmers in the developing world were using Green Revolution seeds, with the greatest use found in Asia, followed by Latin America.

Thanks to the new seeds, tens of millions of extra tonnes of grain a year are being harvested. The Green Revolution resulted in a record grain output of 131 million tonnes in 1978/79 (Fig. 8). This established India as one of the world's biggest agricultural producers.

Fig. 8: Picture depicting the prosperity of Indian farmers due to green revolution

Yield per unit of farmland improved by more than 30% between1947 (when India gained political independence) and 1979. The crop area under high yielding varieties of wheat and rice grew considerably during the Green Revolution.

On the other side, the green revolution has created some problems mainly to adverse impacts on the environment. The increasing use of agrochemical-based pest and weed control in some crops has affected the surrounding environment as well as human health. Increase in the area under irrigation has led to rise in the salinity of the land (Fig. 9). Although high yielding varieties had their plus points, it has led to significant genetic erosion.

Fig. 9: Picture depicting the impact of green revolution on salinity of land

6. FARMERS AND THEIR INDIGENOUS SEED PRACTICES

Indigenous seed practices encompass practically all aspects of crop production, since seed saving is an integral part of cropping activities in indigenous systems. Farmers engaged in the production and multiplication of quality seed deal with asexual propagation, land preparation and soil management, seed and seedling preparation and care, crop and pest management, flowering induction, the enhancement of seed quantity and quality, crop improvement, harvesting or collection, seed processing (drying, threshing, cleaning, and grading), storage, genetic conservation, and quality testing.

Local communities use various indicators to determine favourable times of planting and cropping, such as the different aspects of the sky, sun, moon, stars, bodies of water, and trees, as well as animal behaviour. They have many different ways of preparing the seed for planting; some follow the principle of sympathetic magic, such as filling the mouth with porridge while planting coconut. Seeds may be coated with or soaked in various substances to ensure protection against pests or to promote growth. Sometimes, seeds are dipped in blood (allegedly an offering to spirits, but possibly a form of nutrient enhancement), and fields are prayed over the night before planting, using lighted torches (which may actually kill many of the insect pests attracted to light).

There are various 'organic' methods in use for enriching or altering the soil, managing water, and keeping plants pest-free. Diverse or mixed cropping, specific cropping arrangements, and the use of plants that serve as repellants or attractants are common indigenous strategies for minimizing risks and enhancing stability. In the case of rice, some groups use special arrangements of varieties in a field to manage insects, rats and birds. Mechanical techniques, such as interestingly sculpted bird perches, and insect and rat traps, and biological control strategies are also employed to remedy pest problems.

Plant manipulation to alter or improve crop performance such as grafting, injuring, pruning stems or roots and 'training' plant parts can be traced back to indigenous systems. Plant characteristics may also be altered or preselected by means of a particular index. In papaya, for example, female trees can be obtained by selecting the part of the fruit where the seed is derived and the physical characteristics of the seed or seedling. Indigenous ways to promote flowering and fruiting are available (e.g., smudging, injuring, girdling, pruning, water manipulation), and indigenous harvesting and seed collection techniques have also been developed. As a rule, farmers select the most desirable produce for the next season's crop, which is in itself a form of breeding. In other cases, they practise hybridization, by deliberately growing crops side by side to obtain 'new types', or by hand pollinating female flowers with pollen from another plant. Farmers have also developed techniques of processing and drying (e.g., through the use of special dried leaves as an indicator of dryness). Rice seeds, for example may be sorted by flotation and simply winnowed. Some farmers carry out the harvesting, storage and planting of whole rice panicles. Techniques to ratoon, clone or produce asexually propagated materials are also available for some species that do not readily produce seed.

One of the most important concerns of farmers, seed producers, and seed traders is seed protection during storage. Various methods have been devised to protect the seed from rats, including specially designed storage houses. The kitchen hearth is a common storage place, which generally keeps seeds dry and pest-free, thanks to the smoke from a wood-burning stove. There are also containers designed to keep out moisture (*e.g.*, well-sealed gourds or bamboo tubes), or to keep seeds cool (*e.g.*, moist earthen jars, cave interiors). Desiccants such as charcoal, ash, and lime are also in common use. Botanicals (*e.g.*, lemon grass, glyricidia leaves, hot pepper, and tobacco) are often placed inside containers or mixed with the seed to keep it insect-free. Some of these botanicals are also used during planting. The use of resistant indigenous seeds or varieties clearly plays a major role in the strategy aimed at crop health maintenance, seed longevity and pest management.

Seed quality is another constant concern for farmers, contrary to the belief of many people in the formal sector. There are various ways to determine whether the seed is of good quality, such as germination testing using locally available material (*e.g.*, banana leaf sheath), volume expansion with exposure to dew, flotation, and a test for stickiness which some seed display in the presence of certain substances (ash).

Certain rituals or practices associated with cropping itself border on magic, religious belief, or superstition (*e.g.*, planting sweet potato while naked). However, they serve a definite purpose, by strengthening or institutionalizing the practice, in order to ensure implementation, continuity and sustainability. For those practices that are more directly related to production, but for which no scientific explanation can be offered or which do not work when validated, judgement should be suspended until all possible explanations have been ruled out. The current body of science, especially, when it is dealt with as discrete disciplines or by means of a reductionist approach is admittedly still limited to explaining certain phenomena which people claim they have observed. Some of these practices are backed up by solid scientific evidence, while others will need to be validated in more specific environments, taking into account the ecological, socio-cultural, economic and even political context. Interestingly, some of these indigenous practices, including rituals, are shared by farming communities who are in geographically isolated regions or who are not in contact with each other.

Surveys indicate that 'traditional farming systems' fosters high levels of genetic diversity. Farmers have many reasons to maintain and utilize genetic diversity: i) agronomic criteria (yield, yield stability, resistance to insect pests and diseases, storability etc.), ii) economic criteria (market and home economics) and iii) cultural criteria. In selecting seeds and varieties farmers combine and balance these criteria (Almekinders and Louwaars, 1999).

7. FARMERS AND SEED PROGRAM

The first seed program was established by the first cultivator. Thereafter, every serious farmer has taken steps to assure adequate seed supplies for the crops he cultivates, i.e., developed a "seed program". Steps taken by farmers to assure seed supplies range from the traditional saving of a portion of the production for propagation, through various barter arrangements with other farmers or village merchants, to seed village concept, to the booking of all seed needs with a seed company in advance of the planting season.

Several factors influence the specific seed supply arrangements made by farmers (Delouche and Potts, 1983) *viz.*, production objectives (subsistence or marketing), kinds of crops and area cultivated for each kind, level of production technology employed for each crop kind, availability of superior cultivators, availability of other inputs such as fertilizer and pesticides, water, commodity prices, climatic conditions and the availability of cash or credit. The importance of these factors varies among farmers and among the crop kinds cultivated by a single farmer. Subsistence farmers are usually interested in yield stability (risk aversion), the consumption qualities of the produce (taste preference), and secondary uses of the crop (straw, stalks) as in potentially high yield. They often perceive that the desired crop qualities are more certain in traditional varieties, and follow traditional seed saving practices. Large-scale farmers, on the other hand, often feel it is most cost effective to purchase seed every season from a reputable supplier than to take the time and make the effort to save the seed.

7.1. Farmers and Seed Saving

Though agriculture is the leading sector in most developing countries, for centuries, farmers, particularly the smallholders, have been using their own seed saved from the previous crop, or obtained, usually in exchange for grains or other commodities from neighbouring farmers. These seeds are of varieties described as 'land races', which have been going through a natural mutation process over time to produce new genetic versions of seed varieties to fit each agro-ecological condition. However, the land races are well adapted to the local agro-ecological situations, resistant to factors like disease, insect attack or even moisture stress. They are generally low yielding when compared to improved high yielding varieties of seeds bred in agricultural research stations.

Small-scale farmers usually prefer to use their own seed (Wierema *et al.*, 1993) (Table 3). It is the cheapest, most readily available, and of a variety that the farmer is familiar with. The farmer knows the seed quality, and the seed is available at planting time.

Table 3: Seed source in 1991 of maize and beans plots (% of total number of cases) in Costa Rica, Nicaragua and Honduras

Seed source	Maize			Common Bean		
	Costa Rica	Honduras	Nicaragua	Costa Rica	Honduras	Nicaragua
Farmers' own seed	79	75	81	58	79	72
Other local sources	19	13	12	21	15	14
Formal sources	2	13	6	21	6	13

(Wierema *et al.*, 1993)

7.2. Farmers and Seed Villages

In view of the low replacement rate in crops such as wheat and paddy (8 per cent on an average), the traditional seed supply system is important. Assuming a 5 to 10 per cent annual growth of improved seed sales, traditional supply will be still vital. A method to improve this system of supply must be developed by stimulating the concept of "seed village" involving the village panchayat and women.

In India, women account for at least one half of the agricultural labour. Moreover, a serious constraint faced by small and marginal farmers is non-availability of quality seeds of improved varieties at the proper time and place. Even if the seeds are produced in the village (which is far cheaper) these farmers will find it difficult to retain the produce because they need money. They, therefore, sell it soon after harvest. To alleviate the situation, bank credit may be arranged either to the individual farmers or to village panchayats by hypothecating the seed for sowing.

Alternatively, a cooperative society may be formed in the village for seed production, distribution and sale. The concept of seed village is not new. It started in the sixties, but was forgotten when the dwarf wheat from Mexico was introduced in India. The main task was to convince the farmers about its higher yield potential. Mexican dwarf wheat yielded far more than the Indian wheat at that time which generated a great demand for wheat seed.

In each revenue division of a district, seed production can be organised by the Primary Agricultural Cooperative Societies (PACs) in a cluster of three to four nearby villages to avoid isolation problems and also to have effective monitoring of seed production plots.

The seed should be harvested separately and threshed, cleaned, graded and tested for germination. Care should be taken to avoid physical admixture of cultivars. In each cluster a processing plant of at least 1 ton/hour capacity should be installed or a mobile processing plants for cleaning and grading of the seeds can be taken on rent either by the village panchayat or the State Department of Agriculture. If this is not possible, cleaning can be done manually. Either all or a few selected farmers of the cluster can be given genetically pure seed, which can be multiplied initially under the supervision of experts. One part of seed can be exchanged for 1.5 parts of the grain.

For a crop like paddy, a community nursery can be prepared for transplanting. Training farmers in seed production and testing should form an important element of the programme. Farmers can be taught to produce seeds of oilseeds, pulses, cereals and vegetables.

It is also necessary to prepare a seed map of the country indicating areas suitable for economic seed production of different crops. It is advisable to take up seed production in semi-arid areas under irrigation, preferably in the rabi season. The basic seed map prepared can be modified from time to time indicating new areas for seed production. The seed production areas must be, as far as possible, closer to the states requiring seeds and seed production must be economical.

By involving the village in seed production, the high-yielding varieties of crop plants can be spread faster. It is the quickest way to saturate an entire nation with quality seeds of improved varieties. In this kind of venture, investment is modest when compared to the establishment/expansion of the seed corporation. The concept of exchange of seed for grain can also be developed.

Indigenous knowledge systems, especially indigenous seeds, provide a logical starting point in the search for sustainable alternatives in agriculture. Seed is a basic input in production. As such, it has always played a primary role in development programmes, through the 'HYVs' (high yielding varieties) and associated technology and input packages such as monocropping, chemicals, irrigation, machines, capital, and credit or loans. To promote HYVs, the formal seed sector-both official and private-was operationalized to multiply and make available to the farmers the so-called 'improved' seed.

7.3. Farmers and the formal seed production

Formal seed production follows closely the modern crop production system, which means that it relies heavily on chemicals, external inputs, monocropping, and a limited number of varieties. This approach is known to have led to problems related to the environment (pollution, chemical residue, soil degradation, pest resistance, and genetic erosion); the economy (impoverishment and indebtedness); social structure (inequities); and culture (erosion of tradition, knowledge and tools). It is a highly centralized, 'top-down' and non-participatory approach. Farmers have become mere recipients of institutional seed, rather than participants in the process, from the development of varieties to the actual utilization. As a result, they are dependent on external solutions designed to solve their local problems, and have lost confidence in their ability to help themselves. They see modern, 'hi-tech', or institutionally promoted techniques as superior to their own. To revive the confidence of farmers in their ability to address their own needs, new approaches in both formal and non-formal education, research and extension are needed. Participatory approaches and interactive, dialogic and experiential techniques are an essential part of this process of confidence building and empowerment. Fortunately, there are many local communities throughout the world, which can provide the much-needed indigenous knowledge, practices, techniques, tools, and genetic resources. NGOs and development/rural workers have also put forward various non-conventional approaches, which are useful in bringing about the needed transformation.

In spite of the efforts and resources poured into the formal seed sector, the informal sector (farmers, grassroots NGOs, etc.) has remained the major source of seed (85% of requirements) in many Third World countries. Along with the recognition of the value of indigenous agricultural knowledge, there is tremendous potential for the informal seed sector in the development of a national seed industry. This could very well provide the foundation for sustainable agricultural systems. Unfortunately, the seed industry is often seen in terms of its commercial aspect – which is used as the measure of development and welfare – rather than in terms of equity or environmental protection.

8. FARMERS AND TAILOR MADE SEEDS

8.1. Terminator technology

Now, there is a technology available for the farmers suitably called the 'Terminator technology', which is designed to genetically switch off a plant's ability to germinate a second time.

What is the purpose of switching off seed germination? Many believe, it is purely a business idea by forcing farmers to buy a fresh supply of seeds each year; many of whom are in the developing world and cannot afford to do this.

The traditional practice, tried and tested for thousands of years, of saving seeds for the next harvest comes under threat due to a U.S patent on this technology to prevent "unauthorized seed-saving" by farmers.

Supporters of the terminator technology believe that the idea of it is for the protection of corporations from unscrupulous farmers. Control of seed germination helps prevent growers from pirating their technology. If crops remain fertile, there is a chance that farmers could use any saved transgenic seed from a previous season. This would result in poorer profits for companies. This then becomes a battle over a farmer's (traditional) right and a corporation's right. In India, this is a cause for some concern as scientists fear for the livelihood of 400 million farmers and for food security in the country. However, even without such terminator technology, under patent laws in Canada, U.S. and a number of other industrialized nations, it is illegal for farmers to re-use patented seed.

8.2. Designer seeds

For millennia, farmers have battled insects, microorganisms, and weeds that destroy or compete with their crops; threatening their families with starvation. Indeed, many major events in history have resulted from devastating plant disease epidemics or insect infestations. The Irish potato famine of the mid-1800s, which was caused by the fungus *Phytophthora infestans*, killed more than a million people and prompted a massive Irish emigration to the United States.

In hopes of preventing crop plant destruction by pests, ancient Romans made sacrifices to their various gods. Modern farmers use other techniques in their attempts to kill pests, including spraying pesticide and herbicides. They also make use of improved management practices and benefit from traditional breeding techniques to strengthen their crops. Some of the newer methods, however, have substantial costs and disadvantages. Excessive ploughing can cause soil erosion, for instance and pesticides and herbicides can pollute both soil and water as well as contribute to species extinction.

Thanks to recent advances in the genetic engineering, or bioengineering, of plants, farmers are now beginning to have at their disposal crop seeds that are genetically endowed not only to resist damage from insects but also to be resistant to herbicides. These bioengineered seeds have the potential to revolutionize agriculture and improve environmental quality by making it possible to reduce the use of pesticides and keep ploughing to a minimum.

Like most scientific innovations that have had significant effects on society, bioengineered seeds did not emerge solely from the efforts of researchers to improve pest or weed control. Rather, they were the by-product of earlier researchers curiosity about such basic science questions as: How do bacteria cause plant tumors? How do some viruses protect plants from other viruses? What enables some bacteria to kill insects?

8.2.1. A new breed of seeds

A farmer handles seeds of cotton while contemplating the freshly tilled fields that are ready to receive them. Although the seeds look just like the ones he planted last year, they are far from ordinary. These seeds will yield nearly 10 per cent more cotton and will need less than half as much insecticide as last year's standard seeds. This improvement stems from an additional gene that scientists have inserted in the seeds' genetic material, or genome. The gene, which originated in bacteria and is not found naturally in plants, encodes a toxic protein that kills two of the prime predators of cotton plants, bollworms and budworms. These caterpillars destroy millions of dollars worth of cotton each year, and are the main reason why more than half the insecticide used worldwide is sprayed on cotton plants.

Although the farmer might view the bioengineered seeds in his hands as merely the outcome of recent progress in plant breeding efforts by the company that produced them, they are actually the result of more than 50 years of research by many scientists. Incrementally, these scientists paved the way for isolating the genes that protect a particular organism from pests and for transferring these genes into a wide variety of plants. They could not have known that the answers to their fundamental questions about living organisms would lead to practical results, such as these bioengineered crop seeds.

8.2.2. Limits of traditional breeding

Farmers have noticed for centuries that some individual plants in a given species manage to survive disease or epidemics of insects relatively unscathed, while their neighbours succumb to infection or insect predation. In 1905, Sir Roland Biffen of Cambridge, England, wondered whether healthy plants inherited pest resistance, just as they might inherit the tendency to be tall or short. His experiments on two varieties of wheat showed that the ability to resist infection by a rust fungus was indeed inherited; a discovery that intensified attempts by farmers and plant breeders to produce varieties of pest resistant crop plants.

Similar attempts continue today and primarily involve screening a large number of plant varieties to identify those that are resistant to particular pests. Resistant varieties are then crossed with those desirable for other reasons, for example, because they produce more grain per acre. Careful selection and repetitive crossing of progeny can eventually generate varieties that are both high yielding and resistant to particular pests. But the process is extremely time consuming; it can take more than 15 years to bring a new variety to market.

One challenge encountered in traditional breeding is that generally only closely related species of plants can be cross-bred. If no varieties are naturally resistant to a particular fungus or insect, traditional breeders have no way to create resistance to that fungus or insect. Furthermore, breeders frequently face a situation in which a resistance gene is closely linked to a gene that adversely affects the quality of a crop, that is, where the two traits are always inherited together. In the early 1990s, despite plant breeders' best attempts to improve the pest and disease resistance of cotton, corn, rice, and other crops, farmers worldwide lost about one-fourth of their crops to pests and diseases.

8.2.3. The quest for desirable genes

A tool for introducing genes into plants is useful only if scientists have found genes that they want to transfer. Researchers were quick to see the value of the insect-killing bacteria. By 1938, the first commercial insecticide containing *Bt* hit the market in France, where it was mainly used to kill flour moths. During the next 50 years, other insecticide sprays were developed that contained *Bt* but the products had several limitations; rain washed away the insecticide, for example, and sunlight rapidly broke it down. Furthermore, many pests were not susceptible to the *Bt* spray, and some that were susceptible were able to avoid contact with it, because they fed mainly on roots, inside the stems, or on other parts of crop plants that are inaccessible to sprays. Given those limitations and the availability of more effective chemical insecticides, *Bt* insecticides were used only by niche markets in agriculture and forestry.

That situation changed in the 1980s, as many insects grew increasingly resistant to the commonly used insecticides, and as scientists and the public became aware that many of these chemicals are harmful to the environment. Bt insecticides affect only specific insect pests and do not persist in the soil or on leaves. As a result, they are generally considered environmentally benign. Thus, a number of commercial, government, and academic research laboratories embarked on research aimed at increasing the effective uses of *Bt* insecticides.

A critical piece of missing information was how *Bt* kills insects. When the first *Bt* insecticides were commercialized in the 1930s and 1940s, researchers knew only that they killed insects, not what the mechanism was. By the 1950s, a series of experiments by several research groups revealed that proteins produced by *Bt* bacteria were lethal to particular insect species. Over the next 20 years, several different strains of *Bt* bacteria were discovered, and each strain was found to produce specific proteins toxic to different groups of insect. By 1980, dozens of studies had made it clear that the different proteins produced by different strains of *Bt* bacteria determined which groups of insects would be killed.

Researchers then zeroed in on identifying the genes associated with the production of *Bt* proteins. Information about the genes was gathered by a pair of microbiologists looking into why the *Bt* genes triggered production of their toxic protein only when *Bt* bacteria started to produce spores. In 1981, Helen Whiteley and Ernest Schnepf, then at the University of Washington, discovered that the insecticidal proteins were found in a crystal-like body that was produced by the bacteria. They used recombinant DNA techniques to isolate a gene that encodes for an insecticidal protein. By 1989, more than 40 *Bt* genes, each responsible for a protein toxic to specific groups of insects, had been pinpointed and cloned by various researchers.

8.2.4. *A. tumefaciens* meets *Bt* genes

The stage was now set to develop plants that were resistant to insects. When both the *A. tumefaciens* vector for gene transfer and cloned *Bt* genes became widely available in the mid-1980s, a number of researchers realized that the two could be combined to modify crop plants so that they produce *Bt* proteins and thereby protect themselves from insect pests. Such plants would get around many of the limitations of *Bt* insecticides. Insects that had once hidden from *Bt* sprays would find the *Bt* toxins in whatever part of the plant they bit into, and sunlight or rain would not affect the persistence or potency of the toxins.

Genetic engineers also could transfer genes for several different proteins, so that plants would be protected from several different predators.

8.2.5. Viral resistance

Bt cotton, which reached the market in 1996, was one of the first bioengineered crop seeds to become available commercially. Since then, around a dozen other crop plants have been genetically modified to resist pests and diseases, many with genes derived from basic research findings reported decades earlier. Genetically engineered virus-resistant strains of squash and cantaloupe, for example, would never have made it to farmers' fields if plant breeders in the 1930s had not noticed that plants infected with a mild strain of a virus do not succumb to more destructive strains of the same virus. That finding led plant pathologist Roger Beachy, then at Washington University in Saint Louis, to wonder exactly how such "cross-protection" worked; did part of the virus prompt it?

Since 1992, researchers have pinpointed and cloned several of the genes that make selected plants resistant to certain bacterial and fungal infections; some of these genes have been successfully inserted into crop plants that lack them. Many more infection-resistant crops are expected in the near future, as scientists find more plant genes in nature that make plants resistant to pests. Plant genes, however, are just a portion of the arsenal; microorganisms other than *Bt* also are being mined for genes that could help plants fend off invaders that cause crop damage.

8.2.6. Herbicide resistance

If pests and disease can be deadly to crop plants, weeds are a constant scourge. Weeds compete for nutrients, water, and sunlight and can reduce potential yield by as much as 70 per cent. A standard way for farmers to deal with weeds is to spray their fields with several herbicides, each targeted to a specific type of weed. But many herbicides can damage crops as well as weeds and they can also cause air and water pollution. Farmers also use tilling to kill weeds before planting, or they spray fields with more environmentally benign broad-spectrum herbicides before the emergence of a new crop; but these practices can subject fields to erosion by wind and water.

The concept of 'genetically modified seeds' has recently offered farmers an alternative for weed control. The approach involves spraying fields with broad-spectrum herbicides after herbicide-resistant crops have sprouted and are holding down the soil.

In 1983, researchers at Calgene and Monsanto succeeded in isolating and cloning the genes that produce EPSP synthase. Scientists at Monsanto then modified the gene so that the enzyme it produced was no longer sensitive to glyphosate and used the *A. tumefaciens* vector to introduce the modified gene into crop plants. The new tomatoes and other crop plants produced an EPSP synthase that made them resistant to damage from glyphosate, as reported in 1985. In 1996, the first glyphosate-resistant soybean, cotton, canola, and corn seeds were made available to farmers. Plants have also been engineered to tolerate other types of herbicides, many of which are "friendly" to the environment. These herbicide-resistant crops can be used to support farming practices that protect soil and water quality. Once such crops have sprouted and are holding down the soil, farmers can spray their fields with glyphosate to kill weeds.

8.3. Seed pelleting

The idea behind seed pelleting is to take a seed that is comparatively light in density and oddly shaped and turn it into a seed with a pellet around it that makes that seed heavier, more uniform, and as perfectly round as possible. A planter will be able to plant that seed/pellet in the most accurate and precise way that is mechanically possible.

8.3.1. A Science – selecting the powder and adhesive

Creating a pellet is both a science and an art. Most of the science comes when selecting the powders and adhesive used to create the pellet around the seed. The materials should form a pellet structure that is porous enough to provide immediate oxygen to the germinating seed, and absorb water in a way that is most beneficial to the seed. It is necessary sometimes to have a mixture of four or more powders in order to achieve the desired characteristics and also end up with the desired pellet density.

The adhesive is also very important. It needs to be strong enough to hold the powder together, without making the pellet so hard that it slows the germination process. You can also go one step further by putting together the right combination of adhesives and powders to create a pellet that actually splits open much like an egg cracks open. This splitting action is said to lead to even faster oxygen availability and thus less stress on the germinating seed.

8.3.2. The Art – Making the pellet

There are a couple of ways to create a pellet around a seed, but today the most common method was taken from the candy industry.

In this method, seed is placed in a large rolling pan (Plate 1) like the one pictured in the photograph. The pan is continuously turning and tumbling the seed during this process. The pelleter sprays water until the seed surface is wet.

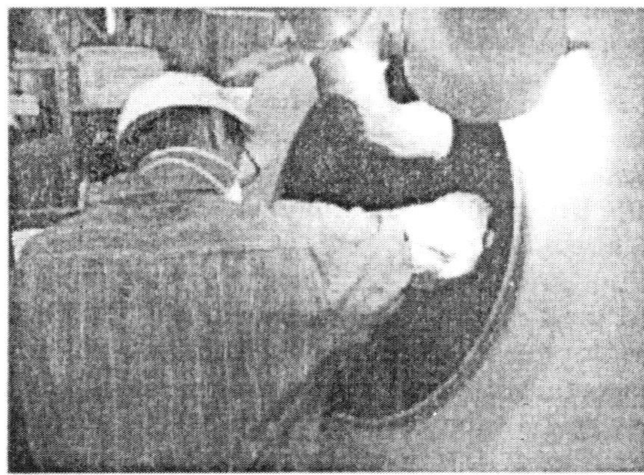

Plate 1: Seed pelletizer

After that, he sprinkles the pelleting powder over the wet, tumbling seed. The powder sticks to the wet seed. The pelleter then sprays adhesive until the seed and powder is again wet. During the process the pellets are removed from the rolling pan several times in order to run them over a sizing screen so that smaller pellets can be built up, while any larger pellets wait outside the pan. This is done until all pellets are exactly the same standard size. Knowing when and how much adhesive to add, and when and how much powder to add is an art that takes some time and practice to perfect. For example, if you get the seed too wet in the beginning the seed sticks to each other and you get two seeds per pellet. Not enough adhesive makes for a pellet that is too soft.

Several companies are starting to use machinery that is more mechanized, using less art, but the method described is still the most common pelleting method used in the industry today.

8.4. Seed coating

A technology once relegated to small scale farming in remote regions of the world has advanced in scope to the point where it is now ready for global agricultural use. Seed coatings technology has come to age in modern agriculture, not only for its ability to protect high value seeds, but also for its ability to hold pesticide and fungicide where it is needed most: at the point where a seed germinates in the ground, and during storage. Seed coating technology is a natural extension of the traditional pesticide broadcast application, where emulsifiable concentrate, wettable powder and suspension concentrate formulations are dispersed at the time of planting. Seed coatings perform the same function as traditional formulations, serving to protect the seed against insect or nematode attack. But insecticidal seed coatings have their advantages. First, there is a cost savings associated with the elimination of at least one pesticide application and second there is a minimal risk of toxic chemical exposure by targeting crop protection chemical exactly at the point of insect attack. Another major advantage of advanced seed coating technology is its potential to deliver the desired effect of active ingredients at a low dosage level. The science of combining a seed coating with an active, however, becomes complex when one considers all of the potential interactions.

8.4.1. The science of seed coating technology

In a paper titled Application of Latex Emulsion Polymers in Seed Coating Technology, Mr. Herbert explains that through its relationship with National Starch & Chemical, Uniqema has made available a series of off-the-shelf emulsion polymers specifically for use in seed coating formulations—having completed the testing and research required to prepare polymers that are applicable in various seed coating technologies. Mr. Herbert noted that "all seed coating product development begins with the establishment of performance criteria or performance targets, and with seed coatings, the performance criteria begins with abrasion resistance of the coating, flow ability of the coated seeds through the planter, and the assurance of no negative effects on seed germination. Crop protection chemical manufacturers may also include efficacy;

that is, the minimum concentration of active ingredient required to control the targeted pest."

The technical team at Uniqema outlines its "triangle tool" that was developed to find out the optimum ratio of components required in any successful seed coating formulation. The report further explains the experimental design approach; a three-component blending process that takes the form of a triangle. Blending along the left side of the triangle determines the polymer requirement, blending along the right side determines the diluent requirement, and the lines parallel to the bases determine the ratio of polymer to diluent required for a given total solids (polymer + diluent) addition.

8.4.2. A formula for success

The advancement of seed coating technology is nothing less than a winning formula for the seed industry, pesticide manufacturers, the consumer and the environment. "Seed coating technology is a step in the right direction for everyone," says Rich Herbert. "The technology gives farmers in developing countries access to higher quality," more effective pesticides that they could not previously afford with the absence of traditional application equipment.

"We tend to view food as a commodity here in the United States but in the other parts of the world it is a precious resource, where one drought or insect infestation can lead to famine. Seed coatings can help to alleviate this situation," said Mr. Herbert, "but more than that, seed coating technology is a lifesaver for the developing world." The benefits of advanced seed coating technology go beyond the third world and beyond the agricultural industry. "In fact, seed coating technology offers farmers the opportunity to reduce the amount of pesticide that enters the environment because it is concentrated in a small area rather than spread over a large one," concludes Mr. Herbert. To find out more about the advantages of seed coating technology in large scale and small-scale agriculture, contact the crop protection specialists of Uniqema in New Castle, Delaware, USA.

8.5. Artificial seeds

Artificial seeds were first introduced in the 1970s as a novel analogue to the plant seeds. The production of artificial seeds is especially useful for plants which do not produce viable seeds. It represents a method to propagate these plants. Artificial seeds are small sized and this provides further advantages in storage, handling, shipping and planting.

8.5.1. What makes an artificial seed?

Artificial seeds can be produced by encapsulating a plant propagule in a matrix which will allow it to grow into a plant. Plant propagules may consist of shoot buds or somatic embryos that have been grown aseptically in tissue culture. In culture, these plant propagules can easily grow into individual plants as we have the capacity to control its growth using chemicals provided in the culture media.

In the production of artificial seeds, an artificial endosperm can be created within the encapsulation matrix. The encapsulation matrix is a hydrogel made of natural extracts from seaweed (agar, carageenan or alginate), plants (arabic or tragacanth), seed gums (guar, locust bean gum or tamarind) or microrganisms (dextran, gellan or xanthan gum). These compounds will gel when mixed with or dropped into an appropriate electrolyte (copper sulphate, calcium chloride or ammonium chloride). Ionic bonds are formed to produce stable complexes. Useful adjuvants such as nutrients, plant growth regulators, pesticide and fungicide can be supplied to the plant propagule within the encapsulation matrix. In most cases, a second coat covering the artificial endosperm is required to simulate the seed coat.

8.5.2. Advantages of artificial seeds

Artificial seeds have the potential for providing an inexpensive plant delivery system. The process of planting can be easily mechanised and this allows the direct delivery of tissue cultured plant propagules to the field. It also provides rapid bulking up for the production of individual genetically engineered plants.

9. CONCLUSION

Quality seed means seed or any propagating material possessing genetic purity, physical purity, good germination and seedling vigour, optimum moisture content, freedom from diseases and pests. The ideal is to minimize pest and disease losses. Concomitantly, one could consider 'Seed Security' as part and parcel of 'Food Security' whereby agricultural development is expected to play a major role in order to alleviate the problem of food insecurity among the poor people. Yet, seeds are such an integral part of agriculture we tend to take them for granted, or overlook their crucial roles in agricultural production and crop improvement. In the future, keeping in mind the importance of seed in agriculture, horticulture and forestry, scientists should try their level best to provide quality seeds, at least to the farmers who are afforded to pay for that.

REFERENCES

Almekinders, C. J. M. and N. P. Louwaars. 1999. Farmers' Seed Production New approaches and practices. Intermediate Technology Publications, London, U. K. pp. 1-58.

Delouche, J. C. and H. C. Potts. 1983. The importance of seed in agriculture and the need for a seed program. **In:** Seminar on Improved Rice Seed Production in West Africa, Freetown, Sierra Leone, September 5-9, 1983, pp. 1-20.

Wierema, H., C. Almekinders, L. Keune and R. Vermeer. 1993. La produccion campesina en Centroamerica. Los sistemas Locales de semilla. IVO, Tilburg, The Netherlands.

2

Seed Development and Maturation

R. Jerlin[1], K. Raja[2], S. Kavitha[2], M. Kathiravan[1] and P. Murugesan[3]
[1]*Agricultural Research Station, Bhavanisagar - 638 457, Tamil Nadu*
[2] *Central Institute for Cotton Research , Coimbatore , Tamil Nadu 641 003*
[3]*National Research Centre for Oil palm, Pedavegi – 534 450,Andhra Pradesh*

1. INTRODUCTION

A seed is the simplest, since is a ripened or matured ovule. It is both the culmination of the activities of one plant generation and the start of a new generation. It contains a plant in miniature, the embryo, with the potential for growth and development into an adult plant. However, the embryo is not expressing that potential, since its development has been arrested by drying. How does this remarkable structure arise? To answer this, it is necessary to study the sequential development until it becomes a unit of dispersal. The important events involved in seed development and maturation include, sporogenesis, pollination, fertilization and development of the fertilized ovule by cell division, accumulation of reserve food material and loss of moisture content.

2. SPOROGENESIS

2.1. Megasporogenesis

The seeds of angiosperms originate from meristematic tissue of the ovary wall called ovule primordial. In species with simple ovaries, these primordial are usually located near the suture of the ovary wall where the carpel is fused. In species with more than one carpel, or with polycarpellate ovaries, the seeds form at the fusion of the carpels or along the septa, or central carpel axes, depending on the type of placentation (Fig 1). In some fruits (*e.g.*, tomato), a well-developed placenta arises from which many ovule primordial develop.

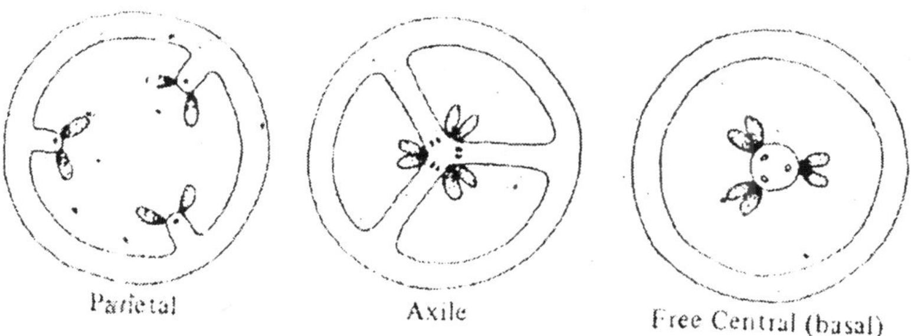

Fig.1: Type of ovule placentation

Within nucellus, or specialized tissue of the carpel, one cell, known as the *archesporial cell*, develops special characteristics that distinguish it from adjacent cells. As this cell increases in size, its nucleus becomes larger and its cytoplasm grows more dense in preparation for cell division. The first division results in a *megaspore mother cell* and a *parietal* cell. Usually the parietal cell remains undivided and soon deteriorates; however, in some species, it undergoes further division and contributes to seed formation. The megaspore mother cell is diploid (2n), having the same number of chromosomes as the parent plant. However, it soon undergoes a two-step cell division known as *meiosis* (Fig.2). This process gives rise to four megaspores, each having one-half the chromosome complement of the mother plant; these are thus haploid (1n) cells. Normally, only one megaspore is functional, while the other three degenerate.

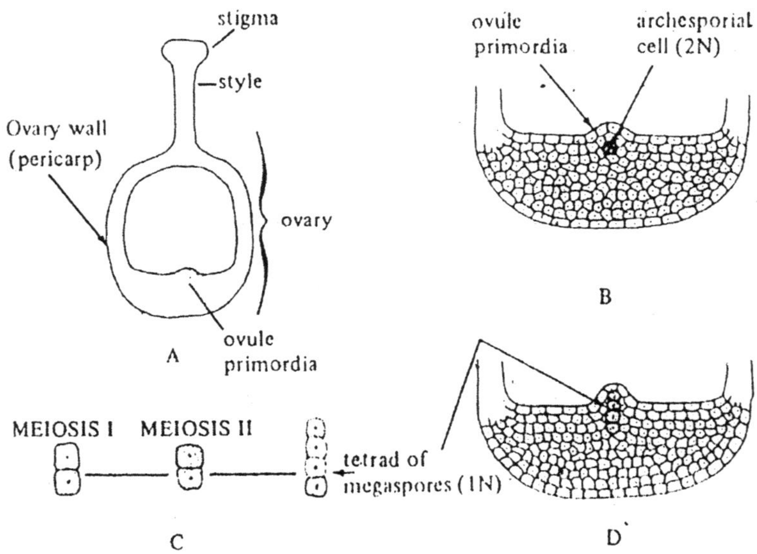

Fig.2: Megasporogenesis. (A) Location of the ovule development, (B) Cutway section of the lower region of the ovary wall, showing origin of the archesporial cell; note that it is larger than surrounding cells, having larger nucleus and dense cytoplasm, (C) Cell division during megasporogenesis, (D) Cutway section of lower part of the ovary, showing location of the four megaspores, three of which normally degenerate

2.1.1. Megagametogenesis

The development of the female gametophyte, or embryosac, from the functional megaspore is known as *megagametogenesis*, which is a process of successive nuclear divisions within an enlarging cell that becomes the *embryo sac*. Three successive free nuclear division (*mitosis*) occur (Fig.3), culminating in eight haploid (1n) nuclei. Soon these nuclei arrange themselves within the enlarging embryo sac and cell walls form, resulting in three antipodal cells at one end, two *polar nuclei* (without cell walls) near the center, and the *egg apparatus* (composted of the egg between two *synergid* cells) at the other end. After the two polar nuclei fuse to form a diploid (2n) nucleus, the resulting seven-cell structure is known as the mature female gametophyte (embryo sac), or megagametophyte, which is ready to receive the mature male gametophyte.

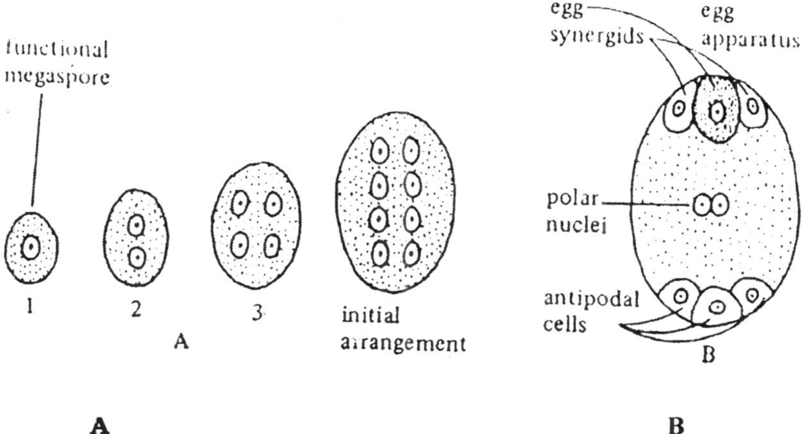

A **B**

Fig.3: Megagametogenesis. (A) Three normal mitotic nulear divisions leading to one large cell enclosing eight nuclei. Later cell walls enclose the nuclei and the entire structure becomes the female gametophyte, or embryosac, (B) mature female gametophyte

This describes the normal embryo sac development as it occurs in most species. Variations to this pattern occur in certain species, especially in the polar nuclei and antipodal development. With few exceptions, the egg apparatus development is as described. The egg cell comprises most of the egg apparatus. It is a complete cell containing a haploid (1n) nucleus with surrounding cytoplasm enclosed in a thin wall, or *follicle.* The egg cell is positioned near the small opening (*micropyle*) of the ovule formed by the surrounding integuments. A small vacuole may be present near the point of attachment away from the micropyle.

2.1.2. The developing ovule

Ovule development (Fig.4) occurs within the ovary, which provides a location for the nurture and development of the female gametophyte, its sexual fusion with the male gametophyte, and embryo development, survival, and eventual regrowth. Ovule growth begins as a small outgrowth within the *nucellus.* As megasporogenesis and megagametogenesis continue, the region of the nucellus that is to become the ovule enlarges and differentiates into definite morphological characteristics. Secondary outgrowths, or collars (*integuments*), soon appear around the periphery of the nucellar outgrowths and envelop it. These usually consist of the inner and outer integuments and ultimately become the *testa* (seed coat) of the mature ovule.

The developing ovule is commonly attached to the placenta by the *funiculus.* The scar on the ovule where the funiculus detaches at maturity is known as the *hilum.* The point where the integuments meet at the nucellar apex is the *micropyle,* and the region of integumentary origin and attachment, usually opposite the micropyle, is the *chalaza.* Between the chalaza and the hilum of many species is an area known as the *raphe*. The raphe may be visible on the seed coat of some species.

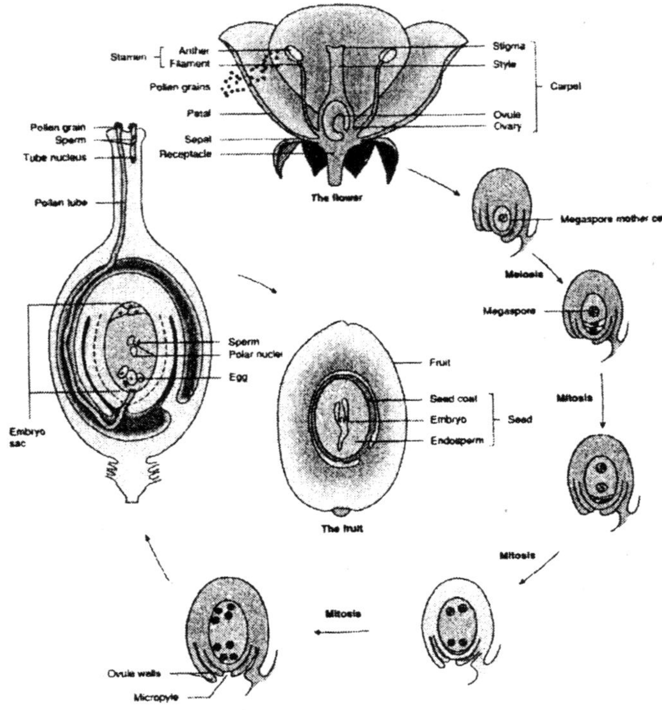

Fig.4: Ovule development

2.1.3. The Nucellus

The nucellus provides tissue for the origin and nurture of the female gametophyte, from the archesporial cell to the mature megagametophyte. It originates from ovary tissue and provides the site of archesporial cell origin. Subsequently, part of it becomes trapped within the integuments as the ovule continues to develop. Normally, no further growth occurs, and the nucellus is at least partially consumed, since it supplies nutritive support to the developing embryosac. However, in some species it undergoes considerable development and contributes substantially to the storage tissue as the perisperm. Examples of species with well-developed perisperm are sugar beet (*Beta vulgaris*) and leafy spurge (*Euphorbia esula*).

2.1.4. Integuments

The nature and thickness of the integuments vary considerably among species, depending on their role in contributing to embryo sac and ovule development. In Apiaceae, the inner integument is completely absorbed and only two or three cellular layers of the outer integument persist. In Asteraceae, most cells of both integuments are absorbed, leaving only a thin layer of crushed integumentary tissue on the inner side of the pericarp. Practically no integumentary tissue remains in the fully developed corn caryposis, and in *Symplocarpus*, both integuments and endosperm and completely consumed by the developing embryo, leaving it naked inside the pericarp.

Integumentary outgrowths

Two types of integumentary outgrowths may occur in certain species, giving rise to special structures not found in most seeds. A third integument, or *aril,* may either arise from the base of the nucellus or split off from the outer integument. *Elymus,* for example, has a well-developed aril. Another type of integumentary outgrowth, a *caruncle*, arises as a proliferation of the outer integument in the region of the micropyle. Seeds of *Euphoribia esula* have a well-developed but fragile caruncle that extends back over the seed and appears to have no function. Still another type of appendage arises from the seed coat over the area of the raphe in some species (*e.g., Stylophorum* and *Trillium*) and is known as the *strophiole.*

Integumentary tapetum

In some species, the cells of the inner integument serve as nutritional support for the developing embryo sac and later harden and act as a protective layer for the ovule). In *Lobelia,* the cells of the inner integument take on a pronounced radial elongation and become binucleate before becoming hardened as the *integumentary tapetum.*

2.1.5. Micropyle

The micropyle is an integumentary pore or opening in the ovule through which the pollen tube grows to fertilize the egg cell of the female gametophyte. The micropyle may assume one of several configurations, depending on the closure of the inner and outer integuments (Fig.5).

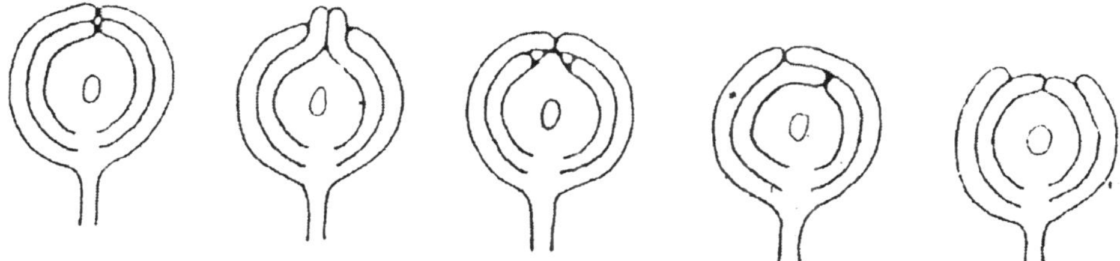

Fig.5: Types of the microphyle arrangements showing the different closer of the inner and outer integuments

(Source: Maheswari, 1950)

2.1.6. Epistase

Epistase is the development of well-defined nucellar or integumentary tissue in the micropylar region of the seed of certain species. In *Castalia* (water lily) and *Costus* (spiral flag) species, epidermal cells of the nucellus proliferate and form a plug beneath the micropyle, which remains after the rest of the nucellar tissue is gone. Cells adjacent to the micropyle may show a marked radial elongation. Another type of epistase, an *operculum*, develops when cells of the integument proliferate and form a tightly compacted micropylar plug, as in species of *Lemna* and *Acorus*.

Mature ovules are classified into five different types based on their arrangement within the ovary (Fig.6). The difference in arrangement begins at the time of archesporial development and becomes well defined by the time of embryo sac maturity. Ovule types have been determined for most well-known plant species and serve as a means of plant classification. The effect of the ovule arrangement is often visible externally. For example, the relative position of the hilum (funicular detachment scar), chalaza, and micropyle of many legumes can be easily seen.

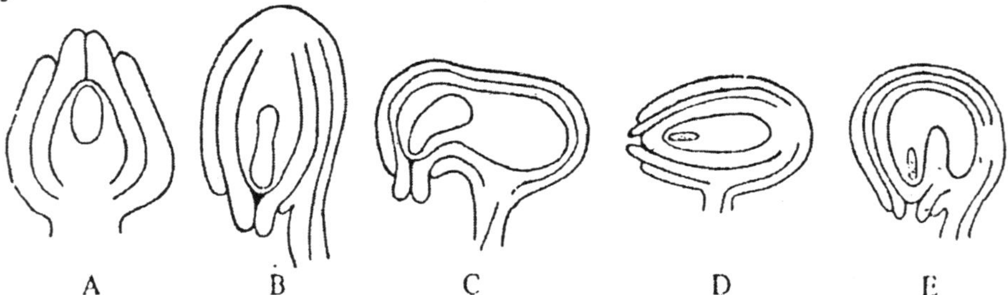

Fig.6: Types of the ovule as seen in the vertical longitudinal section. (A) atrotropus , (B) anatropus, (C) campylotropus, (D) hemianatropus, (E) amphitropus (Maheswari, 1950)

2.2. Microsporogenesis and microgametogenesis

The period of flower development when the stigma is ready to receive the pollen is known as *anthesis*. Pollen is usually produced in four sacs, or *microsporangia* (Fig.7), of the anther, although occasionally fewer sporangia may occur. Within the sporangia, certain cells become the *microspore mother cells* and undergo a two-step reduction division (meiosis), or microsporogenesis, to yield four *microspores*, each of which is haploid (1n). Each of the four microspores is normally functional and undergoes two divisions, known as *microgametogenesis*, giving rise to a microgametophyte, or mature pollen grain.

3. POLLINATION AND FERTILIZATION

The mature anthers dehisce and release pollen grains (haploid microspores). When pollen grains are transferred from an anther to the stigma of the same flower the process is called self pollination or autogamy. If they are transferred to the stigma of another flower, cross-pollination or allogamy is said to have occurred. Self-pollination occurs in those plants where bisexual flowers achieve anther dehiscence and stigma receptivity simultaneously. The majority of angiosperms bear chasmogamous flowers. In some plants, flowers do not open before pollination such flowers is called *cleistogamous*, and this is the most efficient floral adaptation for promoting self pollination.

Crosspollination is ensured in plants which bear unisexual flowers. In bisexual flowers also self pollination may be prevented by selfsterility, *dichogamy* (maturation of male and female organs at different times), *herkogamy* (where the structure of male and female sex organs proves a barrier to self pollination) and

heterostyly (where flowers are of different types depending on the length of the style and stigma and pollination occurs only between 2 dissimilar types).

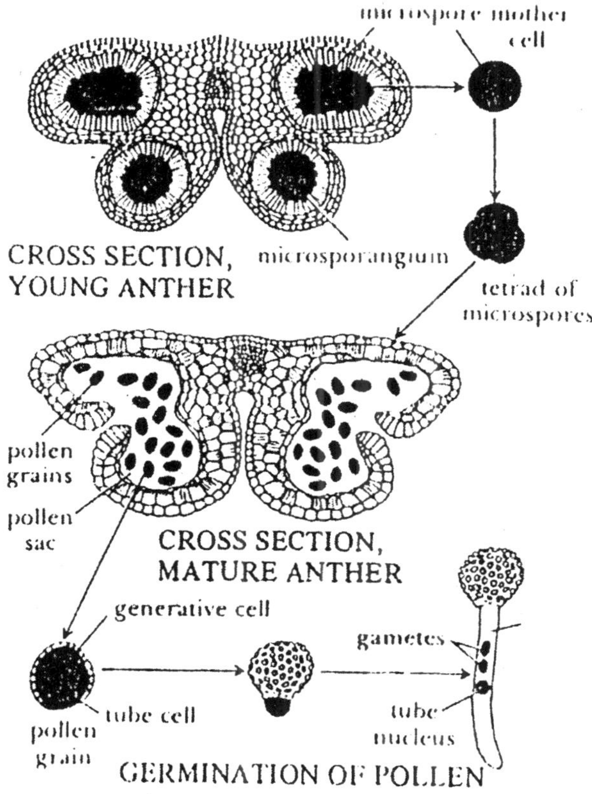

Fig7: Microsporogenesis

The important self pollinated crops are wheat, rice, barely, mungbean and cowpea and cross pollinated are maize, rye, forage legumes and vegetables like carrot, cauliflower and onion. There is yet another category of crops called often cross pollinated crops such as cotton and pigeon pea where there may be 10-40% cross pollination. Different agents bring about the dissemination of pollen grains. The agents may be abiotic including wind (*anemophily*) and water (*hydrophily*) or biotic including insects (*entomophily*) and bats (*cheiropterophily*).

After landing on the stigma, the pollen grain germinates and pollen tube grows through the style (Fig. 8a and b). The surface of the stigma secretes substances, which may provide optimum conditions for pollen germination. The pollen tubes secrete the pectinase which dissolve intercellular substances of the style tissue. After traversing the style, the pollen tube enters embryosac of the ovule. The embryosac consists of 8 cells. The end near the micropyle has the egg apparatus, which consists of egg cell and 2 synergids. There are 2 polar nuclei in the centre and the chalazal end has 3 antipodal cells.

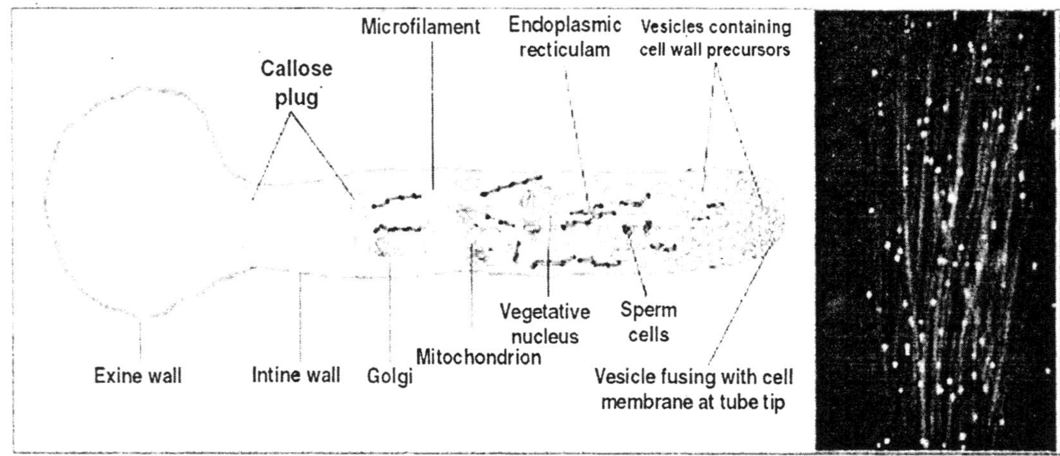

Fig. 8a: Pollen tube growth

Source: Bewley *et al.* (2000) Reproductive development. In: B Buchanan, W Grussem and R Jones, (eds.) Biochemistry and Molecular Biology of Plants. American Society of Plant Physiologists)

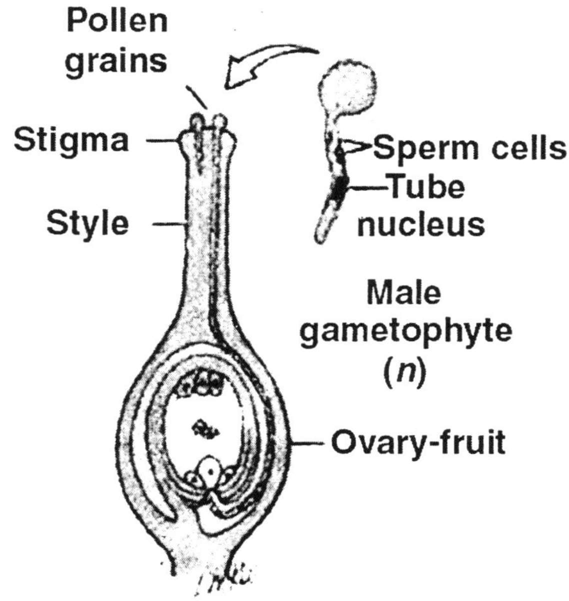

Fig. 8b: Pollen tube growth and fertilization

In angiosperms, fertilization involves the participation of 2 male nuclei (double fertilization). One fuses with the egg nucleus to form the diploid zygote and the other with 2 polar nuclei to produce a triploid nucleus, which is the primary endosperm nucleus (Fig.9).

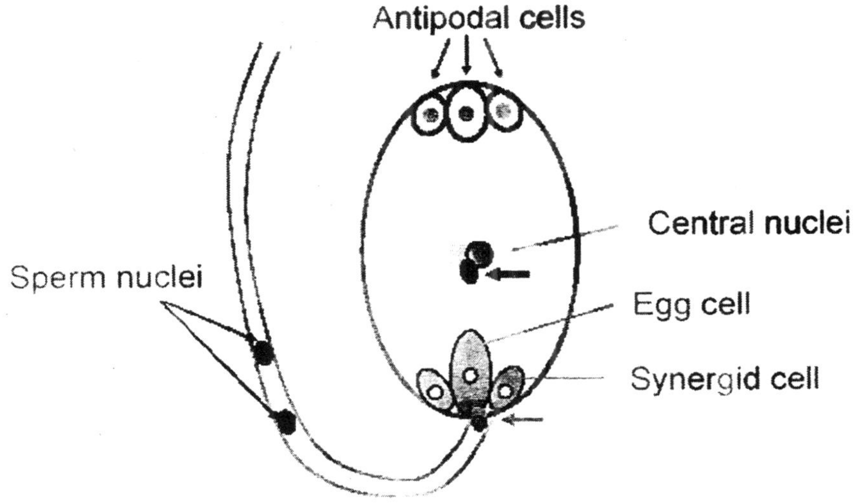

Fig. 9: Double fertilization

4. EMBRYO DEVELOPMENT

The first division of the zygote is transverse in dicots and it results in a small apical cell and a large basal cell. Cell ca divides vertically forming 2 juxtaposed cells and cb undergoes a transverse division forming 2 superimposed cells. These results in a 'T' shaped, 4 celled proembryo. Cell ci divides transversely giving rise to n and n'. These 2 cells divide further resulting in a row of 3 or 4 cells, forming suspensor. Cell m and its derivatives undergo vertical divisions forming a group of 4 to 6 cells. This group divides by oblique-perclinal wall forming a set of inner cells and a row of outer cells. The inner cells form the initials of the root apex and the outer cells form the root cap. The 2 cells formed as a result of the division of ca again divide vertically forming quadrant. Each cell of the quadrant divides transversely and thus an octant containing 2 tiers of cell l and p is formed.

The cells of the octant undergo vertical division resulting in a globular proembryo. Periclinal divisions occur in the peripheral cells of the globular proembryo that delimit an outer layer, the dermatogen. The tier l gives rise to cotyledons and shoot apex and l forms hypocotyl-radicle axis. Certain deviations from the above pattern of embryo development are found in different plants. Different types of embryogeny are distinguished depending on the plane of division of the apical and the extent of contribution of the basal cell towards embryo development (in some plants cb remains undivided and does not take part in embryo development at all). In monocotyledons, the cell cb remains undivided and develops into a haustorial of the suspension. Cell ca divides into

2 by a transverse division. The terminal cell of these 2 by repeated divisions in different planes gives rise to a single cotyledon.

The embryo development in grasses is different from that of other monocotyledons. A dorsiventral symmetry is established as a result of the peculiar oblique position of cell walls early in the embryogeny. The single cotyledon is reduced to absorptive scutellum and additional structures like coleptile and coleorrhiza are formed (Fig.10). Murugesan (2003) studied the structural change in embryo during the development and maturation in ash gourd and found that the cotyledons developed the ability to expand first (20 DAA), followed by embryonic axis development (30-60 DAA). During embryonic axis development, hypocotyls expanded first, followed by root and shoot meristems (Plate 1).

The late globular stage occurred at 5 DAA. Early stage of cotyledenary development was observed at 15 DAA. At 20 DAA, complete cotyledenary structure had developed. During the subsequent stages (upto 50 DAA), cotyledon thickening and embryo differentiation took place. Mature embryo could be observed beyond 50-60 DAA. From over mature seed stage (80 DAA), radicle cells (Plate 2) were highly enlarged and multiple growth was observed in hypocotyl axis. Cell enlargement in all other stages was proportionate to their increased growth stages. In cluster bean, seed development the cotyledons developed the ability to expand first followed by embryonic axis development (20-30 DAA). During the embryonic axis development, hypocotyls expanded first followed by root and shoot meristems (Renuga Devi, 2003).

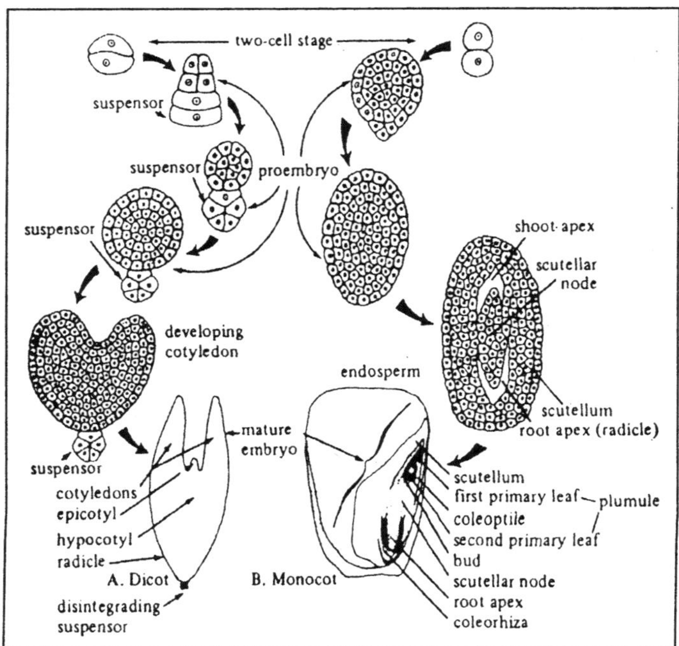

Fig.10: Embryo development: (A) dicotyledonous species, (B) monocotyledonous species

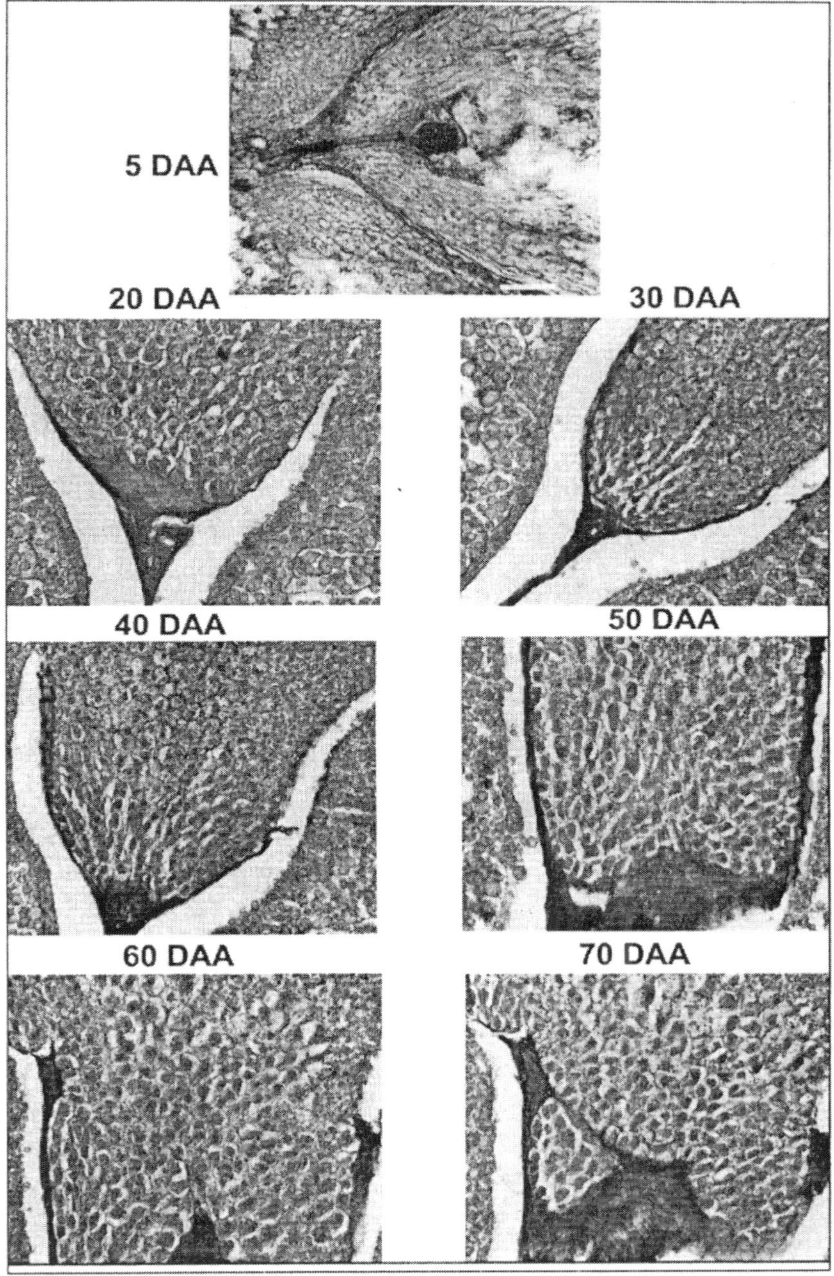

Plate 1: Structural changes in the embryo at different stages of development and maturation in ashgourd

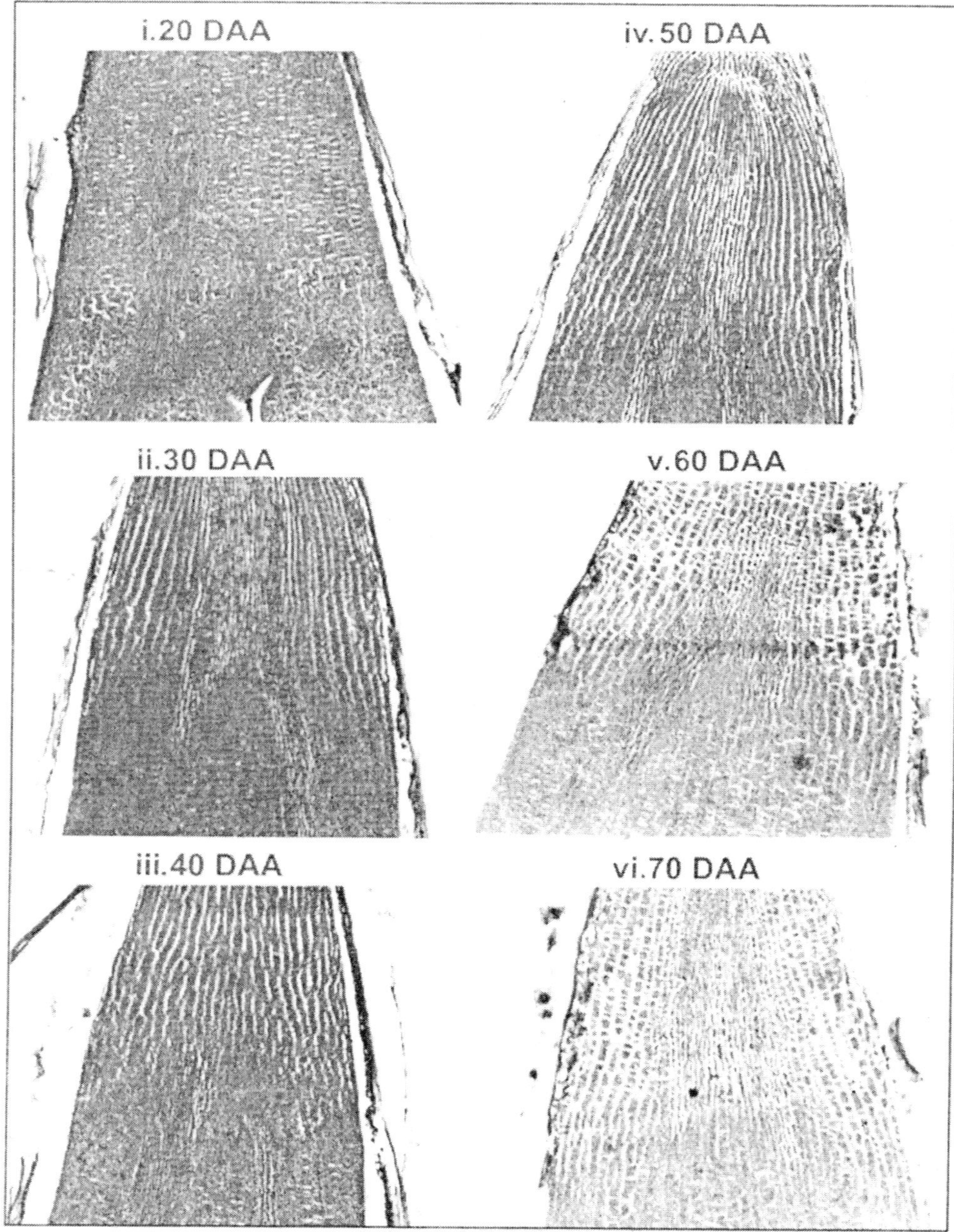

Plate 2: Structural changes in the radicle at different stages of development and maturation in ashgourd

5. ENDOSPERM DEVELOPMENT

There are three types of endosperm development (a) nuclear—where the endosperm nucleus undergoes several divisions prior to cell wall formation, *e.g.*, wheat apple, squash (b) cellular - in which there is no free nuclear phase, and (c) helobial where the free nuclear division is preceded, and is followed by cellularization as in some monocots. During the course of seed development, reserve food materials are accumulated in the endosperm from the adjacent tissues. In endospermic dicot seeds, endosperms are retained as a permanent storage tissue. In non-endospermic dicot seeds, endosperm reserves are depleted and occluded by the developing embryo. The reserves are then reorganized in the cotyledons, which in turn act as the source of stored reserved food for embryo after germination. A part of the endosperm is depleted in cereals during embryo maturation and this lies as a layer between the starchy endosperm and scutellum. (Fig. 11)

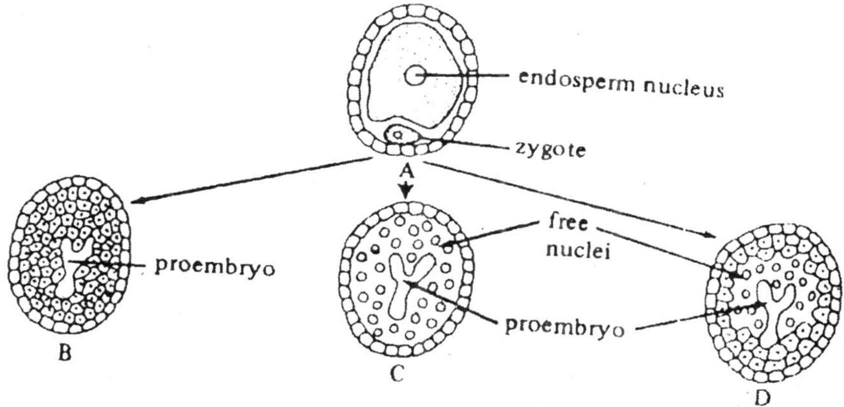

Fig.11: Endosperm development: (A) Ovule after fertilization , showing primary endosperm nucleus and zygote , (B) Cellular endosperm, (C) Nuclear endosperm, (D) Helobial endosperm

6. SEED COAT DEVELOPMENT

Integument's of the ovule undergo marked reorganization and histological changes during maturation to form seed coats. In bitegmic ovules (which have 2 integument's), the seed coat may be derived from both the integument's or from the outer integument only; the inner integument may disintegrate. The histological studies (Murugesan, 2003) on seed coat development of mature seed of ash gourd (Plate 3) reveals that seed coat of over matured seed is visible as non structured cell walls and at the peak maturity stage, multiple layers of cell wall was visible with maximum expansion. Elongated tabular or rod shaped thickening in epidermal cell running in the radial direction from the inside to out side is visible in the seed coat of matured seeds of ash gourd, which was absent in the seed coat of immatured seeds. In immature seed, seed coat layers were so compact that it was difficult to by growing embryo.

7. SEED MATURATION AND DEVELOPMENT

Seed development is the processes, which lead a fertilized ovule to a matured seed. The course of seed development exhibits a sigmoid curve in most of the plant species and a double sigmoid in exceptional cases. The sequential events occurring during the development a seed are summarized in figure 12a and b.

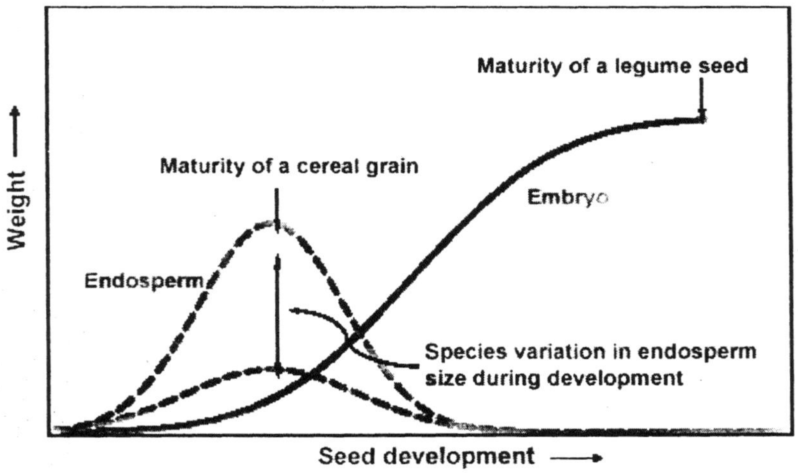

Fig. 12a: Endosperm/embryo size relationships during seed development

Seed Development

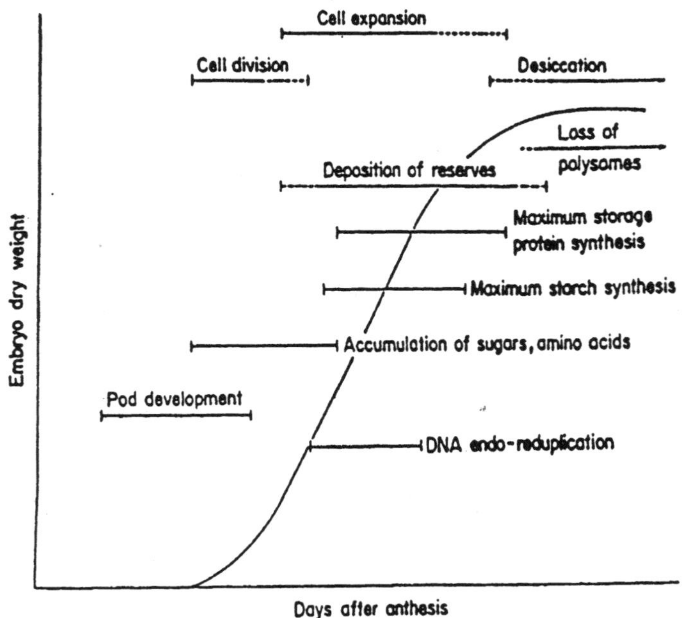

Fig. 12b: Cotyledon development and the deposition of reserves in pea

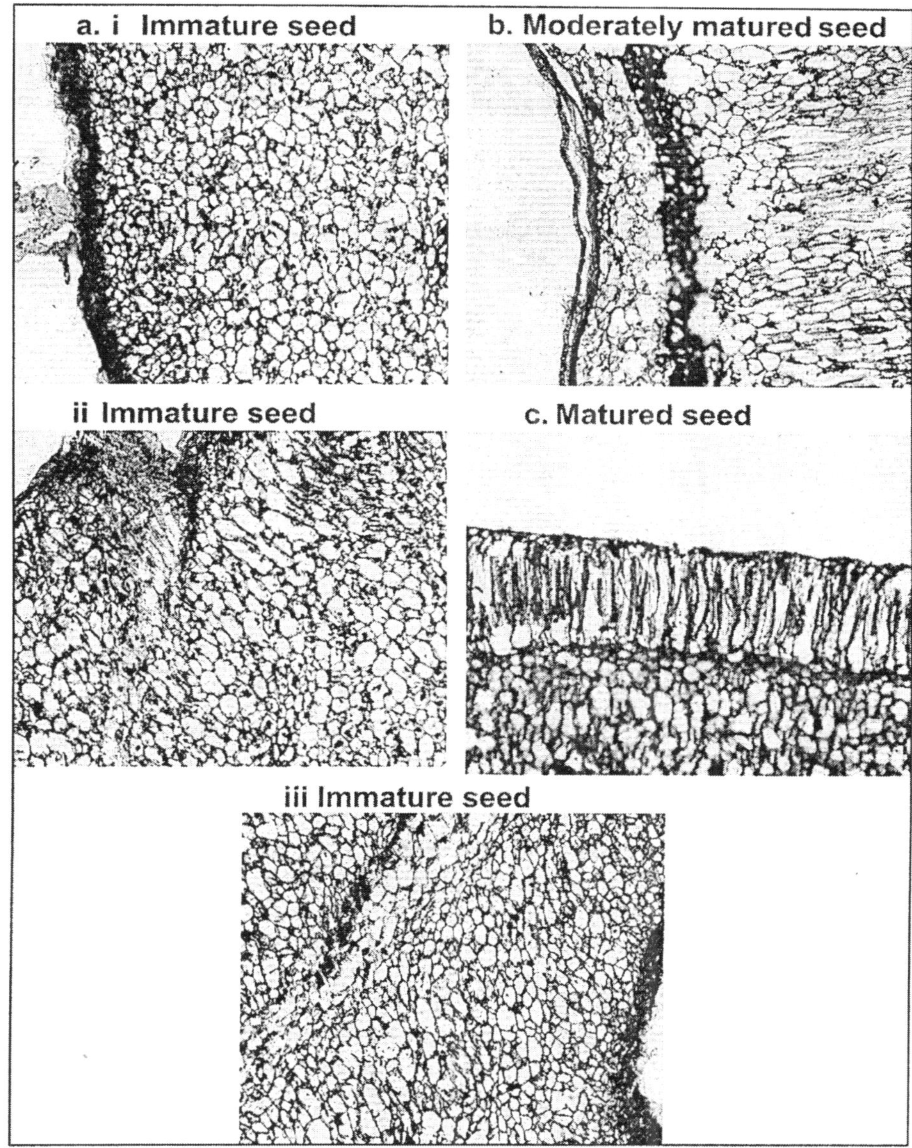

Plate 3: Structural changes in seed coat at different stages of development and maturation in ashgourd

The differentiated, but immature embryo develops by a series of cell divisions from the fertilized egg. Then, cell division activity declines and cell expansion takes place. At this stage, the DNA endoreduplication occurs. In the seed development sequence, pod development preceeds seed development, i.e. the development of the pod and for the fruit wall occurs first and with the culmination of pod and/or fruit wall development, the actual seed development begins. In respect to seed development, the seed coat develops first followed by the embryo. The accumulation of different chemical constituents such as sugars, amino acid, starch, storage protein and their deposition in cells occur in a sequential way with genetic programming. A simplified projection is depicted in the Fig.13.

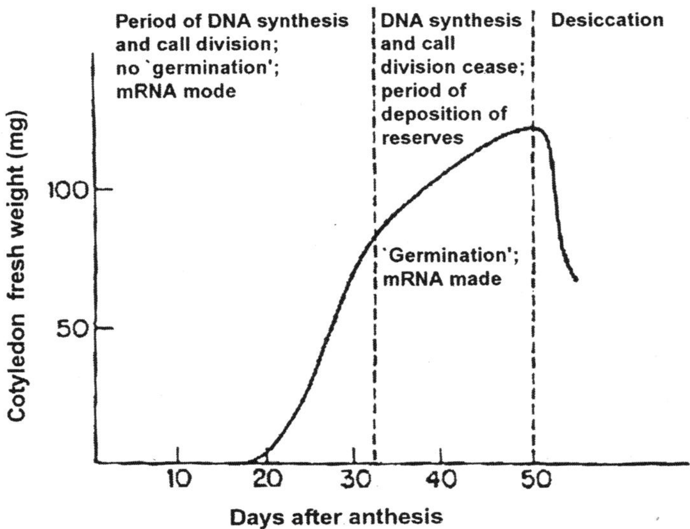

Fig. 13: Seed development in cotton showing time of synthesis of messenger (mRNA) molecules which are used in germination

When the embryo reaches the full size i.e. the deposition of reserves is completed, the seed is matured physiologically. From that time onwards, the moisture content of the seed recedes at a faster rate. Even in this, an orderly and sequential process has been observed in almost all the species.

8. PHYSIOLOGICAL MATURITY

The culmination of the developmental phase accompanied by simultaneous accumulation of food reserves in the seed marks the potential maturity of the seed known by the term physiological maturity. It is at this stage, the seed possess maximum vigour and viability potential. Subsequent to that, no accumulation of reserves take place in the seeds and it is virtually cut off from the mother plant by forming an abscission layer in the funicle, the attachment of the seed with the pod. In certain seeds like maize, sorghum, etc, it is marked by a visible layer at the funicular region called the black layer formation indicating the physiological maturity. In other seeds, it will be accompanied by a colour change in the pod from green to yellow or light brown depending upon the kind of seed. Thereafter, the changes that are occurring in the seed is only receding of moisture or loss of moisture which is only a physical process. At physiological maturity, the moisture content of the seed usually may range from 25 to 35 per cent depending upon the species.

Beyond physiological maturity, the most significant biological event namely, the senescence or deterioration starts in. Depending upon the environmental conditions prevailing at that stage, the quality of the seed will be decided even when it is in the mother plant. Under unfavorable or hostile conditions, the seed may deteriorate considerably even before harvest. In contrast the rate of deterioration may be very little or almost nil under favourable situations. In the technological point of view, it is always desirable to harvest the seeds at physiological maturity, but for the difficulty in dehydrating the seeds from high moisture to a safe moisture level of 8 to12 per cent depending upon the species. This explains why such seed development and maturation studies are important for different crops to resolve quality seeds for sowing.

9. BENCHMARKS IN SEED MATURATION

Seed maturation refers to the morphological, physiological and functional changes that occur from time of anthesis until the seeds are ready for harvest (Khattra and Singh, 1995). Maturity is the critical and the most important factor that determines the size, quality, planting value and storability of the seed, that are directly related with the level of maturity and the seeds harvested at optimum maturity will be possessing maximum germination and vigour (Jerlin *et al.,* 2001). Karuna and Aswathaiah (1990) stated that a considerable knowledge on the stage of seed maturity related with seed quality and performance is essential for higher production of quality seeds of crop plants and the stage of maturity could be optimized based on the physical, physiological and visual indices (Andrews, 1996). Several of the bench marks in the seed maturation process are of special significance in seed production, and thus, should be of interest to seedsmen.

9.1. Physical and physiological changes in fruit and seed

9.1.1. Seed Size

Agricultural Crops

Increasing seed size is, of course, the most obvious change that occurs during maturation. Maximum size is usually attained substantially before the seed reaches physiological maturity. After maximum size is reached, the maturing seed decreases in size (shrinks) as the drying down process continues. In some species, changes in seed size are not easily observed or are deceiving. The hulls of rice seed for example, are full size at the time of pollination. The developing grain between the hulls is not visible so that seed size changes are not observed. Immature and mature rice seed therefore, differ not in over-all size, but rather in the size of the grain. Immature rice seed have incompletely developed grains, hence, are less dense and are removed in processing by aspiration. The same situation occurs in other kinds of seed in which the covering (hulls, pericarp or seed coat) develops the characteristic shape independently of the developing embryo or grain, *e.g.*, cotton, sunflower, castor been, rice.

The coincidence of physiological maturity with maximum seed size, volume, length and breadth was reported by several scientists in different crops belonging to the family fabaceae *viz.*, Jayabarathi (1982), Shabir-Ahamed (1989) ; Gore *et al.*(1997) in soybean ; Paramasivam (1990) and Patrick Jasper (1998) in peas ; Rajasekaran (1997) in rice bean ; Rakesh Seth *et al.* (1999) in fodder cowpea ; Arul Prabu (1998) in pole bean ; Ushamythili (1982) and Kathiravan (2001) in lab-lab and Vijayakumar *et al.* (2002) in field bean. On tracing the development and maturation pattern of cluster bean (Renuga Devi,2004), the pod characters revealed that fresh weight of pod, dry weight of pod, pod length and width observed a steady increase with advances in maturity status upto 45 days and reached the maximum at 50 days after anthesis (DAA). But at 55 days, a slight reduction was observed with length and width of pod, while the reduction in fresh weight was 11 per cent from 50 to 55 days after anthesis (Plate 4).

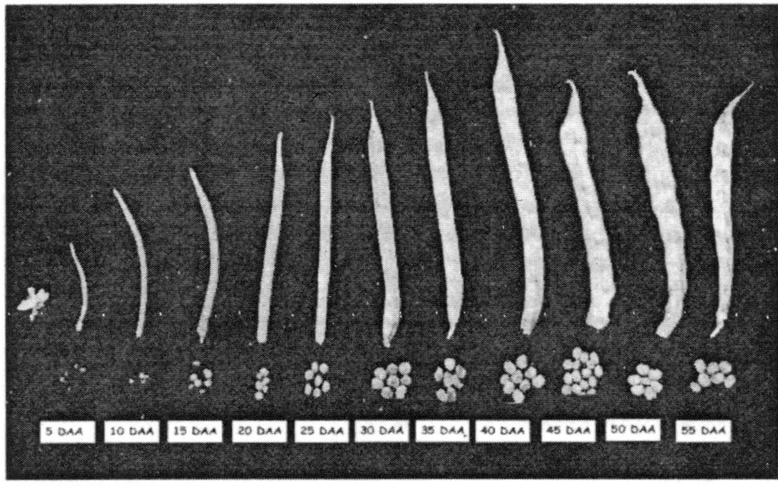

Plate4: Development and maturation of pod and seed

Sai Babu *et al.* (1984) reported that the maximum 1000 seed weight was recorded when the seed was harvested at 30 days after flowering and 25 days after flowering respectively in rice. Seed harvested at 35 days after anthesis showed maximum 100 seed weight in sorghum (Sai Babu and Hussaini, 1984). Sarada *et al.* (1994) observed maximum 100 seed weight in seed from cobs harvested at 45 days after flowering in maize, while it was least in seed from cobs harvested at 30 days after flowering. Seed test weight was the highest in soybean when seeds harvested at 46 days after flowering (Gore *et al.*, 1997).

Horticultural crops

In cucurbits, seed cavity volume to fruit volume ratio decreased with increasing length of the fruit. Pumpkin had the lowest ratio of seed yield to fruit weight (Nerson *et al.*, 2000).

Changes in fruit morphological characters *viz.*, fruit, seed cavity diameter, colour, size and shape and seed parameters namely fresh weight, during fruit growth and maturation of melon cultivars were reported by Miccolis and Saltveit (1991) in order to predict harvestable maturity. Leek (*Allium porrum* L.) attained maximum seed growth rate at 115 days after anthesis. Developing cucurbits seeds from 4 to 9 weeks after anthesis showed marked increase in embryo weight, size, volume and oil content as per the report by Ockenden *et al.* (2001). Fruit weight, length and number of seeds of developing okra cv. Parbhani kranti fruit were studied to determine the optimum stage of seed harvest by Samnotra *et al.* (2002) and they reported that the difference of above growth parameters in spring summer and rainy season crops. In case of ash gourd (Murugesan, 2003), fruit, circumference, length, diameter and pulp thickness of ash gourd were found to increase from anthesis to maturity (Fig. 14). Regarding fruit weight, it was maximum at 60 DAA in summer season and 70 DAA in rainy season. The descriptive characteristics of Ash gourd fruit and seed at different stages of development and maturation is depicted in Table 1 and 2; Plate 5; Fig.15 and 16. Woyke and Lazecka (1994) observed that seeds of *Anethum graveolens* harvested between 37 and 63 days after anthesis gave maximum 1000 seed weight.

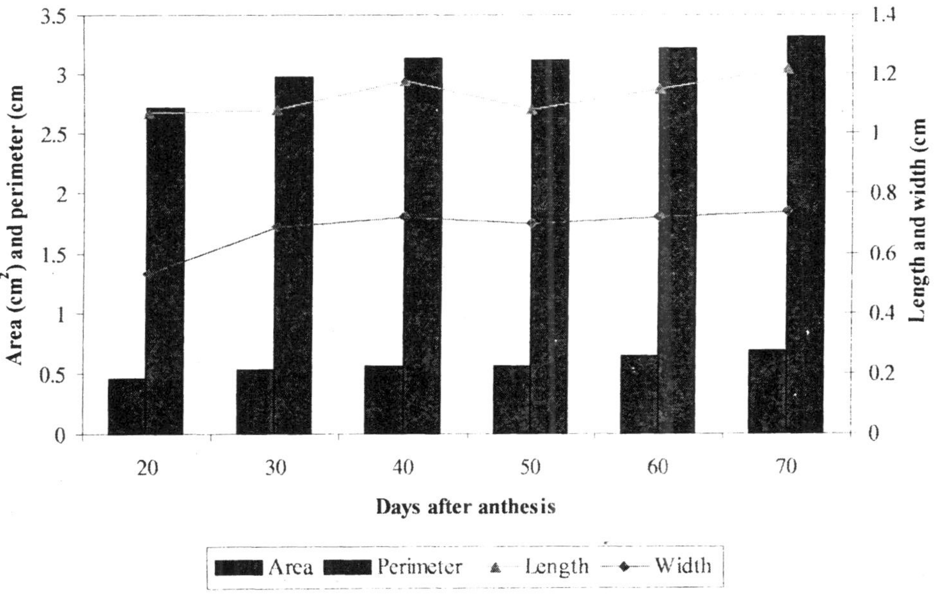

Fig.14: Area, perimeter, length and width of seed matured in image analysis system at different stages of development and maturity in ashgourd varieties

Variation in pod and seed colour was observed during the seed development of *Mucuna pruriens* (Jeeva, 2003). The pods were green at initial stage (5 days after anthesis) and gradually turned to blackish yellow at 45 days after anthesis, while the seed reen (15 days after anthesis) to white (45 days after anthesis)

Table 1: Descriptive characteristics of fruit

a. Summer season

Days After Anthesis	*Descriptive characteristics of fruits*
5-10	Light green fruits with scattered white spot. dark green at pedicel end, persistent corolla, highly pubescent
10-15	Prominent white spots mostly at stylar end in fruit; fruits are light green in colour, pedicel dark green, ashy coat at pedicel end, pubescent.
15-20	Ashy coating covered half portion of rind, intensity of pubescent decreases, pedicel thickened.
20-25	Ashy coat covers entire fruit, but with less intensity except ground portion, pubescent
25-30	Ashy coat covers entire rind surfaces and become thicker and sticky, pubescent reduced.
30-35	Rind of fruit became harder, pubescent at less intensity, dark green pedicel.
35-40	Ashy coat became very thick and sticky, prominent longitudinal groves formed.
40-50	Pubescent disappears, depression formed at stalk end indicating start of maturity, grooves become prominent; pedicel became light green, no longer preferred for vegetable purpose.
50-60	Pubescent disappeared completely, pedicel slightly dried, fruit rind very hard. If we give protracted cut, crack forms in rind, pithy pulp, maximum peduncular depression
60-70	Summer crop senescence in the form of yellow leaves started, pedicel completely dried, ashes become powdery, total crop duration approximately 170 days
b. Rainy season	
5-10	Light green fruits with scattered white spot, dark green at pedicel end, persistent corolla, highly pubescent
10-15	Early ash coating not observed as noticed in summer, which was delayed
15-20	Ashy coating begins here and remaining characters are in agreement with summer crop
20-25	Ashy coat covers entire fruit, but with less intensity except ground portion, pubescent
25-30	Ashy coat covers entire rind surfaces except few patches
30-35	Rind of fruit became harder, pubescent at less intensity, dark green pedicel, pubescent reduced.
35-40	Rinds become slightly hard, but not equivalent to summer, prominent longitudinal groves just developed
40-50	Ashy coat intensity increased, sticky and pulp is still looks tender
50-60	No pithiness as in summer crop, pedicel becomes light green, resistant felt while cutting the fruit pulp,if we give protracted cut, crack not formed in the rind as observed in summer,maximum peduncular shoulder depression
60-70	Angular, light green pedicel, depressed peduncular shoulder, thick ashy coat fruit longevity continued
70-80	Total crop duration continued beyond 180 days, fruits are still intact with vine but gets fungal attack in matured fruits

Table 2: Descriptive characteristics of seed

a. Summer season

Description	Days After Anthesis	Seed characteristics in summer
Pre-embryonic	5	Globular embryo, developing ovary
Pre-cotyledon	10	Heart shaped embryo, remnant of seed coat and water filled cotyledon and distinctly separate.
Early-cotyledon	15	Small papery cotyledon, seed coat develops, seed tightly packed in cavity, creamy seed, liquid cotyledon.
Enlargement I	20	Small embryo, cotyledon develops and thickens, developing seed coat, groves developed in margin of seed, seeds can be removed but attached through funicle
Enlargement II	25	Prominent grooves at seed margin, cotyledon thickened, and seeds could be removed without much effort, slightly yellow seed coat.
Filling	30-35	Cotyledon thickening, seeds easily come out from placenta, they became harder
Filling & Maturation I	40-50	Cotyledon tightly folded, air space noticed in seed cavity
Filling & Maturation II	50-60	Seeds attain maximum width and thickness, fully developed with prominent margin
Maturation III	60-70	Light yellow and maximum size in summer, completely detached from placenta, desiccation started, seeds freely comes out

b. Rainy season

Days after Anthesis (DAA)	Seed characteristics
5-40	No marked difference observed between summer and rainy
40-50	No air space in seed cavity, still seeds attached with placenta
50-60	Seeds attain maximum width and thickness, fully developed with prominent margin.
60-70	No desiccation, but seeds detached from placenta
70-80	Air space visible, but still seeds are fresh and continued to grow, desiccation was delayed,

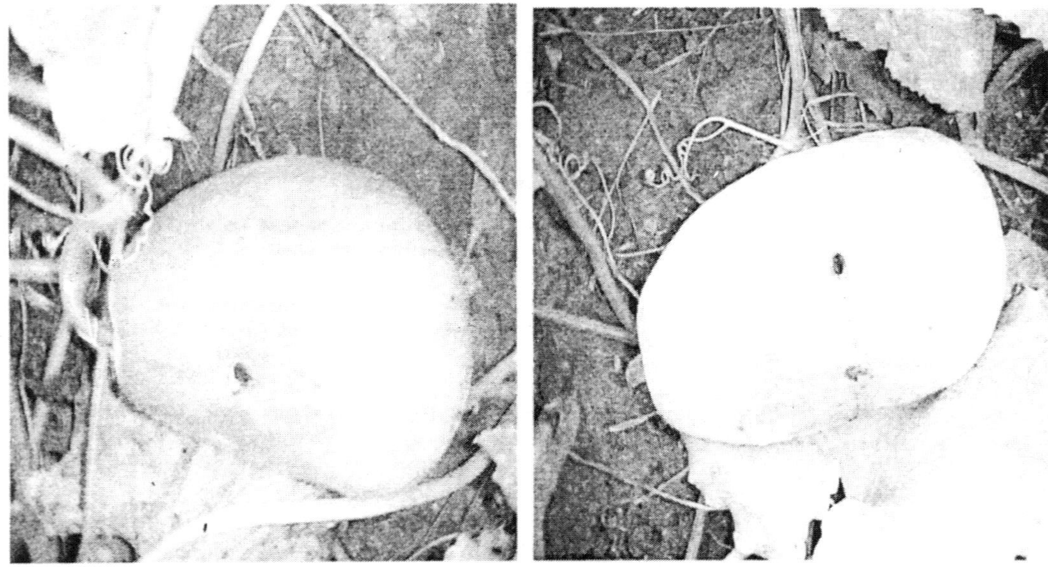

20 DAA 70 DAA

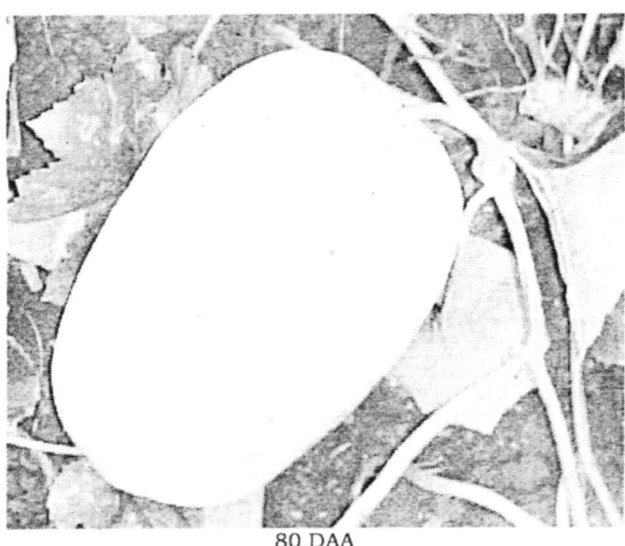

80 DAA

Plate 5: Fruit development and maturation stages in Ash gourd (Murugesan, 2003)

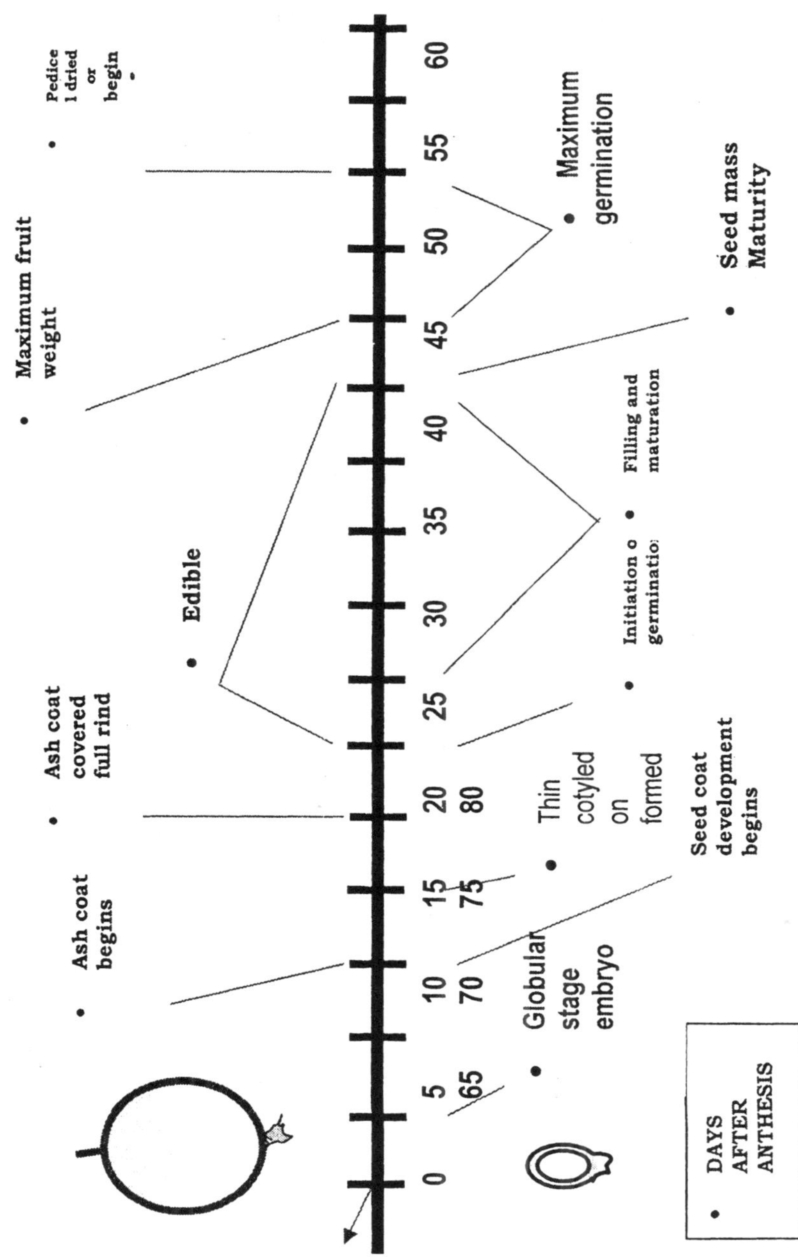

Fig. 15: Comparison of descriptive characteristics of ash gourd fruit and seed during development and maturation in summer season **(Courtesy P. Murugesan, 2003)**

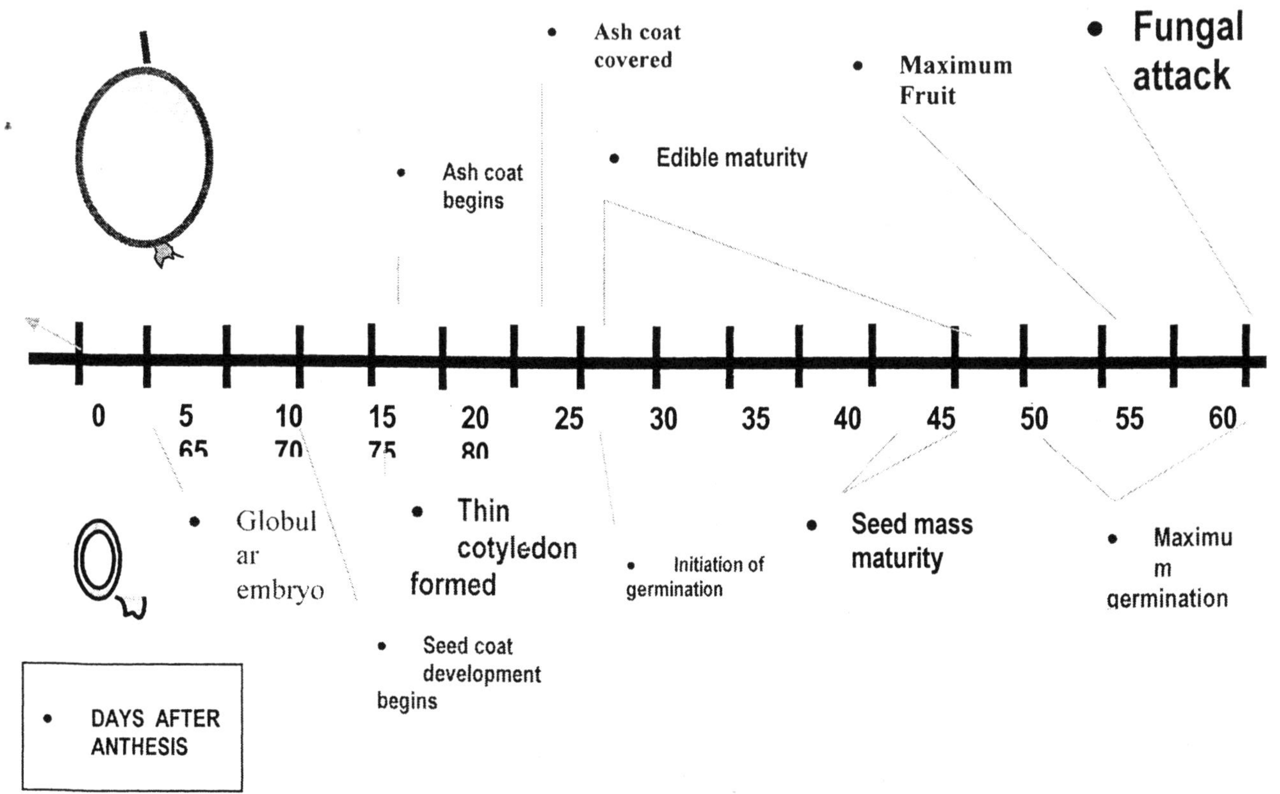

Fig.16: Comparison of descriptive characteristics of ash gourd fruit and seed during development and maturation in rainy season

(Courtesy P. Murugesan, 2003)

9.1.2. Moisture Content

The moisture content of the unfertilized ovule is somewhat above 80% (wet weight basis). Following fertilization, moisture content of the developing seed increases for a few days, then begins to decrease rather slowly until physiological maturity is attained, after which it decrease at a more rapid rate until an equilibrium is established with the field environment at 12 to 18% moisture content.

The time required for seed to decrease in moisture content from the 80% or higher level a few days after fertilization to 12-18% varies with the kind and variety of seed, and climatic conditions (Table 3).

Table 3: Seed moisture content at physiological maturity

Kind	Moisture content
Corn	30-35%
Cotton	50 - 55%
Cowpea	50-55%
Rice (in hulls)	30-33%
Sorghum	30-35%
Soybean	30 - 50%
Red gram	65 – 70%
Wheat	40%
Barley	40%
Oat	38 – 42 %
Safflower	20 – 30%

Bishnoi (1974) stated that maturation process proceeds with water loss at various degrees depending upon the atmospheric conditions, while Manohar and Sachan (1974) described the loss of moisture in maturity stage as an inherent phase of seed development. Khattra and Singh (1995) also expressed that accumulation of drymatter with loss of moisture, as the characteristic feature during seed development and maturation in any crop. The pod moisture content cluster bean (Renuga Devi, 2004) decreased from 82 to 25 per cent with periods of maturation (15 to 55 DAA) (Fig.17 and 18). With advancement in maturation, the decrease in moisture content of pod was at faster rate (Manohar, 1970; Shanmugaraj, 1978; Sabir-Ahamed, 1989) which could be due to the replacement of osmotic material by starch and other large molecules with low hydration capacity.

In case of *Ocimum* spp (Swapna Mova, 2003), the seed moisture content initially recorded was 96.24, 98.01 and 97.90 per cent, which decreased subsequently to 19.61, 16.69 and 17.29 per cent respectively,in *O. basilicum* (22 DAA) (Fig.19), *O. sanctum* (18 DAA) and *O. gratissimum* (30 DAA) at physiological maturity.

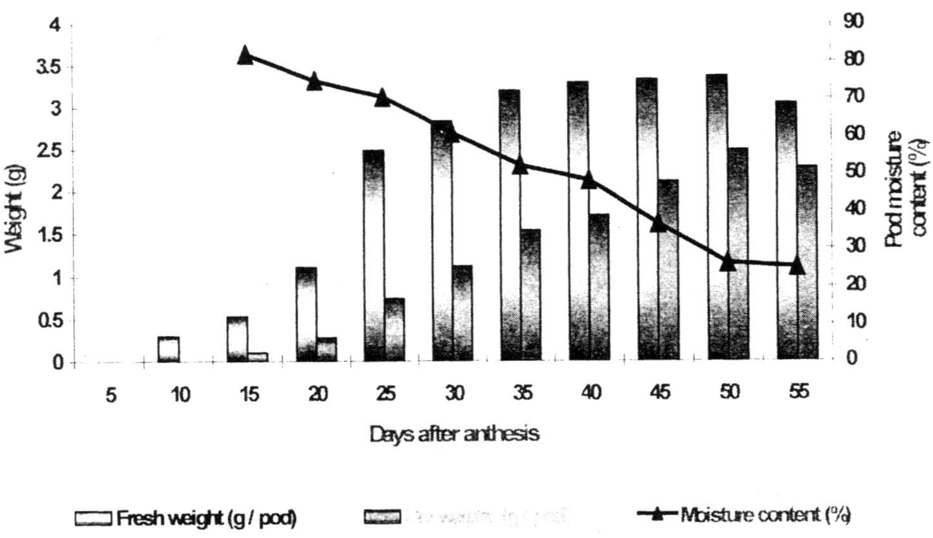

Fig.17: Changes in pod moisture content, fresh and dry weight of cluster bean during seed development and maturation

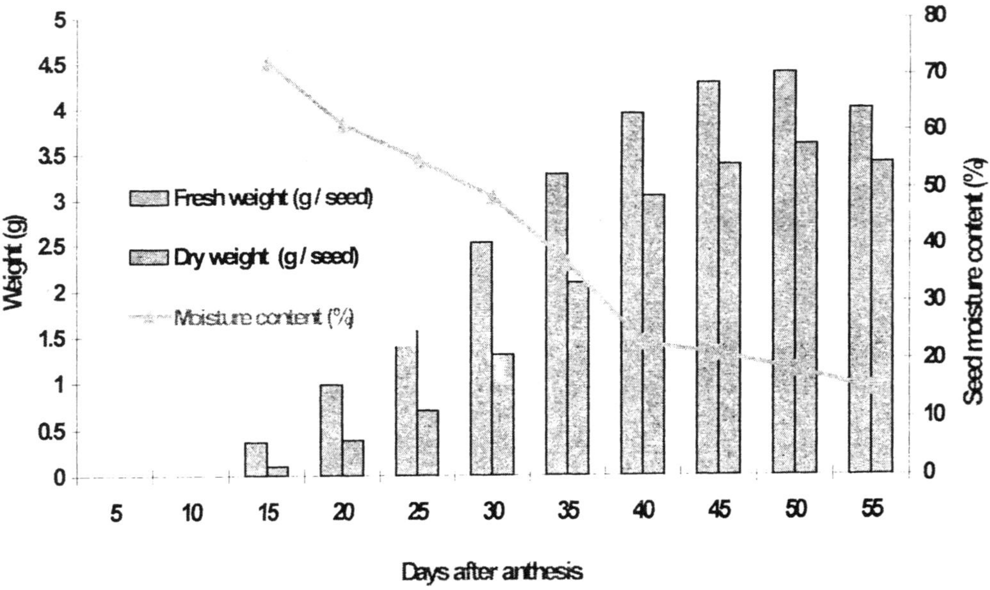

Fig.18: Changes in moisture content, fresh and dry weight of cluster bean seeds during seed development and maturation

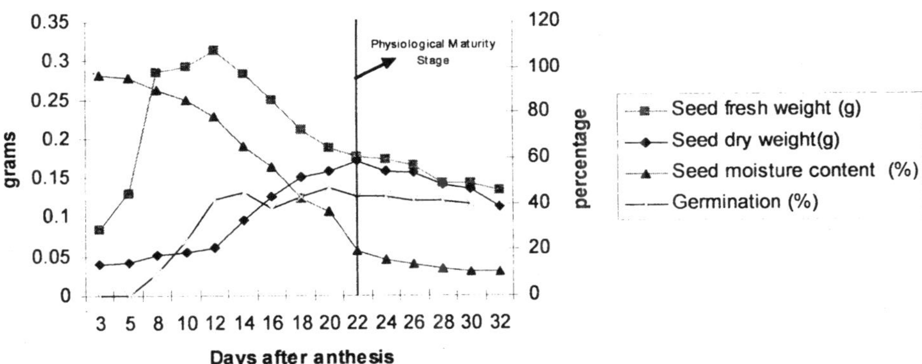

Fig.19: Changes in moisture content, fresh and dry weight and germination of *ocimum basilicum* during seed development and maturation

The loss of moisture content of ash gourd (Murugesan, 2003) seed with time was linear over the range from 59.0 to 43.0 per cent in summer season (Fig. 20) and 70 to 49 per cent in rainy seasons during initial, 20 DAA to mass filling stage, respectively.

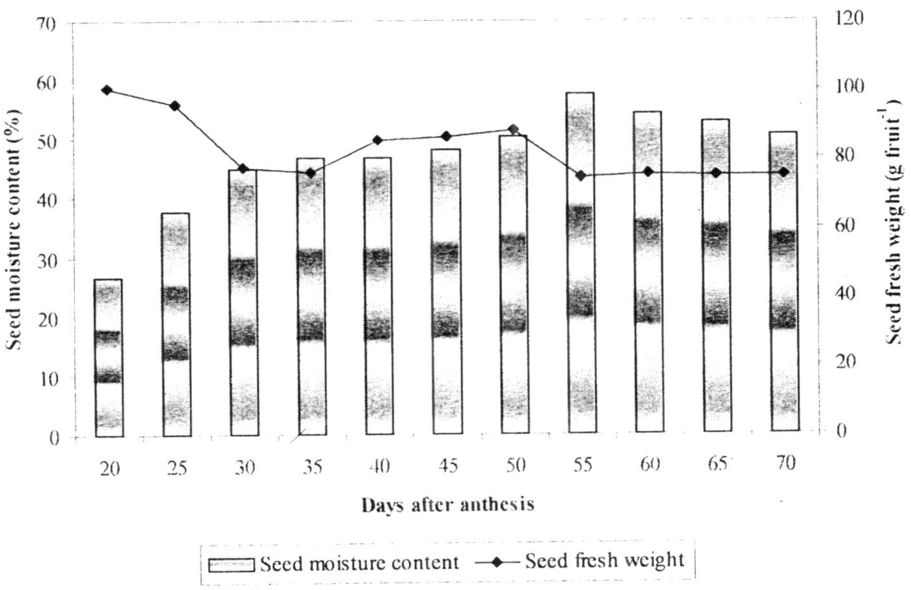

Fig.20: Changes in moisture content and fresh weight of ashgourd during seed development and maturation.

9.1.3. Seed dry weight

Dry weight of the developing seed begins to increase from the time of fertilization – slowly at first, then much more rapidly, and finally more slowly as the point of maximum dry weight is approached. Dry matter accumulation in seed is a most important process. In crops harvested for their grain or seed, the dry matter accumulated is yield. In terms of the reproductive function of seed, the point of maximum dry weight or dry matter accumulation is taken as physiological or functional maturity. Put another way, seed reaches physiological or functional maturity at the time maximum dry weight is attained. Seed quality measured in terms of viability and vigour is the highest at physiological or functional maturity. From this high point, quality can only decrease, and usually does even before harvest. The number of days at which the maximum dry weight was reached for different crops is featured in Table 4.

Table 4: Days after bloom required for maximum dry weight to be attained

Kind	Days after bloom
Rice	28 – 30
Corn	36 – 40
Sorghum	33 – 45
Wheat	26 – 30
Barley	26
Cotton	50 - 55
Safflower	20 - 30

Seed moisture content at the time of physiological maturity, i.e., maximum dry weight, is usually too high for harvest. Since moisture content of seed at physiological maturity is too high for harvest, the seed remains on the plants in the field until they dry down sufficiently for harvest, i.e., harvest maturity stage. During this post-maturation pre-harvest period, the seed are very much influenced by climatic conditions. Rains, heavy dews, warm temperatures, high humidities, all promote deteriorative processes in seed and reduce quality. Field deterioration is, perhaps, the most serious problem confronting the cotton, soybean and peanut seed producers. For these reasons, management must strive to prevent any unnecessary delays in harvest after the seed reach the harvest maturity stage.

Agricultural crops

Bhaskaran (1995) observed higher dry weight was recorded in rice seed harvested at physiological and 3 days after physiological maturity, while the delayed harvest recorded lower dry weight. Maximum fresh weight and dry matter accumulation occurred in CMS lines of pearlmillet harvested at 28 to 25 days after pollination (Apparao *et al.*, 1993). Dry matter content of seed was the highest in soybean when seeds harvested at 46 days after flowering (Gore *et al.*, 1997). Ajayi and Fakorede (2000) observed that harvesting of maize cv. obasupa 1 at 45 days after silking registered higher dry weight over harvested the same before full maturity. In rice (Raja, 2003), the maximum dry weight was accumulated at 30 days after flowering (Fig.21).

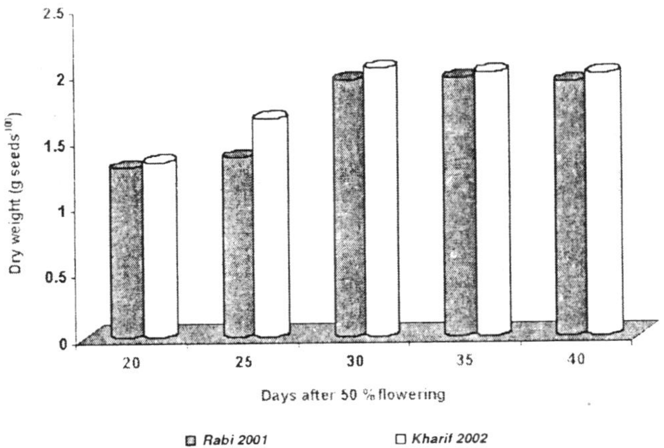

Fig.21: Dryweight accumulation in rice hybrid CORH-2

On tracing the development and maturation pattern of clusterbean (Renega Devei, 2004) the dry weight of pod observed a steady increase with advances in maturity status upto 45 days and reached the maximum at 50 days after anthesis (DAA). But at 55 days, a slight reduction was observed in fresh weight from 50 to 55 DAA (Fig. 17). The fresh weight and dry weight of seed were also increased with maturation upto 50 DAA (Fig. 18).

Horticultural crops

Varatharajan (1979) observed highest weight fruit at 42 days after anthesis in ribbed gourd cv. CO 1. Showalter (1961) had confirmed that in watermelon, the fruit after reaching the maximum weight started loosing weight accompanied by loss in moisture content and this dehydration was an inherent phase of development. Chandrasekaran (1979) reported that peak fruit weight reached 24 days after anthesis in cv. Coimbatore long bottle gourd. Anderson (1985) reported that maximum rate of increase in weight of *Cucurbita pepo* occurred between 11 days after pollination and 11 days previous to beginning of the ripening period. In cucurbits, seed cavity volume to fruit volume ratio decreased with increasing length of the fruit. Pumpkin had the lowest ratio of seed yield to fruit weight (Nerson *et al.*, 2000). Seeds of arona beans (*Phaseolus vulgaris* L.) harvested at 49 days after the beginning of flowering recorded higher drymatter accumulation at physiological maturity (Chamma *et al.*,, 1980). In snapbean, dryweight was the maximum when the seed moisture content decreased to 45-50 per cent (Demir and Ellis (1992). Arulprabhu (1998) revealed that pole bean seeds attained physiological maturity at 40 days after flowering, when the seed moisture content was around 39 per cent and at this stage dry weight of the seed was maximum. Kathiravan (2001) reported that seeds of lablab cv. CO 1 harvested at 49 days after anthesis recorded maximum drymatter. In ash gourd (Murugesan, 2003), the maximum fresh and dry weight (mass filling stage) occurred at 55 DAA in summer (Fig.20) and it was extended upto 60 DAA in rainy season (Fig.22). In subsequent stages of development and maturation, there were no much change occurred in rainy season for weight of both fresh and dry seed, whereas in summer season drastic reduction of fresh weight was observed.

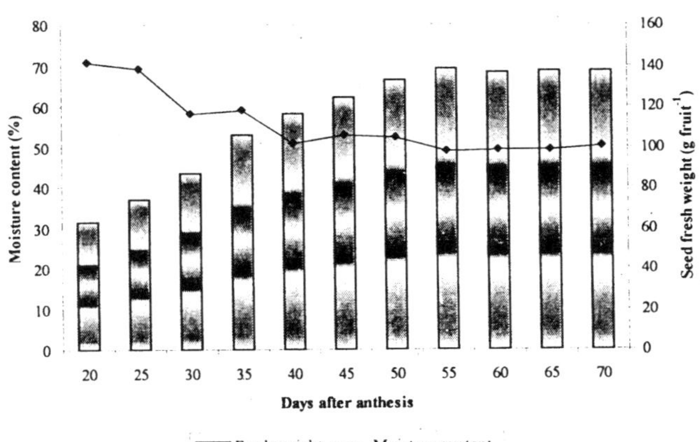

Fig. 22: Seed moisture content and fresh weight at different stages of seed development and maturation in ash gourd

Table 5: Physical properties of fruit and seed during development and maturation in *Solanum nigrum*

Stages	Fruit			Seed recovery (per cent)	Seed		
	Fresh weight (mg fruit^{-10})	Dry weight (mg fruit^{-10})	No. of seeds fruit^{-1}		Fresh weight (mg seeds^{-100})	Dry weight (mg seeds^{-100})	Moisture content (per cent)
5 DAA	41.54	15.71	58	1.15	7.90	3.10	65.53
10 DAA	87.55	27.88	67	2.12	9.80	5.40	61.27
15 DAA	123.30	36.89	55	2.99	15.10	8.90	58.20
20 DAA	151.7	42.26	61	3.42	21.20	12.10	52.23
25 DAA	173.8	55.70	68	3.98	29.20	18.30	46.11
30 DAA	214.8	79.73	60	4.12	31.90	20.80	41.42
35 DAA	246.8	102.80	73	4.82	36.80	25.60	38.48
40 DAA	307.8	140.60	66	5.98	38.20	28.70	36.80
45 DAA	281.1	137.70	68	6.38	34.81	27.90	31.16
CD (P=0.05)	12.09	7.86	NS	0.26	3.637	4.702	6.50

DAA - Days after anthesis

The pod and seed fresh and dry weights of *Mucuna pruriens* (Jeeva, 2003) increased from 5 days after anthesis to 45 days after anthesis and thereafter they showed a declining trend. The seed fresh weight reached the maximum of 0.313g at 12 DAA in *O.basilicum* 0.060g at 14 DAA in *O. sanctum* and 0.109g at 22 DAA in *O. gratissimum* (Swapna Moova, 2003). The fresh and dry weight of fruits increased progressively from the day of anthesis to 40 DAA and thereafter it decreased in *Solanum nigrum* (Gunasekaran, 2003). A steep rise in fruit fresh weight and dry weight was observed between five and 20 DAA. At this stage of development, the fresh weight of fruits increased from 41.54 mg fruit^{-10} to 151.7 mg fruit^{-10} and dry weight of fruits increased from 15.71 mg fruits^{-10} to 42.26 mg fruits^{-10} (Table 5).

9.1.4. Germination and Vigour

Seed attain the capability for germination long before they reach physiological maturity. For example, in Mississippi Lee soybean seed are capable or germination about 38 days after pollination while maximum dry weight (physiological maturity) is reached at 60-65 days; corn seed germinate nearly 100% about 25 days after pollination, while physiological maturity is only reached at 55-60 days. Many of the cereal seeds, *e.g.*, wheat, barley, sorghum, rice, are capable of germination 8-12 days after anthesis or pollination.

Although immature seed are capable of germination, vigour is low and the seedling produced are noticeably smaller and weaker than those produced by physiologically mature seed. Numerous studies have shown that maximum seed / seedling vigour is attained at approximately the same time as maximum dry weight or physiological maturity. Decreasing seed moisture content, increasing dry weight and maximum dry weight, capability for germination, maximum germinability and vigour are the seed maturation "benchmarks" of greatest interest and significance for seeds men. These benchmarks are considered in somewhat greater detail for several kinds of seed in the following sections.

Agricultural crops

Tekrony *et al.* (1980) found that in soybean, seed vigour declined rapidly with the delay in harvest after harvestable maturity. Sai Babu *et al.* (1984) reported that the maximum germination was recorded when the seed was harvested at 30 days after flowering and 25 days after flowering, respectively in rice. Seed harvested at 35 days after anthesis showed maximum germinability in sorghum (Sai Babu and Hussaini, 1984). According to Shaik Mohammad *et al.* (1984) the optimum time to harvest of the EC-68414 cultivar of sunflower was 90 days after sowing for higher germination. Kharb *et al.* (1993) stated that the pigeonpea seeds attained maximum germination potential when harvested at 45 to 49 days after anthesis. Sarada *et al.* (1994) observed maximum germination in seed from cobs harvested at 45 days after flowering in maize, while it was least in seed from cobs harvested at 30 days after flowering. According to Bhaskaran (1995) higher germination per cent was recorded in rice harvested at physiological and 3 days after physiological maturity, while the delayed harvest recorded lower germination. Germination percentage and vigour index were the highest in soybean when seeds harvested at 46 days after flowering (Gore *et al.*, 1997). Basave gowda and Gowda (1999) observed higher germination percentage of sorghum hybrids CSH-1 and CSH-5 when seeds harvested at 44 and 38 days after anthesis, respectively. In rice hybrid CORH 2, the highest germination and vigour were recorded in seeds harvested at 30 – 35 days after 50 % floweing (Raja, 2003). The seeds of cluster bean (Renega Devi, 2004) recorded the germination of 20 per cent and vigour index of 115 at 20 DAA and it increased with advances in maturity and reached a maximum germination of 96 per cent and vigour of 3055 at 50 DAA which coincide well with the accumulation of maximum fresh and dry weight of seed (Table 6).

Table 6: Seed germination and vigour of cluster bean during seed development and maturation

Days after anthesis	Germination (%)	Vigour index
5	-	-
10	-	-
15	-	-
20	20	115
25	24	326
30	36	749
35	52	1350
40	76	2219
45	92	2815
50	96	3055
55	96	3005

Horticultural crops

In tomato, seeds harvested at 10 days after mass maturity showed improved germination as reported by Doijode (1983). Kanwar and Saimbhi (1987) could obtain matured seeds with high germination at 35 days after anthesis in Okra under Punjab condition. Kwon and Bradford (1987) reported that maximum viability and vigour occurred in tomato 15 days after mass maturity. The on set of both germinability occurred either just before or at mass maturity (53 DAA) when moisture content reached 51-53% in pepper during seed development as reported by Demir and Ellis (1992). Sundralingam (1995) obtained good quality seeds harvested at physiological maturity, which occurred seventh week after anthesis in carrot. Baruah and Paul (1997) concluded that okra pods could be harvested 35 to 42 days after flowering to obtain higher germination and vigour. Reddy and Rolston (1997) indicated that quality of coriander seed was maximum at 37 to 42 days after peak flowering. Barbedo *et al.* (1999) found that cucumber seeds maintained higher germination when harvested at 30-35 days after flowering. Menaka (2000) obtained maximum germination and seedling vigour of amaranthus seeds harvested at 25 days after anthesis when physiological maturity coincided. Cent per cent viability and germination were reported in the tea seeds prior to shedding which occur approximately 15-16 months after anthesis. (Bhattacharya *et al.*, 2002). Murugesan (2003) observed higher germination and vigour of ashgourd seeds harvested at 65 and 80 days after anthesis in summer and rainy seasons, respectively. In case of *Mucuna pruriens,* the developing seeds started to germinate at 20 days after anthesis.

The reason for low germination at early stage might be due to imperfect development of seeds and its essential structures. Maximum germination of 96 per cent was recorded at 45 days after anthesis. The developing seed was able to germinate at 10 DAA in *O. basilicum* (Fig.19), 8 DAA in *O. sanctum* and 12 DAA in *O. gratissimum*. The germination increased gradually from 10 per cent (10 DAA) to a maximum of 47 per cent (22 DAA) in *O. basilicum*, 1 per cent (8 DAA) to a maximum of 41 per cent (20 DAA) in *O. sanctum* and from 2 per cent (12 DAA) to 86 per cent (30 DAA) in *O. gratissimum* (Swapna Moova, 2003). In *Solanum nigrum* (Gunasekaran, 2003) , the maximum germination and vigour were attained, when seeds harvested at 40 – 46 days after anthesis (Fig. 23).

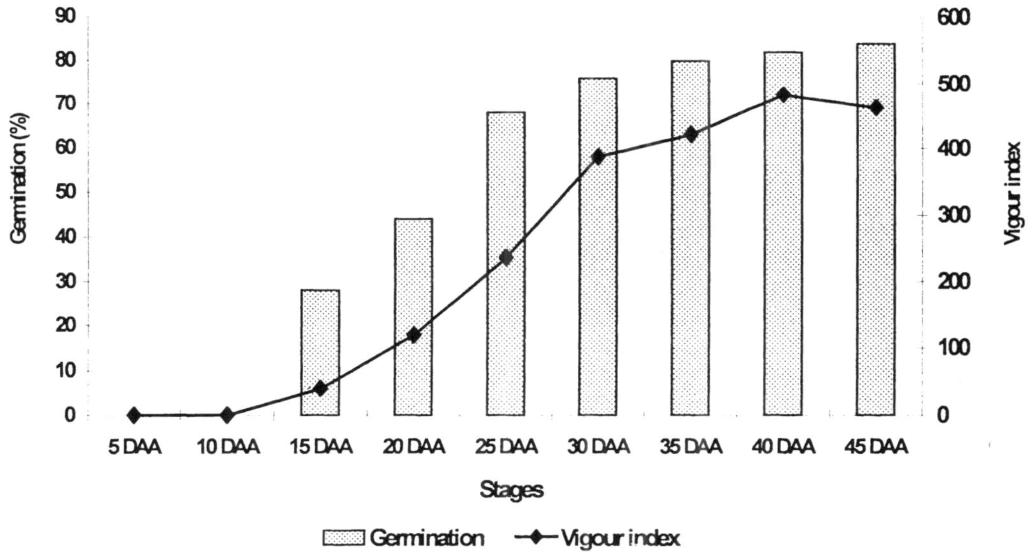

Fig.23: Seed germination and vigour during different stages of seed development in *Solanum nigrum*

9.1.5. Biochemical changes

The late events in seed maturation include biochemical formation of storage sugars, proteins and hormones. The increase in the thickness of the embryo during seed development reflects an increase in the storage reserves especially protein and lipid. Negligible change in the DNA and RNA of the testa occurs during seed development (Fig. 24), since they are nucleotides of the nucleus and cytoplasm, and any marked increase in their occurrence is reflection of cell division. In contrast, DNA and RNA increase rapidly with increased cell division during early embryo and endosperm growth, but level off with increased cell expansion (Fig. 25).

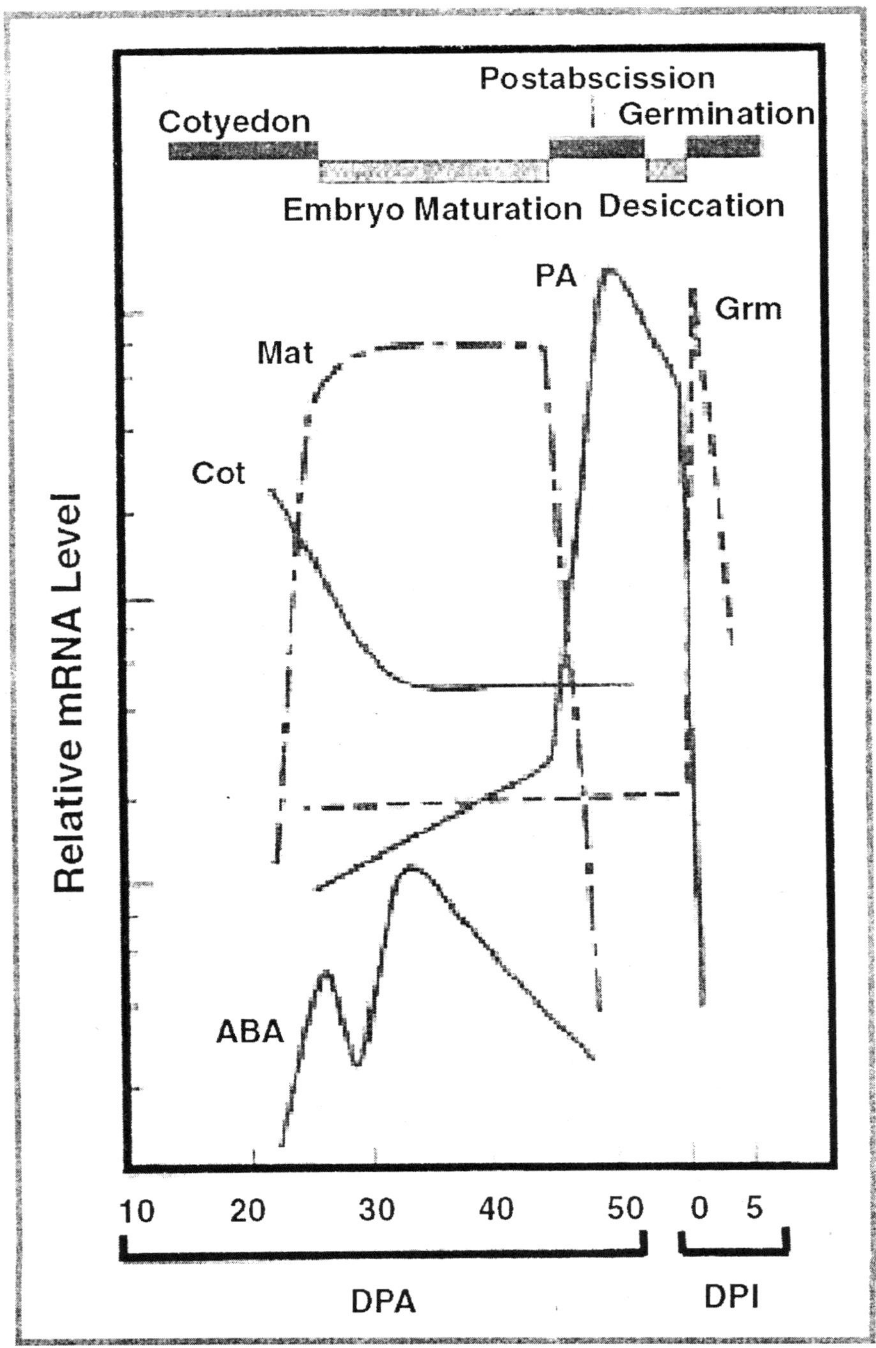

Fig. 24: Gene expression during seed development

Source: Hughes and Galau (1989), Genes and Devl. P: 358 – 369

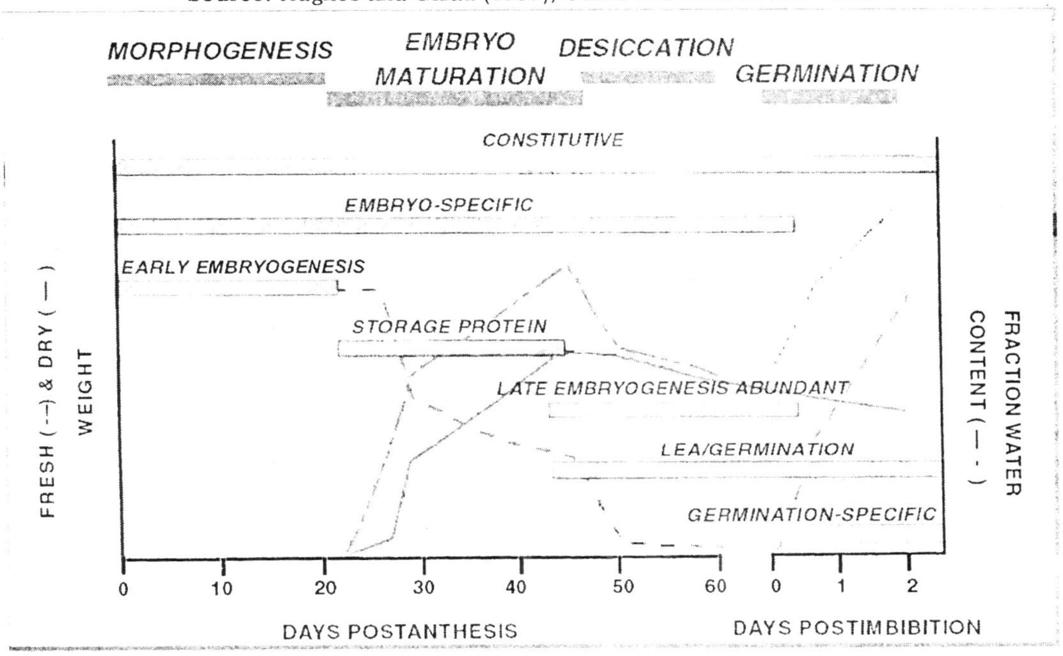

Fig. 25: Gene expression during seed development

Source: (1997) in Larkins and Vasil, Cellular and Molecular Biology of Plant Seed Development

Agricultural crops

Bhaskaran (1995) observed higher protein content, lower electrical conductivity, free sugars and free amino acids were recorded in rice seed harvested at physiological and 3 days after physiological maturity, while the delayed harvest recorded lower values of protein and higher values of electrical conductivity, free sugars and free amino acids.

In rice hybrid CORH 2 (Raja, 2003), the maximum electrical conductivity in seed leachate was registered at 20 days after 50 per cent flowering. The electrical conductivity decreased with advance in seed maturity and it was the minimum at 40 days after 50 per cent flowering indicating that the seed coat integrity increased with increase in maturity time.

The free amino acids and free sugars content increased from 6.46 and 7.36 $\mu g\ ml^{-1}$ in the seed leachate of 20 days after 50 per cent flowering to 7.08 and 9.95 $\mu g\ ml^{-1}$ of seed leachate of 25 days after 50 per cent flowering and thereafter, there was a reduction in free amino acids and free sugars content in seed leachate till 40 days after 50 per cent flowering. The result clearly indicated that the seed coat integrity started increasing from 30 days after 50 per cent flowering to 40 days after 50 per cent flowering.

At early stages, the sugars and amino acids are soluble form and they are converted to some insoluble form and stored when the seeds attain maturity

(Gurusamy, 1996) and this is the probable reason to increase and/or decrease of free amino acids and free sugars at early and later stages of seed maturation, respectively. The maximum of total sugars (0.679 mg g^{-1}), starch (73.15 per cent) and protein (8.69 per cent) were recorded at 30 days after 50 per cent flowering. However, there was a marginal reduction in total sugars, starch and protein content of seeds after 30 days of flowering. The probable reason for increase in total sugars, starch and protein was accumulation of more dry weight (Rajasekaran, 2001).

Dasgupta and Bewley (1982) reported that immature seeds of *Phaseolus vulgaris* failed to germinate without desiccation, which is associated with protein synthesis and desiccation plays a role in permanently suppressing developmental protein synthesis and in inducing germination protein synthesis and the desiccation of developing seeds was characterised by the accumulation of a set of mRNA and related proteins called late embryogenic abundant proteins in the desiccation stage of seeds. But Khattra and Singh (1995) reported that water saturation deficit occur during desiccation which increased with progressive maturity and was positively correlated with soluble sugars and free amino acids. In cluster bean (Renuga Devi, 2003), the protein content was maximum at 50 Days after anthesis (23.8 per cent) which was maintained thereafter upto 55 DAA.

Increase in dehydrogenase activity was positively correlated with germination of developing seeds (Gurusamy, 1996). Seeds harvested at physiological maturity recorded higher alpha-amylase activity than later stage .As the maturity of seeds advanced, the dehydrogenase (Fig. 26).

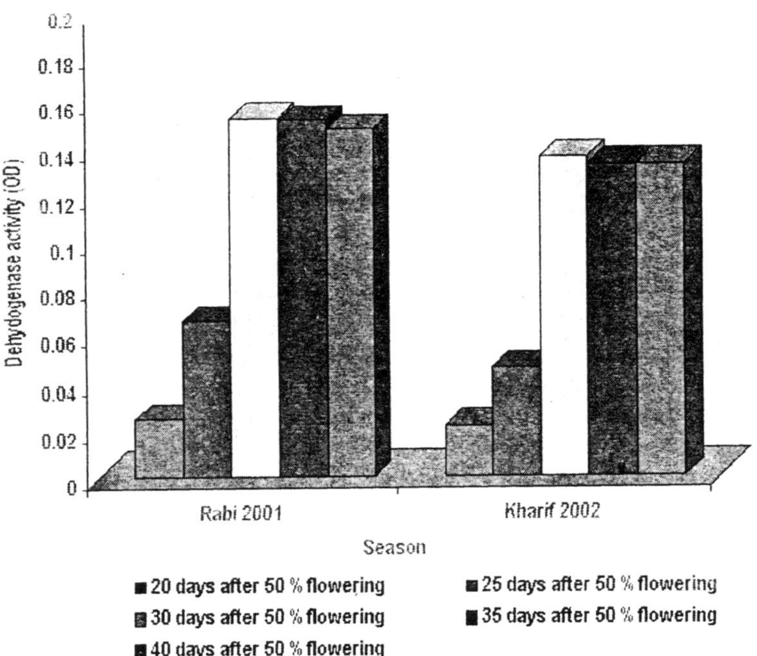

Fig. 26: Influence of stage of harvest on dehydrogenase activity of rice hybrid CORH2

Alpha-amylase and ATPase enzymes activity (Fig. 27) also increased and they reached the maximum at 30 days after 50 per cent flowering and thereafter, a marginal decrease in this enzyme activity was noticed.

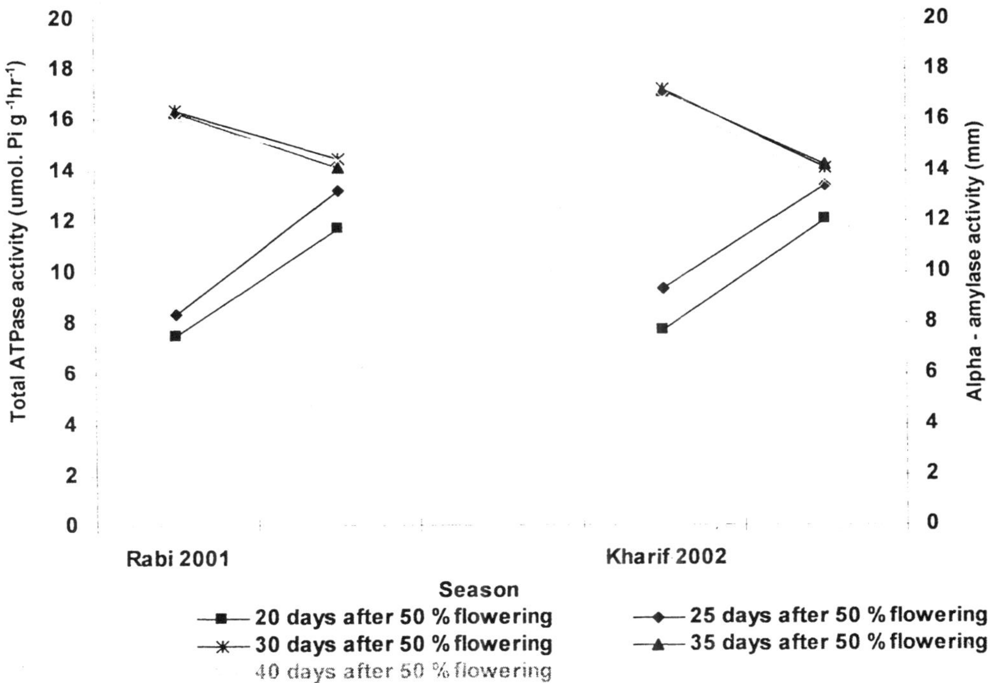

Fig. 27: Influence of stage of harvest on ATPase and alpha amylase activity of rice hybrid CORH2

Horticultural crops

Chittiraichelvan and Shanmugavelu (1977) opined that papaya seeds attained maximum accumulation of seed protein at 135th day after anthesis. Crouch and Sussex (1981) detected both 17s, 12s protein during early embryo development and added that accumulation of the 17s protein stopped when the water content of embryo began to decline, but accumulation of the 12s protein continued until seed maturity in *Brassica napus* L. Immature seeds *Phaseolus vulgaris* failed to germinate without desiccation which is associated with protein synthesis and desiccation plays a role in permanently suppressing developmental protein synthesis and in inducing germination protein synthesis (Dasgupta and Bewley, 1982). The synthesis of soluble protein in developing castor bean endosperm (in the 30 – 33 kDa range) commences by 20 days and continues upto 50 days.

Other proteins (eg, 45 and 46 kDa polypeptides) are synthesized exclusively during late maturation (40 to 50 days of development) (Kermode *et al.*, 1986). Changes in fruit weight, length girth along with biochemical changes in chilli were studied during fruit and seed development in chilli by Rama Rao *et al.*, (1990) and reported that peak seed quality achieved when pod colour changed to yellow.

Blackman *et al.* (1991) indicated that desiccation of developing seeds was characterized by the accumulation of a set of mRNA and related proteins called late embryogenic Abundant (LEA) proteins in the desiccation stage of pea. Sreenivasulu and Amritphale (2000) reported changes in protein composition in cellular membranes of seed parts in cucumber seeds treated with ethanol during imbibition. According to Dasgupta and Mandal (1993) the major classes of storage protein (12s and 2s) were appreciably synthesized two weeks after flowering during seed development in cauliflower. The accumulation of protein followed more or less sigmoidal pattern in ash gourd seed (Murugesan, 2003), which starts from 30 days after anthesis and continued upto 70 day after anthesis in summer crop while rainy crop had protein accumulation beyond 70 days after anthesis and continued upto 80 days after anthesis. The protein content of *Mucuna pruriens* seed was the maximum at 15 days after anthesis and thereafter decreased upto 55 days after anthesis (Table 7). The higher protein content at early stages of maturity might be due to the synthesis and deposition of protein in seed, which were important metabolic steps during early phases of seed development (Jeeva, 2003).

Table 7: Changes in protein content during seed development and maturation of *Mucuna pruriens* L.

Stages in days after anthesis	Protein content (%)
5	-
10	-
15	35.37
20	34.43
25	34.28
30	33.71
35	31.51
40	30.34
45	29.88
50	29.37
55	29.07
Mean	26.18
SEd	0.042
CD (P=0.05)	0.086

In *Solanum nigrum* (Gunasekaran, 2003), the maximum electrical conductivity of seed leachate (0.102 dSm^{-1}) was registered at five DAA (Table 8). As the maturity advanced, it decreased and reached the minimum (0.021 dSm^{-1}) at 45 DAA indicating that membrane integrity was maximum at this stage. More electrical conductivity recorded at earlier stages might be due to poor integrity of cell membrane and immature nature of embryos. Protein accumulation (Table 8) increased gradually and reached maximum at 40 DAA.

Table 8: Physiological properties of seed during development and maturation in *Solanum nigrum*

Stages	Electrical conductivity (dSm^{-1})	Protein content (%)
5 Days after anthesis	0.102	5.72
10 Days after anthesis	0.091	8.81
15 Days after anthesis	0.079	9.16
20 Days after anthesis	0.064	9.84
25 Days after anthesis	0.048	10.91
30 Days after anthesis	0.036	11.38
35 Days after anthesis	0.029	12.83
40 Days after anthesis	0.024	13.52
45 Days after anthesis	0.021	13.10
SEd	0.004	0.15
CD (P=0.05)	0.009	0.32

REFERENCES

Ajayi, S.A. and M.A.B. Fakorede. 2000. Physiological maturity effects on seed quality, seedling vigour and mature plant characteristics of maize in a tropical environment. *Seed Sci.&Technol.,* **28**: 301-319.

Anderson, A.P. 1985. The grand period of growth in a fruit of *Cucurbita pepo* determined by weight. *Mines Bot. Studies,* **1**: 238-279.

Andrews, C.H. 1996. Some aspects of pod and seed development in soybean. Ph.D. Thesis, Mississippi State Univ. State College.

Apparao, S., N. Kameswararao and M.H. Mangesha. 1993. Germinability and seedling vigour of pearlmillet seeds harvested at different stages of maturity. *Crop Res.,* **32**: 141-145.

Arul Prabu, C. 1998. Studies on certain aspects of seed production and .storage in polebean *(Phaseolous vulgaris* L.). M.Sc. (Ag.) Thesis, Tamil Nadu Agricultural University, Coimbatore.

Arul Prabu, C. 1998. Studies on certain aspects of seed production and storage in polebean *(Phaseolous vulgaris* L.). M.Sc. (Ag.) Thesis, Tamil Nadu Agricultural University, Coimbatore.

Barbedo, C.J., A.S.C. Barbedo, J. Nakagawa and O. Sato. 1999. Effect of fruit age and post harvest period of cucumber on stored seeds. *Pesquisa Agropecuaria Brasileira,* **34**(5): 839-847.

Baruah, G.K.S. and S.R. Paul. 1997. Seed development and maturation studies in Okra (*Abelmoschus esculentus*). *Annals of Agriculture Research,* **18**(3): 367-368.

Basave gowda and S.J. Gowda. 1999. Effect of stages of harvest on seed germination during storage in hybrid sorghum. *Crop Res.,* **18**(3): 368-69

Bhaskaran, M. 1995. Studies on the influence of biotic and abiotic factors on seed yield, quality and storability of rice (*Oryza sativa* L.). Ph.D.(Ag.) Thesis, Tamil Nadu Agricultural University, Coimbatore.

Bhattacharya, A., P.K. Nagar and P.S. Ahuja. 2002. Seed development in *Camellia sinensis* (L.) O. Kuitze. *Seed Science Research,* **12**: 39-46.

Bishnoi, V.R. 1974. Physiological maturity of seeds in *Triticale hexaploid* L. *Crop Sci.,* **14**(6): 819-821.

Blackman, S.A., S.H. Wettlaufer, R.L. Obendorf and A.C. Leopold. 1991. Maturation proteins associated with desiccation tolerance in soybean. *Plant Physiology,* **96**: 868-874.

Chamma, H.M.C.P., B. Choudury and C. Singh. 1980. Maturation of seeds of Aroana beans (*Phaseolus vulgaris* L.) and its influence on the storage potential. *Seed Sci.& Technol.,* **18**(2): 371-382.

Chandrasekaran, 1979. Studies on the development maturation and storage of seed in bottle gourd. M.Sc. (Ag.) Thesis, Tamil Nadu Agricultural University, Coimbatore.

Chittiraichelvan, R. and K.G. Shanmugavelu. 1977. Studies on the physico-chemical changes, during growth and development of CO 2 papaya (*Carica papaya* L.) seeds. *Seed Research,* **5**(1): 32-36.

Crouch, M.L. and I.M. Sussex. 1981. Development and storage – protein synthesis in *Brassica napus* L. embryos *in vivo* and *in vitro. Planta,* **153**: 64-74.

Dasgupta, J. and J.D. Bewley. 1982. Desiccation of axes of *Phaseolus vulgaris* during development of switch from a development pattern of protein synthesis to a germination pattern. *Plant Physiology,* **70**: 1224-1227.

Dasgupta, S. and R.K. Mandal. 1993. Compositional changes and storage protein synthesis in developing seeds of *Brassica campestris. Seed Sci. & Technol.,* **21**: 291-299.

Demir, I and R.H. Ellis. 1993. Changes in potential seed longevity and seedling growth during seed development and maturation in marrow. *Seed Science Research,* **3**: 247-257.

Demir, I. and R.H. Ellis. 1992. Development of pepper (*Capsicum annum*) seed quality. *Annals of applied biology,* **121**: 385-399.

Doijode, S.D. 1983. Studies on vigour and viability of seeds at different stages of fruit development

Gore, S.V., R.B. Patil and G.R. Wankhade. 1997. Effect of maturity period and harvesting time on seed quality in soybean (*Glycine max* (L.) Merill) cultivars. *Seed Res.,* **25**(1): 45-49.

Gore, S.V., R.B. Patil and G.R. Wankhade. 1997. Effect of maturity period and harvesting time on seed quality in soybean (*Glycine max* (L.) Merill.) cultivars. *Seed Res.,* **25**(1): 45-49.

Gunasekaran, 2003. Seed development and maturation, extraction methods, germination test and storage in *Solanum nigrum* Linn. M.Sc. (Ag.) Thesis, Tamil Nadu Agricultural University, Coimbatore

Gurusamy, C. 1996. Studies on the pattern of seed development, maturation and deterioration in cauliflower. (*Brassica oleraceae* L. var. botrytis) cv. Kibo Giant. M.Sc. (Ag.) Thesis, Tamil Nadu Agricultural University, Coimbatore.

Jayabarathi, M. 1982. A study on seed development and maturation, production and storage in soybean. M.Sc.(Ag.) Thesis, Tamil Nadu Agricultural University, Coimbatore.

Jeeva, B. 2003. Seed development and maturation, picking and storage studies in *Mucuna pruriens* (L.) M.Sc.(Ag.) Thesis, Tamil Nadu Agricultural University, Coimbatore

Jerlin, R., P. Srimathi, and K. Vanangamudi. 2001. Seed physiological and maturity indices. **In**: Recent techniques and participatory approaches on quality seed production. (Eds. K. Vanangamudi, A.Bharathi, P. Natesan, R. Jerlin, T.M. Thiyagarajan and S. Kannaiyan) Department of Seed Science and Technology, Tamil Nadu Agricultural University, Coimbatore

Karuna, M.N. and B. Aswathaiah. 1990. Accelerated ageing test to predict the storability of beet root and carrot seeds. *Seeds and Farms,* **16** (9-10): 10-12.

Kathiravan, M. 2001. Seed technological studies in lab lab (*Lablab purpureus* (L.). sweet var. typicus parin) cv. CO1. M.Sc. (Ag.) Thesis, Tamil Nadu Agricultural University, Coimbatore.

Kermode, A.R., J.D. Bewley, J. Dasgupta and S. Mistra. 1986. The transition from seed development to germination : a key role for desiccation ?. *Hort. Science,* **21**(5): 1113-1118.

Kharb, R.P.S., B.P.S. Lather and D.P. Deswal. 1993. Studies on physiological maturity in Early pigeonpea. *Seed Res.,* **21**(1): 45-48.

Khattra, S. and G. Singh. 1995. Indices of physiological maturity of seeds. A. Review. *Seed Res.,* **23**(1): 13-21.

Kwon, O.S. and K.J. Bradford. 1987. Tomato seed development and quality as influenced by pre harvest treatment with ethephon. *Hort. Science,* **22**: 588-591.

Maheswari, P. 1950. AN introduction to the Embryology of the Angeosperm. New Yark. McGraw – Hill Book Company

Manohar, M.S. 1970. Seed development and germination studies on lab lab (*Dolichos lablab* L.). *Indian J. Horticulture,* **27** (182): 86-92

Manohar, M.S. and S.C.P. Sachan. 1974. Pod development and germination in pees (*Pisum sativum* L.) *Veg. Sci.,* **1**: 22-30

Menaka, C. 2000. Seed technological studies in amaranthus (*Amaranthus bicolour*) cv. CO 5. M.Sc. Thesis, Tamil Nadu Agricultural University, Coimbatore

Miccolis, V. and M.E. Saltveit. 1991. Morphological and physiological changes during fruit growth and maturation of seven melon cultivars. *Journal of American society of Horticultural Science,* **116**(6): 1025-1029

Murugesan, P. (2003). Harvesting, post harvest handling and conditioning of fruit and seed in Ash gourd Benincasa hispida (Thunb). (ogn) Ph.D. Thesis, Tamil Nadu Agricultural University, Coimbatore-3.

Nerson, H., H.S. Paris, E.P. Paris. 2000. Fruit shape, size and yield in *Cucurbita pepo. Acta Horticulture,* **510**: 227-230.

Ockenden, I., M. West, J. Domingues and J.N.A. Lott. 2001. Changes in the element composition of globoids and whole embryos in developing seeds of *Cucurbita maxima. Seed Science Research,* **11**: 35-44.

Parmasivam, V. 1990. Studies on development, maturation quality and storability of pea (*Pisum sativum* L.) seeds. M.Sc.(Ag.) Thesis, Tamil Nadu Agricultural University, Coimbatore.

Patrick Jasper. 1998. Studies on seed production and storage aspects of pea (*Pisum sativum* L.) M.Sc. (Ag.) Thesis, Tamil Nadu Agricultural University, Coimbatore.

Raja, K. 2003. Investigations on nursery and main field management techniques for quality seed production of rice hybrid CORH 2. Ph.D. Thesis, Tamil Nadu Agricultural University, Coimbatore.

Rajasekaran, R. 2001. Seed production, processing and storage studies in niger (*Guizotia abyssinica* L. f. Cass.) cv. Paiyur 1. M.Sc.(Ag.), Thesis, Tamil Nadu Agricultural University, Coimbatore.

Rajasekaran, S. 1997. Seed technological studies in ricebean M.Sc.(Ag.) Thesis, Tamil Nadu Agricultural University, Coimbatore

Rakesh Seth, C.H. Padmavathi and L.P. Misra. 1999. Seed quality in relation to degree of seed maturity in fodder cowpea. *Seed Res.,* **27**(1): 66-70

Rama Rao, G., G.V. Hanumantha Rao and T. Sujatha. 1990. Rate of fruit development and drymatter accumulation in chilli cultivars (*Capsicum annum* L.). *South Indian Horticulture,* **38**(4): 191-192.

Reddy, K. and M.P. Rolston. 1997. Achievement of maximum seed yield in coriander. Proceedings of Annual conference Agronomy Society of New Zealand. *Seed Abstract,* **22**(4): 1269. pp. 27 and 37-40

Renugadevi, J. 2004.Investigations on seed management, production, processing and storage of cluster bean (*Cyamopsis tetragonaloba* (L.) Taub. Ph.D., thesis Tamil Nadu Agricultural University Coimbatore -3.

Sabir-Ahamed, A. 1989. Studies on the production of quality seed and storage in soybean (*Glycine purpureus* L.). M.Sc. (Ag.) Thesis, Tamil Nadu Agricultural University, Coimbatore.

Sai Babu, K.G.R.S. and S.H. Hussaini. 1984. Effect of maturity on seed quality in sorghum. *Seed Res.,* **12**(2): 47-50.

Sai Babu, K.G.R.S., S.H. Hussaini, B.M.M. Reddy and P.R. Reddy. 1984. Effect of nitrogen fertilization on maturity.

Samnotra, R.K., V. Tripathi and R. Kaul. 2002. Seed development in Okra (*Abelmoschus esculentus* (L.) Moench) cv. Parbhani Kranti. *Environment and Ecology,* **20**(1): 230-232.

Sarada, P., B. Murali Mohan Reddy, S.H. Hussaini and V. Padma. 1994. Effect of stage of harvest and drying methods on the storability of maize inbred, CM 201. *Seed Res.,* **22**(1): 77-78.

Shaik Mohammad, K.V.L.N. Dutt and G. Chandrasagar. 1984. Note on the effect of time of harvest on germinability of sunflower (*Helianthus annuus* L.) seeds. *Seed Res.,* **12**(1): 95-97.

Shanmugaraj, K. 1978. Seed maturation and storage studies in lab lab (*Lablab purpurens* L.) M.Sc. (Ag.) Thesis, Tamil Nadu Agricultural University, Coimbatore.

Showalter, R.K. 1961. Specific gravity, weight and solid relationship in watermelons. *Proceedings flower statistics Horticultural Science,* **74**: 268-271.

Sreenivasulu, Y. and D. Amritphale. 2000. Changes in protein composition in cellular membranes of various parts of secondary dormant cucumber seeds treated with ethanol. *Seed Science Research,* **10**: 61-70

Sundaralingam, K. 1995. Seed technological studies in carrot (*Daucas carota* L.). M.Sc. (Ag.) Thesis, Tamil Nadu Agricultural University, Coimbatore.

Swapna mova. 2003. Tracing seed maturation pattern and development of seed testing procedure for *ocimum* spp. M.Sc. (Ag.) thesis, Tamil Nadu Agricultural University, Coimbatore.

Ushamythili, M. 1982. A study on seed production processing and storage in lab-lab (*Lablab purpuries* L.) sweet var. typicus prain. M.Sc. (Ag.) Thesis, Tamil Nadu Agricultural University, Coimbatore.

Varatharajan, A. 1979. Studies on maturation and quality of seeds in ribbed gourd and bitter gourd. M.Sc. (Ag.) Thesis, Tamil Nadu Agricultural University, Coimbatore.

Vijayakumar, A., P. Geetharani and T. Thangaraj. 2002. Seed Production, Processing and Storage in Frenchbean cv. Premier. *Seed Tech News,* **32(**1): 41.

Vijayakumar, A., P. Geetharani and T. Thangaraj. 2002. Seed Production, Processing and Storage in Frenchbean cv. Premier. Seed Tech News, **32** (1): 41.

Woyke, H.W. and M. Lazecka. 1994. Effect of the ripening phase on the quality of dill (*Anethum graveolens* L) seeds. *Biuletyn Warzywniczy,* **41**: 5-15.

3

Seed—The Structural Function and Biochemical Variations

P. Srimathi[1], K. Natarajan[2] and K. Vanangamudi[2]

[1]*Forest College and Research Institute, Mettupalayam – 641 301*

[2]*Adhiparasakthi Agricultural College, G.B.Nagar, Kalavai – 632 506*

1. SEED

Botanically seed is defined as a ripened ovule containing an embryo in arrested state of development usually with a food reserve and a protective coat. In seed technological term the part of the plant used for sowing to raise the crop is considered as seed or any propagative material is known as seed. Seeds are also defined as the dispersal and propagation units of the Spermatophyta (seed plants): Gymnosperms (conifers and related class) and Angiosperms (flowering plants). Seeds are mature, fertilized ovules. Ovules are structures of seed plants containing the female gametophyte with the egg cell, all being surrounded by the nucellus and 1-2 integuments. In angiosperms, the double fertilization results in formation of the diploid embryo and the triploid endosperm.

2. SEED FORMATION AND DEVELOPMENT IN ANGIOSPERMS

- After fertilization, the ovary becomes the fruit generally and seeds are within the fruit. The true seed formed by the fusion of male gamete (pollen) and female gamete (egg cell) are attached to the seed adopting a definite pattern that is known as placentation. The placentation, occur in any one of the three forms *viz.*, partial, axile and free central (basal).
- **Fruit:** Strictly, the ripened ovary of a plant and its contents. More loosely, the term is extended to the ripened ovary and seeds together with any structure with which they are combined, *e.g.* the apple (a pome) in which the fruit (core) is surrounded by flesh derived from the floral receptacle (Plate1).
- **Achene:** A small, usually single-seeded, dry indehiscent fruit, *e.g.* lettuce.
- **Caryopsis:** A dry, nut-like fruit typical of grasses, *e.g.* a cereal grain. It is an achene with the ovary wall united with the seed coat.

Plate 1: Teak seed

After fertilization, development of the seed starts in ovule. During fertilization, one sperm nucleus combines with the egg cell and other combines with two polar nuclei (n+n) and forms the store of food, known as endosperm (3n). In some plant species, cell of zygote multiplies and form 2 cotyletons. The food material of endosperm is absorbed by the embryonic cells with the help of a suspensor and is stored in swollen cotyledon. Seed coat is formed from the integuments of ovule. The ovule is attached with ovary through funiculus. At the time of maturity, it leaves a scar on seed known as hilum. The place left open by integument for entrance of pollen tube is known as micropyle just above the hilum.

3. GENOMIC CONSTITUTION OF SEED STRUCTURES

The major structures of seed (irrespective of kind) are seed coat (testa, tegmen, and endosperm) embroyonic axis + cotyledon, embryo and perisperm. The genomic constitution of various seed structures are as follows:

Contribution of parts of seed by male and female gamete and their genomic constitution

PART OF SEED	FEMALE PARENT	MALE PARENT
Embryo (2n)	Egg cell (n)	Sperm nucleus (n)
Cotyledon (2n)	Egg cell (n)	Sperm nucleus (n)
Endosperm (3n)	Polar nuclei (2n)	Sperm nucleus (n)
Testa (2n)	Outer integument (2n)	

Hilum (2n)	Scar of funiculus	-
Micropyle	Microhole	-
Caruncle (2n)	Outer integument (2n)	-
Strophile (2n)	Wart like outer growth (2n)	-
Coleoptile (2n)	Egg cell (n)	Sperm nucleus (n)
Coleorhiza (2n)	Egg cell (n)	Sperm nucleus (n)
Scutellum (2n)	Egg cell (n)	Sperm nucleus (n)
Pericarp (2n)	Fruit wall (2n)	-
Caryopsis (2n)	Fruit wall + seed wall	-
Appendages	Out growth of testa (2n)	

4. VARIOUS STRUCTURES OF ANGIOSPERM SEEDS

Seeds also have minor structures, which vary with seed.

4.1. EXTERNAL

Testa: The seed coat. It is derived from the integuments and its several layers thick.

Tegmen: The inner coat of testa.

Pericarp: The body of a fruit developed from the ovary wall and enclosing the seeds.

Raphe: A cord or ridge of fibrovasculer tissue connecting the base of the nucleus with placenta, the adhesent funicle of anatropous ovule (Type of ovule).

Micropyle: The Micropyle is a canal or hole in the coverings (seed coat) of the nucellus through which the pollen tube usually passes during fertilization. Later, when the seed matures and starts to germinate, the micropyle serves as a minute pore through which water enters. The micropylar seed end has been demonstrated to be the major entry point for water during seed imbibition and germination. During germination, the testa ruptures at the micropylar end and the radicle protrudes through the micropylar endosperm.

Hilum: A scar left on the seed where it was formerly attached to the funicle or placenta.

Caruncle: A wart or protuberance near the hilum of the seed.

Legenostome: The tree apex of the nucellus.

Lens: The contracted tissue, of the tree position of nucellus frequently attached to the base of the lagenostone.

Coleorhiza: The sheath of the monocotyledonous embryo, which pierced by the true radicle.

Aril (Arillus): An expansion of the funicle, arising in the placenta and enveloping the seed.

Areole: A place marked out on a surface such as on the flat sides on the seed coats in seeds of sub family mimoseae and caesalpinaceae .

Appendages: External attachment on the seed that favours dispersal of seed (*e.g.*)

Wings:	*Holoptelia integrifolia*
Arils:	Nutmeg.
Caruncles:	Jatropha, rubber
Hairs:	Salix, *Populus*
Others:	Spines, tubercles, elaisomes.

Hilum and funiculus: Funicular scar on seed coat that marks the point at which the seed was attached via the funiculus to the ovary tissue. The hilum is a large oval scar left when the seed breaks away from its placental connection, the funiculus.

Chalaza: Non-micropylar end of the seed. The base of an ovule, bearing an embryo sac surrounded by integuments.

Raphe: Ridge on seed coat formed from adnate funiculus.

Arillate: General term for an outgrowth from the funiculus, seed coat or chalaza; or a fleshy seed coat.

Aril: Outgrowth of funiculus, raphe, or integuments; or fleshy integuments or seed coat, a sarcotesta. Arils probably often aid seed dispersal, by drawing attention to the seed after the fruit has dehisced, and by providing food as an attractant reward to the disperser. The aril of the nutmeg produces the spice mace and the seed itself is the nutmeg.

Strophiole: Outgrowth of the hilum region which restricts water movement into and out of some seeds. In some hard-coated legume seeds, *e.g. Melilotus alba* and *Trigonella arabica*, a plug covering a special opening - the strophiolar cleft - must be loosened or removed before water can enter, and then only through this region.

Operculum: A little seed lid. It refers to a dehiscent cap of a seed or a fruit that opens during germination.

Carunculate: Seed with an excrescent outgrowth from integuments near the hilum, as in Euphorbia.

Fibrous: Seed with stringy or cord-like seed coat, as mace in *Myristica.*

Funicular: Seed with a persistent elongate funiculus attached to seed coat, as in Magnolia.

Strophiolate: Seed with elongate aril or strophiole in the hilum region.

Elaiosomes: A speciality in the dispersal through animals is that through ants (myrmecochory). Such seeds or fruits bear attachments, the elaiosomes that contain lures and nutrients. Myrmecochory is common with plants that live at the forest soil like violets (*Viola*).

Caruncle: A reduced aril, in the form of a fleshy, often waxy or oily, outgrowth near the hilum of some seeds. Usually it is brightly colored. It acts as an aid to dispersal. Viola seeds have an oily caruncle and are sought and dispersed by ants.

Mucilage: A layer of polysaccharide slime produced by some seeds upon imbibition. Serves in water uptake during imbibition and germination.

4.2. INTERNAL

Fruits: Fruits are mature, ripened ovaries containing seeds. The pericarp ("fruit coat") is diploid maternal tissue.

Testa (seed coat): Outer protective layer of the seed, developed from the integuments of the ovule, diploid maternal tissue

Endosperm: Storage tissue of seeds and is formed by fussion of pollen cell with 2 polar nuclei. It is nonviable in nature. Food storage tissue, triploid (3n), result of double fertilization, 2/3 of the genome is of maternal origin

Embryo: Young sporophyte, diploid (2n), result of fertilization. The mature embryo consists of cotyledons (seed leaves), hypocotyl (stem-like embryonic axis below the cotyledons), radicle (embryonic root). It consists of embryonic axis and cotyledon either one or two.

Embryonic axis: It is growing part of seed the miniature plant consisting of plumule and radial.

Cotyledon: It is the storage tissue of dicots (mostly) and is the part of embryo. It is viable.

Scutellum: It is the colyledon of monocot.

Coleoptile: Covering of miniature plumule.

Plumule: It is miniature part of shoot region of plant

Radicle: It is the miniature part of root region of plant.

Hypocotyl: It is the region below the colyledon above the radicle of the seed.

Perisperm: Diploid maternal food storage tissue originates from the nucellus or unutilized nucellus region of seed, normally seen as papery growth outside the endosperm and inside the seed coat. Only in some species, *e.g. Beta vulgaris, Piper nigrum, Coffea arabica*, many Caryophyllales

Epicotyl: It is region above the cotyledon and below the primary leaf of seed (plumule).

4.3. FUNCTIONS

Seed coat: The seed coat is the only protective barrier for the embryo from the external environment.

Hilum: Seed identification

Raphe: Seed identification

Micropyle: Water absorption

Strophiole: After absolute impermeability is reached the entry of water is under the control of strophiole

Scutellum: It supplies growing embryo with food material absorbed from endosperm through epithelium. The initial synthesis of alpha-amylase and certain proteolases also occurs in scutellum.

Aleuron layer: This layer secrets amylase and proteolytic enzymes, which hydrolyse reserves of endosperm.

5. STRUCTURE OF FEW IMPORTANT SEED SPECIES

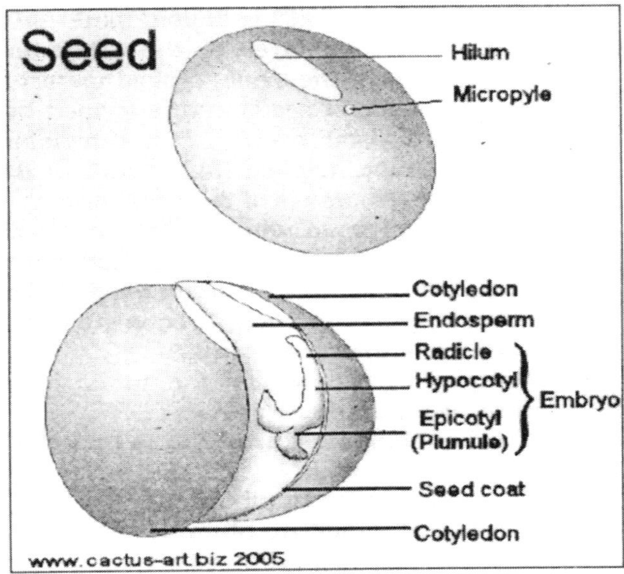

Plate 2a: Structure of cactus seed

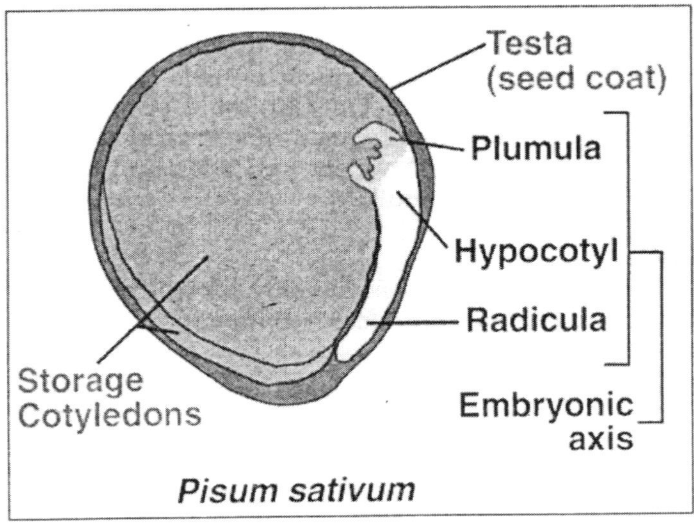

Plate 2b: Structure of *Pisum* seed

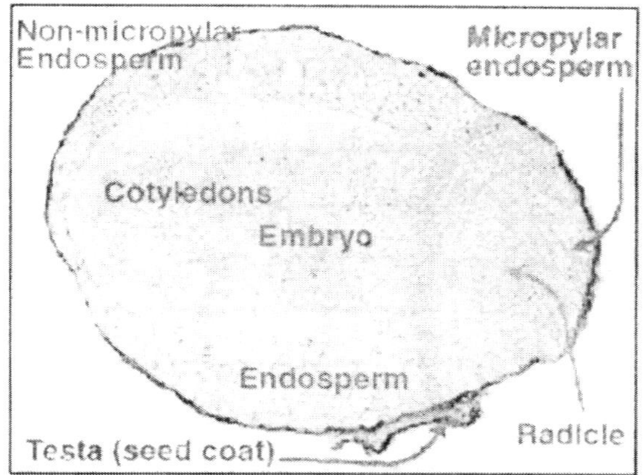

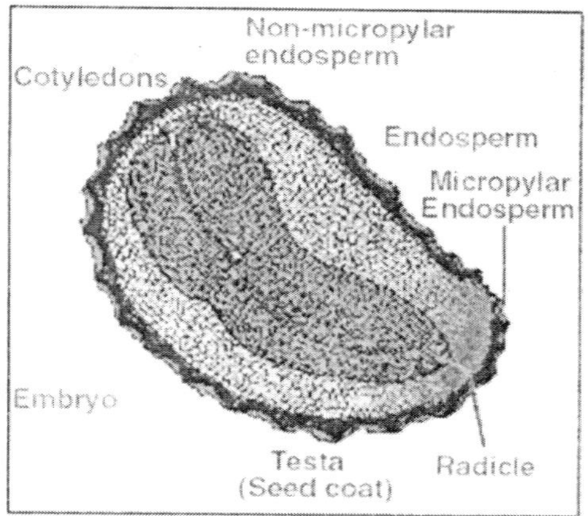

Plate 2c: Structure of *Nicotiana tobacum*

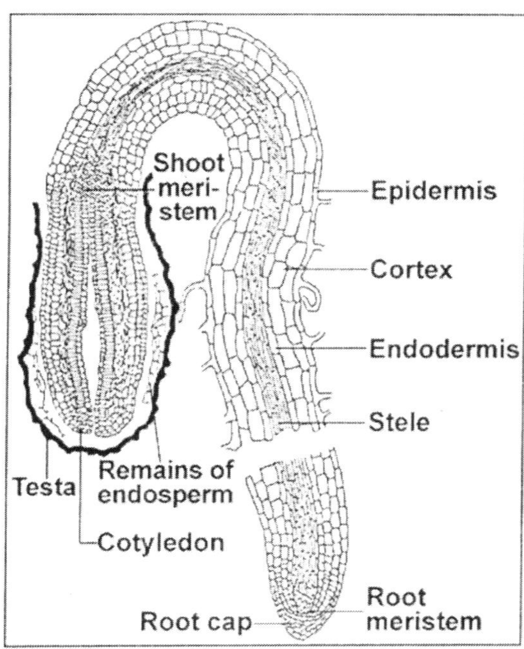

Plate 2d: Structure and germination of tobacco seed and the developmental anatomy of the seedling plant

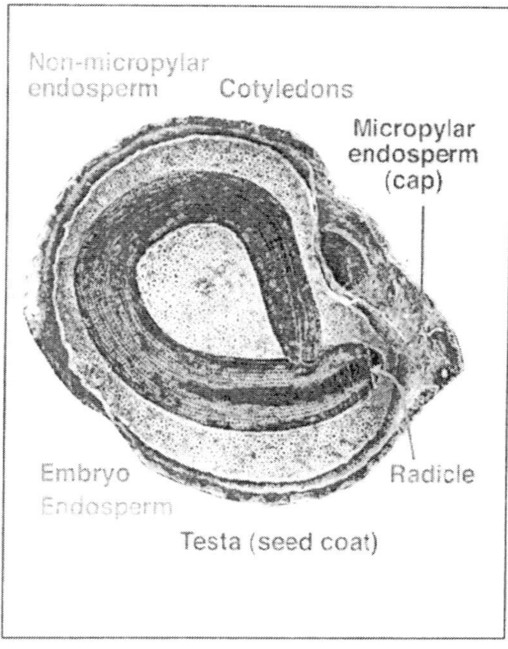

Plate 2e: Structure of tomato and pepper seed

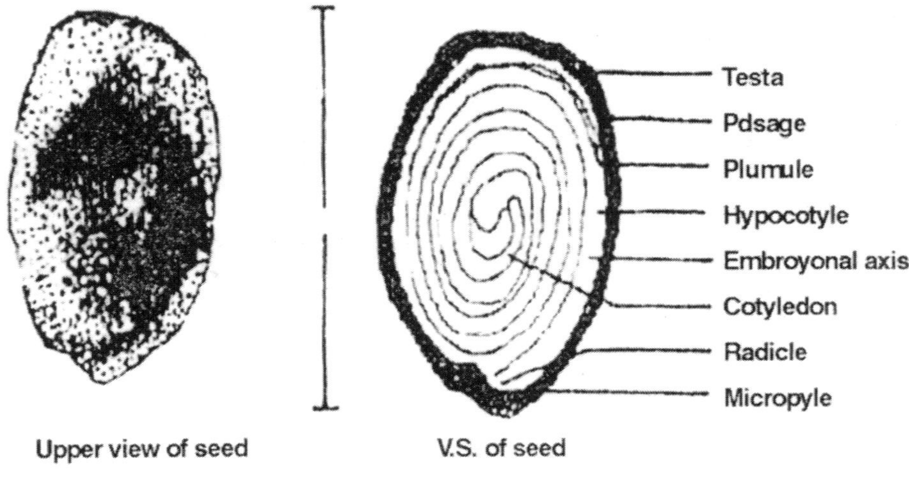

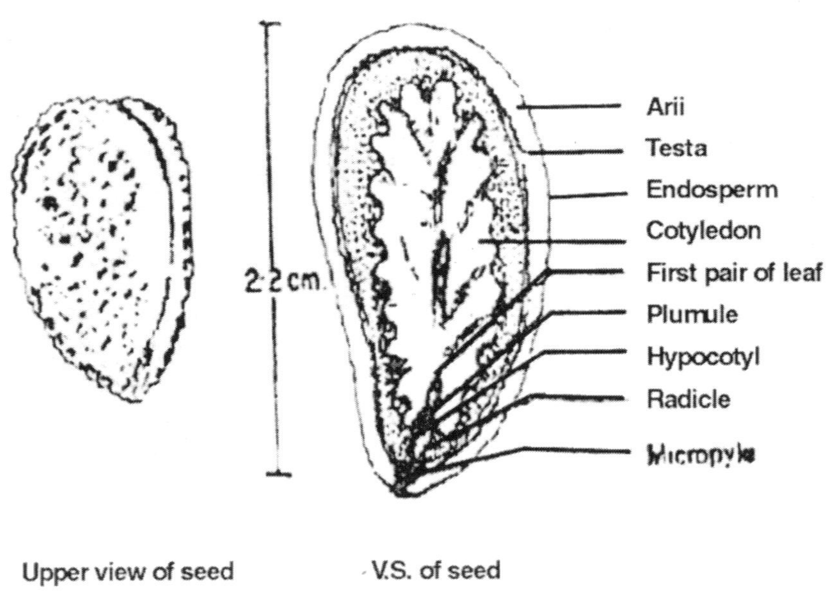

Fig. 1: Structure of *Dedonia viscosa* and castor

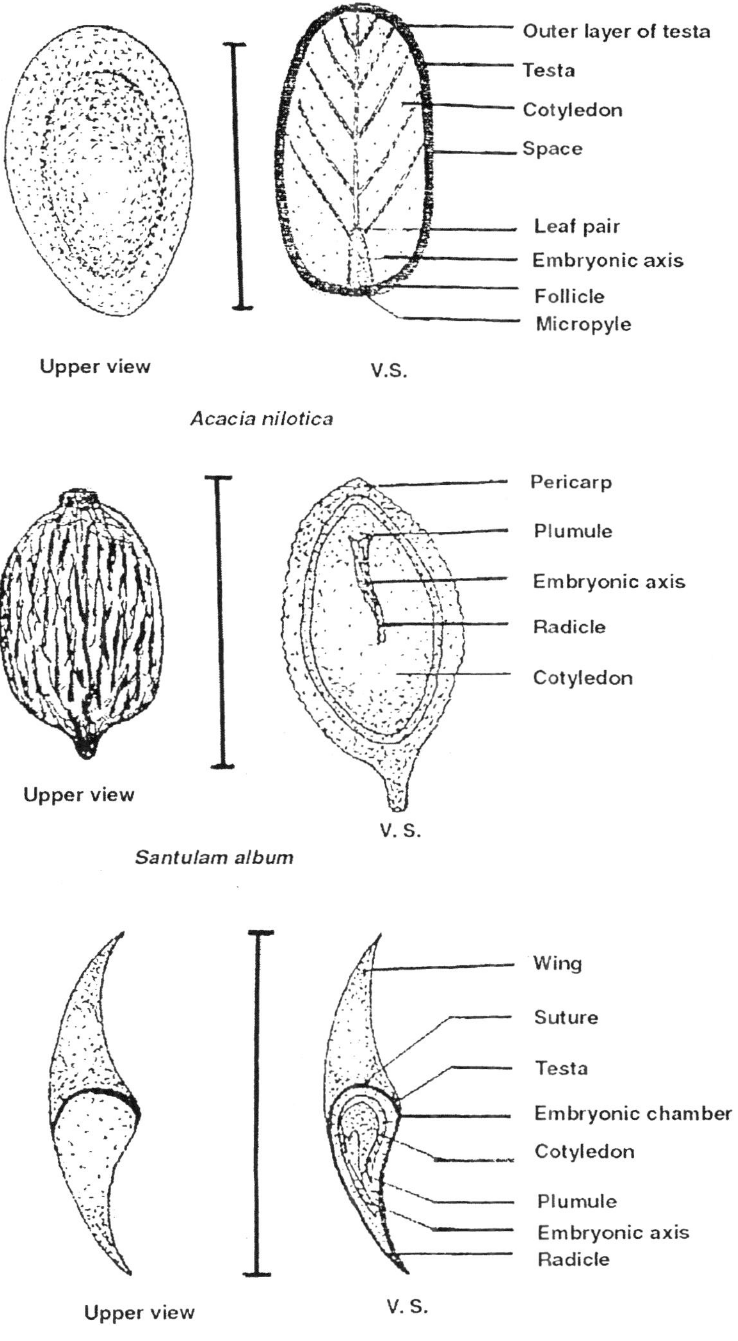

Fig.2: Structure of forest tree species

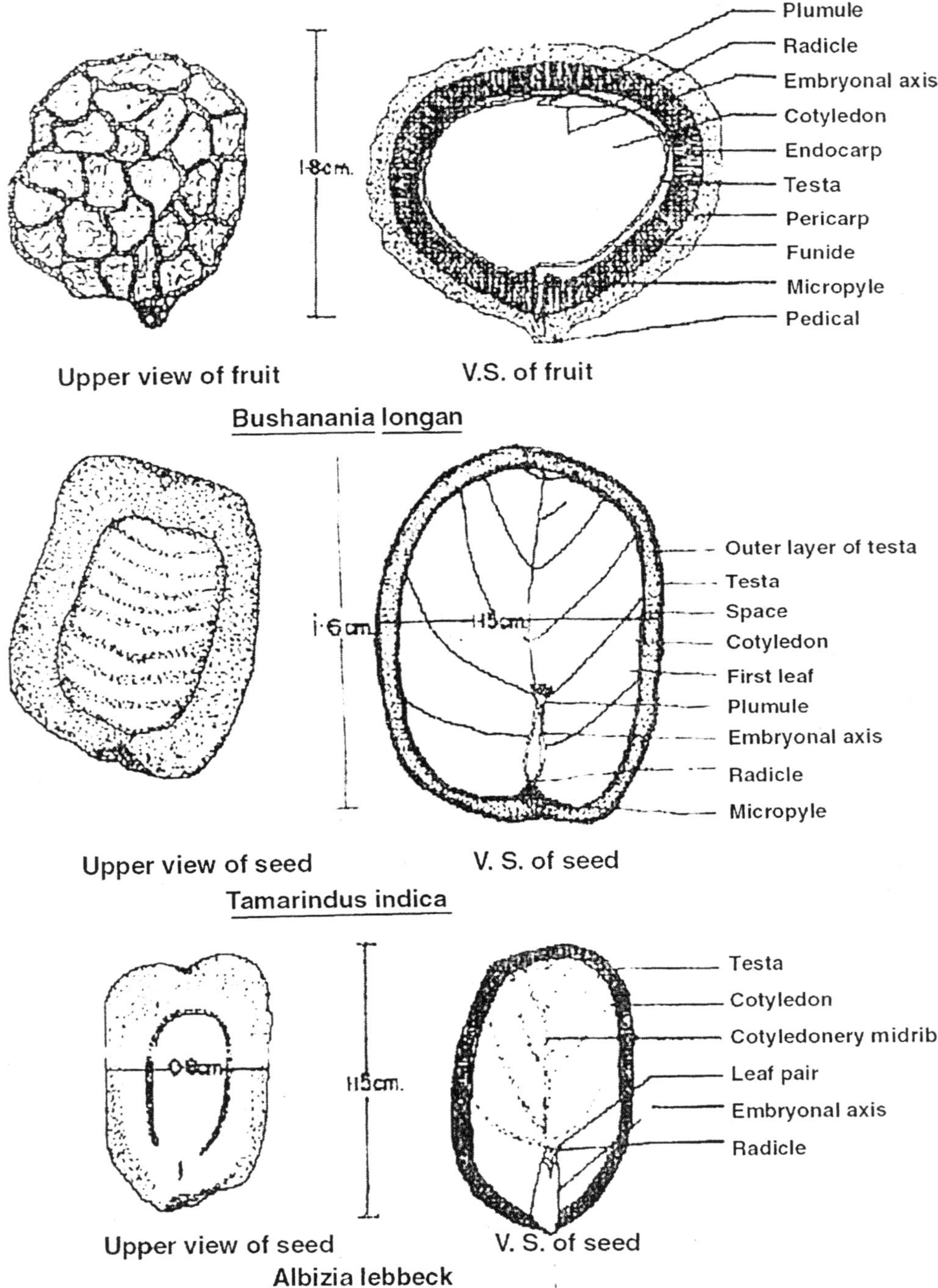

Fig.3: Structure of *Bushanania longan, Tamarindus indica* and *Albizia lebbeck*

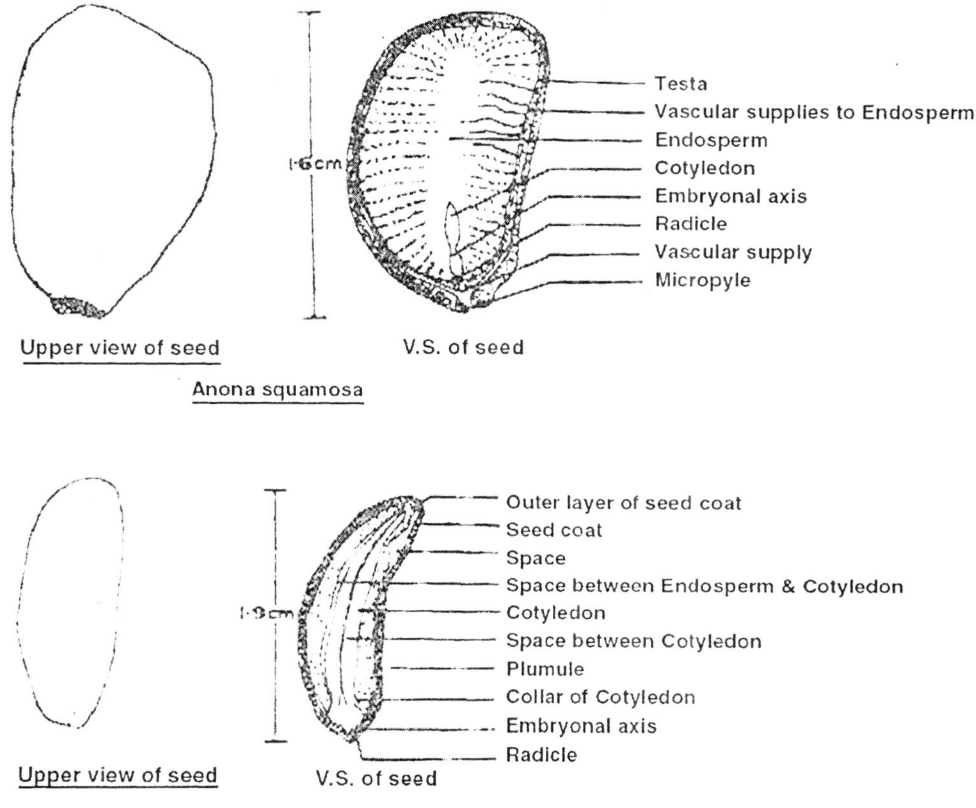

Fig. 4: Structure of *Anona squamosa* and *Delonix regia*

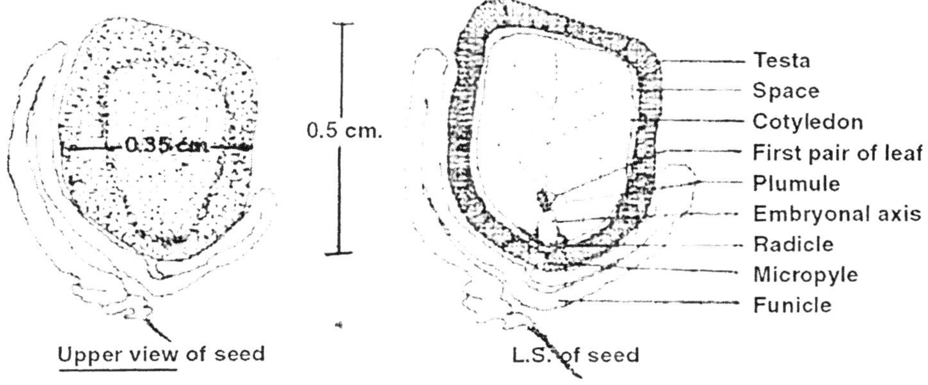

Fig.5: Structure of *Acacia auriculiformis*

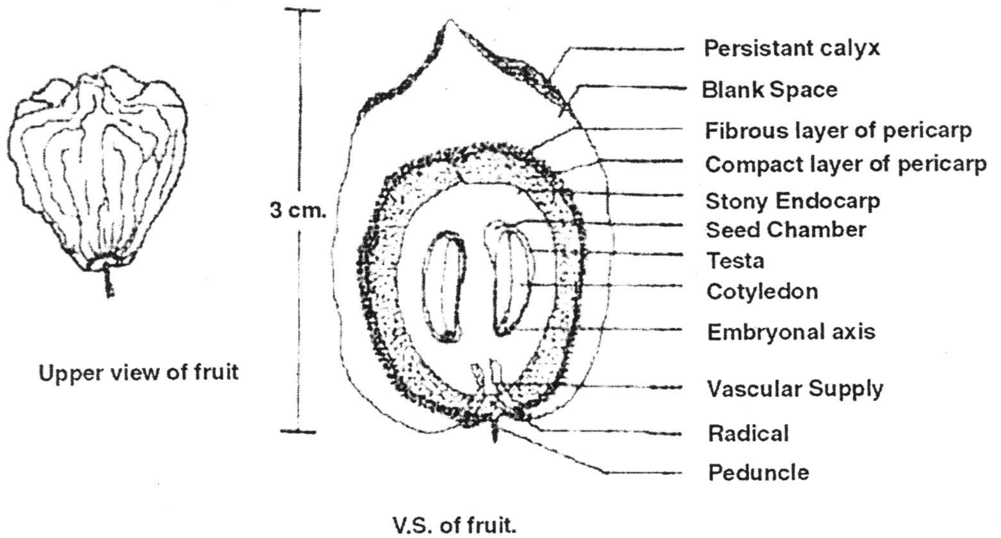

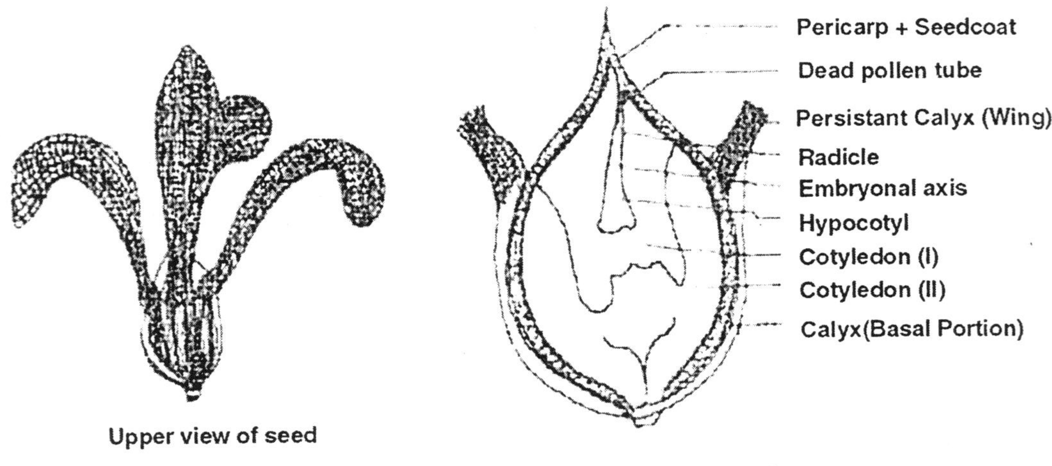

Fig.6: Structure of teak and sal (*Shorea robusta*)

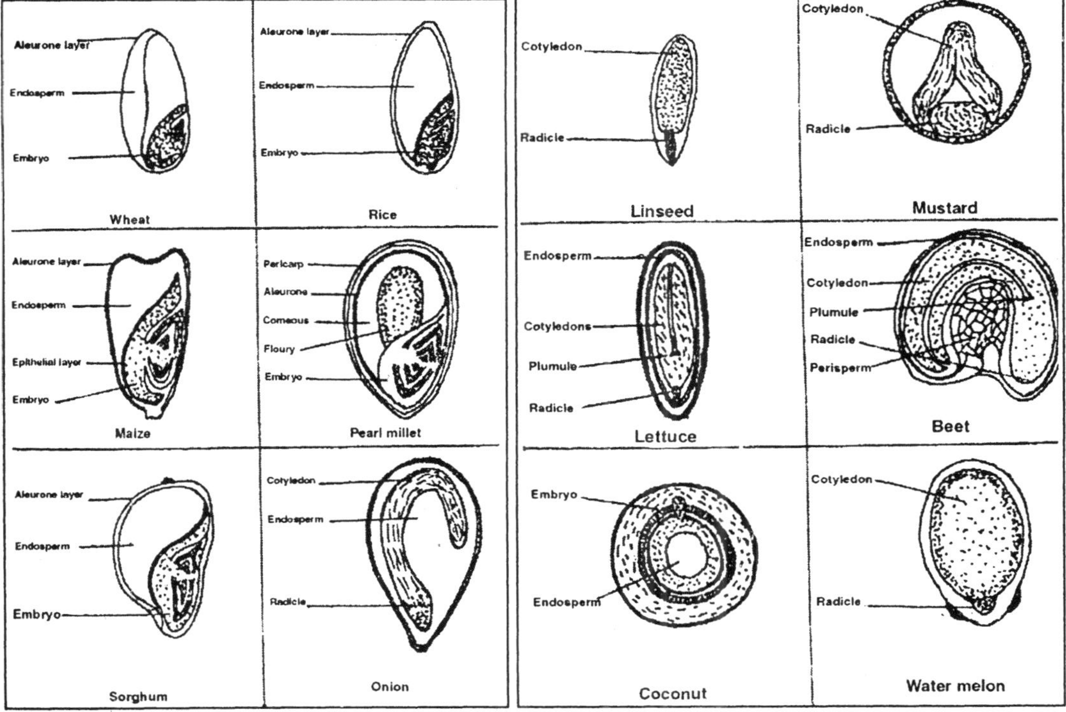

Fig.7: Seed structure of cereals, millets and vegetables

6. CLASSIFICATION OF SEEDS BASED ON SEED STRUCTURE

6.1. Endospermic seeds

The endosperm is present in the mature seed and serves as food storage organ. Testa and endosperm are the two covering layers of the embryo. The amount of endosperm in mature seeds is highly species-dependent and varies from an abundant endosperm layer (*Nicotiana tabaccum*) to a single layer (*Arabidopsis thaliana*) *e.g.* castor, fenugreek (Fig.8)

6.2. Non-endospermic seeds

The cotyledons serve as sole food storage organs as in the case of pea (*Pisum sativum*). During embryo development, the cotyledons absorb the food reserves from the endosperm. The endosperm is almost degraded in the mature seed and the embryo is enclosed by the testa. eg. chickpea, pea, beans (Fig.9.)

6.3. Perispermic seed

Type of endosperm like storage tissue in a mature seed which develops from the nucellus of the parent plant – thus it has the 2n chromosome number.

Examples of species with well developed perisperm tissue include beet and pigweed (Fig. 10).

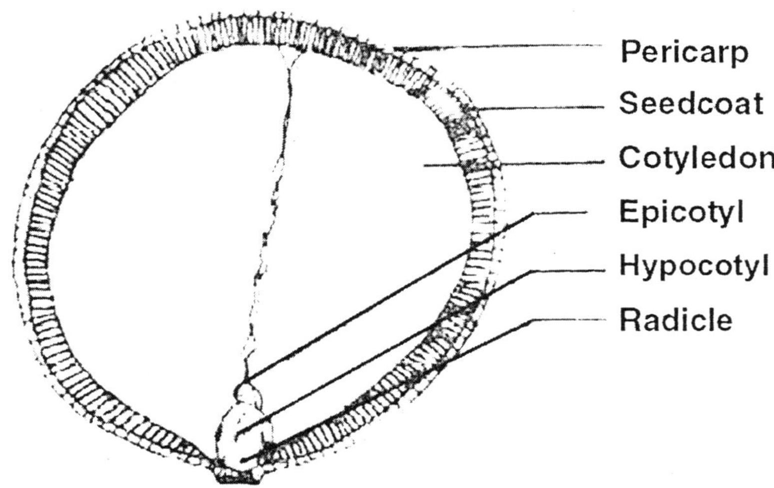

Fig.8: V.S. of Endospermic seed

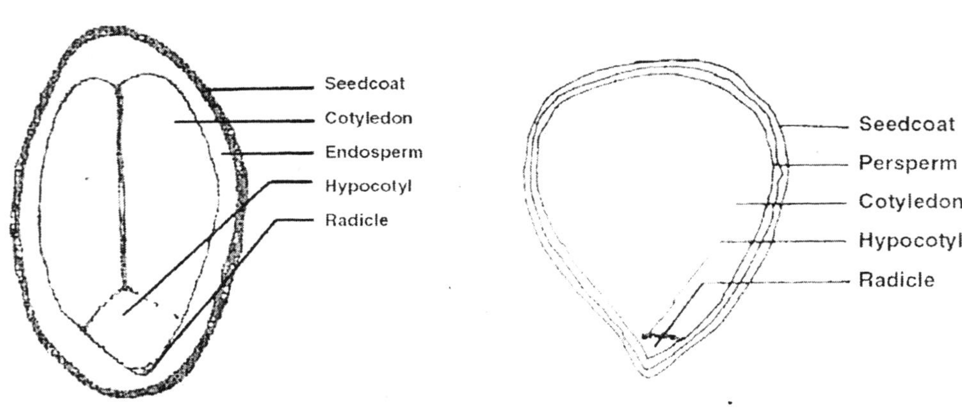

Fig. 9: V.S. of Non endospermic seed

Fig. 10: V.S. of Perispermic seed

7. CLASSIFICATION OF SEED BASED ON COTYLEDON

7.1. Monocotyledons (mostly albuminous) eg. rice, sorghum (Fig.11)

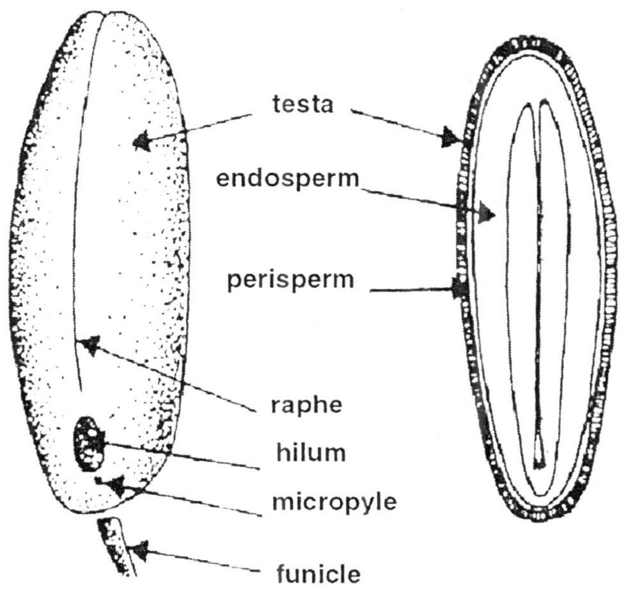

Fig.11: Structure of monocotyledon

7.2. Dicotyledons (both albuminous and nonalbuminous) (*e.g.*) Bean, castor, sunflower (Fig.12).

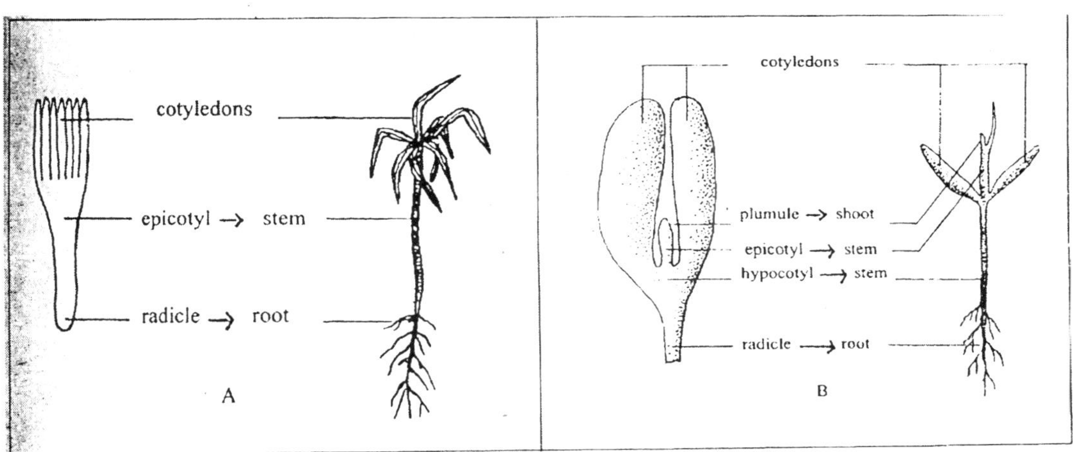

Fig.12: Structure of dicotyledons

8. MORE SPECIAL SEED STRUCTURAL FEATURES

The structure, anatomy and morphology of mature seed model species

Family (clade)	Examples	Description
a. Endospermic seeds		
Cucurbitaceae -Core Eudicots - Rosid clade	Muskmelon (*Cucumis melo*)	In the muskmelon seed, the embryo is surrounded by a perisperm /endosperm envelope. Callose (ß-1,3-glucan) deposition in this envelope is responsible for the apoplastic semi permeability of muskmelon seeds. The perisperm/endosperm envelope is weakened prior to the completion of germination.
Fabaceae - Core Eudicots - Rosid clade	Castor bean (*Ricinus communis*)	Only some legume (Fabaceae) seeds are endospermic, most legume seeds are non-endospermic.
Brassicaceae - Core Eudicots - Rosid clade	Garden Cress (*Lepidium sativum*) Mouse-Ear Cress (*Arabidopsis thaliana*)	Only some brassicaceae seeds are endospermic, most brassicaceae seeds are non-endospermic. Mature seeds of *Lepidium* have 2-3 cell layers of endosperm, while *Arabidopsis* has a single endosperm cell layer. These two species exhibit, as tobacco, a two-step germination process with distinct testa rupture and endosperm rupture. (We have established *Lepidium* as a novel model system for endosperm weakening.)
Solanaceae - Core Eudicots - Asterid clade	Cestroideae subgroup: Tobacco (*Nicotiana tabacum*) other Nicotiana-species Petunia (*Petunia hybrida*) Solanoideae subgroup: Tomato (*Lycopersicon esculentum*) Pepper (*Capsicum annuum*) Datura (*Datura ferox*)	Mature seeds of the solanaceae family usually have an abundant endosperm layer. Well investigated examples are tobacco and tomato, which are model systems in seed biology for the study of endosperm weakening and the regulation of germination by plant hormones and environmental factors. The solanaceae family can be divided into two large subgroups: Cestroideae subgroup of solanaceae (tobacco, petunia): Straight or slightly bent embryos and prismatic to subglobose seeds, two-step germination (distinct testa rupture and endosperm rupture), typically capsules as fruits. Solanoideae subgroup of solanaceae (pepper, tomato):

		Curved embryos and flattened, discoid seeds, no visible distinction beween testa rupture and endosperm rupture, often berries as fruits.
Rubiaceae - Core Eudicots - Asterid clade	Coffee (*Coffea arabica*)	The coffee embryo is enveloped by an endosperm tissue. The fully differentiated embryo lies inside an embryo cavity. The endosperm is surrounded by endocarp, which resembles a seed coat. Endosperm weakening of coffee is inhibited by abscisic acid (ABA) and promoted by gibberellins (GA).
Oleaeceae - Core Eudicots - Asterid clade	*Syringa* species	Low temperature dormancy of *Syringa* seeds is mainly imposed by the mechanical resistance of the endosperm layer.
Asteraceae - Core Eudicots - Asterid clade	Lettuce (*Lactuca sativa*)	Lettuce "seeds" are actually fruits and have 2-3 cell layers of endosperm below the pericarp. Lettuce endosperm weakening has been demonstrated and the hormonal regulation of lettuce seed germination is similar to tobacco.
Apiaceae - Core Eudicots - Asterid clade	Celery (*Apium graveolens*)	Seeds of Apiaceae (Umbelliferae) contain relatively large amounts of living endosperm which completely surrounds a small embryo located at the micropylar end of the seed. Endosperm breakdown has been demonstrated and is promoted by gibberellins.
Ranunculaceae - Basal Eudicots	*Trollius* species	The seeds of basal angiosperms often have underdeveloped embryos that are embedded in abundant endosperm tissue. Two-step germination process with distinct testa rupture and endosperm rupture.
Poaceae and other monocot families - Basal Angiosperms - Monocots	Wheat (*Triticum aestivum*) Barley (*Hordeum vulgare*) Maize (*Zea mays*) Onion (*Allium cepa*)	A typical monocot seed with endosperm is onion (*Allium cepa*), Alliaceae family. In the highly specialized cereal grains/caryopsis (wheat, barley, maize), the endosperm can be divided as starchy endosperm (starch grains, food storage, dead cells, flour) and the aleurone layer (living cell layer surrounding the starchy endosperm). The cereal embryos are highly specialized in their structure.

b. Non-endospermic seeds		
Fabaceae - Core Eudicots - Rosid clade	Pea (*Pisum sativum*) Garden bean (*Phaseolus vulgaris*) Soybean (*Glycine max*)	Most species of the legume family (Fabaceae) including pea (*Pisum sativum*) and diverse beans have non-endospermic seeds. The cotyledons serve as sole food storage organs as in the case of pea (*Pisum sativum*). During embryo development the cotyledons absorb the food reserves from the endosperm completely. In the mature seed, the embryo is enclosed solely by the testa as the only seed covering layer. The regulation by ethylene of pea seed germination and seedling emergence has been studied in detail.
Brassicaceae - Core Eudicots - Rosid clade	Rape (*Brassica napus*) Wild mustard (*Sinapis alba*) Wild radish (*Raphanus sativus*)	Most species of the mustard family (Brassicaceae) including several *Brassica*-species have non-endospermic seeds. The cotyledons serve as sole food storage organs as described for the non-endospermic Fabaceae seeds.
c. Seeds with perisperm		
Amaranthaceae (incl. Chenopodiaceae) - Core Eudicots - Caryophyllid clade	Sugar beet (*Beta vulgaris*) Lambsquarters (*Chenopodium album*)	Perisperm is diploid maternal food storage tissue that originates from the nucellus. It is present in mature seeds of many Caryophyllales (Centrospermae) including the Amaranthaceae (*Beta, Chenopodium*) among the eudicots, but also in basal angiosperms like black pepper (*Piper nigrum*), Piperaceae.

9. SEED FORMATION IN GYMNOSPERMS

In gymnosperm, there is normally a single protective integument which in a typical female cone is partially sured to the ovuliferous scale carrying the paired ovules. Within the integument is the .nucellus which of fertilization, as in angiosperms is clearly separated from the integument only in the region of the micropyle meiosis within the nucellus followed by mitotic division leads to the formation of multicellular haploid tissue the *female gemetophyte*. By the time of fertilization, it has developed much further than the 8 nucleate embryo sac in the angiosperms and has largely displaced the nucellus. At its micropyler end, it is differentiated into one to many *arehegonie* each of which contains a large egg cell.

At fertilization, the pollen tube releases two male nuclei. The resulting zygote later develops into the new archegonium, one of which unites with the egg nucellus diploid embryo.

The second male nucleus aborts in *Pinus,* but many fertilize a second archegonium as in *Cupressus* spp. It never units with female polar nuclei to form a triploid tissue andogous to the endosperm of gymnosperms.

The matured seed consists of some or all of the following structures.

The seed coat or testa – developed from integument, diploid from the female parent.

The perisperm – developed from nucellus and is diploid in nature and is having female genomic constitution. In most species, this is absorbed by the female gametophyte and has disappeared by the time seed is ripened, but it is still recongnisable as a distinct tissue in *Pinus pinea.*

The female gametophytic tissue – It is haploid in nature and serves as food storage organ to nourish the embryo. Its function is the same as the endosperm in angiosperms and it's frequently called by that name.

The embryo – The parts of embryo are radicle, cotyledons, plumule and hypocotyl as in angiosperm, but the cotyledon number varies upto 18.

In angiosperm, more than one archegonium may be fertilized within a single ovule, part in the great majority of cases only one embryo per seed develops to maturity polyembryony in uncommon in most severe.

10. DEVELOPMENT

After fertilization, the female cone increases in size and weight, in moisture content and accumulated food reserves. As the cones approach maturity, the moisture content decreases and the accumulated food reserves more from one cone to seed and the cone become more or less woody.

A membranous flake became detached from the ovuliferous scale and attached to the ripe seed forming a wing in *Pinus* spp. In Junipers, the cone scales grow together to form a fresh berry like fruit, while in *Podocarpus* and *Taxus* each singly borne seed becomes partly enclosed in a brightly coloured cup, the aril. The woody cone is the characteristic features of gymnosperm.

11. SPECIAL FEATURES

➢ Seed maturation period is longer in gymnosperm

Tropical species

Pinus kesiya	: 23 months.
P. oocerpc	: 18 – 21 months
Agothis robusta	: 16 months.
Araucaria cunninghamii	: 24 months.
A. hunsteinii	: 21 – 24 months.
Temperate species.	
Pseudotsurga menziesii	: 5 months.

- Even the unpollinated female fully develops into matured cone with empty seed.
- Parthenocarpy is common in Abies, Juniperus, Lerix, Pieca, Taxus and Thuja.
- Parthenocarpy is rose in pines.
- In most genera, the seed is attached with adnate or articulate wing.

12. STRUCTURAL VARIATION IN GYMNOSPERM

12.1. External

- Gymnosperm seeds develop from naked ovules without enclosing ovaries and poser's integuments or seed coat.
- Seeds are oblong, ovate or ovoid usually acute at the base or triangular.
- The hilum is quite inconspicuous of the base of seeds.
- The seed coat is thin papery or thick or woody.
- Some genera have resin vesicles on or within their seed coat. The presence of resin in the seed coat hands to make the seed sticky.
- In most genera, an adnate or articulate wing is seen. The development of seed coat is closely linked to the formation of the wing. The wing may be thin, papery, but in species thickened at base.
- The integument is 3 layered in some species.
- The seed is dicotyledonous.
- Matured seed is non-endospermous.
- The entire seed at maturity is invested by a fleshy, scarlet cup-like aril which arises as an outgrowth from the base of the ovule. Variation in seed coat character with example.

12.1.1. Genomic nature of gymnosperm seed

1.	Seed coat and perisperm (2n)	:	Derived from pan diploid parent sporophyte.
2.	Endosperm (Nutritive tissue – n)	:	Derived from pan haploid female gametophyte.
3.	Embryo	:	It is fertilized egg in a new sporophyte (diploid chromosome).
4.	Wings seen as trace/scale on the integument	:	*Pinus wallichiana*

12.2. Internal character

- The seed coat is two layered, the outer layer being the hard, brown testa and the inner in membranous segment.
- The major, whole part of the seed consists of only endosperm. At the pointed end, their membranous cap the perisperm representing the remains of nucellus.

- In the centre of the endosperm is called *Corrosion Zone* in which lies a embryo.
- The embryo consists of a pointed radicle (over which the degenerated suspensor forms a mere small cap), large hypocotyl and a minute plumule. Bearing many cotyledons. The cotyledon number will vary with species

  (*e.g.*) *P. roxburghii* - 8-14
  P. gerardiana - 9-12

13. SPECIAL MORPHOLOGICAL CHARACTERS OF SEED WITH EXAMPLE IN GYMNOSPERM SEED

Characters	Seed
Seed	
Oval or oblong seed	*Abies* spp.
Irregularly triangle	*Cedrus* spp.
Small, oblong acute at base	*Pieca* spp.
Seed coat	
Thin and soft	*Abies* spp.
Thin or thick and woody	*Pinus* spp.
With vesicles on or within the seed coat	*Abies* spp *Cedrus deodara*
Appendages	
Presence of wing	Pinaceae
Seed coat extends into a wing which cannot be detached without injury to the seed	*Pinus roxburghii*
The wing easily detachable without any injury as the blade is very short and the base has no effective grasp on the seed.	*P. geradiane*

14. MORPHOLOGICAL SEED STRUCTURAL VARIATION AND ITS INFLUENCE ON SEED QUALITY

Occurrence of one or more distinct type of seed in a single seed lot collected from natural or artificial population is known as polymorphism. The occurrence of polymorphism in seed may either due to phenotype variation or due to genotypic variation normally variation occur within the species for expression of phenotypic characters *viz*; branching, leaf size, shape. Occurrence of inflorescence and formation and fertilization of flowers within the inflorescence. Because of this variation, the seed and seed quality vary with seed which are also expressed through some characters. This variation may also occur in seed even after seed formation and during maturation due to the inflorescence of environmental and nutritional factors. The occurrence of seed within the seed lot either genetically or due to environmentally is known as seed polymorphism or variation in seed. Based on the expression of this polymorphism on or within the seed, the seed polymorphism is divided into four groups.

(1) Morphological polymorphism.
(2) Physiological polymorphism.
(3) Genetic polymorphism.
(4) Position polymorphism.

14.1. Morphological polymorphism

It is the variation expressed by the seeds of a single population based on seed morphological characters. The variation may be due to size, (or) shape (or) weight (or) surface texture (or) colour (or) occurrence of appendages.

14.1.1. Seed size

It is the commonest polymorphism that occurs in any seed. Because variation in size could not be prevented completely a development, as it is decided by position of seed on plant, location of production, health status of plant and changes in the environment (pollination and seed set).

Seed size is positively associated with germination, seedling vigour and plant establishment. This polymorphism could be minimized to low level by processing the seed using standard sieves, and thereby eliminating the bigger and smaller seeds from a population. The homogenity within the seed lot achieved minimizing the seed size variation brings uniformity in seed germination, expression of seedling vigour, synchronize of flowering and the uniform maturation of seed for harvesting.

The applicability of size polymorphism is wider value in seed which are having higher compactness between seed coat and embryo. (*e.g.*) Acacias, Albizias, Teak.

14.1.2. Seed weight

The variation due to weight occur in seed either due to ill filling or due to insect, pathogen damage. It is positively correlated with seed germination but may or may not correlate positively with seed size. This type of polymorphism could be reduced to minimum level by grading the seeds using specific gravity either manually (water and liquid floatation) or mechanically (specific gravity separator). The applicability of seed specific gravity grading will be more in seed posses the genetic character where the seed coat and embryo are loosely attached. (*e.g.*) Casuarina, Emblica, Jatropha.

14.1.3. Seed colour

Polymorphism in this morphological character occur due to improper harvest, biotic damage or by environmental influence.

Seeds of early harvest and delayed harvest, seeds damaged by pathogen and insects, seed exposed to high humidity or higher rainfall express this type of polymorphism.

This has positive association with seed and seed quality characters. Either dull or dark from original characteristic colour are known as off coloured seed. More the off coloured seed in a lot, lesser will be the seed quality characters and this type of polymorphism could be minimized by grading the seeds either manually or mechanically using colour sorter. (*e.g.*) Casuarina, Tamarind

14.1.4. Seed shape

Shape polymorphism will be less in agriculture due to stabilization of characters that occurred due to continuous cultivation. But silvicultural species, the occurrence is wider and shape character is widely influenced by the place of production or due to seed source variation. This polymorphism has strong association with weight and thereby on seed quality characters. Seed shape with heavier or larger size is better in seed quality characters.

(*e.g.*) Neem - Linear, oval
Tamarind - Round, oval, rectangle, triangle, square.
Pungam - Round, oval, flat, bulged.

14.1.5. Surface texture

Occurrence of this polymorphism is due to genetic influence or genetic admixture. Smooth type of peas mixed with wrinkled type of pea seed.

14.2. Physiological polymorphism

Physiological polymorphism is the variation in expression of physiological characters by the seed. This seed lot normally in fresh seeds, the type of polymorphism will be minimum. Occurrence of this polymorphism is influenced by production factors (spacing, weeding, irrigation, infection by pest and diseases), nutritional factor (NPK fertilizer, foliar spray), post harvest seed handling (processing, drying, extraction) and age of the seed in storage. This polymorphism is expressed through variation in germination and vigour characters. It is also influenced by the morphological, genetic and positional polymorphism.

14.3. Genetic polymorphism

This variation is expressed mostly in dormancy characteristics. For example, in seeds of in *Xanthium pennsylvanium* the upper seed is deeply dormant, while the lower seed is shallow dormant. This variation also could be influenced by seed source, wild type and fertility status of the soil. The polymorphic expression would be both in morphological expression and physiological expression.

14.4. Positional polymorphism

Occurrence of this polymorphism is due to the developmental variation. The seed formed in top, middle and bottom position of plant or branch expresses variation in morphological (size) and physiological characters due to the relation between source to sink ratio and flow of nufrients from source to sink.
(e.g) coffee, jamun, ber, amla.

15. CHEMICAL COMPOSITION OF SEED

The important chemical constituent of a seed is its food reserve which mainly includes carbohydrates (starch+sugar), fats and protein. The food reserves are present in varying amounts in different crops (Tables 1, 2, 3).

Table 1: Percent of food reserves of some crop species

Crop	Protein (%)	Fat (%)	Starch (%)	Sugar (%)
Cereals				
Wheat	10-14	1-2	65-75	3-5
Maize	10-12	4-6	70-75	2-4
Jowar (Sorghum)	8-19	3-4	56-69	1-3
Legumes				
Soybean	38-42	18-22	Negligible	5-8
Lentil	25-30	0.5-1	42-45	Not known
Chickpea	20-24	3.1-5.8	41-51	5.1-6.3
Oilseeds				
Sunflower	26-41	40-55	Negligible	2-3

Table 2: Average chemical composition of seeds

Plant	Protein (%)	Fat (Lipid) (%)
Acorn (red oak)	3.2	10.7
Barley (Pacific coast states)	8.7	1.9
Mung bean	23.6	0.2
Navy bean	22.9	1.4
Pinto bean	22.5	1.2
Beechnuts	15.0	30.6
Buckwheat	10.3	2.3
Chickpeas	20.3	4.3
Cottonseed kernel (without hull)	38.4	33.3
Flaxseed	24.0	35.9
Kafir grain	11.0	2.9
Wild mustard	23.0	38.8
Oats	12.0	4.6
Peas	23.4	1.2
Peanut (Groundnut) (without hulls)	30.4	47.7
Rape	20.4	43.6
Rice (rough grain)	7.9	1.8
Rye	12.6	1.7
Soybean	37.9	18.0
Sunflower	16.8	25.9
Vetch	29.6	0.8
Wheat	13.2	1.9

(Morrison, 1961)

Table 3: Chemical composition of different parts of corn seed*

Chemical	Entire seed	Endosperm	Embryo	Pericarp-Testa
Starch	74.0	87.8	9.0	7.0
Sugars	1.8	0.8	10.4	0.5
Oil (Lipid)	3.9	0.8	31.1	1.2
Protein	8.2	7.2	18.9	3.8
Ash	1.5	0.5	11.3	1.0

*Includes only selected chemical components (Earle *et al.*, 1956)

Depending on the major food reserve present in seeds, they are classified as starchy *e.g.*, wheat, maize and as fatty *e.g.*, sunflower, rapeseed. As a rule, protein is never the major component of a seed, but soybean is an exception, and is categorized among the proteinaceous seeds.

A knowledge of the chemical composition of seeds is essential for several reasons: (1) seeds are a basic source of food for both man and animals, (2) they are an important source of medicine and drugs, (3) they contain various antimetabolites that adversely affect human and animal nutrition, and (4) they contain reserve food supplies and growth substances that influence seed germination and seedling vigour, seed storage and longevity, as well as industrial and agricultural uses of seeds.

Because of genetic factors, the chemical composition of seeds varies widely among species and even among varieties. The effectiveness of breeding and selection techniques in increasing the protein and oil content of crop seeds is shown by work with soybeans (Shannon *et al.*, 1972) and corn (Dudley and Lambert, 1968).

15.1. Environmental influences on chemical composition of seeds

Many environmental factors influence the chemical composition of seeds, and because of the interrelationships of these factors, it is sometimes difficult to determine the cause of variability. Cultural practices are known to influence chemical composition of seed (Table 4).

Table 4: Effect of irrigation on the mineral content of Wheat, Barley and Oat Seed

Element	Increase (+) or decrease (-) over controls, %		
	Wheat	Barley	Oats
Nitrogen	-21	-19	-40
Phosphorus	+55	+30	+35
Potassium	+35	+14	+31
Calcium	+155	+41	+22
Magnesium	+32	+ 9	+65

(Greaves and Carter, 1923)

Perhaps the easiest environmental parameter to control in an evaluation of factors that affect seed chemical composition is the mineral nutrition that the mother plant receives. In most instances, seeds that are mineral deficient will perform poorly compared to normal seeds unless they are planted in a soil that is nutritionally adequate and provides the missing essential element or elements. Many studies have been conducted to evaluate the influence of nitrogen, phosphorus, and potassium on seed quality. Some of these are cited below.

Corn plants grown under either high nitrogen fertilization or low plant populations produce seed with higher protein content than those produced under low nitrogen availability or high plant populations. Similar results with rice indicate that when lower seeding rates are used, each rice plant has greater access to available nitrogen, which is then absorbed and supplied to the seeds (DeDatta *et al.*, 1972). Scott (1969) found that excess nitrogen application could have an indirect and detrimental influence on sugar beet seed quality.

16. STORAGE COMPONENTS

16.1. Starch

Seeds are composed largely of metabolically inactive food reserves that are stored until needed during germination. Starch is stored in two related forms, amylose and amylopectin, which are two polymers of D-glucose, one linear and the other branched. Starch is the principal, most widespread, storage carbohydrate of seeds.

Both amylose and amylopectin are hydrolyzed by the enzymes α- and β-amylase during normal metabolism and germination. Alpha-amylase possesses a molecular weight of 60,000 and requires the cation Ca^{++} for activation and stabilization against proteolytic destruction and thermal denaturation of the enzyme. This enzyme attacks both amylase and amylopectin randomly, cleaving both α1,4 and α1,6 linkages. Also the action pattern varies, there is a general amylolysis (amylase catalyzed breakdown) of the starch molecule to large amounts of dextrin molecules to maltose and finally glucose. The enzyme is formed *de novo* in the aleurone of barley and the scutellum of corn. In corn, α-amylase accounts for 90 per cent of the amylolytic activity and β-amylase for the remaining 10 per cent.

Beta-amylase hydrolyzes both amylose and amylopectin from the nonreducing end by breaking the α1,4 linkage and forming maltose. In the case of amylose, almost complete hydrolysis occurs. However, since β-amylase is unable to bypass the branched chain point or α1,6 linkage of amylopectin, this molecule is only partially hydrolyzed, yielding the residual β-limit dextrin. This means that only the outer portions of the amylopectin molecule are attacked by β-amylase. This enzyme is already formed and is present *in situ* in seeds. In wheat kernels, β-amylase is present in an active latent form, apparent chemically bound to the wheat glutenin by disulfide linkages. During germination, the secretion of a substance capable of releasing the glutenin-bound enzyme may occur, accounting for the observed β-amylase activity.

In seeds, most starch is laid down in discrete subcellular bodies called *starch grains* that range from 2 to 100 μ in size within the endosperm. Many starch grains appear to form around a central point, the hilum, around which shells of the polysaccharide are deposited. These shells probably reflect a diurnal periodicity in the synthesis and deposition of starch, since they are absent from starch grains of seeds developing in continuous light under experimental conditions. The shape of the starch grain appears to depend on the amylose content-the less angular, rounded grains having relatively high amylose levels.

Most starch grains are composed of about 50 to 75% amylopectin and 20 to 25% amylose. However, some seeds such as rice may be high in amylose (up to 37%) and are classified as starchy. The difference in appearance between amylopectin and amylose starch forms is illustrated in waxy corn (almost 100% amylopectin) versus normal corn (50 to 75% amylopectin) (Kerr, 1950).

16.2. Hemicellulose

The other major form of carbohydrate storage in seeds is hemicellulose. Usually found in the cell walls of plants, although in certain seeds, they are found as reserve food materials. They are usually found in the thickened tertiary layers of cell walls of the endosperm or cotyledons instead of in the interior region of starch concentration. Hemicelluloses are particularly characteristic of seeds of many of the palm species, such as date palm and South American and Polynesian ivory nut palms. Seeds of the date palm have a small cylindrical embryo embedded in a sizable horny endosperm of hemicellulose.

16.3. Other seed carbohydrates

16.3.1. Mucilages

Physically, they are similar to the gums found in the bark and stems of many plants infrequently may be found in rather large amounts. A well known example is the seed of buckhorn plantain, which is covered with a thick coating of mucilage that becomes sticky when wet and tends to cling to material that it touches. Obviously, this is a seed dispersal mechanism. However, seed conditioners also use the sticky coat characteristic when cleaning buckhorn seed from certain crop seed. Mucilages from buckhorn plantain seed are separated commercially for industrial use.

16.3.2. Pectic compounds

These compounds occur in seeds and in other plant parts mainly as components of the cell wall and the middle lamella. The three principal pectic compounds are pectic acid, pectin, and protopectin. They occur in the primary cell wall and the middle lamella where they bind the cell walls together. When protopectins are converted into pectins, they are instrumental in the softening of ripening fruits.

16.4. Lipid storage in seeds

The occurrence of high lipid concentrations differentiates seeds from other plant organs. High lipid content is usually associated with high protein content (for example, in soybeans, peanuts, cotton). However, in some species, such as certain oaks, it is associated with high carbohydrate levels.

Lipids are plant or animal substances that are insoluble in water, but soluble in ether, chloroform, benzene, or other organic solvents. They are esters of either fatty acids or glycerol or their various hydrolytic products. They are known as glycerides, or, more specifically, *triglycerides*.

The term *lipid* is often used interchangeably with fat and oil, although it is actually the generic term for both. Oils are distinguished from fats only in that they remain in liquid form at ordinary room temperatures, whereas fats remain in solid form. Oils are often designated as fatty oils to differentiate them from essential oils, which are chemically unrelated.

16.4.1. Fatty acids

The fatty acids are constituent of natural fats, and in the free state they resemble fats in physical properties. Free fatty acids are seldom found in plant parts other than in germinating or deteriorating seeds as a result of fat hydrolysis. Fatty acids are saturated or unsaturated, depending on the type of carbon linkage in the molecule. The unsaturated fatty acids contain one or more double bond links, which means the compound can take up hydrogen atoms to form a saturated compound. The unsaturated fatty acids are most common in seeds with oleic (one double bond) and linoleic (two double bonds), accounting for 60% of the weight of all the lipids present in most oil seed crops. Saturated fatty acids are also present and contain an even number (n) of carbon atoms (n usually being between 4 and 24). Palmitic acid (n=14) is the most common saturated fatty acid in oil seeds.

The great majority of seed lipids are simple lipids, which include fats, fatty oils, and waxes. Waxes are simple lipids that are fatty acid esters of some alcohol other than glycerol. Waxes are less soluble in ordinary fat solvents than fats. They can be saponified, but only with difficulty. They are usually found as coverings for the leaves, fruits, and seeds of many plants.

16.4.2. Hydrolysis of lipids

When oilseeds with high lipid content germinate, there is a rapid disappearance of lipid with a concomitant rise in sucrose content. Concurrently, lipase activity rises sharply and participates in the stepwise hydrolysis of triglycerides to diglycerides, monoglycerides, and finally free glycerol and free fatty acids. Fatty acids are further oxidized by both α- and β-oxidation during the course of germination. Most lipids are stored in oil storage bodies called *spherosomes* that range in size from 0.2 to 6.0 μ in diameter. Many of the enzymes essential for fatty acid biosynthesis or for triglyceride hydrolysis are present in the spherosomes.

16.5. Protein storage in seeds

Proteins are nitrogen-containing molecules of huge size and exceedingly complex structure, the greater part of which yield amino acids upon hydrolysis of the peptide bond. Proteins are so important plant life that all physiological reactions of living cells revolve around their physical and chemical properties and those in related compounds. Aside from water, they are the principal component of all protoplasm of both plant and animal cells.

Proteins comprise a valuable food storage component in seeds of most plant species. Soybeans are one of the few species known in which proteins comprise more of the reserve food supply than do fats or carbohydrates. Most of the plants known to have high protein seed are legumes having nitrogen-fixing capacity. However, many high protein species are nonlegumes. Seed storage proteins are less complex than protoplasmic proteins and less likely to be tied up with lipids and other prosthetic groups, although they are structurally similar.

The great majority of seed proteins are metabolically inactive and serve merely as food reserves for use by the growing embryo during germination. Metabolically active proteins constitute a small amount of the total, but they are extremely important to the developing and germinating seed. As enzymes, they catalyze all metabolic processes in digestion, translocation, and utilization of stored food reserves; no growth can occur without them. The *nucleoproteins* are another extremely important form of active protein and are molecules of enormous size (the molecular weight may number in the million). They are formed by protein-nucleic acid linkage and are a huge conglomerate molecule composed of a protein, a pentose sugar, a cyclic nitrogen compound (purine or pyrimidine), and phosphoric acid. If the pentose sugar is deoxyribose, the resulting nucleoprotein is called *deoxyribonucleic acid* (DNA); if the pentose sugar is D-ribose, the nucleoprotein is called *ribonucleic acid* (RNA). Both these forms have been popularized for their role in protein synthesis and their crucial role in the structure and function of chromosomes, genes, and life itself.

Proteins are stored in the seed in units known as protein bodies that are 1-20μ in diameter bounded by a lipoprotein unit membrane. These are somewhat like starch grains in size and shape and are usually a mixture of different proteins. They may be visualized as the kind of units that are deposited in the albuminous aleurone layers of cereal seeds and that, during seed germination, play an important role both as food reserves and as hydrolytic enzymes to aid in starch breakdown.

16.5.1. Seed albumins are soluble in water at neutral or slightly acid pH and are coagulated by heat. Examples are leucosins of cereal grains, legumelin of various pulse seeds, and ricin of rice. These are mainly enzymes.

16.5.2. Seed globulins are soluble in saline solutions, but not in water and are generally more difficult to coagulate with heat than are animal globulins. Their solubility is modified by combined acids and concentration of saline solutions. Globulins are found predominately in dicot seeds such as legumes. Examples are legumin, vignin, glycinin, vacilin, and arachin.

16.5.3. Seed glutelins are soluble in aqueous or saline solutions or ethyl alcohol, but can be extracted with strong acid or alkaline solutions. Glutelins are found in most cereal seeds. Examples are glutenin of wheat and oryzenin of rice.

16.5.4. Seed prolamines are soluble in 70 to 90% ethyl alcohol, but not in water; however, the salts with acids and bases are soluble in water. They are found only in cereal seeds: gliadin of wheat and rye and zein in maize are example. Upon hydrolysis they yield proline, glutamic acid, and ammonia.

In general, glutelins and prolamines form the major components of seed protein that are metabolically inactive and associated with structural architecture. Seed proteins generally have a high nitrogen and proline content and are low in lysine, tryptophan, and methionine.

17. OTHER CHEMICAL COMPOUNDS FOUND IN SEEDS

17.1. Tannins

Although one usually thinks of particularly in the seed coat structures. They are found in the testa of cocoa and bean seed (Bonner and Varner, 1965).

17.2. Alkaloids

Alkaloids are complex cyclic compounds containing nitrogen. Most are white solids; however, nicotine is a liquid at room temperatures. Better-known alkaloids that occur in plants or their seeds are morphine from poppy fruits, strychnine from seeds of *Strychnos nuxvomica*, atropine from the deadly nightshade, and colchicine from meadow saffron. Other extremely common alkaloids are caffeine from coffee and tea, nicotine from tobacco leaves, and theobromine from cocao.

17.3. Glucosides

While most glucosides are found in the vegetative organs of plants, some do appear in seeds. Examples of glucosides in seeds and vegetative plant parts are salicin in bark and leaves of willows; amygdalin in seeds of almonds, peaches, and plums; sinigrin in black mustard; aesculin in horse chestnut seed; and quereitron in the bark of oak (Bonner and Varner, 1965).

Glucosides are formed by a reaction between a sugar (usually glucose) and one or more nonsugar compounds. In their pure state, they are mostly crystalline, colourless, bitter and soluble in either water or alcohol. Some of the glucosides, such as saponin (which comes from tung seed), are highly poisonous to both man and animals.

17.4. Phytin

Phytin is the major form of phosphorus storage in seeds. In cereal grains, phytin is generaly associated with the protein bodies in the aleurone layer and is more or less absent from the protein bodies of the cotyledons. During the germination of seeds, phosphatases that hydrolyze phytin increase several-fold.

The phytase activity is the highest in the scutellum and aleurone layers. Since a large proportion of the seed's phosphate, magnesium, and potassium are present in phytin, much of the seed's subsequent metabolism is dependent on the hydrolysis of phytin and the concomitant release of magnesium and potassium ions. In lettuce seeds, 50% of the total phosphorus is tied up in phytin.

18. GROWTH REGULATORS

18.1. Hormones

The term *hormone* is used to designate certain organic compounds that, when present in minute quantities, exert important regulatory effects on the metabolism of either plants or animals. Many important plant hormones also occur in seeds. They are variously designated as phytohormones, growth hormones, growth substances, and growth regulators.

18.2. Gibberellins

The presence of gibberellins in higher plants was demonstrated by Radley (1956), who was able to induce tall growth in dwarf peas by giving them an extract from normal tall plants. It is now agreed that gibberellins are a normal constituent of green plants, as well as seeds. They appear to have a major physiological role in both seed development and seed germination. They also exert an influence on floral induction and initiation.

The best-known gibberellin is gibberellic acid (GA_3), although at least forty others have been identified. Gibberellins are produced commercially by fermentation of fungal cultures (*Gibberella* spp.).

18.3. Cytokinins

Another group of compounds that occur in seeds and exerts hormonal influence is the cytokinins (kinins). They were first discovered in coconut milk (Van Overbeek *et al.*, 1941). Fifteen years later, kinetin was purified and the chemical structure identified. The first naturally occurring cytokinin to be extracted from seeds was zeatin.

Cytokinins appear to be necessary for cell growth and differentiation; perhaps this is the base for their influence in promoting seed germination. They have also been shown to inhibit senescence in leaves (Richmond and Lang, 1957) and to regulate the flow of chemicals through the plant system (Mothes *et al.*, 1959).

18.4. Inhibitors

There is growing conviction among plant physiologists that dormancy-whether in seeds, buds, tubers, or other plant parts – is regulated by a balance, or interaction between endogenous inhibitors and growth promoters such as gibberellins and auxins. Dormin (Cornforth *et al.*, 1965; Thomas *et al.*, 1965) and abscisin II (Ohkuma, 1965) were preliminary names given to abscisic acid (ABA), which is reportedly instrumental in inhibiting the formation of the leaf abscission layer and (particularly) the onset of winter dormancy in deciduous plants. Another inhibitor that influences seed dormancy is coumarin. Ethylene gas can both inhibit and promote seed germination, and is sometimes considered a hormone (Crocker *et al.*, 1935), as is the growth regulator, maleic hydrazide (Meyer *et al.*, 1960).

18.5. Vitamins

From the standpoint of human physiology, vitamins qualify as growth regulators and cannot be sharply distinguished from hormones, since they are necessary for the diet but are required in small, often minute, quantities. Although the role of vitamins in animal life and their almost universal occurrence in green plants have long been recognized, their role in plant growth is not so well known. Chemically, vitamins represent a very heterogeneous group.

In contrast to animals, which are dependent on green plants for their vitamin requirements, green plants are self-sufficient for their vitamin needs (although certain parts of a plant may depend on other parts for their vitamin supply). All known vitamins or their immediate precursors are synthesized in higher plants. Whether all are necessary in a plant's metabolism is not certain. However, specific roles for certain vitamins have been determined. Thiamine appears to be necessary for the developing embryo and endosperm of seeds in some species. It is also needed for normal root development. The basis for both needs seems to be the role of thiamine in continued cell division, which is rapidly occurring in these parts. In the case of both roots and the developing seed, thiamine is produced in the vegetative plant parts or by the cotyledon and translocated into the areas where needed. Biotin and ascorbic acid are apparently involved in respiration processes in seeds. The role of biotin is not known, but ascorbic acid appears to be responsible for regulating the oxidation-reduction potential during germination.

REFERENCES

Bonner, J., and J.E.Varner. 1965. *Plant Biochemistry*. New York: Academic Press.

Cornforth, J.W., B.V.Milborrow, G.Ryback, and P.F.Wareing. 1965. Identity of sycamore "dormin" with abscisin II. *Nature*, **205**: 1269-1270.

Crocker, W., A.E.Hitchcock, and P.W.Zimmerman. 1935. Similarities in the effects of ethylene and the plant auxins. *Contributions from Boyce Thompson Institute*, *7*: 231-248.

DeDatta, S.K., W.N.Obcemea, and R.K.Jana. 1972. Protein content of rice grain as affected by nitrogen fertilizer and some triazines and substituted ureas. *Agronomy J.*, **64**: 785-788.

Dudley, J.W., and R.J.Lambert. 1968. Genetic variability after 65 generations of selection in Illinois high oil, low oil, high protein, and low protein strains of *Zea mays* L. *Crop Sci.*, **9**: 179-181.

Earle, R.J., J.Curtice and J.E.Hubbard. 1956. Composition of the component parts of the corn kernel.*Cereal Chemistry*, **23**: 507

Greaves, J.E. and E.G. Carter. 1923. The influence of irrigation water on the composition of grains and the relationship to nutrition. *J. Biol. Chem.*, **58**: 531-541.

Kerr, R.W. 1950. *Chemistry and Industry of Starch*, 2d ed. New York: Academic Press.

Meyer, B.S., D.B.Anderson and Bohning. 1960. *Introduction to Plant Physiology*. New York: D.Van Nostrand Company.

Morrison, F.B. 1961. *Feeds and Feeding*, Ithaca, N.Y.: Morrison Publishing Co., 1961, pp. 611-640.

Mothes, K., L.E. Engelbrecht, and O.Kulajewa. 1959. Uber die Wirkung des Kinetins auf Stickstoffverteilung und Eiweiss-synthese in isolierten Blattem. *Flora: Oder Allgemeine Botanische Zeitung*, **147**: 445-464.

Ohkuma, K. 1965. Synthesis of some analogs of abscisin II. *Agricultural and Biological Chemistry*, **29**: 962-964.

Radley, M. 1956. Occurrence of substances similar to gibberellic acid in higher plants. *Nature*, **178**: 1070-1071.

Richmond, A.E., and A.Lang. 1957. Effect of kinetin on protein content and survival of detached *Xanthium* leaves. *Science*, **125**: 650-651.

Scott, R.K. 1969. The effect of sowing and harvesting dates, plant population and fertilizers on seed yield and quality of direct-drilled sugar beet seed crops. *J. Agri. Sci.*, **70**: 373-385.

Shannon, J.G., J.R.Wilcox, and A.H.Probst. 1972. Estimated gains from selection for protein and yield in the F4 generation of six soybean populations. *Crop Sci.*, **12**: 824-826.

Thomas, T.H., P.F. Wareing, and P.M.Robinson. 1965. Action of the sycamore "dormin" as a gibberellin antagonist. *Nature*, **205**: 1270-1272.

Van Overbeek, J., M.E. Conklin, and A.F.Blakeslee. 1941. Factors in coconut milk essential for growth and development of very young *Datura* embryos. *Science*, **94**: 350-351.

4

Physiology of Seed Germination

Mallika Vanangamudi[1], J. Annie Sheeba[1], M. Djanaguiraman[1] and P. Saravananae[2]
[1]*Department of Crop Physiology, Tamil Nadu Agricultural University Coimbatore – 641 003*
[2]*Department of Soil and Crops, Adhiparasakthi Agricultural College G.B.Nagar, Kalavai- 632 506*

1. INTRODUCTION

A seed is essentially a baby in a suitcase carrying its lunch. "Baby" refers to the embryo, or immature plant, that will grow and develop into the seedling and ultimately the mature plant. The dicot embryo consists of (1) radicle (embryonic root) - develops into the root system (2) epicotyl (embryonic shoot) - develops into the stems bearing leaves (3) cotyledons (seed leaves) - serve to store nutrients for germination and/or as an organ to transfer nutrients from the endosperm into the embryo and (4) hypocotyl - embryonic stem that connects the epicotyl and radicle. The embryos of grasses like maize and oats differ slightly. The "suitcase" is the seed coat (or testa) that surrounds the seeds and "lunch" refers to the nutritive source for the germinating seedling. The food for the germinating seedling may be stored in part of the embryo itself, such as the fleshy cotyledons of a bean seed, or it may take other forms including endosperm, which is a special starch-rich storage tissue that surrounds the embryo.

Germination can be defined as the onset of the growth of the embryo, usually after a period of dormancy. The transition of metabolically quiescent seed to an active seedling starts with the imbibitions of water followed by a series of physiological and biological events, such as hydration and enzyme activation, enzymatic degradation of complex metabolites, subsequent mobilization of simple metabolites to the meristematic region, radicle emergence and seedling growth.

2. REQUIREMENTS FOR GERMINATION

Seed is a dry structure with hard seed coat which is impermeable to water and oxygen and acts as a barrier between embryo and these substances. If germination is to occur, the seed coat must become permeable to water and oxygen. Hence, seeds must be placed in proper environmental conditions *viz.*, optimal moisture, proper temperature and ample oxygen. If these conditions are met, germination can take place.

2.1 Moisture

Water is needed to activate the biochemical processes during germination. It is also an important reagent at this stage, being used in the hydrolysis of food stores. The quantity of water taken up by seeds is generally small (not exceeding

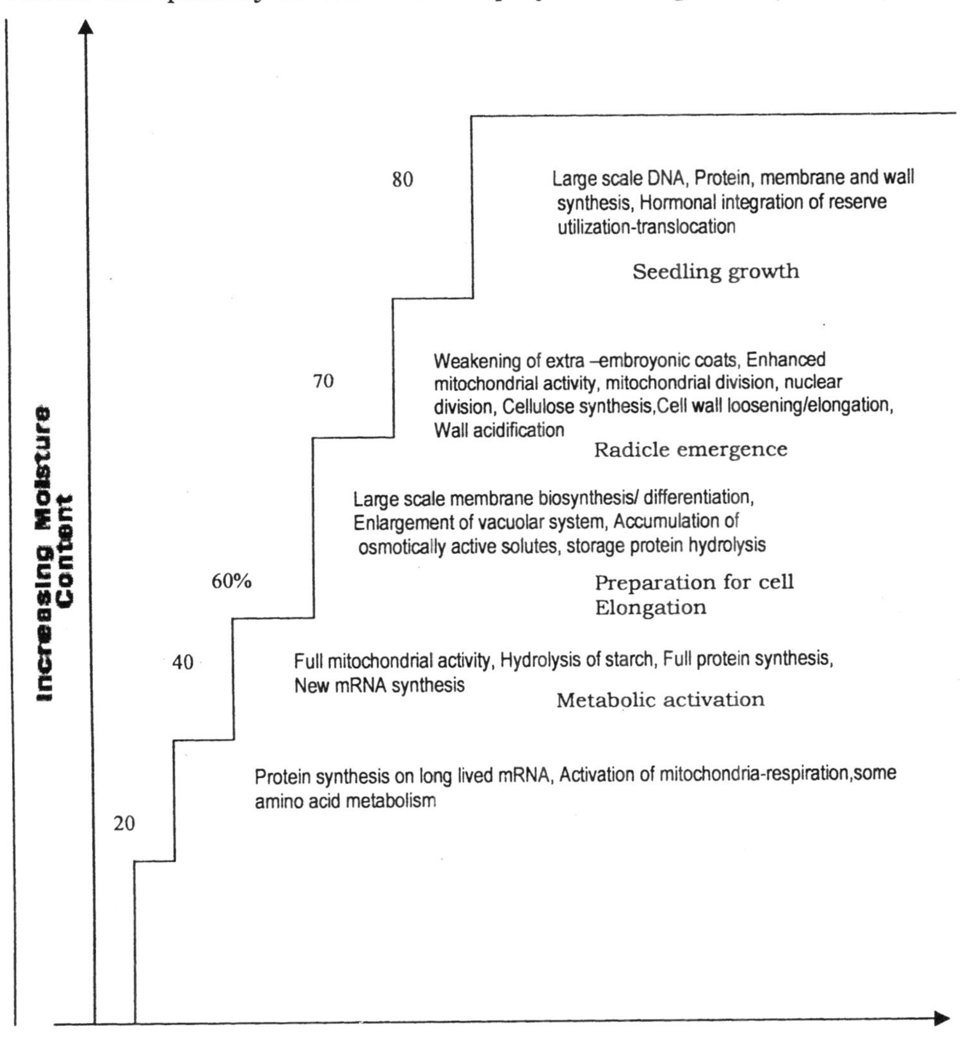

Fig. 1: Some proposed relationships between the major biochemical/ structural events and increases in seed water content, which lead to germination and seedling establishment (Adapted from Obroucheva and Antipova, 1997).

two to three times the seed dry weight). Substantial quantities of water are required for the establishment and subsequent growth of the seedling. It is generally recognized that seed germination is more sensitive to moisture stress than the subsequent seedling development. In normal germination, the germination medium is sufficiently moistened throughout the course of germination. While inadequate moisture in the medium will result in delayed and poor germination, excessive moisture will hinder germination due to decreased aeration. It is therefore important that the germination medium be not so wet that a film of water forms around the seeds, or, when pressing the medium, a film of water forms on the finger. Seeds of some tropical species such as *Paraserianthes falcataria* were found to be sensitive to excessive medium moisture in germination. Some steps associated with water uptake are illustrated graphically in Fig. 1 and show, in a generalized way, some metabolic and other events which take place at particular hydration levels. It shows four stages associated with the germination process: (1) metabolic activation, (2) preparation for cell elongation, (3) radicle emergence, and (4) seedling growth (Obroucheva and Antipova, 1997).

In this model, loss of coat impermeability is associated with hard-coated legumes; nuclear division and cell wall elongation may be synchronous, and precede radicle emergence, or separated in time by the process of radicle emergence. A critical event is the requirement for the embryo to overcome the resistance of encompassing tissues. This is achieved by embryos lowering their water potential, by lowering either the osmotic potential or their turgor. That the concept of seed water potential is a crucial event in germination has now been embodied in a model using the concept of hydro time (Bradford, 1996).

2.2. Temperature

Plants may be classified into basically four groups with regard to the temperatures at which their seeds germinate. The first type is cool temperature tolerant plants which represent most plants that are native to temperate zones. Seed of cool temperature tolerant plants can germinate at temperatures as low as 4.5°C but prefer optimal temperature in the range of 25°-30°C for germination to occur, *e.g.* broccoli, cabbage.

The second group of plants requires cool temperatures for seed germination. Seeds of this type originate from cool season plants, most of which are native to the Mediterranean type climates. These climates usually have hot, dry summers and cool moist winters. In terms of seed germination, it is favourable for seed to germinate in winter under cool moist conditions rather than during the hot, dry summer. Thus, plants of this group posses mechanisms that prevent their seed from germinating when soil temperatures exceed 25°C (77°F), *e.g.* lilac, freesia.

The third group of plants requires warm temperatures in order for seed germination to take place. Seed of this type generally will not germinate unless soil temperatures are 10-15°C and show visible signs of chilling injury, if germination at lower temperatures occurs. Seedlings afflicted with chilling injury may be chlorotic, slow growing and accompanied by disease problems.

Some examples of warm temperature requiring seeds include cotton, beans, egg plant and corn. Most of these plants are tropical in origin and adapted to regions where soil temperatures do not get very low.

The fourth group of plants requires an alternating diurnal cycle of temperatures to initiate germinate. Basically, this involves alternating temperatures on the surface of the soil of warmth provided by sunlight during the day and cooling of soils at night through radiant heat loss, *e.g.* fir, red pine and rose.

Temperature can affect the percentage and rate of germination through at least three separate physiological processes.

(i) Seeds continuously deteriorate and unless in the meanwhile they are germinated, they will ultimately die. The rate of deterioration depends mainly on moisture content and temperature. The Q_{10} for rate of loss of viability in orthodox seeds consistently increases from about 2 at -10°C to about 10 at 70°C.

(ii) Once seeds have lost dormancy, their rate of germination (reciprocal of the time taken to germinate) shows a positive linear relation between the base temperature (at and below which the rate is zero) and the optimum temperature (at which the rate is maximal); and a negative linear relation between the optimal temperature and the ceiling temperature (at and above which the rate is again zero). The optimum temperature for germination rate is typically higher than that required to achieve maximum percentage germination in partially dormant or partially deteriorated seed populations.

(iii) Most seeds are initially dormant. Relatively dry seeds continuously lose dormancy at a rate which is temperature dependent. Unlike enzyme reactions, the Q_{10} remains constant over a wide range of temperature at least up to 55°C, and typically has a value in the region of 2.5-3.8. Hydrated seeds respond quite differently, high temperatures generally reinforce dormancy or may even induce it. Low temperatures may also induce dormancy in some circumstances, but in many species they are stimulatory (stratification response), especially within the range -1°C to 15°C. Small, dormant, hydrated seeds are usually stimulated to germinate by alternating temperatures which typically interact strongly and positively with light (and often also with other factors including nitrate ions). The most important attributes of alternating temperatures are amplitude, mean temperature, the relative periods spent above and below the median temperature of the cycle (thermo period) and the number of cycles.

2.3. Oxygen

Another factor affecting germination is concentration of oxygen available to the seed. Germinating seed are very metabolically active. Respiration is occurring at high rates breaking down storage materials and sugars and converting them to other forms more useful to the seed. All this requires oxygen which is thus necessary for seed germination. Oxygen may be limiting in heavy soils such as clay or flooded conditions which in turn will affect seed germination.

2.4. Light

Another factor affecting germination in some types of seeds is light. Epiphytes are examples of plants that require light in order for seed to germinate. Many epiphytic plants posses seeds that are light sensitive, meaning that if exposed to darkness for extended periods, viability are lost. In other words, light must be present in order for these seeds to germinate. Light may also be a factor in physiological dormancy. Many plants having very small seeds require exposure to light to overcome physiological dormancy. Seed of this type must be on or near the soil surface in order to germinate. Lettuce is an example of a plant that generally needs exposure to light in order to germinate. Conifers need light requirements to satisfy seed germination.

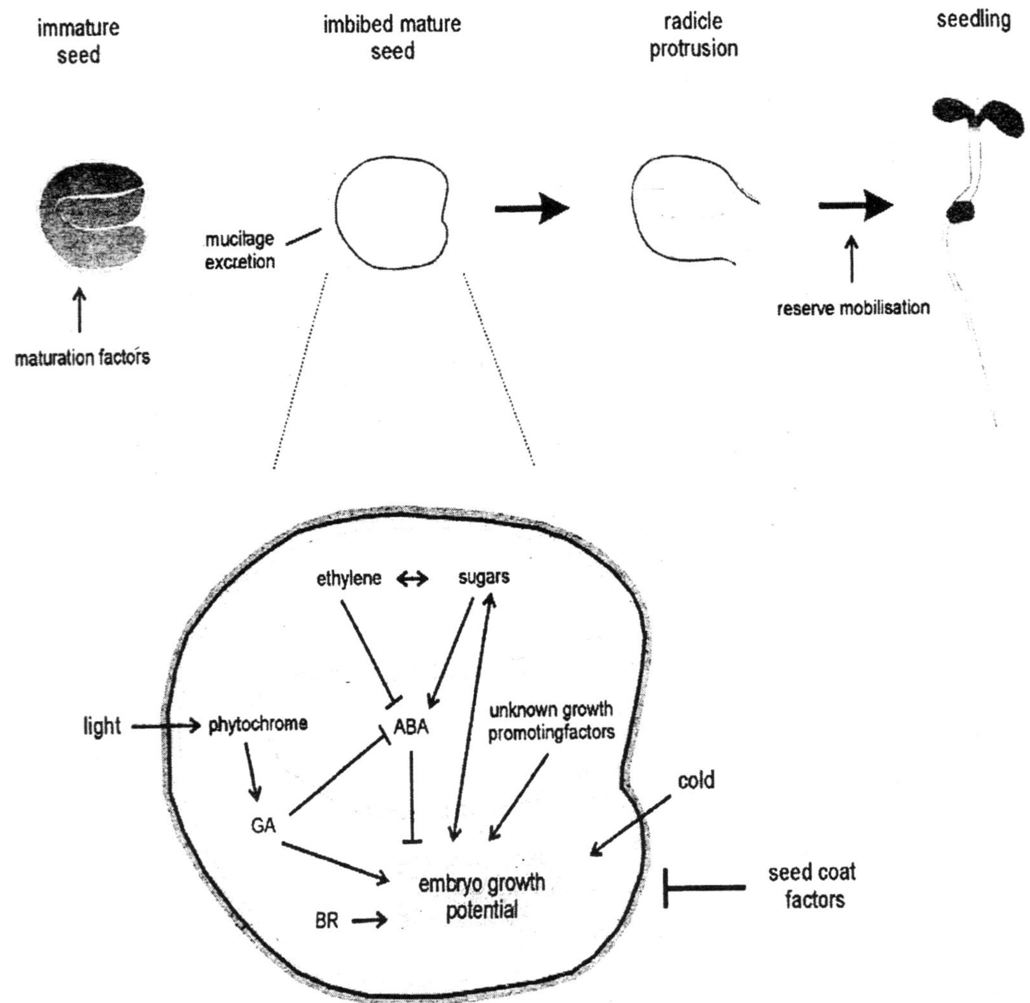

Fig. 2: Overall physiological process during germination

Conversely, there are plants that need darkness in order for their seed to germinate. This is related to soil depth. Cacti and other desert plants are examples and are adapted so that germination only occurs in total darkness deep within the soil. The deeper the seed is in soil the closer it is in proximity to moisture which is a requirement for survival in desert environments.

Photoperiod or changes in day length can also affect germination. Certain seeds, especially those of some woody, temperate species require particular day lengths before germination can occur. The various physiological steps occurring seed germination has been depicted in the following figure (Fig.2), which includes imbibition, mobilization of food reserves and radicle emergence.

3. IMBIBITION AND THE RESUMPTION OF METABOLISM

Uptake of water by a mature dry seed is triphasic (Figure 2), with a rapid initial uptake (phase I) followed by a plateau phase (phase II). A further increase in water uptake occurs only after germination is completed, as the embryonic axes elongate. Because dormant seeds do not complete germination, they cannot enter phase III. The influx of water into the cells of dry seeds during phase-1 results in temporary structural perturbations, particularly to membranes, which lead to an immediate and rapid leakage of solutes and low molecular weight metabolites into the surrounding imbibition solution. This is symptomatic of a transition of the membrane phospholipid components from the gel phase achieved during maturation drying to the normal, hydrated liquid crystalline state (Crowe and Crowe, 1992). Within a short time of rehydration, the membranes return to their more stable configuration, at which time solute leakage is curtailed. How repair to desiccation and rehydration induced damage to membranes and organelles is achieved is unknown. However, during the imbibition of cotton seeds, the amount of N-acetylphosphatidylethanolamine, a phospholipid with membrane stabilizing properties, increases, as does that of the corresponding synthase. These molecules may be involved in maintaining or enhancing membrane integrity (Sandoval *et al.,* 1995). Upon imbibition, the quiescent dry seed rapidly resumes metabolic activity. The structures and enzymes necessary for this initial resumption of metabolic activity are generally assumed to be present within the dry seed, having survived, at least partially intact, the desiccation phase that terminates seed maturation. Reintroduction of water during imbibition is sufficient for metabolic activities to resume, with turnover or replacement of components occurring over several hours as full metabolic status is achieved.

One of the first changes upon imbibition is the resumption of respiratory activity, which can be detected within minutes. After a steep initial increase in oxygen consumption, the rate declines until the radicle penetrates the surrounding structures. At this time, another burst of respiratory activity occurs (Bewley and Black, 1994). Both the glycolytic and oxidative pentose phosphate pathways resume during phase I, and the Kreb's cycle enzymes become activated (Salon *et al.,* 1988). Germinating seeds of many species frequently produce ethanol (Morohashi and Shimokoriyama, 1972). This is often the result of an internal deficiency in oxygen that is caused by restrictions to gaseous diffusion by the structures that surround the seed and by the dense internal structure of most seeds. This oxygen deficiency may result in more pyruvate production than utilization for activities of the Kreb's cycle and electron transport chain. Tissues of the mature dry seed contain mitochondria, and although these organelles are poorly differentiated as a consequence of maturation drying, they contain sufficient Kreb's cycle enzymes and terminal oxidases to provide adequate amounts of ATP to support metabolism for several hours after imbibition (Attucci *et al.,* 1991). During germination of embryos, there appear to be two distinct patterns of mitochondrial development. These patterns, which are particularly obvious in cotyledons, depend on the nature of the stored reserves. In starch storing seeds, repair and activation of preexisting organelles predominate, whereas oil storing seeds typically produce new mitochondria (Morohashi, 1986). For example, the biogenesis of mitochondria in germinating maize embryos (which store oil in the scutellum, although starch is the major endosperm reserve) involves the synthesis of *cytochrome c oxidase* subunits encoded by the organellar genome, which is followed within hours by the synthesis of nuclear-encoded subunits (Ehrenshaft and Brambl, 1990). This observation also implies that the coordinated regulation of mitochondrial and nuclear genomes in plants begins during the early stages of germination.

3.1. Protein synthesis during germination

All of the components necessary for the resumption of protein synthesis upon imbibition are present within the cells of mature dry embryos, although polysomes are absent. However, within minutes of rehydration there is a decline in the number of single ribosomes as they become recruited into polysomal protein-synthesizing complexes. Initial protein synthesis is dependent on existing ribosomes, but newly synthesized ribosomes are produced and used within hours of initial polysome assembly (Dommes and Van de Walle, 1990). Preformed mRNAs are also present within the dry embryo. Some of these are residual messages associated with previous developmental processes (Lane, 1991) and may be used transiently during early germination. Messages encoding proteins that are important during seed maturation and drying, such as late embryogenesis abundant (LEA) proteins, are likely to be degraded rapidly upon imbibition. Conversely, those encoding proteins required during early germination (*e.g.*, ribosomal protein messages; Beltrán-Petia *et al.,* 1995) are replaced by identical messages at later times, with protein synthesis becoming more dependent on the new transcripts with time. The nature of the stored messages in seeds has not been studied extensively, although it is known that they are likely to be present in association with proteins as messenger ribonucleoprotein complexes (mRNPs) (Peumans *et al.,* 1979). An alternative possibility is that stored transcripts remain sequestered within the nucleus (Hammett and Katterman, 1975). New mRNAs are transcribed as germination proceeds.

The majority of these are likely to encode proteins essential for the support of normal cellular metabolism, that is, "growth maintenance" reactions that are not restricted to germination (Bewley and Marcus, 1990). Indeed, no specific protein markers exclusive to germination have been found. Therefore, it is cautioned here that many so-called germination-specific mRNAs reported in the literature encode enzymes integral to the mobilization and conversion of the major stored reserves; these are post germinative events that are important during seedling growth, but they are unrelated to germination *per se* (Fig.3).

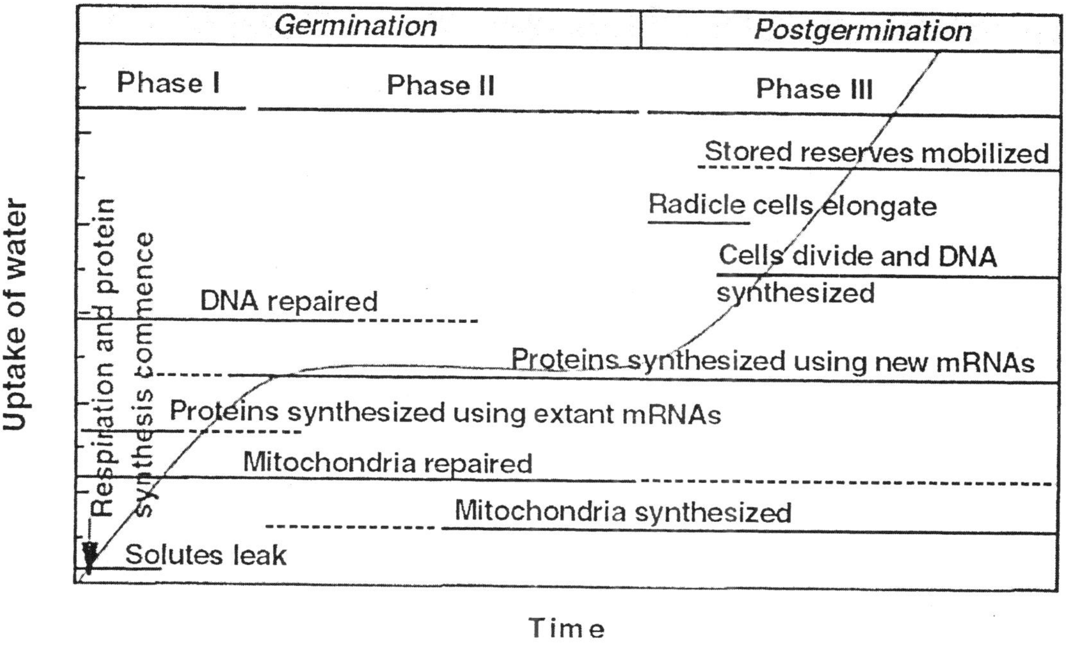

Fig. 3: Time course of major events associated with germination and subsequent post germinative growth. The time for events to be completed varies from several hours to many weeks, depending on the plant species and the germination conditions (Bewley, 1997).

Nevertheless, some changes in embryo mRNA populations and synthesized proteins do occur during germination of several species of monocots (*e.g.*, maize; Sánchez-Martinez *et al.*, 1986), dicots (*e.g.*, peas; Lalonde and Bewley, 1986), and conifers (*e.g.*, loblolly pine; Mullen *et al.*, 1996). The importance of these newly synthesized proteins to the completion of germination of the embryo remains to be elucidated. One promising candidate for a "germination-specific" gene could be the one that encodes the protein germin (Lane, 1991), which is an oxalate oxidase. However, by its nature, the timing of its synthesis, and its potential metabolic functions (Lane, 1994), germin is more likely to be involved with post germinative cell elongation. Thus, the search for specific protein markers with an exclusive role in germination must continue.

3.2. Mobilization of food reserves

3.2.1. Carbohydrates

The major form of stored carbohydrate present in most of the seeds is starch along with other polysaccharides such as hemicellulose, galactomannan, and certain oligosaccharides. Starch is found in amyloplasts. Starch is a polymer of glucose bound by alpha-1,4-glucose bonds. Starch molecules may be branched. Amylose and amylopectin are hydrolytically degraded by α- and β-amylases in the following manner.

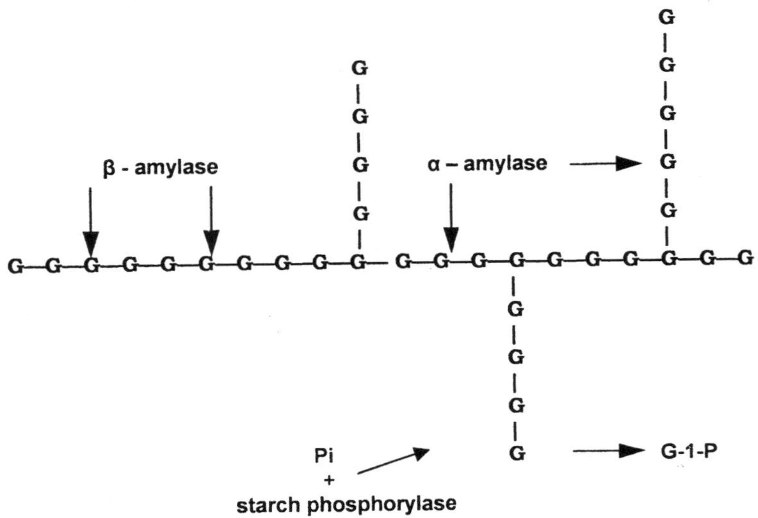

Starch may also be degraded phosphorolytically:

$$\text{Amylose} + \text{Amylopectin} + Pi \xrightarrow[\text{Phosphorylase}]{\text{Starch}} \text{Glucose 1 - P} + \text{limit dextrin}$$

α-Amylase cleaves the internal α-1,4-linked bonds of the glycan chains, releasing shorter amylose chains that are then further hydrolyzed to maltose (a disaccharide) by β-amylase. The majority of β-amylase is already present in the endosperm, but cannot act on starch grains until their initial hydrolysis by α-amylase. Neither α-amylase nor β-amylase can hydrolyze the α-1, 6 bonds in the branch points in amylopectin, resulting in what is called a limit dextrin. Specific debranching enzymes are required to hydrolyze these bonds, releasing additional amylose chains for further degradation. The maltose is converted to glucose by α-glucosidase, and the glucose is converted to sucrose *via* UDP-Glc pyrophosphorylase and sucrose-6-P synthetase. Starch phosphorylase can also release Glc-1-P directly from amylose or amylopectin. The walls of the endosperm cells are also degraded by specific hydrolases. Similar processes are involved in starch mobilization in dicots, although the separation of storage and secreting tissues is not so evident in those where the cotyledons are the storage tissues.

Sucrose is the major form of carbohydrate transport, because sucrose is uniquely resistant to hydrolysis by several enzymes. Glucose-1-phosphate, the product of starch phosphorylation, is a direct substrate for sucrose biosynthesis. Free glucose, however, must first be phosphorylated, namely to glucose-6-P, and then isomerized to G-1-P to be combined with uridine diphosphoglucose (UDPG), which transfers glucose to free fructose cr to fructose 1-P as follows.

$$\text{G-1-P} + \text{UTP} - \xrightarrow{\textit{UDPG}\text{ phosphorylase}} \text{UDPG} + \text{PPi}$$

$$\textit{UDPG} + \textit{Fructose} \xleftrightarrow{\textit{Sucrose synthetase}} \textit{Sucrose} + \textit{UDP}$$

or

$$\textit{UDPG} + \textit{Fructose-1-P} \xrightarrow{\textit{Sucrose}\text{-6=P synthetase}} \textit{Sucrose-6-P} + \textit{UDP}$$

$$\textit{Sucrose- 6- P} \xleftrightarrow{\textit{Sucrose}\text{-6-phosphatase}} \textit{Sucrose} + \textit{Pi}$$

Sucrose can be hydrolyzed to glucose and fructose by β-fructofuranosidase (invertase) whenever needed. Among cereal grains, barley (*Hordeum* sp.) has been thoroughly investigated as to its mobilization of carbohydrates reserves during germination. Barley varieties exhibit significant variation in the degree and timing or response to germination stimuli. The embryo is situated at the base of the kernel (mostly endosperm). The barley embryo consists of the acrospire, nodal and rootlet region, and attached scutellum. The endosperm is surrounded by the aleurone layer, encroaching around the periphery of the scutellum. Bewley and Black (1978) reviewed the metabolic changes associated with mobilization of food under two headings: (1) mobilization of the reserves of the embryo and (2) utilization of the reserves of endosperm. The food reserves of the embryo are quantitatively minor components of total mobilization (Fig. 4).

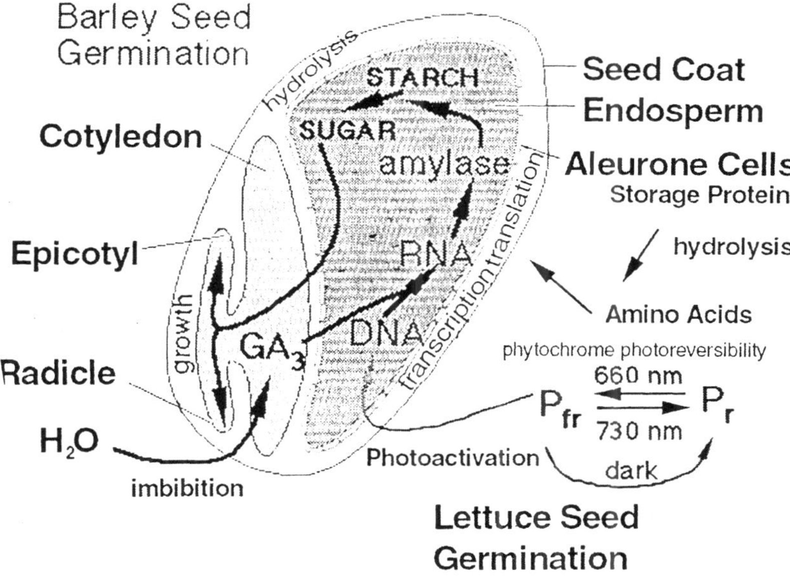

Fig. 4: Mobilization of food reserves

Gibberellins are synthesized by the seed embryo in either the nodal region of the axis or the scutellum. Indole acetic acid is also synthesized by the coleoptile tip soon after imbibition and then diffuses to the scutellar node. This auxin promotes the lignification of cell wall in the center of the node and subsequently plays a role in depositing lignin in the elongated provascular strand of the scutellum. Bewley and Black (1978) listed the chronological sequence of the events leading up to carbohydrate breakdown during the first two days of germination and growth of barley seeds (Table 1). These events pertain to barley seeds imbibed and germinated under nonmalting conditions. In seeds where cell wall galactomannanans are the major carbohydrate reserves, the corresponding hydrolytic enzymes are endo-β-mannanase, β-mannosidase, and α-galactosidase. Endo-β-mannanase cleaves internal bonds in the mannan chains; β-mannosidase is an exo-enzyme that converts these to mannose, and α-galactosidase removes the galactose side chains from the mannan backbone. The released sugars are then converted into sucrose and exported to the seedling.

These events (Table 1) begin after 5 h of imbibition, at which time the vascular system is ready for translocating metabolites from the scutellum to the embryo and possible for the transport of gibberellins synthesized at the node to the scutellum for their release.

Table 1: Chronological events in barley seed (cv. Proctor) during first two days of germination and growth

Time (h)	Events
0	Imbibition of water commences: 'dry' grain moisture content is 12%. Fat droplets, but no starch, observed in scutellum; β-amylase present.
1	Moisture content of embryo is 40%
5	Lignification of scutellar node and vascular tract commences.
12	Starch granules observed in scutellum; shrinkge of fat deposits.
16	Moisture content of endosperm is 34%
17	Translocating of synthesized gibberellin to aleurone layer; coleorhiza ruptures coat
22-24	Dissolution of intermediate layer begins; scutellar epithelial cells elongate: endo-β-glucanase and α-amylase released from aleurone layer
22-29	Dissolution of starch grain in endosperm observed
24	Isocitrate lyase present in scutellum
48	Axis accumulates starch

(Source: From Seeds Hand book, Desai *et al.*, 1997)

The gibberellin thus released diffuses into the aleurone layer to stimulate the synthesis and release of a number of hydrolytic enzymes into the endosperm. The GA released from the embryo induces the aleurone layer to undergo a series of metabolic changes, which result in the release of α-amylase and other hydrolases into the endosperm for the degradation of food reserves stored in it.

3.2.2. Lipids

The mobilization of lipid reserves involves a specialized organelle called the glyoxysome. In the glyoxysomes lipases in the oil bodies release fatty acids from the triacylglycerols, and the fatty acids are then transported to the glyoxysome. Glycerol is metabolized through glycolysis after its oxidation to triose phosphates. The fatty acids are further oxidized through the β-oxidation pathway by sequentially removing two carbon atoms in the form of acetyl CoA. The later enters the TCA cycle for its complete oxidation to CO_2 and water. The lypolysis of fats to fatty acids and glycerol in seeds takes place in fat-storing oil bodies, whereas fatty acids oxidation and synthesis of succinate *via* glyoxylate cycle occurs is glyoxysomes (Fig. 5). Succinate is converted to oxaloacetate in the mitochondria. Oxaloacetate is then metabolized to sucrose in the cytoplasm.

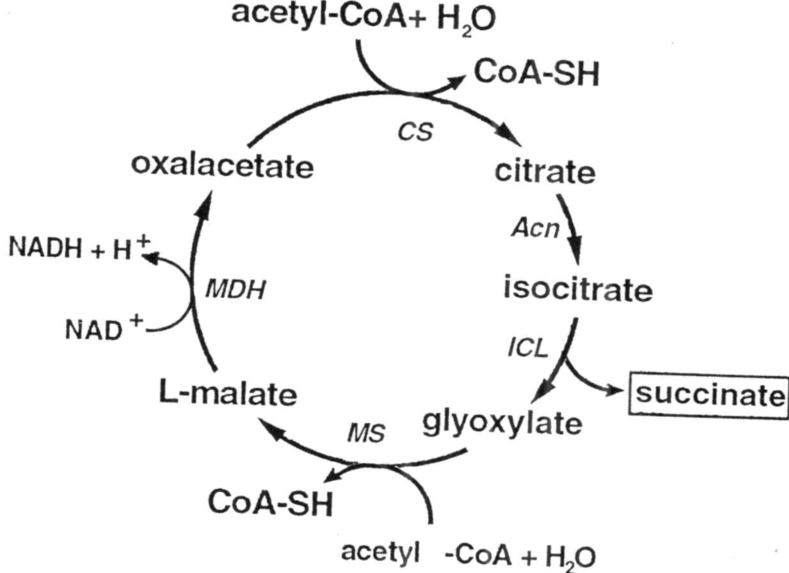

Fig. 5: The glyoxylate cycle. In higher plants, the glyoxylate cycle is an independent pathway, apparently disengaged from Krebs cycle. CS, citrate synthase; Acn, aconitase; ICL, isocitrate lyase; MS, malate synthase; MDH, malate dehydrogenase

3.2.3. Proteins

Plants accumulate protein reserves in developing seeds. The major amount of these reserves consists of specific storage proteins (Shewry and Casey, 1999), like globulins, which predominate in dicotyledonous seeds, or prolamines, which are the major storage proteins of most cereals. Accumulation of protein reserves takes place during the middle and late maturation stages when seeds act as nitrogen sinks in the plant.

They are deposited in membrane bounded compartments, the protein bodies (PB), which are generated either directly from the endoplasmic reticulum (ER) as in maize, or from the protein storage vacuoles (PSV) as in the majority of species (Muntz, 1998). During this deposition period storage protein degradation and turnover are negligible (Madison *et al.*, 1981) indicating that the proteins are protected against premature proteolytic attack. Storage proteins are mobilized during seed germination and subsequent seedling growth (Fig.6). Storage protein mobilization has been predominantly investigated in seeds of cultivated plants such as grain legumes, rape, sunflower, and cereals which, during domestication, underwent selection for increased protein accumulation in the cotyledon storage tissues, as in dictoyledonous seeds, or in the endosperm, as in cereal grains. In these tissues, storage protein mobilization only becomes measurable after the radicle has protruded and germination is finished. Inactive precursors of proteinases are synthesized on the rER, from where they are transported through the end membrane system into the PB. Here, enzyme activation takes place and protein breakdown is initiated (Müntz, 1996). The only alternative model has been presented for buckwheat cotyledons where a stored metalloendopeptidase starts globulin breakdown (Belozersky *et al.*, 1990).

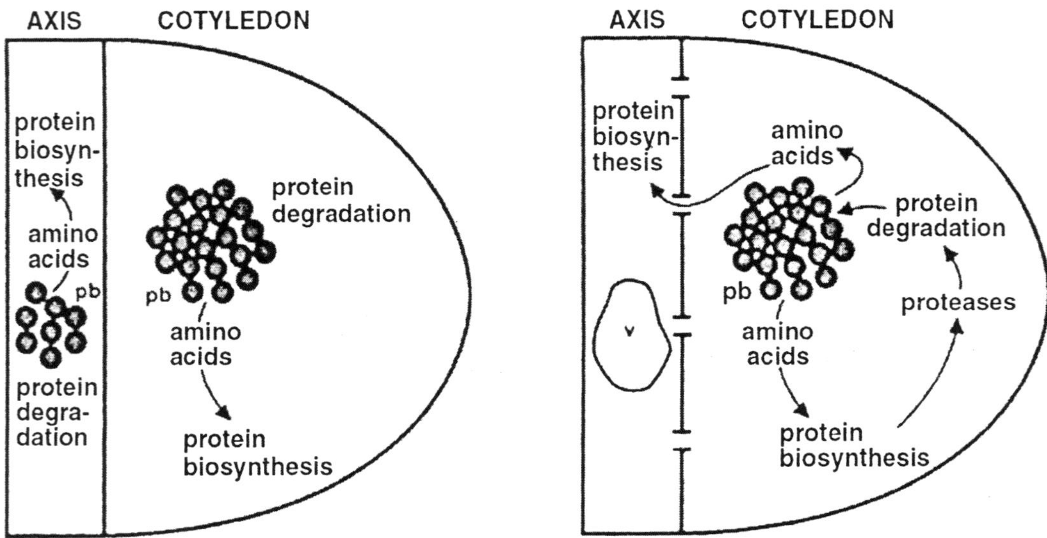

Fig. 6: Storage protein mobilization in embryonic axis and cotyledons during germination (stage 1) and after germination (stage 2). In stage 1, endogenous sources supply amino acids for protein biosynthesis in both organs whereas, in stage 2, the cotyledons are the source of amino acids for the embryonic axis which now acts as a sink (Schlereth *et al.*, 2000).

Storage proteins are not only accumulated and mobilized in specific storage tissues, but also in axis organs like the radicle and the embryonic shoot (Tiedemann *et al.*, 2000). During germination small amounts of storage proteins are broken down in limited regions. This degradation might be mediated by stored proteinases. It is not known whether storage protein mobilization in embryonic axes is mediated by *de novo* formed or by stored proteinases, neither it is known to what extent early protein biosynthesis during germination in axis nor storage tissue (Bewley, 1997) depends on free amino acids released by protein mobilization. Reviewers always remark that endo- and exopeptidases are present in the PB from the dry seeds of many plants (Müntz, 1996). Papain like cysteine proteinases (CPRs) were found in PB of dry cucumber cotyledons (Hara-Nishimura *et al.*, 1982). Recently, a new CPR family was detected with strict specificity for Asn in the P1 position at the cleavage site. These so called legumains (Hara-Nishimura *et al.*, 1998) most probably have polypeptide processing functions in vacuoles. Vacuolar processing enzymes (VPE) from the PSV of maturing seeds are classified as ßVPE. Legumains of the latter group like proteinase B from vetch (Becker *et al.*, 1995) and legumain-like proteinase (LLP) from kidney bean (Senyuk *et al.*, 1998) also participate in storage globulin breakdown in cotyledons of seedlings after germination. Both ßVPE and proteinase B-like legumains are localized in PB of seeds (Becker *et al.*, 1997). ßVPEs are formed in the developing seeds at the time of storage protein deposition. They are present in PB of dry seeds and their activity declines only slowly following germination (Okamoto *et al.*, 1996). The PB of dry buckwheat cotyledons contains a metalloproteinase (MPR) as well as an aspartic proteinase and a carboxypeptidase(s) (CP) (Elpidina *et al.*, 1991). An aspartic proteinase was also found in the PB of scutellum cells in dry barley grains (Marttila *et al.*, 1995). Carboxypeptidases are the only exopeptidases found in PB whereas aminopeptidases are localized in the cytoplasm of storage tissue cells (Shutov and Vaintraub, 1987). Small amounts of endopeptidases are present in protein bodies of embryonic axes and cotyledons of dry seeds from dicotyledonous plants also play an important role in the initiation of storage protein mobilization during early germination (Muntz *et al.*, 2001).

3.2.4. Phytic Acid

Phytic acid (*myo*-inositol hexa*kis*phosphate) is the major storage form of phosphorus in plant seeds. In maize grains, phytate phosphorus represents up to 88% of total phosphorus. It is a compact source of phosphate and mono (K^+) and divalent cations (Ca^{++} and Mg^{++}). During germination, stored reserves are used as a source of nutrients by the plant seedling. Phytic acid is degraded by the activity of phytases to yield inositol and free phosphate. Adequate levels of phosphorus are critical to the growth and development of all organisms for a range of functions such as macromolecular structure, energy generation and metabolic regulation. The demand for phosphorus increases dramatically during periods of rapid cell growth and division, such as seed germination. Phosphorus is stored in plant seeds as phytic acid (*myo*-inositol hexa*kis*phosphate) during seed development (Raboy, 1997). In soybeans (*Glycine max* L. Merr.), phytic acid is deposited in protein bodies (protein storage vacuoles) as a complex of chelated minerals and protein known as phytin (Prattley and Stanley, 1982). Maximal phytic acid levels are achieved at seed maturity immediately preceding desiccation (Raboy and Dickinson, 1987) and represent approximately 1% of the total weight in dry seeds. Upon seed germination, hydrolysis of phytin reserves provides nutrients for the rapidly growing seedling.

Dephosphorylation of phytic acid to a series of *myo*-inositol esters and inorganic phosphate is catalyzed by the enzyme phytase (Loewus and Murthy, 2000). Phytases have been purified from the seeds of various monocot and dicot species (Wodzinski and Ullah, 1996). In soybean seed germination, a pronounced increase in phytase activity accompanies a concomitant decrease in phytic acid, with maximal phytase activity attained at approximately 10 days after germination (Gibson and Ullah, 1988).

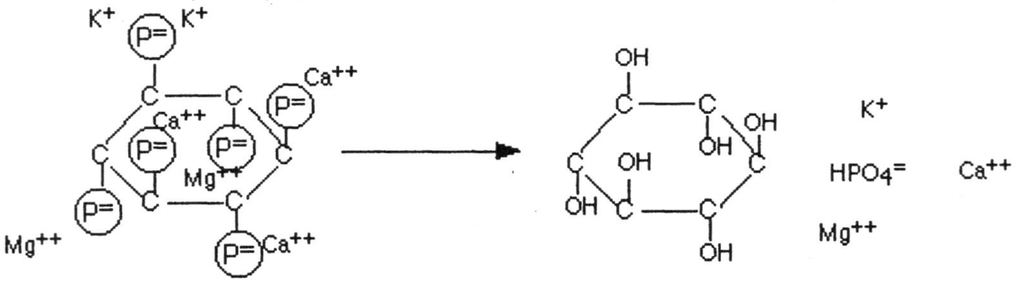

Inositol + Free phosphate

Phytin ⟶ Inositol + Free phosphate

4. RADICLE EXTENSION AND THE COMPLETION OF GERMINATION

With few exceptions, radicle extension through the structures surrounding the embryo is the event that terminates germination and marks the commencement of seedling growth. This extension may or may not be accompanied by cell division. Two discrete phases of DNA synthesis occur in the radicle cells after imbibition (Fig.1). The first takes place soon after imbibition and probably involves the repair of DNA damaged during maturation drying and rehydration as well as the synthesis of mitochondrial DNA. DNA synthesis associated with post germinative cell division accounts for the second phase (Osborne and Boubriak, 1994). Extension of the radicle is a turgor-driven process that requires yielding of walls in those cells of the embryonic root axis that lie between the root cap and the base of the hypocotyl (Cosgrove, 1997). There are three possible reasons for the commencement of radicle growth. One possibility is that late during germination, the osmotic potential $(\psi\pi)$ of the radicle cells becomes more negative because of the accumulation of solutes, perhaps as a result of the hydrolysis of polymeric reserves present within the radicle cells themselves. The decrease $(\psi\pi)$ would lead to increased water uptake, and the resulting increase in turgor would drive cell extension. However, there is no consistent evidence for changes in cellular $(\psi\pi)$ during germination (Bradford, 1996). A second possibility is that extensibility of the radicle cell walls allows for their elongation. Whether the mechanisms by which cells of the radicle become more extensible differ from those in other tissues is not known. Cell wall loosening may result from the cleavage and rejoining of xyloglucan molecules that tether adjacent cellulose microfibrils, which would permit expansion by microfibril separation. The activity of xyloglucan endotransglycosylase (XET), an enzyme capable of reversibly cleaving xyloglucan molecules, increases in the apical region of maize seedling roots during their elongation (Wu *et al.*, 1994), but this increase occurs after germination is completed. Alternative candidates for cell wall-loosening proteins are the expansins, which have the ability to disrupt the hydrogen bonds between cell wall polymers (*e.g.*, matrix polysaccharides and cellulose microfibrils). Expansins have been strongly implicated in the expansion of cucumber hypocotyls (McQueen-Mason and Cosgrove, 1995). However, neither of these cell wall-loosening proteins has been reported in germinating seeds. Moreover, both XET and expansin activities in seedlings appear to be enhanced by auxin, which is generally regarded as ineffective in promoting seed germination, and XET activity in maize seedling roots is also enhanced by abscisic acid (ABA), a potent inhibitor of embryo radicle elongation. A third possibility is that the seed tissues surrounding the radicle tip weaken, thus allowing the tip to elongate. Because there are no changes in cell *($\psi\pi$)* before radicle growth commences, it is axiomatic that the turgor potential ($\psi\pi$) of the radicle cells is sufficient to drive their elongation if there is little or no restraint exerted by the surrounding structures. In many germinating seeds, including those of rape, the testa splits during imbibition, and it is only the rigidity of the radicle cell walls that restrains growth (Schopfer and Plachy, 1985). As the walls yield during the initial stages of radicle elongation, there is a decline in cell ψp. Conversely, in other seeds, $\psi\pi$ alone is insufficient to drive wall extension, and there is a severe constraint on radicle cell growth imposed by the surrounding structures. In lettuce, tobacco, and tomato seeds, the endosperm is the constraining structure,whereas in muskmelon it is the perisperm. A reduction in the resistance of these enclosing structures is necessary for germination to be completed. Measurements have revealed a decline in the mechanical resistance of the structures covering the embryo root cap at the time of radicle emergence in the endosperm of pepper seeds (Watkins and Cantliffe, 1983) and the perisperm of muskmelon seeds (Welbaum and Bradford, 1990). This decline in resistance is likely to be achieved by cell wall hydrolases, such as hemicellulases, produced within and secreted by the endosperm itself.

Seedling growth can be of two types: epigeous germination or hypogeous germination. Both refer to the position of the cotyledons during germination. In epigeous germination (*epi* - Latin meaning above or beyond), the cotyledons are pushed above the soil surface. Beans and other legumes are examples. In hypogeous germination (*hypo* - Latin meaning under), the cotyledons as well as most of the seed remains underground with only the shoot emerging from the soil surface (Fig. 7).

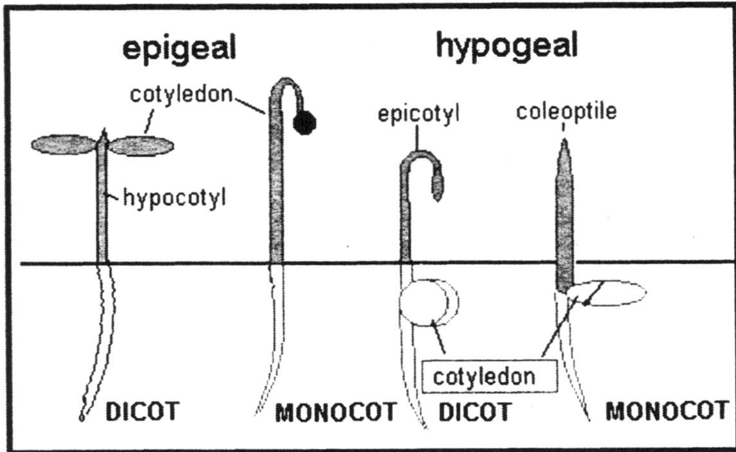

Fig.7: Epigeal and hypogeal germination

In dicots, the primary root emerges through the seed coats, while the seed is still buried in the soil. The hypocotyl emerges from the seed coats and pushes its way up through the soil. It is bent in a hairpin shape as it grows up. The two cotyledons protect the epicotyl structures, the plumule from mechanical damage. Once the hypocotyl arch emerges from the soil, it straightens out. This response is triggered by light. Both red light, absorbed by phytochrome and blue light, absorbed by cryptochrome can do the job. The cotyledons spread apart exposing the epicotyl, consisting of two primary leaves and the apical meristem. In many dicots, the cotyledons not only supply their food stores to the developing plant but also turn green and make more food by photosynthesis until they drop off (Fig.8).

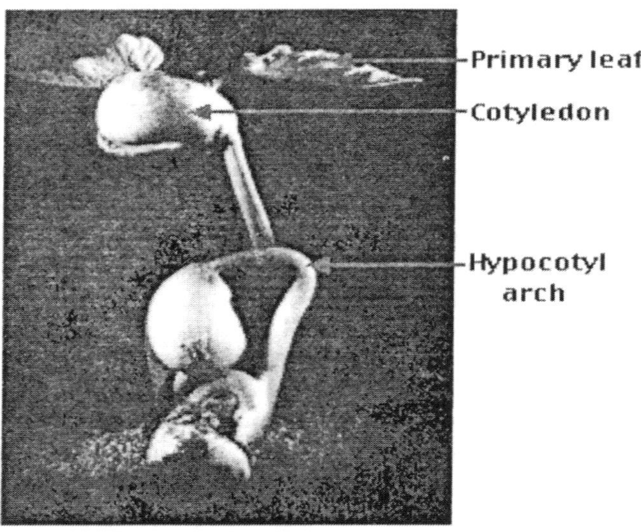

Fig. 8: Germination in dicots

In monocots when grass seeds like corn (maize) or oats germinate, the primary root pierces the seed (and fruit) coverings and grows down; the primary leaf of the plant grows up. It is protected as it pushes up through the soil by the coleoptile—a hollow, cylindrical structure. Once the seedling has grown above the surface, the coleoptile stops growing and the primary leaf pierces it (Fig.9).

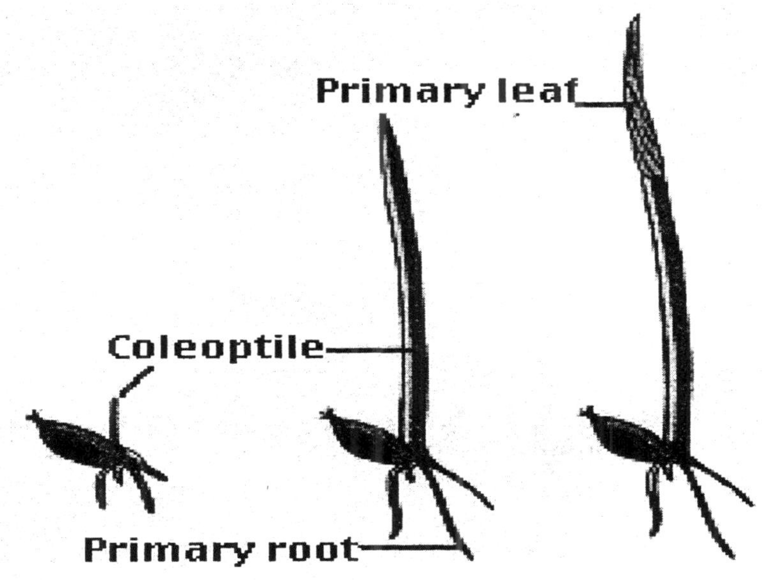

Fig. 9: Germination in monocots

5. ROLE OF PLANT HORMONES IN SEED GERMINATION

5.1. Gibberellins

Gibberellin (GA) is an essential phytohormone that controls many aspects of plant development, including seed germination, leaf expansion, stem elongation, flowering, and seed development (Davies, 1995). Various studies have suggested that these GA mediated growth responses are regulated in part by the modulation of cellular GA concentrations and by altering the ability of cells to respond to this hormone (Richards *et al.,* 2001).

GAs are tetracyclic diterpenoids synthesized from geranylgeranyl diphosphate. The biosynthesis pathway that converts geranylgeranyl diphosphate to biologically active GAs, such as GA_1 and GA_4, has been well studied (Hedden and Kamiya, 1997). Recent studies have started to identify some regulatory factors that control this pathway (Yamaguchi and Kamiya, 2000).

Seed dormancy and germination are complex physiological processes that are controlled by a range of developmental and external cues (Koornneef *et al.,* 2002). Germination begins with water uptake by seeds and terminates with the initial elongation of the embryonic axis (Bewley, 1997). Completion of germination is visible by the emergence of the radicle. Numerous studies have shown that GA plays an essential role in promoting the germination of Arabidopsis seeds (Fig. 10) (Ogawa *et al.,* 2003). This is evident by the defects in seed germination of severe GA-deficient mutants such as *ga1-3* and *ga2-1* (Koornneef and Van der Veen, 1980). The inhibitory effect of GA biosynthesis inhibitors, such as paclobutrazol and uniconazole, on seed germination (Jacobsen and Olszewski, 1993) suggests that newly synthesized. GAs is required after imbibition for radicle emergence. These lines of evidence demonstrate that the level of GA is a critical determinant for seed germination. However, it remains unclear how endogenous GA concentrations are modulated during Arabidopsis seed germination.

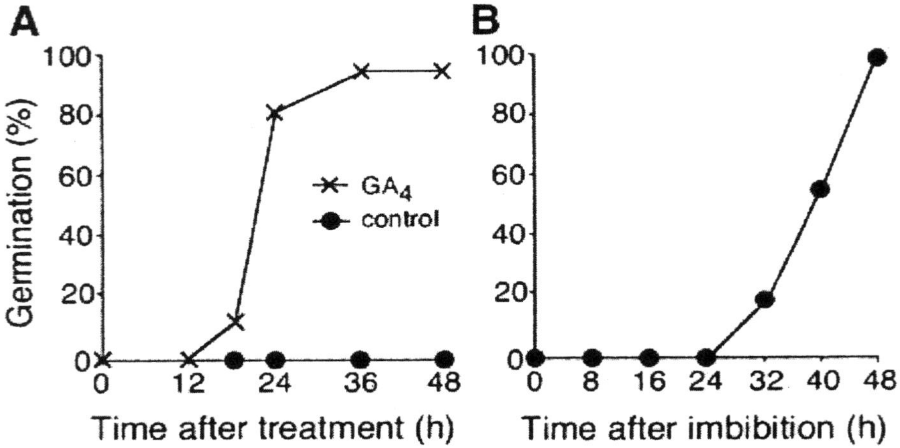

Fig. 10: (A) Germination of *ga1-3* seeds. The seeds were stratified at 4°C in the dark for 48 h and then incubated at 22°C for 24 h under continuous white light before treatment with 50 μM GA_4 or a mock solution (control). (B) Germination of wild-type seeds at 22°C under continuous white light. (Ogawa *et al.,* 2003)

At least two roles for GA have been proposed in stimulating germination in small-seeded plant species such as tomato and Arabidopsis (Yamaguchi and Kamiya, 2002). First, GA is necessary to overcome the mechanical restraint conferred by tissues that surround the embryo, such as the aleurone and the testa. In Arabidopsis, embryos of nongerminating GA-deficient mutants can develop into dwarf plants when the surrounding tissues are removed mechanically (Telfer *et al.,* 1997). Therefore, one of the roles of GA in the induction of radicle protrusion appears to be to weaken the tissue that surrounds the embryo (Fig.10). Second, GA also increases the growth potential of the embryo, as indicated by the reduced growth rate of GA-deficient embryos (Groot and Karssen, 1987). Several GA-inducible genes related to cell wall loosening have been identified in tomato seeds, including those encoding endo-β-mannanase (Nonogaki *et al.,* 2000), xyloglucan endotransglycosylase/hydrolase (Chen *et al.,* 2002), expansin (Chen *et al.,*

2001), and β-1,3-glucanase and chitinase (Wu *et al.,* 2001), some of which are expressed specifically in the micropylar endosperm cap around the radicle.

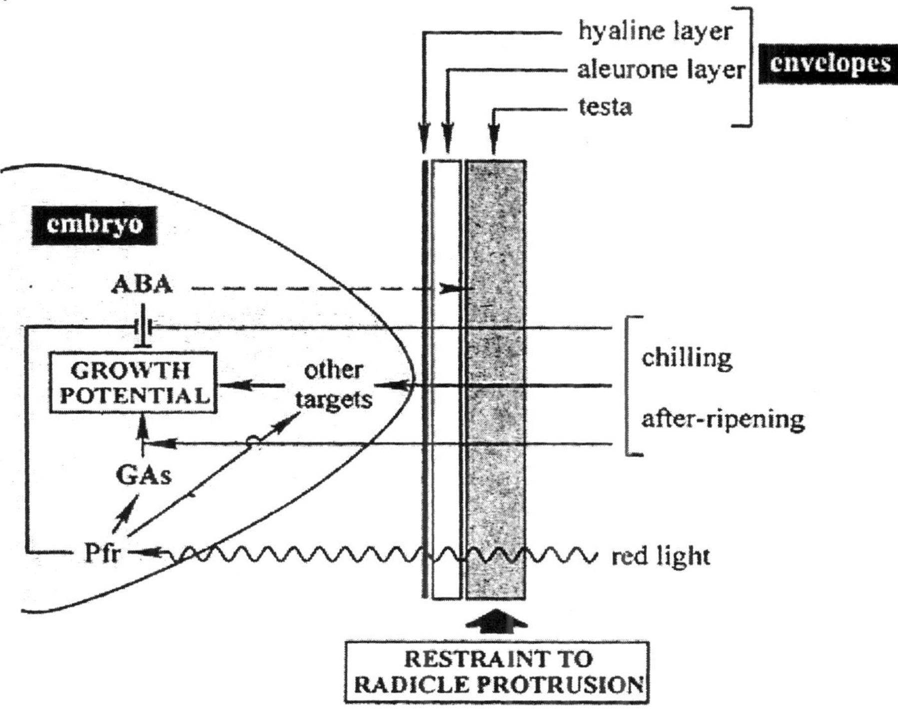

Fig.11: Schematic presentation of the interactions between the envelopes and the embryo affecting dormancy breaking and germination of Arabidopsis seeds.

Germination occurs when the growth potential of the embryo is sufficient to overcome the restraint to radicle protrusion imposed by the seed envelopes. When the restraint of the testa is weakened by mutations, the growth potential threshold required for germination is lowered. The sharp arrows and the blunt arrows stand for a promotive and an inhibitory action, respectively. Dashed arrow indicates leaching of ABA through the testa. (Debeaujon and Koornneef, 2000). Genetic studies have identified several GA-signaling components in Arabidopsis, some of which play a role in GA-induced seed germination, as shown by their loss of function mutant phenotypes (Olszewski *et al.,* 2002). RGA-LIKE1 (RGL1) and RGL2, members of the DELLA subfamily of GRAS proteins, and SPINDLY, a Ser/Thr *O*-linked *N*-acetylglucosamine transferase, act as negative regulators of GA-dependent seed germination (Wen and Chang, 2002). By contrast, SLEEPY1 (SLY1) is proposed to be a positive regulator in the GA response pathway (Steber *et al.,* 1998) in germinating seeds. The importance of defining specific GA responses is emphasized by the fact that germination in GA-deficient non germinating seeds can be rescued when the abundance of or the sensitivity to, other hormones is modulated. Abscisic acid (ABA) plays a critical role in the maintenance of seed dormancy and inhibits germination (Leung and Giraudat, 1998). ABA biosynthesis mutants were first isolated as suppressors of the nongerminating GA deficient mutants (Koornneef *et al.*, 1982).GA levels are increased just before radicle emergence and that a unique set of genes is upregulated or downregulated in response to exogenous GA treatment. GA-responsive genes constitute subsets of genes with different temporal patterns of expression during wild type seed germination. The diverse nature of GA-responsive genes also is suggested by *in situ* hybridization analysis, which indicates that at least a fraction of GA-regulated genes are expressed predominantly in non-GA producing cell types.

5.2. GA distribution in germinating cells and its response

The distribution of GA-regulated genes within a seed is largely unknown. It has been shown that two GA biosynthesis genes, *AtGA3ox1* and *AtGA3ox2*, are expressed predominantly in the cortex and endodermis of the embryonic axis in germinating Arabidopsis seeds (Yamaguchi *et al.,* 2001). Because these are the major genes encoding GA 3-oxidases that catalyze the final step to produce active GAs during seed germination, it is clear that active GAs are synthesized mainly in these cell types at this stage. *At-XTH5* transcripts were detected mainly in the embryonic axis, including cell types exhibiting *AtGA3ox1* expression. However, unlike *AtGA3ox1*, *At-XTH5* expression appeared abundant in the radicle. Upregulation of *AtCP1* (encoding a putative Cys proteinase) by GA_4 was most evident in the epidermis and provasculature of the embryonic axis and the aleurone layer, where *AtGA3ox* transcripts were not detected (Ogawa *et al.,* 2003). GA_4 induced transcript accumulation of the *PROTODERMAL FACTOR1* (*PDF1*) gene (Abe *et al.*, 1999) was observed mainly in the epidermis of the embryonic axis. In germinating cereal grains, active GAs synthesized in the embryonic epithelium have been shown to cause the induction of α-amylase in the aleurone (Kaneko *et al.,* 2002). A similar embryo aleurone interaction was seen in Arabidopsis, provided that the expression of *AtGA3ox* genes is restricted to the embryo (Yamaguchi *et al.,* 2001), whereas *AtCP1* expression responded to exogenous GA_4 in the aleurone.

5.3. GA-ABA interaction in germinating cells

There are at least three possible mechanisms by which GA downregulates ABA upregulated genes: (1) GA reduces ABA levels by affecting ABA biosynthesis; (2) GA negatively regulates the ABA response pathway; and (3) GA and ABA signals are targeted independently to distinct *cis*-regulatory sequences of a single gene. In germinating lettuce seeds, exogenous GA_3 treatment has been shown to reduce endogenous ABA levels when the seeds are incubated in the dark after far-red light treatment that inhibits germination (Toyomasu *et al.,* 1994). ABA contributes primarily to the initial downregulation of the ABRE-containing genes, which then are downregulated further to nearly undetectable levels by active GAs that increase at later times during wild-type seed germination. An early decrease in the ABA level and a late increase in the active GA level after seed imbibition also have been reported in germinating barley embryos (Jacobsen *et al.,* 2002).

5.4. Role of GA in cellular activity and modulation of metabolic pathways

5.4.1. Cell elongation and Cell division

Among GA-upregulated genes identified, there are some classes of genes implicated in cell elongation, Aquaporin, a class of intrinsic membrane proteins, facilitates water uptake into symplast to increase turgor for cell expansion. Xyloglucan endotransglycosylase/hydrolase (XTH) loosens the cell wall by hydrolyzing cell wall components (Rose *et al.*, 2002), whereas expansin does it by disrupting hydrogen bonding (Lee *et al.*, 2001). Although the roles of pectin methylesterase (PME) have not been elucidated fully, this enzyme also may be involved in cell elongation by modifying cell wall pectin (Micheli, 2001). These cell wall–loosening activities may be associated with weakening of the tissue surrounding the embryo and/or the facilitation of embryo growth (Fig. 12, Table 2). GA-dependent expression of cell wall related enzymes during germination already has been reported for other species (Chen *et al.*, 2002) majority of expansin and PME gene family members were upregulated by GA_4, suggesting that these cell wall–modifying activities are dependent in large part on GA during Arabidopsis seed germination (Fig. 12).

Table 2: GA-responsive genes implicated in cell elongation

Protein name[a]	Total[b]	Microarray[c]	Germination[d]	GA Up[e]	GA Down[f]
Expansin	26	12	10	7	0
XTH	33	24	12	5	1
PME	67	32	9	7	0
Aquaporin	38	24	18	5	1

[a] PME, pectin methylesterase; XTH, xyloglucan endotransglycosylase/hydrolase.

[b] Number of genes identified in the Arabidopsis genome (Arabidopsis Genome Initiative, 2000).

[c] Number of genes represented on the microarray.

[d] Number of genes whose mRNAs were detectable within 32 h after imbibition of wild-type seeds.

[e] Number of GA-upregulated genes. [f] Number of GA-downregulated genes

(Ogawa *et al.*, 2003).

Most cells in mature dry seeds are in G1-phase (Georgieva *et al.*, 1994). Micro-array studies indicated that classes of cell cycle related genes, notably those involved in the transition from G1- to S-phase, were upregulated after GA_4 treatment (Ogawa *et al.*, 2003). These include genes that encode D-type cyclin (At1g70210) and the DNA replication licensing factor MCM protein (At1g44900 and At2g16440) (Vandepoele *et al.*, 2002). GA also unregulated a gene that encodes replication protein A (RPA; At2g24490). The GA-induced expression of *OsRPA1* (which encodes an RPA) in internodes of deepwater rice has been reported (Van der Knaap *et al.*, 1997). Because the expression of the

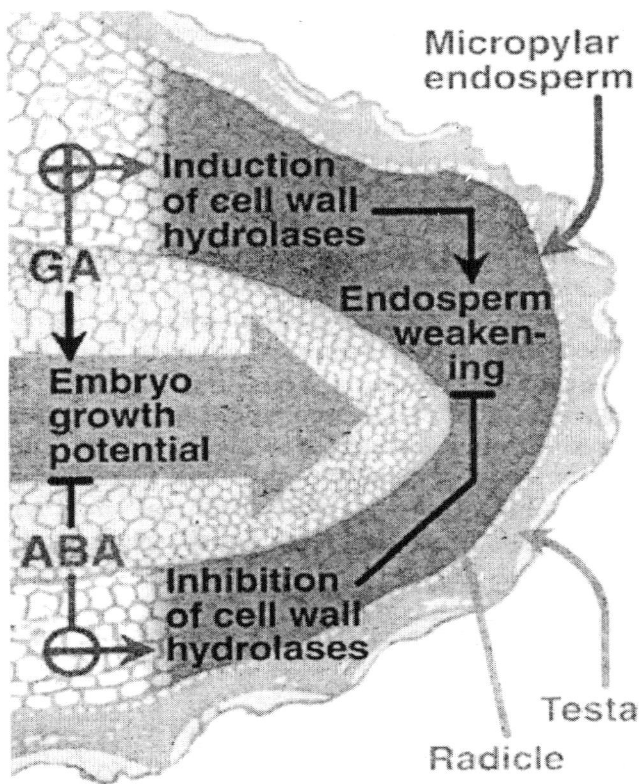

Fig. 12: The model shows the micropylar endosperm and the radicle tip of a tobacco seed. Gibberellins (GA) promote the induction of cell wall hydrolases and thereby promote endosperm weakening and endosperm rupture. Abscisic acid (ABA) inhibits the induction of cell wall hydrolases and thereby inhibits endosperm weakening and endosperm rupture. GA promotes and ABA inhibits the embryo growth potential. Endosperm weakening involves several cell-wall hydrolases and other molecular mechanisms like ROS.

majority of genes implicated in the transition from G2- to M-phase was not GA dependent, we speculate that the GA signal for cell division is targeted mainly to G1 cells to initiate DNA replication. Transcriptional changes for both cell elongation and cell division took place before radicle appearance. However, cell elongation related genes tended to be upregulated earlier than genes related to cell division. Role of GA in seed germination is illustrated in Fig. 13 by Taiz and Zeiger (2002).

1. GA_1 from the embryo first binds to a cell surface receptor.

2. The cell surface GA receptor complex interacts with a heterotrimeric G-protein, initiating two separate signal transduction chains.

3. A calcium-independent pathway, involving cGMP, results in the activation of a signaling intermediate.

4. The activated signaling intermediate binds to DELLA repressor proteins in the nucleus.

5. The DELLA repressors are degraded when bound to the GA signal.

6. The inactivation of the DELLA repressors allows the expression of the *MYB* gene, as well as other genes, to proceed through transcription, processing, and translation.

7. The newly synthesized MYB protein then enters the nucleus and binds to the promoter genes for α-amylase and other hydrolytic enzymes.

8. Transcription of α-amylase and other hydrolytic genes is activated.

9. α-Amylase and other hydrolases are synthesized on the rough ER.

10. Proteins are secreted via the Golgi.

11. The secretory pathway requires GA stimulation via a calcium–calmodulin-dependent signal transduction pathway.

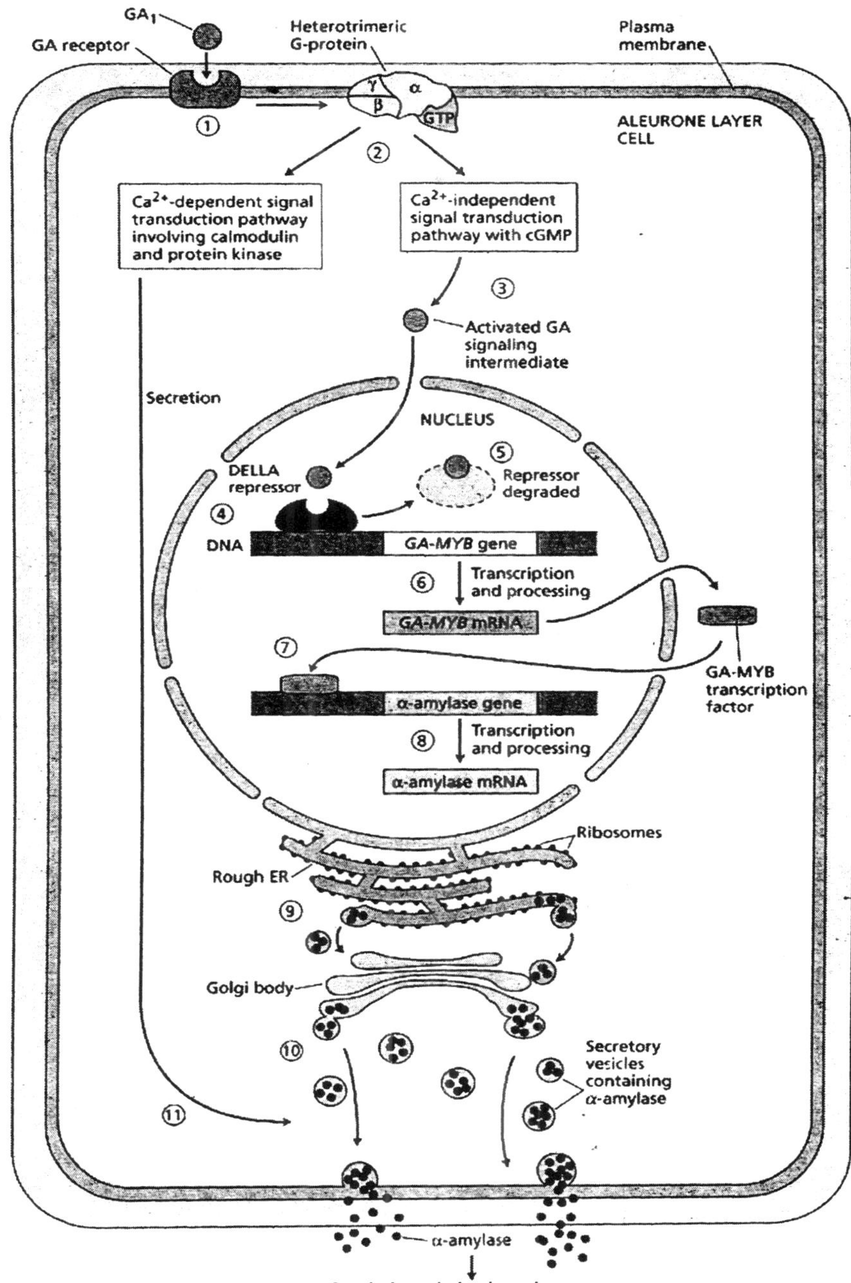

Fig. 13: Composite model for the induction of α-amylase synthesis in barley aleurone layers by gibberellin. A calcium-independent pathway induces α-amylase gene transcription; a calcium-dependent pathway is involved in α-amylase secretion. DELLA proteins are nuclear repressors of plant gibberellin (GA) responses (Taiz and Zeiger, 2002).

5.4.2. Role of ethylene and auxin in seed germination

Ethylene acts positively on seed germination in Arabidopsis (Kepczynski and Kepczynska, 1997). Exogenous ethylene treatment induces the germination of GA-deficient mutant seeds (Karssen *et al.*, 1989). Some

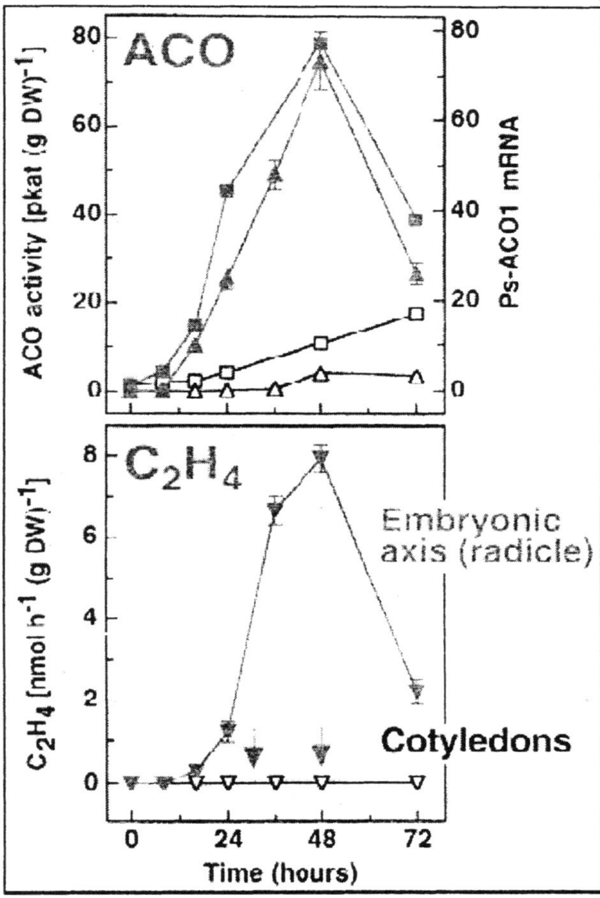

Fig. 14: Tissue specificity of ethylene biosynthesis in germinating pea seeds. An early onset and sequential induction of ACC biosynthesis, Ps-ACOl mRNA and ACO activity accumulation and ethylene production were localized almost exclusively in the embryonic axis, but not in the cotyledons of germinating pea seeds. Within the embryonic axis, ethylene biosynthesis and responsiveness are localized to the elongation and differentiation zones of the radicle (Petruzzelli *et al.,* 1999).

ethylene-related genes are affected by exogenous GA_4 treatment in imbibed *ga1-3* seeds. A gene that encodes a 1-aminocyclopropane-1-carboxylic acid oxidase (ACO), which catalyzes ethylene formation, was upregulated by exogenous GA_4. GA_4 may increase ethylene production in imbibed Arabidopsis seeds. Transcripts of ethylene-inducible genes, such as

HOOKLESS1 (*HLS1*) (Lehman *et al.*, 1996), and a member of the ethylene receptor gene family, *ERS1* (Hua *et al.,* 1998), also were upregulated by GA_4 treatment, which supports the idea that GA activates ethylene synthesis and/or response. Increased ethylene evolution accompanies seed germination of many species including *P.sativum.* Ethylene promotes ethylene biosynthesis during pea seed germination by positive feedback regulation of 1-aminocyclopropane-1-carboxylic acid oxidase (ACO). An early onset and sequential induction of ACC biosynthesis, Ps-ACO1 mRNA and ACO activity accumulation and ethylene production were localized almost exclusively in the embryonic axis, but not in the cotyledons of germinating pea seeds (Fig. 14). Ethylene independent signalling pathways regulate the spatial and temporal pattern of ethylene biosynthesis, whereas the ethylene signalling pathway regulates high level ACO expression in the embryonic axis, and thereby enhances ethylene evolution during seed germination (Fig. 14). Ethylene sensitive pea class I β-1,3-glucanase (ßGLU I) is induced exclusively in the embryonic axis, but not the cotyledons of germinated seeds. Distinct ethylene and tissue specific regulation of β1,3-glucanases and chitinases were found during pea seed germination. Ethylene biosynthesis and responsiveness are localized to the elongation and differentiation zones of the pea radicle. Ethylene indution of Ps-ACO1 and βGlu I are localized in this tissues and ethylene induces root hair formation and elongation. Calcium is required for ethylene dependent responses involving ACO in radicle tissues of germinated seeds (Petruzzelli *et al.,* 2003).

The roles of auxin in seed germination have been unclear; the expression of several auxin related genes was modulated by exogenous GA_4 treatment. Polar auxin transport is controlled by auxin influx and efflux carriers, which pump auxin into and out of cells, respectively (Muday and DeLong, 2001). Several genes that encode (putative) auxin carrier proteins are upregulated in response to exposure to GA_4 at relatively early time points. Two genes that encode CYP79B2 and CYP79B3 were determined to be upregulated by exogenous GA_4. Recent work has demonstrated that these CYPs are necessary not only for indole glucosinolate synthesis, but also for the formation of indoleacetic acid through indole-3-acetaldoxime (Zhao *et al.,* 2002). Thus, GA activity may be targeted to enhancing auxin biosynthesis as well. Although direct evidence for this idea remains to be found, GA activity results in substantial changes in auxin level and transport during seed germination. Hydrolysis of IAA-containing proteins or other types of high molecular mass IAA conjugates, which are abundant in Arabidopsis seeds (Walz *et al.,* 2002), might also contribute to the free IAA pool during germination. Hydrolysis of IAA containing proteins are carried out by auxin conjugate hydrolases (Rampey *et al.,* 2004).

5.4.3. Brassinosteroids

Brassinosteroids (BRs), a group of over 40 naturally occurring plant steroid hormones found in a wide variety of plant species (Schumacher and Chory, 2000) are also involved in the control of germination in Arabidopsis. It is suggested that the BR signal is required to reverse ABA induced dormancy and to stimulate germination (Steber and McCourt, 2001). BRs could overcome the lack of germination of the *sleepy1* mutant, probably by bypassing its GA requirement. Two BR mutants *DET2* and *BRI1* show reduced germination, but eventually germinate without BR, indicating that in contrast to GAs, BRs are not absolutely required for germination (Steber and McCourt, 2001). These results indicate that seed germination is determined by the net effect of multiple hormones and suggest the presence of crosstalk among actions by GA, ABA, ethylene, and BR.

6. ROLE OF PHYTOCHROME IN SEED GERMINATION

Phytochrome, stand for "plant colour", was originally coined to describe the proteinaceous pigment that controls reversible seed germination of lettuce (cv. Grand rapids) by red (R) and far-red (FR) light. Red light promotes seed germination, whereas subsequent far-red light treatment abolishes red light induction of seed germination. The germination response of lettuce seeds repeatedly treated with R/FR cycles is determined by the last light treatment.

Thus R/FR reversibility is a characteristic feature of this response. In addition, the law of reciprocity applies to this response, i.e., the response is dependent on the total amount of photons received irrespective of the duration of light treatment. This type of responses is now defined as the low fluence responses (LFRs) (Mancinelli, 1994). LFRs also induce other transient responses, such as changes in ion flux, leaf movement, chloroplast rotation and changes in gene expression (Haupt and Hader, 1994). Besides the R/FR reversible LFRs, there are two other modes of phytochrome action. The very low fluence responses (VLFRs) are activated by extremely low light intensities of different wave lengths (far-red, red and blue), such as light induced expression of the *LHCB* gene and light induction of seed germination, and the high irradiance responses (HIRs) which depend on prolonged exposure to relatively high light intensities. HIRs are primarily responsible for the control of seedling de-etiolation (*e.g.* inhibition of hypocotyl elongation and promotion of cotyledon expansion) under all light qualities (Casal *et al.*, 1998).

Pr ——————> Pfr ——————> responses

Inactive form	Active form
Red absorbing	Far red absorbing
(600-690 nm)	(720-780 nm)

7. SEED GERMINATION AND SEEDLINGS DE-ETIOLATION

At least three phytochromes have been documented to be involved in control of Arabidopsis seed germination: phyA, phyB and phyE. phyA is responsible for the photoirreversible VLFR responses triggered by a broad spectrum of irradiations (ultraviolet, visible and far red light), while phyB controls the R/FR photoreversible effects of low fluence response (Shinomura *et al.,* 1996). Recently, it was found that phyE also plays a role in the control of seed germination by FR light. Either phyE is directly involved in the photoperception of FR for this response or the action of phyA in mediating seed germination requires the presence of phyE (Hennig *et al.,* 2002).

Phytochrome mediates germination in photoblastic lettuce seeds by an upregulation of active GA biosynthesis. Expression of genes encoding GA 20-oxidases, producing GA_{20}, the immediate precursor of GA_1, is induced after an initial 3-h imbibition in the dark, and a red-light treatment converts Pr to Pfr, which upregulates gene expression of GA 3β-hydroxylase (Fig. 15). This results in an increase in active GA_1, which induces germination (Toyomasu *et al.,*1998).

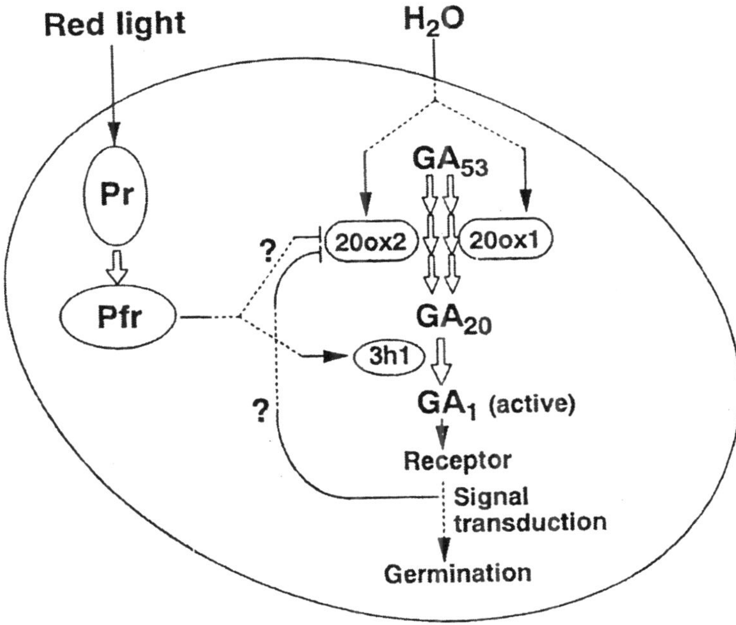

Fig. 15: Model of the mechanism of germination of photoblastic lettuce seed. (Toyomasu *et al.*, 1998).

Following seed germination, light acts to inhibit hypocotyl elongation and promotes cotyledon expansion. Under far-red light, phyA is probably the only active photoreceptor (Whitelam *et al.*, 1993). In white or red light, although phyB plays a major role, the long hypocotyl and reduced cotyledon expansion phenotype of its null mutants is enhanced by mutations in phyA or phyD, suggesting that multiple photoreceptors sense white light during de-etiolation. The contribution of phyE to seedling de-etiolation seems negligible (Devlin *et al.*, 1998). Interestingly, a recent high-resolution kinetic analysis of the growth of Arabidopsis seedlings revealed that the red light inhibition of hypocotyl elongation is controlled by a sequential and coordinated action of phyA and phyB. PhyA contributes to the initial hypocotyl growth inhibition (first 3 hr of irradiation), while phyB functions in the later phase (Parks and Spalding, 1999).

Studies with Arabidopsis *phyC* overexpressors suggested that *phyC* likely plays a role in primary leaf expansion. The more accurate assessment of *phyC* function became possible until recent identification of *phyC* loss of function mutants. *phyC* mutant seedlings exhibited a partial loss of sensitivity to red light, with longer hypocotyls and smaller cotyledons than wild type seedlings. Further, hypocotyl of *phyA phyC* double mutant is significantly longer than that of *phyC* single mutant, suggesting that *phyA* and *phyC* acts redundantly to inhibit hypocotyl elongation under red light. Consistent with this, *phyAphyBphyDphyE* quadruple mutant still retains some weak responses to red light, suggesting that *phyC* has some roles in mediating red light signaling (Franklin *et al.,* 2003). However, *phyBphyC* double mutants de-etiolated similarly to *phyB* single mutants, suggesting that *phyC* function is dependent on *phyB* (Monte *et al.,* 2003).

8. SIGNAL CROSS TALK DURING SEED GERMINATION

Seeds integrate many intrinsic signals to control germination (Koornneef *et al.,* 2002). It has been repeatedly shown that GA induces germination and that abscisic acid (ABA) antagonizes the GA effect (Lovegrove and Hooley, 2000). Seed reserve mobilization in the gramineae is coordinated to a large extent by aleurone cells, which are regulated by the opposing actions of the plant hormones gibberellin (GA) and abscisic acid (ABA). GA is perceived at the aleurone plasma membrane (Gilroy and Jones, 1994) and induces transcription of α-amylase genes. This effect is overcome by ABA. Functionally important DNA sequence elements in the promoters of GA and ABA regulated α-amylase genes have been identified, and genes encoding proteins that bind to some of these have been cloned. GA regulated *myb* gene, *GAmyb,* encodes a protein that binds to the GA response element of high pI α-amylase promoters. *GAmyb* is thought to be a component of the GA response pathway and has been shown to transactivate a high pI α-amylase promoter. It is not known how GA regulation of *GAmyb* and α-amylase gene transcription is linked to GA perception at the plasma membrane. However, recent evidence has suggested the possible involvement of protein phosphatases, cGMP, and inositol 1,4,5-trisphosphate at early stages in GA signal transduction. Seed germination of GA synthesis mutants can be rescued by the application of GA and ABA synthesis and insensitive mutants lack the ABA inhibition of GA-induced germination as well as being viviparous (Leon Kloosterziel *et al.,* 1996b). Brassinosteroid (BR) probably acts downstream of GA because BR is able to rescue germination of GA deficient (Steber and McCourt, 2001) and GA response (Steber *et al.,* 1998) mutant seeds. The inhibitory effect of high concentrations of sugars on germination may occur *via* ABA. The evidence supporting this comes from measurements of ABA in glucose (Glc) treated seedlings (Arenas-Huertero *et al.,* 2000) and from the observation that ABA synthesis and response mutants are insensitive to Glc (Rook *et al.,* 2001). Ethylene controls the Glc inhibition of germination (Zhou *et al.,* 1998). Evidence supporting this includes the observations that high concentrations of ethylene antagonize the Glc repression of germination (Ghassemian *et al.,* 2000). Epistasis analysis between various sugar and ethylene-insensitive mutants predict that ABA functions downstream of ethylene consistent with the conclusion that ethylene blocks Glc-induced ABA synthesis (Ghassemian *et al.*, 2000). The mechanism by which these signals are integrated to control germination is only now being addressed (Ullah *et al.*, 2002). One possible mechanism involves coupling of and cross talk between signals by heterotrimeric G proteins (Fig. 16).

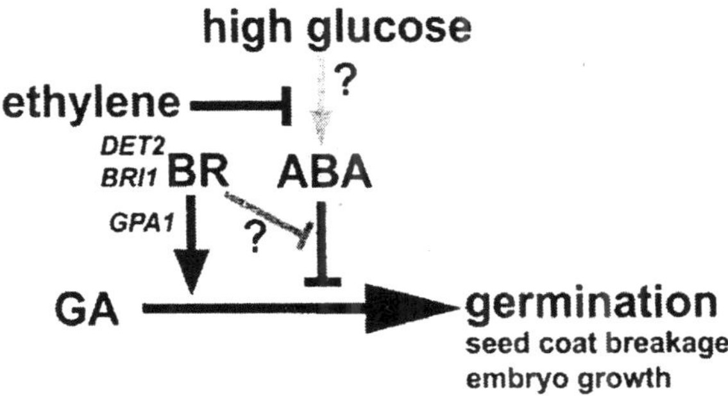

Fig. 16: DET2 plays an important role in senescence. It encoded a protein that encode enzymes involved in the synthesis of BRs. DET2 catalyzes the conversion of 3-dehydro-$\Delta^{4,5}$-campesterol to 3-dehydrocampestanol in the BR biosynthetic pathway. *BRI1* has been shown to encode a protein that has sequence homology to transmembrane leucine-rich repeat (LRR) receptor kinases (Chory and Li, 1997). BRI1 as a critical component of the BR-receptor complex, which likely transduces the BR signal to downstream targets through its kinase domain. GPA 1 codes for alpha subunit of the G protein in Arabidopsis.

A heterotrimeric G protein may mediate many of the signals that seeds use to control germination (Assmann, 2002). Heterotrimeric G proteins are associated with the cytoplasmic face of the plasma membrane of a variety of eukaryotic cells and transduce information from cell surface G protein coupled receptors to effector proteins of signal transduction pathways (Neer, 1995). The α-subunits of heterotrimeric G (Gα) proteins transduce signals from G protein coupled receptors to effector proteins, accompanied by the GTPase-catalyzed hydrolysis of GTP (Fig. 17).

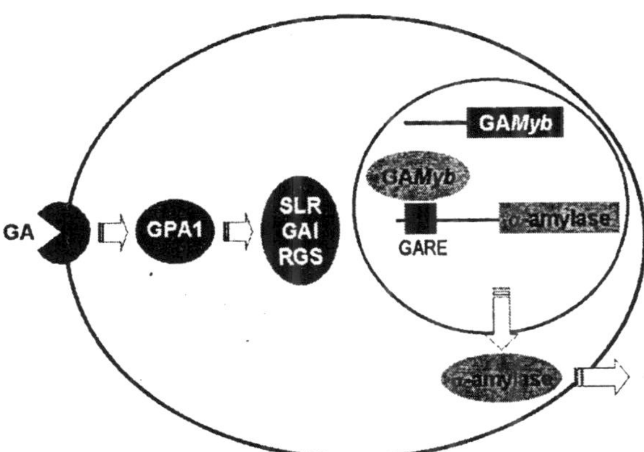

Fig. 17: RGS is the regulator of G protein signalling proteins. RGS protein regulates a molecular on-off switch called a G protein alpha (Gα) subunit, which is important for cellular processes. GAI and SLR are regulatory genes of GA signaling pathway.

GPA1 couples BR potentiation of GA-stimulated germination. Black arrows indicate pathways tested for GPA1 coupling in signal control of seed germination. The results here suggest that only BR signaling is coupled by GPA1 because the BR response and biosynthetic mutants, *bri1* and *det2*, respectively, lack BR rescue of GA-dependent germination and because *gpa1* mutants are completely insensitive to BR with regard to rescue of GA dependent germination. Grey arrows indicate potential pathways not yet ruled out or in. BR may also control ABA sensitivity or levels. The mechanism of sugar inhibition is unclear. On the one hand, results here show that the inhibition by high sugar on germination is dependent on ABA synthesis, but a sugar-induced increase in ABA was not detected in seeds.

Mutational and molecular analyses in Arabidopsis have identified several of the elements in the ethylene signal transduction pathway (Johnson and Ecker, 1998). Ethylene is perceived by a family of receptors related to ETR1, and ethylene binding inhibits the signalling activities of these receptors (Hua and Meyerowitz, 1998). In the absence of ethylene, the receptors activate CTR1, which regulates negatively the downstream signalling components, including EIN2 (Alonso *et al.,* 1999). *ein2-45* mutation was isolated as a suppressor of the ABA-resistant seed germination phenotype of *abi1-1*. Furthermore, in the wild-type *ABI1* background, *ein2-45* mutant seeds were also supersensitive to ABA. This phenotype is not specific to the *ein2-45* allele because *ein2-1* mutation similarly enhances ABA sensitivity of seed germination and that the ABA-supersensitive *era3-1* mutant described by Cutler *et al.,* 1996 is also an *ein2* mutant allele. Conversely, like *ein2-1* (Guzman and Ecker, 1990), *ein2-45* and *era3-1* mutants were insensitive to ethylene, as shown by their failure to develop the characteristic triple response evoked by ethylene (or ACC) in dark-grown seedlings. In addition to its role as a positive regulator of ethylene signaling, *EIN2* is a negative regulator of ABA sensitivity in seeds. *EIN2* mutants have also been recovered in screens for mutants resistant to cytokinins (Cary *et al.,* 1995) or inhibitors of auxin transport (Fujita and Syono, 1996) and in screens for delayed leaf senescence (Oh *et al.,* 1997). None of the other ethylene-insensitive loci was recovered in these screens. Furthermore, expression of the EIN2 C-terminal domain in *ein2* mutant plants restored responsiveness to jasmonic acid and paraquat-induced oxygen radicals but was not sufficient to induce the triple response (Alonso *et al.,* 1999). Hence, EIN2 has been proposed to lie at the crossroads of multiple hormone response pathways (Alonso *et al.,* 1999). However, EIN2 is not the sole ethylene signalling element that controls ABA responsiveness in seeds. As did in the *ein2* mutations, the ethylene-insensitive *etr1-1* mutation in the ETR1 receptor enhanced ABA sensitivity in seeds. Conversely, seeds of the constitutive ethylene response mutants *ctr1* were less sensitive than the wild type to ABA. Hence, the entire pathway defined by ETR1, CTR1, and EIN2 impinges on ABA transduction in seeds.

It is currently unclear whether the ETR1–EIN2 cascade mediates ethylene signalling exclusively. For instance, various observations suggest that ETR1 (Hua and Meyerowitz 1998), and possibly a whole subset of the ethylene receptors (Gamble *et al.,* 1998), may have additional, ethylene-independent functions.

Nevertheless, the simplest interpretation of the present data is that the enhanced dormancy of *etr1-1* and *ein2* seeds actually results from impaired ethylene signalling, and hence that endogenous ethylene regulates negatively the dormancy of Arabidopsis seeds. Ethylene has long been suspected to play a role in seed dormancy, but direct evidence has been scarce (Kepczynski and Kepczynska, 1997). Results of studies with biosynthesis inhibitors were ambiguous because in many cases these inhibitors markedly lowered ethylene production without affecting seed germination (Kepczynski and Kepczynska, 1997). However, studies with 2, 5-norbornadiene, a competitive antagonist of ethylene, indicated that endogenous ethylene is necessary for the germination of dormant seeds from *Chenopodium album* (lambquarters) (Machabee and Saini, 1991) and *Amaranthus retroflexus* (Kepczynski *et al.,* 1997). Similarly, endogenous ethylene regulates negatively the degree of seed dormancy in Arabidopsis. Under experimental conditions endogenous ethylene content in the wild type was at or close to saturating values for its action on seed dormancy because constitutive activation of the ethylene pathway in the *ctr1* mutants or treating wild-type seeds with 10 μM ACC barely decreased dormancy (Beaudoin *et al.,*2000).

Seed dormancy mutants of Arabidopsis can be classified into two groups depending on whether or not they are affected in ABA action (biosynthesis or sensitivity) (Leon-Kloosterziel *et al.,* 1996a). As discussed above, the *etr1-1* and *ein2* mutations markedly enhanced the sensitivity of seed germination to exogenous ABA. Furthermore, epistasis analyses between *ein2-45* and severely ABA-insensitive mutants indicated that ethylene regulates negatively the degree of seed dormancy by decreasing the sensitivity to endogenous ABA. *Etr1-1* seeds were somewhat less dormant but were more sensitive to exogenous ABA than *ein2-1* seeds. Conversely, *abi1-1 ein2-45* double-mutant seeds were more dormant but less sensitive to exogenous ABA than wild-type L*er* seeds. It is unclear whether these few discrepancies suggest that ethylene has in addition some ABA-independent effect on seed dormancy or whether they simply result from the fact that inhibition of seed germination by exogenous ABA mimics only imperfectly the action of endogenous ABA on seed dormancy (Steber *et al.,* 1998).

Ethylene regulates seed dormancy largely by counteracting the effect of ABA (Fig.17). In this respect, there are many similarities between the roles of ethylene and gibberellin (GA). GA is absolutely required for the germination of wild-type Arabidopsis seed because GA-deficient mutants such as *ga1* fail to germinate (unless supplied with exogenous GA) (Koornneef and Van der Veen, 1980). In contrast to the *ga1* single mutant, however, double mutants that combine *ga1* with ABA-deficient or ABA-insensitive mutations do not require exogenous GA to germinate (Steber *et al.,* 1998). Finally, the GA-deficient *ga1* and GA-insensitive *sly1* mutations increase the sensitivity of seed germination to ABA inhibition. *ga1* and *sly1* mutants were isolated by Steber *et al.,* 1998 as suppressors of *abi1-1* in a screen very similar to the one in which we recovered *ein2-45.* Hence, both endogenous GA and ethylene appear to control seed germination by decreasing ABA responsiveness. GA and ethylene can replace each other to a certain extent because germination of the *ga1* mutant could be induced by exogenous ethylene (Karssen *et al.,* 1989), and conversely, a GA treatment stimulated germination of the *etr1-1* mutant (Bleecker *et al.,* 1988). The roles of endogenous GA and ethylene are unlikely to be completely redundant, however, because mutations that impair the action of either of these two hormones lead to tangible alterations in seed germination (Fig. 18).

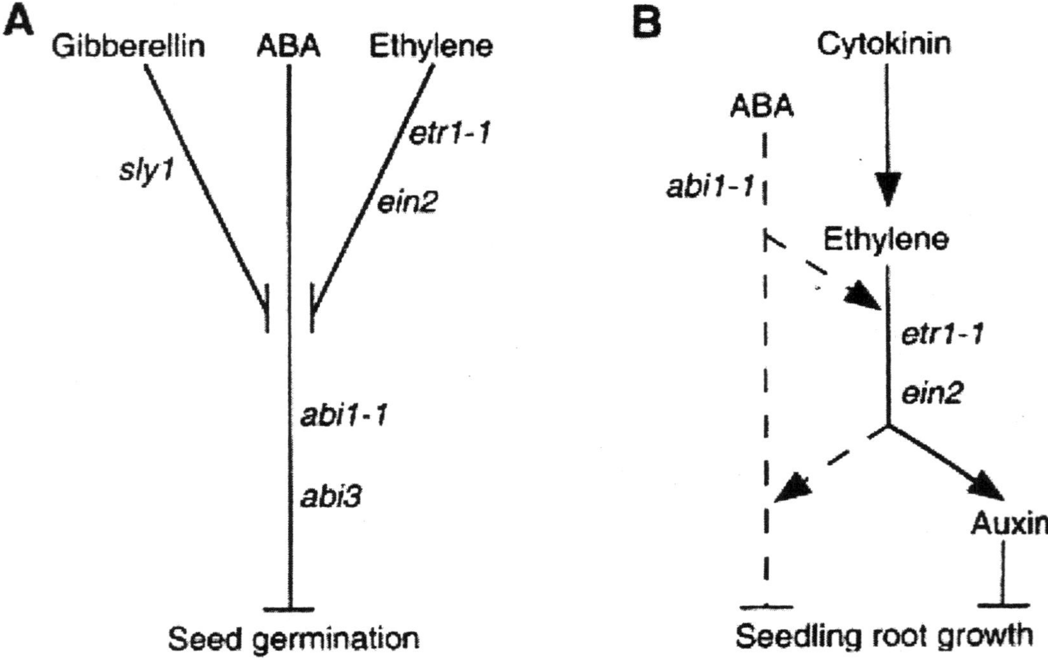

Fig. 18: Schematic representations of the interactions between the ABA and Ethylene signalling cascades.

ABI3 is the gene encoding a protein orthologous to the maize Viviparous 1 protein. The *ABI1* gene encodes homologous protein serine/threonine phosphatases 2C. The *ETR1* gene encodes a protein with sequence similarity to bacterial two component histidine kinases (Chang *et al.*, 1993). The *EIN2* gene product is novel and weak homology to the integral membrane domain of the Nramp family of metal transporters (Alonso *et al.,* 1999).abi1,abi3 are ABA insensitive mutants with reduced seed dormancy.abi1 and abi3 are ethylene insensitive mutants. The ABA-insensitive (*abi*) mutants *abi1* to *abi5* are able to germinate in the presence of ABA concentrations that are inhibitory to the wild type (Nambara *et al.,* 1995).

(A). Regulation of seed germination: Ethylene and GA signaling cascades promote seed germination by decreasing sensitivity to endogenous ABA in imbibed seeds. The *abi1-1* and *abi3* mutants are ABA insensitive, the *sly1* mutant is GA insensitive, and the *etr1-1* and *ein2* mutants are ethylene insensitive. The positions of the intersections between the ABA, ethylene and GA pathways are speculative.

(B). Inhibition of seedling root growth: Inhibition of root growth in response to increased amounts of ethylene is largely mediated through internal accumulation of auxin. Cytokinin inhibits root growth by stimulating ethylene biosynthesis. ABA inhibition of root growth is apparently not mediated by an increase in ethylene biosynthesis but requires an active ethylene signaling cascade. Either the basal activity of the ethylene pathway is required to synergize ABA action or ABA stimulates the ethylene signaling cascade.

Primary seed dormancy is established on the mother plant during the late-stages of seed development (seed maturation). Studies on ABA-deficient mutants clearly demonstrated that embryonic ABA is required for the onset of dormancy during Arabidopsis seed maturation (Leon-Kloosterziel *et al.,* 1996a). Because ABA content markedly decreases at the end of seed maturation, Karssen *et al.* (1989) postulated that after the onset of dormancy, endogenous ABA may not be required for its maintenance. However, recent experiments in barley (Wang *et al.,* 1995) and in *Nicotiana plumbaginifolia* (Grappin *et al.,* 2000) indicate that the maintenance of dormancy in imbibed seeds is actually an active process involving *de novo* ABA synthesis. Ethylene and GA do not seem to regulate the induction of dormancy during seed maturation and are thought to act mainly during imbibition to break dormancy and trigger germination (Kepczynski and Kepczynska, 1997). The observation that seeds of ethylene-insensitive (this study) and GA-insensitive (Steber *et al.,* 1998) mutants are supersensitive to exogenous ABA suggests that ethylene and GA may counteract directly the action of ABA in dormancy maintenance by inhibiting ABA signalling in imbibed seeds.

Whereas, *etr1-1* and *ein2* seeds were supersensitive to ABA, however, the roots of these mutants were cross-resistant to ethylene and ABA. Treatment with exogenous ethylene (or ACC), or constitutive activation of the ethylene cascade by the *ctr1* mutation, inhibits the growth of seedling roots. Ethylene inhibits auxin transport in both shoots and roots, and evidence suggests that ethylene inhibition of root growth is largely mediated through internal auxin accumulation (Marchant *et al.,* 1999). Roots of the ethylene-insensitive mutants *ein1* (allelic to *etr1*) and *ckr1* (allelic to *ein2*) are resistant to cytokinin (Cary *et al.,* 1995) because cytokinin inhibits root growth by stimulating ethylene biosynthesis (Cary *et al.,*1995) (Fig. 18). Whereas *etr1-1* and *ein2* roots were resistant to both ABA and ethylene, *abi1-1* root was resistant to ABA but had normal sensitivity to ethylene. This suggested that ethylene acts "downstream" of ABA. However, ABA-treated seedlings did not display any detectable change in ethylene production or in abundance of the ethylene responsive *GST2* mRNA. Hence, unlike cytokinin, ABA inhibition of root growth is apparently not mediated by an ABA-induced stimulation of ethylene production. Nevertheless, the decreased ABA sensitivity of *etr1-1* and *ein2* roots shows that ethylene signaling is required for efficient ABA inhibition of root growth. This implies either that ABA can stimulate the ethylene signaling pathway or that the seedling's content of endogenous ethylene at rest has a synergistic effect on root growth inhibition by ABA (Fig.18). Whether ABA inhibition of root growth requires auxin action, however, is currently unclear.

CONCLUSION

Seed germination involves a series of biochemical and physiological events that includes imbibition, reserve mobilization, radicle emergence and seedling growth. Plant hormones and their proportionate concentrations also decide seed dormancy and germination. The interrelationship of various hormones or the signal crosstalk during seed germination might be helpful to unravel further information occurring during seed germination.

REFERENCES

Abe, M., T. Takahashi and Y. Komeda. 1999. Cloning and characterization of an L1 layer-specific gene in *Arabidopsis thaliana. Plant Cell Physiol.*, **40**: 571–580.

Alonso, J.M., T. Hirayama, G. Roman, S. Nourizadeh and J.R Ecker. 1999. EIN2, a bifunctional transducer of ethylene and stress responses in Arabidopsis. Science, **284**: 2148-2152.

Arabidopsis Genome Initiative 2000. Analysis of the genome sequence of the flowering plant *Arabidopsis thaliana. Nature* ,408: 796–815.

Arenas-Huertero, F., A.Arroyo, L. Zhou, J.Sheen and P.Leon. 2000. Analysis of *Arabidopsis* glucose insensitive mutants, *gin5* and *gin6*, reveal a central role of the plant hormone ABA in the regulation of plant vegetative development by sugar. *Genes Dev.*, **14**: 2085-2096.

Assmann,S.M. 2002. Heterotrimeric and unconventional GTP binding proteins in plant cell signaling. *Plant Cell*, **14**:355–373.

Attucci, S., J.P. Carde, P.Raymond, V.Saint Gès, A.Spiteri and A.Pradet. 1991. Oxidative phosphorylation by mitochondria extracted from dry sunflower seeds. *Plant Physiol.*, **95**:390-398.

Beaudoin, N.C., F. Serizet, J. Gosti, and J.Giraudat. 2000. Interactions between abscisic acid and ethylene signaling cascades. *Plant Cell*, **12**: 1103-1115.

Becker, C., V.I Senyuk, A.D Shutov, V.H Nong, J. Fischer, C. Horstmann and K. Müntz. 1997. Proteinase A, a storage-globulin-degrading endopeptidase from vetch (*Vicia sativa* L.) seeds, is not involved in early steps of storage protein mobilization. *European J. Biochem.*, **248**: 304–312.

Becker, C., A.D Shutov, V.H. Nong, R.Jung, C. Horstmann, J. Fischer, N.C. Nielsen and K.Mountz. 1995. Purification, cDNA cloning and characterization of proteinase B, an asparagine-specific endopeptidase from germinating vetch (*Vicia sativa* L.) seeds. *European J. Biochem.*, **228**: 456–462.

Belozersky, M.A., Y.E Dunaevsky and N.E Voskoboynikova. 1990. Isolation and properties of a metalloproteinase from buckwheat (*Fagopyrum esculentum*) seeds.*Biochemical Journal*, **272**: 677–682.

Beltran-Petia, E., A.Ortiz-López, and E.Sanchez de Jiménez. 1995. Synthesis of ribosomal proteins from stored mRNAs early in seed germination. *Plant Mol. Biol.*, **28**: 327-336.

Bewley, J.D.1997. Seed germination and dormancy. *Plant Cell*, **9**: 1055–1066.

Bewley, J.D. and M. Black. 1978. Physiology and biochemistry of seeds in relation to germination,vol,1. Springer - Verlag ,Berlin.

Bewley, J.D. and M.Black. 1994. Seeds: Physiology of Development and germination (New York: Plenum Press).

Bewley, J.D. and A.Marcus. 1990. Gene expression in seed development and germination. *Prog. Nucleic Acid Res. Mol. Biol.*, **38**:165-193.

.Bleecker,A.B, M.A.Estelle, C.Somerville, and H.Kende. 1988. Insensitivity to ethylene conferred by a dominant mutation in *Arabidopsis thaliana. Science*, **241**:1086-1089

Bradford, K.J. 1996. Population-based models describing seed dormancy behaviour: Implications for experimental design and interpretation. In Plant Dormancy, G.A. Lang, ed (Oxford, UK:CAB International), pp. 313-339.

Cary, A.J., W. Liu and S.H. Howell. 1995. Cytokinin action is coupled to ethylene in its effects on the inhibition of root and hypocotyl elongation in *Arabidopsis thaliana* seedlings. *Plant Physiol.,* **107**:1075-1082.

Casal, J.J., R.A. Sánchez, and J.F. Botto. 1998. Modes of action of phytochromes. *J. Exp. Bot.*, **49**:127–138.

Chang C., S.F. Kwok, A.B. Bleecker, and E.M Meyerowitz. 1993. Arabidopsis ethylene-response gene ETR1: similarity of product to two-component regulators. *Science* **262**:539-544.

Chen, F., P.Dahal, and K.J.Bradford. 2001. Two tomato expansin genes show divergent expression and localization in embryos during seed development and germination. *Plant Physiol.*, **127**: 928–936.

Chen, F., H. Nonogaki, and K.J. Bradford. 2002. A gibberellin-regulated xyloglucan endotransglycosylase gene is expressed in the endosperm cap during tomato seed germination. *J. Exp. Bot.,* **53**: 215–223.

Chory, J., and J. Li. 1997.Gibberellins, brassinosteroids and light-regulated development, *Plant Cell and Environ.,* **20**: 801–806.

Cosgrove, D.J. 1997. Relaxation in a high-stress environment: The molecular bases of extensible cell walls and cell enlargement. *Plant Cell,* **9**: 1031-1041.

Crowe, J.H., and L.M.Crowe. 1992. Membrane integrity in anhydrobiotic organisms: Toward a mechanism for stabilizing dry seeds. In Water and Life, G.N. Somero, C.B. Osmond, and C.L. Bolis, eds (Berlin: Springer-Verlag), pp. 87-103.

Cutler, S., M.Ghassemian, D.Bonetta, S.Cooney, and P. McCourt. 1996. A protein farnesyl transferase involved in abscisic acid signal transduction in *Arabidopsis. Science,* **273**:1239-1241.

Davies, P.J. 1995. Plant Hormones: Physiology, Biochemistry and Molecular Biology. (Dordrecht, The Netherlands: Kluwer Academic Publishers).

Debeaujon and Koornneef, M. 2000. Gibberellin requirement for Arabidopsis seed germination is determined both by testa characteristics and embryonic abscisic acid. *Plant Physiol.* **122**: 415–424.

Desai, B.B., P.M. Kotecha, and D.K. Salunkhe. 1997. Seeds Hand Book. Marcel Dekker,INC.

Devlin, P.F., S.R. Patel, and G.C. Whitelam. 1998. Phytochrome E influences internode elongation and flowering time in *Arabidopsis. Plant Cell,* **10**:1479–1487.

Dommes, J., and C.Van de Walle. 1990. Polysome formation and incorporation of new ribosomes into polysomes during germination of the embryonic axis of maize. *Physiol. Plant.*, **79**: 289-296.

Ehrenshaft, M., and R. Brambl. 1990. Respiration and mitochondrial biogenesis in germinating embryos of maize. *Plant Physiol.*, **93**:295-304.

Elpidina, E.N., N.E Voskobyonikova, M.A Belozersky and Y.E Dunaevsky. 1991. Localization of a metalloproteinase and its inhibitor in the protein bodies of buckwheat seeds. *Planta,* **185**: 46–52.

Franklin,K.A., S.J.Davis, W.M.Stoddart, R.D.Vierstra,and Whitelam. 2003.Mutant analyses define multiple roles for phytochrome C *Arabidopsis* photomorphogenesis: *Plant Cell,* **15**: 1981–1989.

Fujita, H., and K. Syono. 1996. Genetic analysis of the effects of polar auxin transport inhibitors on root growth in *Arabidopsis thaliana. Plant Cell Physiol.*, **37**:1094-1101.

Gamble, R.L., M.L.Coonfield and G.E Schaller. 1998. Histidine kinase activity of the ETR1 ethylene receptor from Arabidopsis. *Proc.Natl.Acad.Sci.,USA* **95**:7825-7829.

Georgieva E.I, G.Lopez-Rodas, A. Hittmair, H. Feichtinger, G. Brosch, and P. Loidl. 1994. Maize embryo germination. *Planta,* **192**:118–124.

Ghassemian, M., E. Nambara, E. Cutler, H. Kawaide, Y.Kamiya, and P. McCourt. 2000. Regulation of abscisic acid signaling by the ethylene response pathway in *Arabidopsis. Plant Cell,* **12**: 1117-1126.

Gibson, D..M. and A.H.J. Ullah. 1988. Purification and characterization of phytase from germinating soybean seeds. *Arch Biochem Biophys.*, **260**: 503-513.

Gilroy, S., and R.L Jones. 1994 .Perception of gibberellin and abscisic acid at the external face of the plasma membrane of barley (*Hordeum vulgare* L.) aleurone protoplasts. *Plant Physiol.*, **104**: 1185-1192

Grappin, P., D.Bouinot, B.Sotta, E.Miginiac, and M. Jullien. 2000. Control of seed dormancy in *Nicotiana plumbaginifolia*: Post-imbibition abscisic acid synthesis imposes dormancy maintenance. *Planta*, **210**:279-285.

Groot, S.P.C., and C.M. Karssen. 1987. Gibberellins regulate seed germination in tomato by endosperm weakening: A study with gibberellin-deficient mutants. *Planta*,**171**:525–531.

Guzman,P., and J.R. Ecker. 1990. Exploiting the triple response of Arabidopsis to identify ethylene-related mutants. *Plant Cell*, **2**:513-523

Hammett, J.R., and F.R. Katterman. 1975. Storage and metabolism of oly (adenylic acid)-mRNA in germinating cotton seeds. *Biochemistry*, **17**: 3250 – 3256.

Haupt, W., and D.P. Häder. 1994. Photomovement: In Photomorphogenesis in plants, R.E. Kendrick, and G.H.M. Kronenberg, eds, (Dordrecht, The Netherlands:Kluwer Academic Publishers), pp. 707–732.

Hara-Nishimura, I., T. Kinoshita, N. Hiraiwa, and M.Nishimura. 1998. Vacuolar processing enzymes in protein-storage vacuoles and lytic vacuoles. *J. Plant Physiol.*, **152**:668–674.

Hara-Nishimura,I., M. Nishimura, H. Matsubara and T.Akazawa. 1982. Suborganeller localization of proteinase catalysing the limited hydrolysis of pumpkin globulin.*Plant Physiol.*, **70**: 699–703.

Hedden, P., and Y. Kamiya. 1997. Gibberellin biosynthesis: Enzymes, genes and their regulation. *Ann. Rev. Plant Physiol. Plant Mol. Biol.*, **48**: 431–460.

Hennig, L., W.M. Stoddart, M. Dieterle, G.C. Whitelam, and E. Schafer. 2002. Phytochrome E controls light-induced germination of *Arabidopsis*. *Plant Physiol.*, **128**: 194–200.

Hua J. and E.M. Meyerowitz. 1998. Ethylene responses are negatively regulated by a receptor gene family in *Arabidopsis thaliana*. *Cell*, **94**:261-271

Hua, J., Sakai, H., Nourizadeh, S., Chen, Q.G., Bleecker, A.B., Ecker, J.R., and Meyerowitz, E.M. 1998. *EIN4* and *ERS2* are members of the putative ethylene receptor gene family in Arabidopsis. *Plant Cell*, **10:** 1321–1332.

Jacobsen, J.V., D.W Pearce, A.T.Poole, R.P.Pharis, and L.N.Mander. 2002. Abscisic acid, phaseic acid and gibberellin contents associated with dormancy and germination in barley. *Physiol. Plant.*, **115**: 428–441.

Jacobsen, S.E. and N.E. Olszewski. 1993. Mutations at the *SPINDLY* locus of Arabidopsis alter gibberellin signal transduction. *Plant Cell*, **5**: 887–896.

Johnson, P.R., and J.R.Ecker. 1998. The ethylene gas signal transduction pathway: A molecular perspective. *Ann. Rev. Genet.*, **32**:227-254.

Kaneko, M., H. Itoh, M. Ueguchi-Tanaka, M.Ashikari, and M.Matsuoka. 2002. The amylase induction in endosperm during rice seed germination is caused by gibberellin synthesized in epithelium. *Plant Physiol.*, **128**:1264–1270.

Karssen, C., S.Zagorski, J.Kepczynski and S.Groot. 1989. Key role for endogenous gibberellins in the control of seed germination. *Ann. Bot.*, 63: 71–80.

Kepczynski, J., and E.Kepczynska. 1997.Ethylene in seed dormancy and germination. *Physiol. Plant.*, **101**: 720–726.

Kepczynski, J., M. Bihun, and E. Kepczynska. 1997. Ethylene involvement in the dormancy and germination of *Amaranthus* seeds. In A.K.Kanelis, C.Chang, H.Kende and D.Grierson, eds. Biology and Biotechnology of Plant Hormone Ethylene. Dordrecht, The Netherlands, Kluwer Academic Publishers. pp.113–122.

Koornneef, M., L. Bentsink and H. Hilhorst. 2002. Seed dormancy and germination. *Curr. Opin. Plant Biol.*, **5**: 33-36.

Koornneef, M.and J.H.Van der Veen. 1980. Induction and analysis of gibberellin sensitive mutants in *Arabidopsis thaliana* (L.) Heynh. *Theor. Appl. Genet.*, **58**: 257–263.

Koornneef, M., M.L.Jorna, D.L.C.Brinkhorst-van der Swan, and C.M. Karssen. 1982. The isolation of abscisic acid (ABA) deficient mutants by selection of induced revertants in non-germinating gibberellin sensitive lines of *Arabidopsis thaliana* (L.) Heynh. *Theor. Appl. Genet.*, **61**:385–393.

Lalonde, L.and J.D. Bewley. 1986. Desiccation during germination and its effects on the messenger RNA population. *J. Exp.Bot.*, **37**: 754-764.

Lane, B.G. 1991. Cellular desiccation and hydration: Developmentally regulated proteins, and the maturation and germination of seed embryos. *FASEB J.*, **5**:2893- 2901.

Lane, B.G. 1994. Oxalate, germin, and the extracellular matrix of higher plants. *FASEB J.*, **8**: 294-301.

Lee, Y., D.Choi and H. Kende. 2001. Expansins: Ever-expanding numbers and functions. *Curr. Opin. Plant Biol.*, **4**: 527–532.

Lehman, R.Black, and J.R.Ecker. 1996. *HOOKLESS1*, an ethylene response gene, is required for differential cell elongation in the Arabidopsis hypocotyl. *Cell*, **85** : 183–194.

Léon-Kloosterziel,K.M., M.Alvarez Gil, G.J.Ruijs, S.E.Jacobsen, E.Olszewski, S.H.Schwartz, J.A.D.Zeevaart and M. Koornneef. 1996a.Isolation and characterization of abscisic acid-deficient *Arabidopsis* mutants at two new loci. *Plant J.*, **10**:655-661.

Léon-Kloosterziel, K.M., G.A.van de Bunt, J.A.D.Zeevaart, and M. Koornneef. 1996b. Arabidopsis mutants with reduced seed dormancy. *Plant Physiol.*, **110**:233-240.

Leung, J. and J.Giraudat. 1998. Abscisic acid signal transduction. *Annu. Rev. Plant Physiol. Plant Mol. Biol.*, **49**: 199–222.

Loewus F.A. and P.P.N. Murthy. 2000. *Myo*-Inositol metabolism in plants. *Plant Sci.*, **150**: 1-19.

Lovegrove and R. Hooley. 2000. Gibberellin and abscisic acid signalling in aleurone. *Trends Plant Sci.*, **5**: 102-110

Machabée,S. and H.S. Saini. 1991. Differences in the requirement for endogenous ethylene during germination of dormant and non-dormant seeds of *Chenopodium album* L. *J. Plant Physiol.*, **138**:97-101.

Madison, J.T, J.F Thompson and A.E.Muenster. 1981. Turnover of storage protein in seeds of soybean and pea. *Ann. Bot.*, **47**: 65–73.

Mancinelli, A.L. 1994. The physiology of phytochrome action: In Photomorphogenesis in plants, R.E. Kendrick, and G.H.M. Kronenberg, eds, (Dordrecht, The Netherlands: Kluwer Academic Publishers), pp. 211–269.

Marchant, A., J.Kargul, S.T.May, P.Muller, A.Delbarre, Perrot-Rechenmann, and M.JBennett. 1999. AUX1 regulates root gravitropism in Arabidopsis by facilitating auxin uptake within root apical tissues. *EMBO J.*, **8**:2066-2073.

Marttila, S., B.L Jones and A.Mikkonen. 1995.Differential localization of two acid proteinases in germinating barley (*Hordeum vulgare*) seed. *Physiol Plant.*,**93**: 317–327.

McQueen-Mason, S. and D.J. Cosgrove. 1995. Expansin mode of action on cell walls. Analysis of wall hydrolysis, stress relaxation, and binding. *Plant Physiol.*, **107**: 87-100.

Micheli, F. 2001. Pectin methylesterases: Cell wall enzymes with important roles in plant physiology. *Trends Plant Sci.*, **6**: 414–419.

Monte, E., J.M. Alonso, J.R. Ecker, Y. Zhang, X. Li, J. Young, S. Austin-Phillips, and P.H. Quail. 2003. Isolation and characterization of *phyC* mutants in *Arabidopsis* reveals complex crosstalk between phytochrome signaling pathways. *Plant Cell*, **15**: 1962–1980.

Morohashi, Y. 1986. Patterns of mitochondrial development in reserve tissue of germinated seeds: A survey. *Physiol. Plant.*, **66**: 653-658.

Morohashi, Y.and M.Shimokoriyama. 1972. Physiological studies on germination of *Phaseolus mungo* seeds. II. Glucose and organic-acid metabolisms in the early phases of germination. *J. Exp. Bot.*, **23**: 54-61.

Muday, G.K., and A. DeLong. 2001. Polar auxin transport: Controlling where and how much. *Trends Plant Sci.*, **6**:535–542.

.Mullen, R.T, J.E King and D.J. Gifford. 1996. Changes in mRNA populations during loblolly pine (*Pinus* taeda) seed stratification, germination and postgerminative growth. *Physiol. Plant.*,**97**:545-553.

Müntz, K. 1996. Proteases and proteolytic cleavage of storage proteins in developing and germinating dicotyledonous seeds. *J.Exp. Bot.*, **47**: 605–622.

Müntz, K.1998.Deposition of storage proteins. *Plant Molecular Biology*, **38**: 77-99.

Müntz,K., M.A.Belozersky, Y.E.Dunaevsky A.Schlereth and J.Tiedemann. 2001. Stored proteinases and the initiation of storage protein mobilization in seeds during germination and seedling growth. *J.Exp. Bot.*, **52**:1741-1752.

Nambara, E, K.Keith, P.McCourt and S.Naito. 1995. A regulatory role for the *ABI3* gene in the establishment of embryo maturation in *Arabidopsis thaliana*. *Development*, **121**:629-636.

Neer, E.J. 1995. Heterotrimeric G proteins: Organisers of transmembrane signals. *Cell*, **80**: 249–257.

Nonogaki, H., O.H.Gee, and K.J. Bradford. 2000. A germination specific endo-mannanase gene is expressed in the micropylar endosperm cap of tomato seeds. *Plant Physiol.*, **123**:1235–1245.

Ogawa, M., A.Hanada, A.Yamauchi, Y.A.Kuwahara, and Yamaguchi, S. 2003. Gibberellin biosynthesis and response during Arabidopsis seed germination. *Plant Cell,* **15**: 1591-1604.

Oh, S.A., J.H.Park, G.I.Lee, K.H.Paek, S.K.Park, and H.G.Nam. 1997. Identification of three genetic loci controlling leaf senescence in *Arabidopsis thaliana. Plant J.*, **12**:527-535.

Obroucheva, N.V., O.V. Antipova. 1997. Physiology of the initiation of seed germination.*Russs. J .Plant Physiol.*, **44**: 250–264.

Okamoto,T.,Y.MiuraIzu,S.I.Ishii,T.Minamikawa. 1996. Asparaginylendopeptidasen developing and germinating legume seeds: immunological detection and quantitation. *Plant Sci.*, **115**: 49–57.

Olszewski,N., T.P. Sun, and F.Gubler. 2002. Gibberellin signaling: Biosynthesis, catabolism, and response pathways. *Plant Cell,* **14** : 61–80.

Osborne D.J. and I.I.Boubriak. 1994. DNA and desiccation tolerance. *Seed Sci. Res.*, **4**: 175-185.

Parks, B.M., and E.P. Spalding. 1999. Sequential and coordinated action of phytochrome A and B during *Arabidopsis* stem growth revealed by kinetic analysis. *Proc. Natl. Acad. Sci. USA.,* **96**: 14142–14146.

Petruzzelli, L., C. Kunz, R. Waldvogel, F. Meins Jr., G. Leubner-Metzger. 1999. Distinct ethylene and tissue specific regulation of ß-1,3-glucanases and chitinases during pea seed germination. *Planta,* **209**: 195-201.

Petruzzelli, L., M. Sturaro, D. Mainieri, G. Leubner-Metzger. 2003. Calcium requirement for ethylene-dependent responses involving 1-aminocyclopropane1-carboxylic acid oxidase in radicle tissues of germinated pea seeds. *Plant Cell and Environ.*, **26**: 661-671.

Peumans, W.J., L.I.Caers, and A.R.Carlier. 1979. Some aspects of the synthesis of long-lived messenger ribonucleoproteins in developing rye embryos. *Planta,* **144**:485-490.

Prattley, C.A. and D.W. Stanley. 1982. Protein-phytate interactions in soybean: I. Localization of phytate in protein bodies and globoids. *J. Food Biochem.*, **6**: 243-253.

Raboy, V. 1997. Accumulation and storage of phosphate and minerals. *In* B A Larkins, IK Vasil, eds, Cellular and Molecular Biology of Plant Seed Development. Kluwer Academic Publishers, Dordrecht, The Netherlands, pp 441-477.

Raboy, V. and D.B Dickinson. 1987. The timing and rate of phytic acid accumulation in developing soybean seeds. *Plant Physiol.,* **85**: 841-844.

Rampey, S., M.LeClere, K.Kowalczyk, G.Ljung, Sandberg and B. Bartel. 2004. A family of auxin-conjugate hydrolases that contributes to free indole-3-acetic acid levels during Arabidopsis germination. *Plant Physiol.*,**135**: 978-988

Richards, D.E., K.E.King, T.Ait-Ali, and N.P.Harberd. 2001. How gibberellin regulates plant growth and development: A molecular genetic analysis of gibberellin signaling. *Annu. Rev. Plant Physiol. Plant Mol. Biol.*, **52**: 67–88.

Rook, F., F. Corke, R. Card, G. Munz, C. Smith and M.W. Bevan. 2001. Impaired sucrose-induction mutants reveal the modulation of sugar-induced starch biosynthetic gene expression by abscisic acid signaling. *Plant J.*, **26**: 421-433.

Rose, J.K., J.Braam, S.C.Fry, and K.Nishitani. 2002. The XTH family of enzymes involved in xyloglucan endotransglucosylation and endohydrolysis: Current perspectives and a new unifying nomenclature. *Plant Cell Physiol.,* **43**: 1421– 1435.

Salon, C., P.Raymond, and A.Pradet. 1988. Quantification of carbon fluxes through the tricarboxylic acid cycle in early germinating lettuce embryos. *J. Biol. Chem.*, **263**: 12278-12287.

Sánchez-Martinez, D., P.Puigdom Bnech, and M.Pagès. 1986. Regulation of gene expression in developing Zea *mays* embryos. Protein synthesis during embryogenesis and early germination of maize. *Plant Physiol.*, **82**: 543-549.

Sandoval, J.A., H.Huang, D.C.Garrett, D.A.Gage, and K.D.Chapman. 1995. N-Acylphosphatidylethanolamine in dry and imbibing cotton seeds. Amounts, molecular species, and enzymatic synthesis. *Plant Physiol.*, **109**: 269-275.

Schopfer, P.and C.Placy. 1985. Control of seed germination by abscisic acid. III. Effect of embryo growth potential (minimum turgor pressure) and growth coefficient (cell wall extensibility) in *Brassica napus* L. *Plant Physiol.,* **77**: 676-686.

Schlereth, C.Becker, C.Horstmann, J.Tiedemann and K.Müntz. 2000. Comparison of globulin mobilization and cysteine proteinases in embryonic axes and cotyledons during germination and seedling growth of vetch (*Vicia sativa* L.). *J. Exp.Bot.*, **51**: 1423–1433.

Schumacher, K. and J. Chory. 2000. Brassinosteroid signal transduction: still casting the actors: *Curr. Opin. Plant Biol.*, **3**: 79–84.

Senyuk, V.I, V.Rotari, C. Becker, A. Zakharov, C .Horstmann, K. Müntz and I.A Vaintraub. 1998. Does an asparaginyl-specific endopeptidase trigger phaseolin degradation in cotyledons of kidney bean seedlings? *European Journal of Biochemistry*, **258**: 546–558.

Shewry,P.R. and R.Casey. 1999. Seed proteins. Dordrecht: Kluwer Academic Publishers.

Shinomura T., A. Nagatani, H. Hanzawa, M.Kubota, M.Watanabe, and M.Furuya. 1996. Action spectra for phytochrome A- and B-specific photoinduction of seed germination in *Arabidopsis thaliana. Proc. Natl. Acad. Sci.* USA, **93**: 8129–8133.

Shutov,A.D.and I.A Vaintraub. 1987. Degradation of storage proteins in germinating seeds. *Phytochemistry,* **26**: 1557-1566

Steber, C.M, and P.McCourt. 2001. A role for brassinosteroids in germination in Arabidopsis. *Plant Physiol.*, **125**: 763–769.

Steber, C.M., S.E.Cooney and P.McCourt. 1998. Isolation of the GA-response mutant *sly1* as a suppressor of *ABA-1* in *Arabidopsis thaliana. Genetics*, **149**:509-521.

Taiz, L., and E Zeiger. 2002. Plant Physiology, Third Edition., University of California, Santa Cruz, and University of California, Los Angeles, published by Sinauer Associates.

Telfer, K.M.Bollman and R.S.Poethig. 1997. Phase change and the regulation of trichome distribution in *Arabidopsis thaliana. Development*, **124**: 645–654.

Tiedemann, J., B. Neubohn and K.Müntz. 2000. Different functions of vicilin and legumin are reflected in the histopattern of globulin mobilization during germination of vetch (*Vicia sativa* L.). *Planta,* **211**: 1–12.

Toyomasu,T., H.Yamane, N.Murofushi, and Y. Inoue. 1994. Effects of exogenously applied gibberellin and red light on the endogenous levels of abscisic acid in photoblastic lettuce seeds. *Plant Cell Physiol.*, **35**:127–129.

Toyomasu, T., H. Kawaide, W. Mitsuhashi, Y. Inoue, and Y.Kamiya. 1998. Phytochrome regulates gibberellin biosynthesis during germination of photoblastic lettuce seeds. *Plant Physiol.*, **118**: 1517-1523.

Ullah, H., J,G, Chen, S.Wang, and A.M.Jones. 2002. Role of a heterotrimeric G protein in regulation of Arabidopsis seed germination. *Plant Physiol.*, **129**: 897-907.

Van der Knaap, E., S.Jagoueix, and H.Kende. 1997. Expression of an ortholog of replication protein A1 (RPA1) is induced by gibberellin in deepwater rice. *Proc. Natl. Acad. Sci. USA* ,**94**: 9979–9983.

Vandepoele, K., J.Raes, L.De Veylder, P.Rouze, S.Rombauts and D.Inzé. 2002. Genome-wide analysis of core cell cycle genes in Arabidopsis. *Plant Cell,***14**: 903–916.

Walz,A., S. Park, J.P Slovin, J. Ludwig-Müller, Y.S. Momonoki and J.D. Cohen. 2002. A gene encoding a protein modified by the phytohormone indoleacetic acid. *Proc Natl. Acad. Sci. USA.*, **99**: 1718–1723.

Wang, M., S.Heimovaara-Dijkstra and B.Van Duijn. 1995. Modulation of germination of embryos isolated from dormant and nondormant barley grains by manipulation of endogenous abscisic acid. *Planta,* **195**:586-592.

Watkins, J.T., and D.J.Cantliffe. 1983. Mechanical resistance of the seed coat and endosperm during germination of *Capsicum annuum* at low temperature. *Plant Physiol.*, **72**: 146-150.

Welbaum, G.E. and K.J.Bradford. 1990. Water relations of seed development and germination in muskmelon (*Cucumis melo* L.). V. Water relations of imbibition and germination. *Plant Physiol.*, **92**:1046-1062.

Wen, C. K., and C.Chang. 2002. Arabidopsis *RGL1* encodes a negative regulator of gibberellin responses. *Plant Cell,* **14**: 87–100.

Whitelam, G.C., E. Johnson, J. Peng, P. Carol, M.L. Anderson, J.S. Cowl, and N.P. Harberd. 1993. Phytochrome A null mutants of *Arabidopsis* display a wild-type phenotype in white light: *Plant Cell,* **5**: 757–768.

Wodzinski, R.J. and A.H.J. Ullah. 1996 Phytases. *Adv. Appl. Microbiol.*, **42**: 263-302

Wu, C.T., G. Leubner-Metzger, F.Meins Jr., and K.J. Bradford. 2001. Class Iβ-1,3-glucanase and chitinase are expressed in the micropylar endosperm of tomato seeds prior to radicle emergence. *Plant Physiol.*, **126**: 1299–1313.

Wu, Y., W.G.Spollen, R.E.Sharp, P.R.Hetherington, and S.C.Fry. 1994. Root growth maintenance at low water potentials.Increased activity of xyloglucan endotransglycosylase and its possible regulation by abscisic acid. *Plant Physiol.*, **106**:607-615.

Yamaguchi, S., and Y.Kamiya. 2000. Gibberellin biosynthesis: Its regulation by endogenous and environmental signals. *Plant Cell Physiol.*, **41**: 251–257.

Yamaguchi, S. and Y.Kamiya. 2002. Gibberellins and light-stimulated seed germination. *J. Plant Growth Regul.,* **20**: 369–376.

Yamaguchi, S., Y.Kamiya, and T.P Sun. 2001. Distinct cell-specific expression patterns of early and late gibberellin biosynthetic genes during *Arabidopsis* seed germination. *Plant J.*, **28**: 443–453.

Zhao, Y., A.K. Hull, N.R.Gupta, K.A.Goss, J.Alonso, J.R.Ecker, J.Normanly, J.Chory and J.L.Celenza. 2002. Trp-dependent auxin biosynthesis in *Arabidopsis*:Involvement of cytochrome P450s CYP79B2 and CYP79B3. *Genes Dev.,* **16**: 3100–3112.

Zhou, L., J.C Jang, T.L Jones and J. Sheen. 1998. Glucose and ethylene signal transduction crosstalk revealed by an *Arabidopsis* glucose-insensitive mutant. *Proc. Natl. Acad. Sci. USA,* **95**: 10294-10299.

5

Seed Germination under Various Abiotic Stresses

J. Annie Sheeba[1], M. Djanaguiraman[1], Mallika Vanangamudi[1], K. Vanangamudi[2] and G. Kayalvizhi[2]

[1]Department of Crop Physiology, Tamil Nadu Agricultural University Coimbatore – 641 003
[2]Department of Plant Breeding and Genetics, Adhiparasakthi Agricultural College G.B.Nagar, Kalavai- 632 506

1. INTRODUCTION

Seed germination and seedling growth are critical life history stages often subject to high mortality rates. Seeds and seedlings may be less stress tolerant than adults or may be exposed to the more extreme environmental fluctuations at or near the soil surface. In either case, the ability to successfully negotiate this "regeneration niche" can be a strong determinant of species distributions in harsh environments. In coastal and inland saline habitats, salinity and water availability (flooding and/or drought) are the abiotic stresses thought to primarily limit species distributions (Ball and Pidsley, 1995).

Abiotic stress, particularly drought and salinity, is a complex phenotypic and physiological phenomenon in plants. Evaluating tolerance is made more complex by variation in sensitivity to salt during the life cycle. For example, it has long been known that grain yield in rice is much more depressed by salt than in vegetative growth and germination is relatively salt resistant. In tomato, tolerance at germination is not correlated with the ability to grow under salt stress: both are controlled by different mechanisms, although it is possible to find genotypes with similar tolerance at germination and during vegetative growth.

2. SALINITY

Germination is a complex phenomenon involving many physiological and biochemical changes and leading to the activation of embryo. Salinity, as an abiotic hazard, induces numerous disorders in seeds during germination. It either inhibits germination at higher levels or induces a state of dormancy at lower level. It first reduces imbibitions of water because of the lowered osmotic potential of the medium. Second, it causes toxicity that is, it changes enzymatic activity, hampers protein metabolism, upsets plant growth regulators balance, and reduces the utilization of seed reserves. It may elicit changes at ultra structural, cellular and tissue or even at organ levels.

Salinity interacts with certain plant and environment factors during germination. Among the plant factors, seed coat, dormancy, seed age, seed polymorphism, and seedling vigour are prominent. Environmental factors include temperature, light, availability of water and oxygen. Efforts have been made to ameliorate the adverse effects of salinity on germination by employing certain chemical and biochemical agents. Gibberellic acid initiates germination by breaking salt-induced dormancy, whereas kinetin stimulates. Similarity, polyamines, thiourea, amino acids, betaines and sugars have been successfully used to accomplish a higher germination rate and seedling growth. The role of calcium has been well documented in the mitigation of ionic toxicity and regulation of membrane processes during germination. Moreover ammonium, nitrate, potassium, and magnesium have also proved their worth in germination and seedling development.

2.1. Process of germination

Germination of a viable seed starts with the imbibition of water and terminates with emergence of embryonic tissue. This involves the hydration of

Table 1: Effect of different types of salinity on the process of germination of seeds and propagules

Germination	Salinity applied	Response elicited	Reference
Imbibition	NaCl	Reduced hydration of embryo and cotyledon.	Petruzzelli *et al.*, (1991)
Active metabolism	$NaCl + Na_2SO_4 + CaCl_2$ $MgCl_2$	Disaggregation of intermembrane particles : leakage of solutes; reduced mobilization of reserves; inhibited activities of carbohydrates and fatty acid metabolism enzymes; altered pattern of protein synthesis, production of osmotically active solutes	Rogers *et al.*, (1995)
Emergence and elongation of embryonic tissues	NaCl	Delayed and reduced emergence of radicle and plumule and reduced elongation of embryonic tissues	Small and Gutterman, (1992)
	CaCl	Reduced emergence of seedlings	Wahid *et al.*, (1997)
	$Na_2SO_4 + CaCl_2 + MgCl_2 + NaCl$	Inhibited radicle emergence	Katerji *et al.*, (1994)
	$Nacl + Na_2SO_4 + MgCl_2 + CaSO_4$ (Ψ from - 2.5 to - 15.0 bars)	Delayed emergence and final suppression of embryonic tissues growth	Niazi *et al.*, (1987)
	$NaCl + CaSO_4$	Reduced emergence of radicle and plumule elongation of embryonic tissues	Sinha and Gupta, (1982)
	Sea water	Delayed emergence of seedlings	Tipton, (1988)
	$NaCl + Na_2SO_4 + MgSO_4$	Inhibited growth of seedlings	Huang and Redman, (1995)
Establishment of seedlings	NaCl	Enhanced seedling mortality and reduced seedling growth	Rogers *et al.*, (1995)
	$NaCl + Na_2SO_4 + MgCl_2 + CaSO_4$	Reduced growth, vigor and establishment of seedlings.	Sinha and Gupta., (1982)

proteins, subcellular structural changes, respiration, synthesis of macromolecules and cell elongation. Whereas, Chong and Bible (1995) have regarded growth of embryonic tissues as an important step in the completion of germination, some workers have given due emphasis to the establishment of seedlings under stressful conditions. The latter seems crucial under salinity, as without a successful crop stand simple emergence of embryonic tissues will prove futile. The processes of germination are greatly influenced by the nature and extent of salinity and, above all, on the behaviour of seeds. For a better understanding of the adverse effects of salinity, the process of germination have been categorized into four events (a) imbibition, (b) active metabolism, (c) emergence and elongation of embryonic tissues and (d) establishment of seedlings (Table 1).

2.1.1. Imbibition of water

Hydration of stored materials is the initial step for the onset of germination. The osmotic component of salinity poses a strong inhibitory effect on the hydration of the embryo, cotyledon, and endosperm. It is independent of types of salinity and growth media whatsoever, as use of any salt induces an osmotic effect.

2.1.2. Active metabolism

The ions are inevitably taken up by seeds, during exposure to salinity, which cause toxicity to various physiological and biochemical processes. The activities of enzymes are hampered, leading to the altered and reduced synthesis of micro and macro molecules and their reduced mobilization to the developing tissues. The synthesis of new proteins in response to saline stress was observed in wheat embryo. This pattern of synthesis was attributed to the specific effect of ions on the activities of enzymes. Salinity also causes the accumulation of soluble sugars, free proline, and soluble proteins. These metabolites may prove to be beneficial to the germination; first by reducing osmotic inhibition and second by providing substrates for the growth of embryonic tissues.

2.1.3. Emergence and elongation of embryonic tissues

Seed reserves are utilized in the growth of the embryo and the elongation of young tissues and involve the turnover *de novo* synthesis of macromolecules. Germinating seeds in saline media exhibit a lowered and delayed production of radicle and plumule. Sodium chloride affects the emergence of young tissues more adversely than other salinities.

2.1.4. Establishment of seedlings

A successful crop stand depends on the establishment of young seedlings. Prolonged exposure to substrate salinity results in an extremely poor stand caused by seedling mortality. This may be more pronounced in the case of glycophytes owing to their high sensitivity to salinity. A good stand of crop was achieved in *Sorghum halepense* under mild salinity owing to the rapid rate of germination. Plants having higher seedling vigor also show better stand under salinity.

2.2. Toxicity effect

Germination and salinity interaction is often studied on the premise that it has dual action; that is, osmotic and toxic actions. Attempts to separate both the components of salinity by using isotonic solutions of salts and non permeating osmotica yielded conflicting data. Some regard the osmotic effect as the crippling factor, whereas the majority consider ion toxicity as being a noxious component or that both the component are equally detrimental to germination. Wahid *et al.* (1998) reported that incubation of seeds in salt solution followed by reduced germination in water gave credence to the major role of ion toxicity.

Increasing salinity generally reduces germination in glycophytes and to a lesser degree in halophytes (Khan and Ungar, 1997). Salts can affect seed germination either by restricting the supply of water (osmotic effect) or causing specific injury through ions to the metabolic machinery (ionic effect) (Ungar, 1991). Under natural conditions, however, these basic effects additionally interact with soil moisture content (volume of water relative to soil). As soil moisture declines, soil total water potential (Ψ_w) declines due to both soil matric potential (Ψ_m) and covarying soil Ψ_s, which reflects increased concentration of solutes (Hardegree and Emmerich, 1990). Seed of halophytes under natural conditions are subjected to saline stress and usually it is dominated by NaCl. However, other chloride, sulfate and carbonate salts and their interaction plays a significant role in affecting seed germination.

Studies on the effect of various chloride and sulfate salts on the germination of halophytes revealed that the salts exhibited some osmotic effects but no specific ion effect (Pujol *et al.*, 2000), while, others reported both osmotic and ionic effects (Mohammed and Sen, 1990). Seed germination under isotonic solutions of PEG 6000 and NaCl had similar effect on the seed germination of halophytes (Bajji *et al.*, 2002). High salinity could be injurious to most glycophytes, however, seeds of halophytes usually recover completely when saline stress is removed indicating an osmotic effect (Hardegree and Emmerich, 1990). Many halophyte seeds have the ability to maintain seed viability for extended period of exposure to hypersaline conditions with other physicochemical factors and germinate when conditions are favourable (Khan and Ungar, 1997). Seeds of some halophytes when pretreated with salinity show the priming effect of salinity on germination, while others showed no effect of salinity and recover immediately after salinity stress is removed and still other halophytes failed to germinate when exposed to high salinity (Khan and Ungar, 1997). *Suaeda fruticosa* showed a high recovery percentage when transferred to distilled water. Potassium chloride, potassium sulfate, sodium chloride and sodium sulfate had no effect on the germinability of the seeds and all of them germinated to control level when transferred to distilled water except for potassium chloride where recovery percentages were about 20% higher than non-treated control (Mohammed and Sen, 1990). Exposure to potassium chloride had a priming effect and germination was significantly increased in comparison to control. There was no specific ion toxicity and that an osmotic effect limited germination as in other halophytes (Pujol *et al.,* 2000) (Table 2).

Table 2: Seed germination (Mean ± S.E.) of *Chenopodium glaucum* when treated with different salt solutions. Different letters indicate significant difference among the data in same column at P= 0.05. Bonferroni test. (Duan *et al.*, 2004)

Salt concentration	Germination (%)					
	NaCl	SES	Na_2CO_3	Na_2SO_4	$MgSO_4$	$MgCl_2$
0.0	72.5	72.5	72.5	72.5	72.5	72.5
0.1	65.0	68.8	62.5	56.9	73.8	48.8
0.2	44.4	46.3	41.3	16.9	53.8	11.3
0.3	12.5	16.3	4.4	0.6	36.3	0.0
0.4	2.5	4.4	0.0	0.0	21.3	0.0
0.6	0.0	0.0	0.0	0.4	0.0	0.0

Seeds of *C. glaucum* showed 73% germination in non-saline control (Table 2). Seed germination decreased with increase in NaCl, Na_2CO_3, Na_2SO_4, $MgSO_4$, $MgCl_2$ salts and salt solutions extracted from soil. $MgCl_2$ inhibited germination more than other salts and only 11% seeds germinated at 0.2 M, whereas 21% seeds germinated at 0.4 M $MgSO_4$ solution. Little seed germination was observed in 0.6 M salt treatments (Table 2). Inhibition of *C. glaucum* seed germination at high salt concentration was mostly due to osmotic effects while some ionic effects were noted at lower salt concentrations. The inhibition of different salt solutions to halophyte seed germination varied greatly. Ungar (1996) reported germination of *Puccinellia festucaeformis* seeds by various salts was in the order of $CaCl_2$, $MgCl_2$ > NaCl > KCl > $NaNO_3$ > KCl > $MgSO_4$. Ungerminated seeds from all salt treatments when transferred to distilled water germinated to levels similar to non-saline control. Iso-osmotic solutions of NaCl and PEG have similar effect on seed germination at similar concentrations (Fig. 1) and no seed germinated at or above -1.8 MPa PEG. When recovery of germination from various salts and PEG were compared at -1.8 MPa and -2.7 MPa PEG, 88% seeds germinated at -1.8 MPa in all salts; however, a few seeds germinated when pretreated with -2.7 MPa Na_2CO3 and PEG (Fig. 2). Comparison of recovery of germination when treated with various concentrations of NaCl and PEG had a similar pattern showing about 60% germination at the highest osmotic potential. It was also reported that after the stress was removed, the seed germination and early seedling growth were higher than those of the untreated seeds (Yuying *et al.*, 1999). *Chenopodium glaucum* seeds showed a priming effect when treated with high salt concentrations. Seeds of many halophytes have the ability to maintain viability for extended period of exposure to hyper-saline conditions with other physicochemical factors and germinate when conditions are favourable (Khan and Ungar, 1997). Seed of some halophytes when pretreated with salinity showed a priming effect of salinity on germination, while others showed no effect of salinity and recovered immediately after salinity stress was removed. Other halophytes failed to germinate when exposed to high salinity (Khan and Ungar, 1997).

2.3. Metabolism of stored materials

Seeds, whether from monocots or dicots, comprise of storage tissue (endosperm and cotyledons, respectively) and an embryo. The nature and extent of stored materials may be different in different species. Major stored materials include proteins, sugars, and oils, whereas nucleic acids, plant growth regulators, nitrogenous compound other than proteins, sugars, and some nutrients may form a small component. Salt stress hampers the metabolism of **stored materials and the growth of the embryo.**

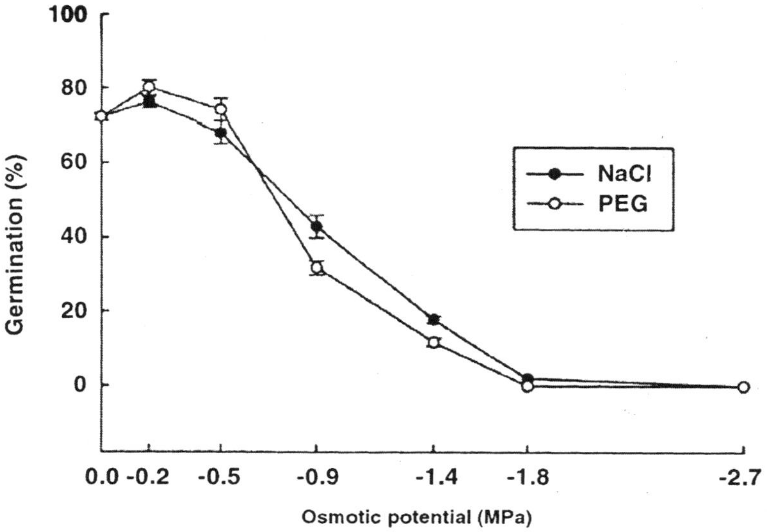

Fig. 1: Seed germination of *Chenopodium album* in different concentrations of NaCl and PEG solutions.

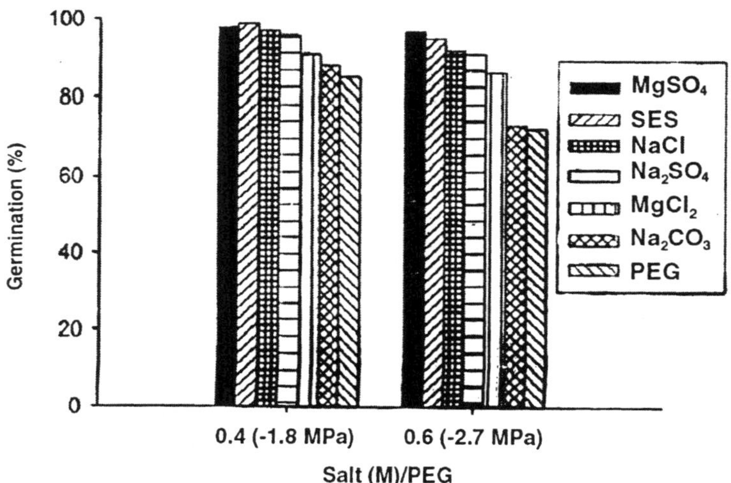

Fig. 2: Relative germination percentages of seed treated with 0.4/0.6 M salt or -1.8/-2.7 Mpa concentrations.

At the onset of germination, synthesis of enzymes and changes in the metabolic pattern are initiated, but salt stress either alters it or does not permit the synthesis of specific metabolisms required for germination. Application of salinity hampers the utilization and mobilization of materials required for the production of seedlings by affecting the enzymatic activities of seed essential for these reactions (Table 3).

Table 3: Physiological and biochemical changes in germinating seeds under salinity

Metabolic activity	Effect of salinity	Reference
Proteins	Toxic to the protein phosphatase and protein kinase specific activities; strong inhibition of methionine and leucine uptake and incorporation into proteins chain; *de novo* synthesis of eight new heat-stock proteins during stress and seven during recovery; a major change in protein phosphorylation-dephos phorylation and a depression in seedling growth	Ramgopal, (1990)
Carbohydrates	Reduction in endospermic α-amylase activity in a concentration dependent manner; promotion of cotyledonary α-glycosidase activity; decrease in reducing and non reducing sugars; accumulation of sugars as osmotica	Corchette and Guerra, (1986)
Nucleic acids	Slow rise in activity or inhibited *de novo* synthesis of RNase; inhibition of incorporation of precursor into nucleic acids reflecting their suppressed biosynthesis, decrease in DNA level throughout germination period; condensed chromatin material	Gomes Filho, (1983)
Lipid	Reduction in the activity of glyoxysomal enzymes; reduction in utilization of total lipids and triacylglycerol	
Polyamine	Induction of putrescine synthesis with the activation of arginine decarboxylase; no effect or spermine; significant increase in spermidine; accumulation of total polyamines	Basu and Ghosh, (1991)
Other compounds (nitrogenous and non-nitrogenous)	Increase in betaine aldehyde dehydrogenase; rise in betaine content; progressive accumulation of free proline and sugars and soluble proteins; decrease in the level of soluble amino nitrogen	Yokoishi and Tanimoto, (1994)

2.3.1. Proteins

Salinity creates an impact on the activities of the enzymes for protein metabolism. Protease, which catalyzes the turnover and solubilization of proteins to soluble nitrogen in seed, is largely inhibited by salinity. It interferes with the incorporation of (H) leucine and (S) methionine during protein not synthesized otherwise. Ramagopal (1990) found the synthesis of qualitatively and quantitatively different eight new proteins in germinating barley embryo under salt stress and seven during recovery.

Sadeghian and Yavari (2004), who stated that seedling growth was severely diminished by water stress in sugar beet. Moreover, distinct genetic differences were found among the cultivars with respect to germination and seedling growth subjected to NaCl and PEG. At the same water potential, the lower mean germination time in NaCl than in PEG could be explained by more rapid water uptake in NaCl solutions. Khajeh-Hosseini *et al.* (2003) found faster germination in NaCl in soybean. Delgado *et al.* (1994), who studied the effects of salt stress on growth and nitrogen fixation by pea, faba bean, common bean and soybean plants, indicated that pea was the most sensitive legume among them. Root and shoot length, and seedling fresh and dry weight were decreased by increasing NaCl and PEG concentrations. Consequently, seedling growth was inhibited in pea. The effects of salt and drought stresses at the water potentials of -2, -4, -6 and -8 bars induced by NaCl and PEG 6000 (polyethylene glycol 6000) each, on germination and early seedling growth, were investigated for 3 pea cultivars (Bolero, Sprinter and Utrillo). Electrical conductivity (EC) values of the NaCl solutions were 4.5, 8.8, 12.7 and 16.3 dS m^{-1}. The germination results revealed that the genotypes significantly differed for salt and drought stress. Bolero appeared to be more tolerant to salt stress, but Sprinter gave higher values under drought stress. Both NaCl and PEG inhibited germination and seedling growth in all cultivars, but the effects of NaCl compared to PEG were less on germination and seedling growth. All cultivars were able to germinate at all NaCl levels without a significant decrease in germination, while a drastic decrease in germination was recorded at -6 bars of PEG. It was concluded that inhibition in germination at equivalent water potentials of NaCl and PEG was mainly due to an osmotic effect rather than salt toxicity (Okcu *et al.*, 2005). The growth of *Pancratium* seedlings in terms of fresh weight of the shoot was reduced by salt-stress, but this reduction was alleviated in part by the addition of exogenous proline. Measurements of dry weight and water content of seedlings indicated that proline did not improve the dry weight yield of salt stressed seedlings. Therefore, exogenous proline could only maintain higher water content under salt stress and this is further emphasized by measurements that showed that the water content of salt stressed seedlings was higher in the presence of proline. In the roots, the water content decreased in response to salt stress, however, exogenous proline kept it comparable to that of the control in salt stressed roots. The ability of exogenous proline to maintain higher water content in severely stressed seedlings might be attributed to its contribution to osmotic adjustment both directly by increasing the internal proline content and indirectly by increasing the internal contents of other amino acids.

The effect of salt-stress on protein content depended on the concentration of NaCl. At lower levels of NaCl, there was an increase in protein content, but higher concentrations caused it to decline in both shoot and root. This suggests that the initial response to water stress involves increased protein synthesis that is prevented when the stress becomes too severe. Exogenous proline increased the protein content of the shoot and root at all levels of salt stress and, interestingly, also in non-stressed plants (Fig. 3). Changes in growth and protein contents as a result of exogenous proline correlated with increase in the internal content of proline suggesting that proline was taken up into the roots and transported to the shoots (Fig. 3). Proline may perform this adaptive effect by working as an additional nitrogen source and/or decreasing proteolytic activity. It would be predicted that plants under stress would have a powerful protein turnover machinery to degrade stress damaged and environmentally regulated proteins.

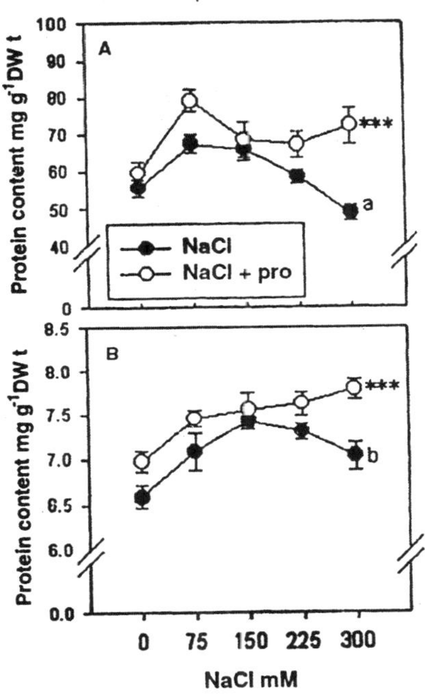

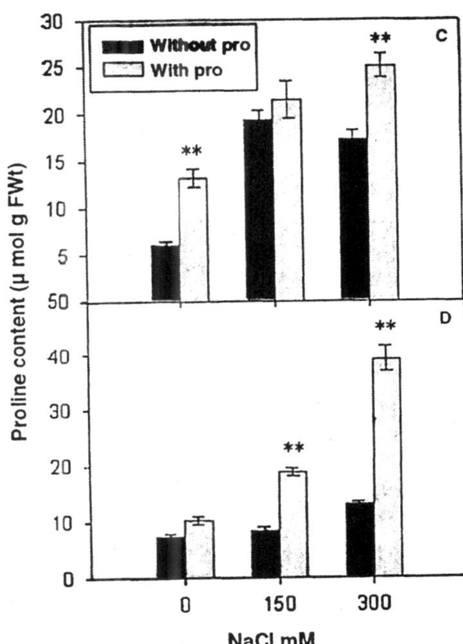

Fig 3: Effect of salt stress on protein and proline content of *Pancratium* seedlings. (A) Shoot, three asterisks indicates significant from 'a' at P=0.001. (B) Root, three asterisks indicates significant from 'b' at P=0.001. Proline contents of shoot (C) and root (D) of *Pancratium* seedlings after treatment with NaCl and proline. Values are internal proline content. Two asterisks indicates significant from the corresponding black bar at P=0.01.

Catalase is involved in scavenging hydrogen peroxide that is produced during photorespiration and β-oxidation of lipids in the peroxisomes under normal conditions and accumulated under stressful conditions. The induction of catalase activity under water stress is well documented and a positive relationship has been found between its upregulation and stress tolerance (Ushimaru *et al.,* 2001). The down-regulation of catalase by salt-stress may indicate that the plant is not able to maintain protection against active oxygen under salt-stress particularly at high salt concentrations. The decrease in catalase activity measured under severe water stress has been reported by Schwanz *et al.* (1996). Such a decrease may be due to some stress-induced damage to the enzyme. This may be supported by the ability of proline to maintain higher catalase activity under stress, since proline can work as an enzyme protectant under stress. Like catalase, peroxidase is involved in neutralizing hydrogen peroxide, but at the expense of another substrate being oxidized such as ascorbate. Peroxidase is also reported to be enhanced by water stress and this was positively correlated with water stress tolerance (Ushimaru *et al.,* 2001). The consistent increase in peroxidase activity in the shoots even during severe salt stress and its differential regulation show that peroxidase may be more stable or more important for stress tolerance than catalase. This is probably dictated by the wide range of metabolic processes in which peroxidases are known to be involved, such as lignin biosynthesis and formation of isodityrosine bridges that are believed to crosslink structural protein molecules, in addition to antioxidative activity. The reduction of peroxidase activity by severe salt stress may be due to stress damage that can be partially relieved by exogenous proline. The pattern of expression of dehydrins is now established to be developmentally and environmentally regulated. They are expressed in the late stages of embryogenesis and induced in vegetative tissues in response to different kinds of stress including dehydration, salinity, cold, and heat stress. In barley, one dehydrin was detected in well hydrated barley seedlings, but, upon water stress, many others were newly synthesized (Close *et al.,* 1993). Many dehydrins are constitutively expressed in both stressed and non-stressed plants and that proline was able to increase expression at all stress levels. In *Pancratium,* the expression of the whole set of dehydrin proteins in the shoot and root may indicate that these proteins are all required for normal metabolism, but more amounts are required during water stress. However, the pre-existence of dehydrins in the non-stressed tissues may provide more protection upon water stress. The quantitative and qualitative regulation of dehydrins by salt stress suggests quantitative differences in the requirement of individual dehydrins to cope with the stress and that the protection against stress damage may be attained by the coordinated action of the whole set of dehydrins rather than by the action of individual proteins (Bartels and Salamini, 2001). The induction of dehydrins by proline indicates that it might be a component of the regulatory process(s) that leads to the accumulation of dehydrins as can be deduced from its ability to induce dehydrins of the shoot and root to levels that are comparable to these under stress conditions. Similar results have been found by Garcia *et al.* (1997), who found that the salT gene of rice is induced by proline and it was found previously that this gene is induced by NaCl, drought and ABA (Claes *et al.,* 1990). From these results it is concluded that proline could act as a component of signal transduction pathways that regulate stress responsive genes in addition to its previously described osmoprotective roles, thereby improving the tolerance to salt stress.

2.3.2. Carbohydrates

Carbohydrates (as starch) constitute a major bulk of storage material in some seeds (*e.g.,* caryopsis). Amylases mainly regulate the metabolism of carbohydrates, and their activity is greatly attenuated by salinity. The activity of α-amylase is reduced under salinity in a concentration-dependent manner, depressing the growth of seedlings (Lin and Kao, 1996). Greater salt tolerance of sorghum during germination was attributed to the enhanced activity of α-amylase. Salt treated lentil seeds indicated no variation in different solute contents; however, the activity of α-galactosidase increased and caused the accumulation of sucrose, galactose, and mannose in the embryonic tissues. An accumulation of osmotically active sugars and proline was noted in different plants which played an important role during and after relief from salinity.

2.3.3. Nucleic acids

The most important factor in nucleic acid metabolism is the synthesis and activation of ribonuclease (RNase). Salinity delays the *de novo* synthesis and/or activation of RNase in *Vigna* cotyledons due to its toxic effect. A slight increase in the cotylendonary RNA level during the first day of germination was noted, but it decreased subsequently upto 7 days; however, DNA decreased continuously during this period. Petruzzelli *et al.* (1991) suggested that the suppression of nucleic acid biosynthesis in wheat embryo was due to salt induced inhibition of the incorporation of precursors into nucleic acid.

2.3.4. Lipids

Glyoxysomal enzymes are responsible for the metabolism of stored lipids. Salinity exerts an inhibitory effect on glyoxysomal catalase, malate synthase, and iso citrate lyase, decreasing the levels of triacylglycerol, diacylglycerol, and monoacylglycerol and increasing the levels of free fatty acids and polar lipids.

2.3.5. Polyamines

Polyamines have recently gained importance in the escape of seedlings from the adverse effect of salinity. They promote seedling growth by the production of ethylene-forming enzymes. Lin and Kao, (1995) reported an increase in the level of spermidine under salinity but a low level of putrescine in the shoot and roots of rice seedlings. Accumulation of putrescine and spermidine, with the activity of arginine decarboxylase in rice seedlings, plays a specific role in salt tolerance owing to production of ethylene (Basu and Ghosh, 1991).

2.3.6. Other organic compounds

Various endogenous compounds are differentially metabolized during germination and seedling growth. The glycine betaine, a compatible solute, either disappears, exhibits no change, or accumulates as a result of salt stimulated betaine aldehyde dehydrogenase activity and rescues the seed from the adverse effect of salt. Similarly, the rise in the level of free proline and soluble sugars of seeds or seedlings also plays a beneficial role (Wahid *et al.*, 1998).

2.3.7. Seed nutrients

The higher content of seed nutrients is of vital importance for germination, but salinity suppresses their role in the metabolism of seed and the production of seedlings. During germination of sorghum caryopses, a higher content of potassium, calcium, phosphorus, and nitrogen was partitioned into the plumule and radicle as a strategy of tolerance to salinity. Guerrier, (1986) attributed the reduced salt tolerance of tomato to its inability to accumulate and transport lower amounts of calcium and potassium.

2.3.7.1. Regulation of ions in reeds and seedlings

Exposure of seeds or seedlings to salinity results in the influx of ions with the imbibition of water, which exerts an adverse effect on the growth of embryo. This may lead to a marked decreased in the internal potassium concentration (Petruzzeli *et. al.*, 1991), a vital nutrient for protein synthesis and plant growth. Seedlings exposed to salinity are highly prone to excessive ions, sometimes leading to their death shortly after emergence. The ability of plants to cope with ion toxicity is principally related to the greater transport of ions to shoot. Grasses show a strategy of salt tolerance by storing toxic ions in the mesocotyl up to a certain limit. This has significance in that the epicotyl and hypocotyls avoid ion toxicity, thus ensuring their better growth.

2.4. Structural changes in seeds and seedlings under salinity

Salinity triggers structural changes at various levels of organization in seeds and seedlings (Table 4). At the subcellular level, major changes were found in (a) nuclear chromatin, which was condensed indicating suppressed nucleic acid biosynthesis; (b) formation of small provacuoles instead of single large vacuole, and (c) damaged mitochondrial apparatus and reduced oxygen uptake. Salinity caused the contraction of plasmalemma away from the cell walls, which may be due to disaggregation of intramembranous particles. The cell wall of salt-treated cotton roots and the sorghum mesocotyl became considerably thickened.

Table 4: Salt-induced changes in anatomical characteristics during germination in various tissues

Metabolic activity	Effect of Salinity	Reference
Subcellular	Formation of small provacuoles in coleorhiza cells	Bliss *et al.*, (1986)
	Diffusion and condensation of chromatin material in embryo	Petruzzuli *et al.*, (1991)
	Reduced size of plasmalemma and mitochondria	Petruzzuli *et al.*, (1991)
	Lignification and thickening of cell wall	Wahid *et al.*, (1998)
Cellular	Reduced size of cortical cells in mesocotyl of sorghum	Wahid *et al.*, (1998)
	Induction of endodermis with casparian band and subberin lamellae close to root base	Reinhardt and Rost, (1995)
Tissue	Earlier development and differentiation of secondary xylem in hypocotyls	Reinhardt and Rost, (1995)
	Constriction of cortical tissue of mesocotyl	Wahid *et al.*, (1998)
	Increased lignification of secondary tissues	Valenti *et al.*, (1992)

At cell and tissue levels, the salinity reduced the cortical cell area and as a result the mesocotyl of sorghum was considerably constricted and appeared to act as a repository of ions. Furthermore, there was the induction of exodermis with a casparian band having suberin lamellae close to the root base and in the transition zone of the hypocotyls of cotton (Reinhardt and Rost, 1995). This protected the root from water loss and/or leakage of solutes important for osmotic adjustment. Salinity also stimulated the development and lignification of secondary tissues and enhanced the number of water storage cells in the epidermis and cortical layer of the hypocotyls (Valenti *et al.*, 1992).

2.5. Factors interacting with salinity during germination

2.5.1. Plant factors

2.5.1.1. Seed coat

The seed coat is the first barrier to the entry of water and ions into the seed. The hard and thick seed coat offers resistance to the entry of water into the seed and minimizes the contact of ions with the embryo. It also acts as a buffering agent to ionic toxicity (Poljakoff-Mayber *et al.*, 1994) and enhances germination.

2.5.1.2. Dormancy

One of the primary impacts of salinity is the induction of dormancy in the seed either due to inhibition of the synthesis of nucleic acids or plant growth regulator imbalance (Khan and Rizvi, 1993). Although dormancy carries no consistent relationship with salinity, it is important for the halophytes, since it permits the seed to remain viable for the period until the environment becomes conducive to germination.

2.5.1.3. Seed age

Aging or prolonged storage of a seed affects its germinability. This has been used to test and predict the salt tolerance potential during germination. Smith and Dobrenz, (1987) found a strong negative relationship between salt tolerance and seed age in a sensitive but a significant decline in the solute leachate during imbibition of water in a tolerant alfalfa genotype.

2.5.1.4. Seed polymorphism

The seed size of a species also shows a differential response to salinity. The greater the seed size, the greater is the salt tolerance. Smaller seeds containing a higher amount of toxic ions and a low amount of reserves per unit weight show higher dormancy-delayed germination and reduced weight of seedling (Khan and Ungar, 1984).

2.5.2. Environmental factors

2.5.2.1. Temperature

A slight variation in temperature may change the germination greatly. The adverse effect of high salinity is further aggravated by higher temperature regimes, which may prolong the time taken for emergence of seedling (Khan and Ungar, 1997). However, a synergistic effect of low temperature and high salinity has been noted in halophilic barley seeds.

2.5.2.2. Light

Light has a profound effect on the germination of many species. The light may be effective in breaking the dormancy and promoting germination in some halophyte species. This may result in better establishment of a stand in marginally saline areas (Khan and Ungar, 1997).

2.5.2.3. Water

Salt-induced lowering in the water potential of the germination medium enhances toxicity, whereas, scarcity of water further aggravates it. Major interaction of water stress under salinity conditions includes a differential pattern of protein synthesis (Petruzzelli *et al.*, 1991), delayed emergence of embryonic tissues, and a decrease in final rate and percentage of germination. The supply of water to seeds or seedlings reverses these processes to a great extent, as it minimizes ionic toxicity.

2.5.2.4. Oxygen

Salt-induced dormancy reduces the availability of oxygen to the embryo for metabolic activities. High salinity coupled with hypoxia significantly reduces both emergence and elongation of the radicle and plumule (Wijte and Gallagher, 1997). Anoxia completely restricted the process of germination; however, no specific difference was discernible with respect to salinity and anoxia tolerance. *Spartina alterniflora* could show better tolerance salinity and anoxia than *Phragmites australis*, as the former showed a rapid rate of coleoptile and mesocotyl growth.

2.6. Alleviation of salt stress

Various chemical agents have been employed to ameliorate the adverse effects of salts. The plant growth regulators are the most widely used. Some nitrogenous compounds, sugars, and certain nutrients have also been employed.

2.6.1. Implication of plant growth regulators

Both naturally derived and synthetic plant growth regulators have been employed separately or in combination. The primary action of osmotic inhibition is the reduction in water uptake, and plant growth regulators may offset this inhibition and promote the process of germination. Presoaking in gibberellin after salt stress releases the seed from physiological dormancy, enhances water uptake, mobilizes starch and improves the rate and percentage of germination. Auxins like indole butyric acid and indole acetic acid promote germination and seedling growth better than kinetin by eliminating osmotic effect of salinity.

Bozcuk, (1981) reported that kinetin releases the seed from salt induced dormancy and enhances seed protein synthesis. Kinetin applied to salt stressed seeds promotes germination by enhanced production of pregermination ethylene with the synthesis of 1-aminocyclopropane-1-carboxylic acid. The use of kinetin in combination with gibberellins not only promotes germination but also stimulates the growth of seedlings under salinity. It is likely that salinity suppresses the endogenous level of plant-growth of seedlings under salinity. It is likely that salinity suppresses the endogenous level of plant growth regulators, and their exogenous supply fulfills this requirement for the initiation of germination.

2.6.2. Polyamines

The polyamines compete with the ethylene pathway, as S-adenosylmethionine is a common precursor. The exogenous application of putrescine to seed not only counteracts the adverse effect of salinity, but also induces tolerance up to the seedling stage owing to their *de novo* synthesis in response to salinity. Lin and Kao (1995) found an increase in the level of putrescine by exogenously applying the precursors of putrescine biosynthesis (L-arginine and L-ornithine), but they did not induce a significant mitigation of salt toxicity.

2.6.3. Other organic compounds

Some organic chemicals have also been employed to lessen the adverse effect of salinity. Application of sucrose and glucose partially reverses the salt inhibition of germination. Noor and Khan (1995) reported the efficacy of a low amount of thiourea in breaking salt induced dormancy. The application of glycinebetaine and proline increased the root growth but did not affect the hypocotyl growth. Fusicoccin also counteracts the inhibitory effect of NaCl on the process of germination by cell wall loosening, promoting transport activity, incorporating amino acids in the protein chain and enhancing potassium uptake owing to stimulation of proton efflux. Addition of methionine sulfoximine was found to reduce the seedling growth of salt stressed plants, which was reversed by stimulation of glutamine synthetase and glytamate synthase activity, and with the addition of glutamine and glutamic acid to the growth medium.

2.6.4. Inorganic agents

Calcium has been extensively used to alleviate salt toxicity because of its crucial role in the maintenance of membrane processes, modulation of enzyme activities, and buffering of Na^+ toxicity. Although presoaking of seed in $CaCl_2$ does not initiate germination, its application to a saline medium significantly improves the rate and percentage of germination. In addition, supplying calcium stimulates plumule emergence and root growth as well. Application of ammonium promotes the protease activity in the endosperm, whereas, nitrate enhances the seed germinability owing to the solubilization and availability of catabolites for the synthesis of embryonic structures. Furthermore, the use of potassium and magnesium also counteract the NaCl inhibition of the root growth of rice seedling.

3. FLOODING

For most crops, seed germination is probably the most vulnerable, reflecting both the fast metabolic rate of germinating seeds being coupled with complete inundation. Problems are exacerbated if germination takes place in association with decomposing organic matter such as straw residues from the previous crop. Microbial colonization of such residues depletes the soil of oxygen and nitrate while discharging potentially harmful substances such as unsaturated fatty acids and the growth-inhibiting hormone abscisic acid. Leaking of soluble carbohydrates from seeds is stimulated by anaerobiosis and this also encourages fungal pathogens such as *Gliocladium roseum.* Seeds of rice are exceptional in being able to germinate in flooded soil without oxygen. But the germination is abnormal, since only the coleoptile, but not the root or the mesocotyl or leaves emerge from the embryo without oxygen. Elongation by the emerging coleoptile is stimulated under these conditions by a combination of lack of oxygen, carbon dioxide and ethylene.

Once germinated, stages in subsequent development of crop plants influence susceptibility to flooding injury. Small cereal seedlings with their shoots below ground are highly susceptible (*e.g.*, winter wheat), but thereafter tolerance rises until early reproductive stages when susceptibility again increases (*e.g.* in soybeans). Germinating seeds have been identified as being especially vulnerable to damage from soil waterlogging and the associated anaerobiosis and injury from chemically reduced metabolites such as organic acids and respiratory intermediates such as acetaldehyde. It is therefore, not surprising that pre-treating seeds with an oxidizing coat has been tried as an insurance against such effects. Coatings include calcium hydroxide and calcium peroxide. There are companies selling products that promise to improve plant growth by using calcium peroxide to 'inject' oxygen into the soil. A more mundane approach in situations where germinating seeds risk contact with undegraded crop residues in wet conditions, is to rotovate and dices these residues or removed them before sowing, possibly by burning, if legislation allows this. Flooding may also increases the incidence of soil-borne fungal diseases. Germinating seeds are particularly vulnerable to fungal colonization (*e.g.*, *Gliocladium roseum*). Infection of alfalfa, vegetables and trees by *Phytophthora* (wilting), *Pythium* (damping-off) and anaerobic bacteria (*e.g. Pseudomonas putida*) are common problems in practical farming. However, it is not always clear whether injury is principally the result of the microbial infection or of the direct effects of flooding and if infection follows rather than precedes injury.

Seeds of higher plants have an absolute oxygen requirement for germination. However, germination and seedling development under anoxia have been reported in rice, *Echinochloa* species, *Erythrina caffra* and the water plant. *Echinochloa* species exhibit an ability to vent most of the ethanol produced during anaerobic germination to the external media and the seedlings are very tolerant of ethanol (Kennedy *et al.,* 1987). Although mitochondria in other species including rice degenerate or exhibit abnormalities under anoxia exposure, anaerobic tolerant *E. phyllopogon* (Stev.) Koss. (synonymous with *E. crusgalli* var. *oryzicola* Ohwi, and *E. oryzicola* Vasing.) developed intact mitochondria even under anaerobic conditions (Kennedy *et al.,* 1990). Indeed, ATP was generated anoxically by the continued functioning of part of the TCA cycle (Kennedy *et al.,* 1987). The pentose phosphate pathway and lipid biosynthesis were also operative under anoxia. Under anoxic stress, plants shift metabolism from aerobic respiration to anaerobic fermentation. Most of the anaerobic stress proteins (ASPs) identified are enzymes associated with glycolysis and fermentation (Drew, 1997). Indeed, continuous glycolytic flux and production of NAD by alcoholic fermentation are essential to maintain formation of ATP under anaerobic conditions. The activities of enzymes associated with glycolysis and alcoholic fermentation during anaerobic germination of formosensis and praticola. Interestingly, an initial and also an intervening enzyme of the pathway, sucrose synthase and aldolase, respectively (Type A enzymes), exhibited similar patterns in their timing and magnitude of change under both aerobic and anaerobic conditions. Likewise, the last part of the pathway, PDC, ADH and ALDH (Type B enzymes) also exhibited similar patterns (Fig. 4). Enolase exhibited an intermediate response between type A and type B enzymes in the degree of induction by anoxic stress and the different pattern on day 1 under aerobic conditions (formosensis, induced; praticola, not induced). In maize seedlings, no common patterns of change in the transcriptional level of glycolytic and fermentative enzymes were observed under hypoxia or anoxia (Andrews *et al.,* 1994). On the contrary, in rice seedlings, two types of mRNA accumulation were observed in genes associated with glycolytic and fermentative pathway under submergence stress. Although in *E. crusgalli* (including formosensis and praticola), induction patterns of the enzymes at the transcriptional level have not been investigated yet, two types of induction patterns were recognized at the enzymatic level. Each type of enzyme (A and B) may be regulated by an identical mechanism such as a common transcription factor and/or signal transduction pathway.

4. DROUGHT

Soil water supply is an important environmental factor controlling seed germination. If the water potential is reduced, seed germination will be delayed or prevented depending on the extent of its reduction. Water stress affects practically every aspect of plant growth and metabolism. Plant responses to water deficit depend upon various factors such as duration and degree of stress, growth stage

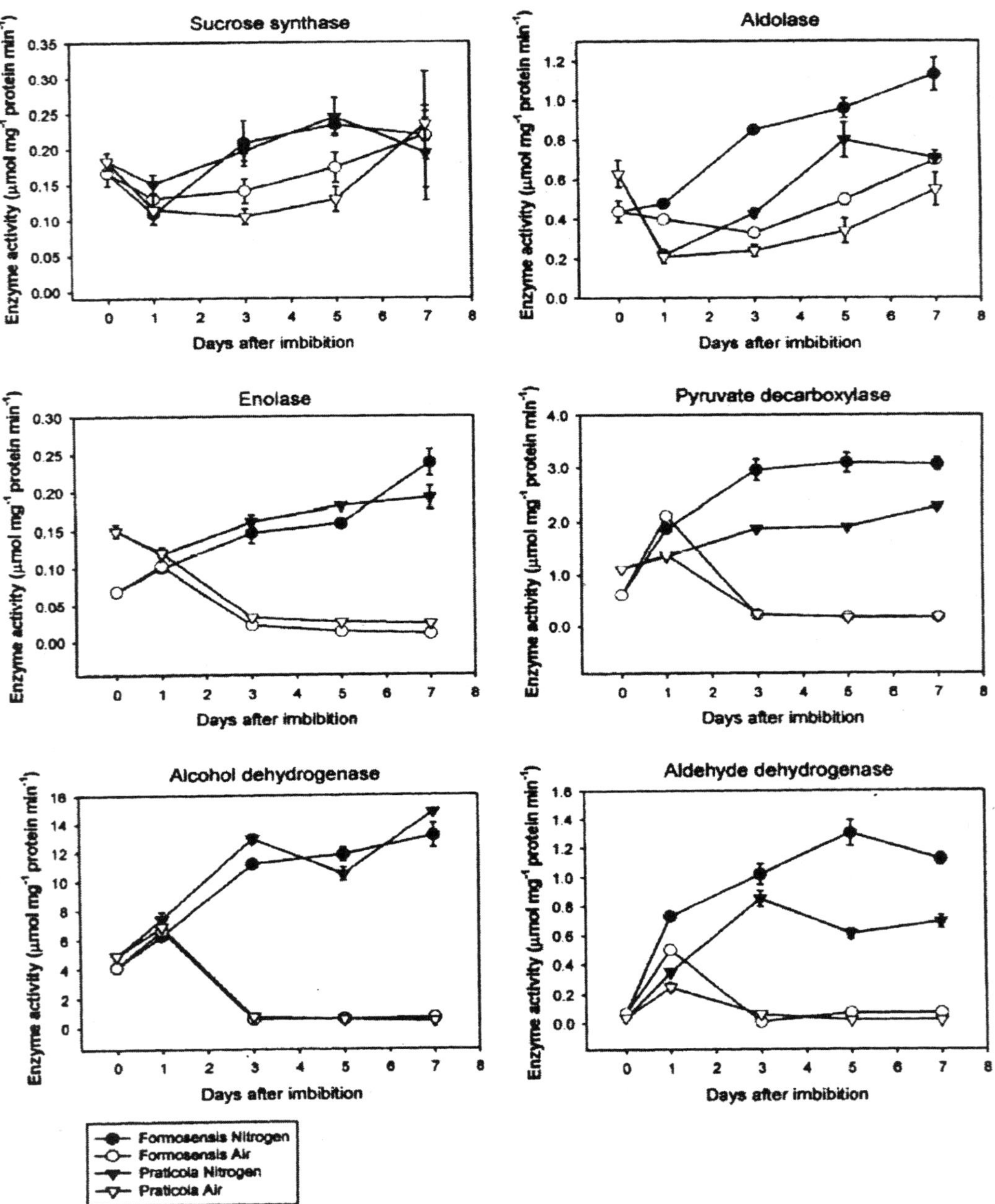

Fig. 4. Enzyme activities during germination under aerobic and anaerobic conditions in formosensis and praticola. Shoot (anaerobic) or root (aerobic) emergence was initiated on day 3 in each species, except for praticola under nitrogen in which germination did not occur. Each activity value represents an average of three measurements of the same sample. Standard deviation values are indicated when larger than the symbol.

and time of stress exposure. Due to their sedentary mode of life, plants resort to many adaptive strategies in response to different abiotic stresses such as high salt, dehydration, cold and heat, which ultimately affect the plant growth and productivity (Gill *et al.*, 2003). Against these stresses, plants adapt themselves by different mechanisms including change in morphological and developmental pattern as well as physiological and biochemical responses (Bohnert *et al.*, 1995). Adaptation to all these stresses is associated with metabolic adjustments that lead to the modulation of different enzymes (Ehsanpour and Amini, 2003). Among these enzymes are phosphatases, which are believed to be important for many physiological processes, including regulation of soluble phosphorous (Pi) (Yan *et al.*, 2001). Phosphatases are traditionally classified as being acid and alkaline depending on their optimum pH for enzyme activity, above and below pH 7.0 (Barret-Lennard *et al.*, 1982). Free soluble phosphate reserves plays vital role in energy transfer, metabolic regulation, important structural constituent of biomolecules like phytin bodies in the ungerminated seeds, protein and nucleotide phosphorylation (Ehsanpour and Amini, 2003). Although, some abiotic stresses like salt, osmotic and water have been reported to increase phosphatase activities by maintaining a certain level of inorganic phosphate in plant cells (Olmos and Hellin, 1997), the exact role of phosphatases in the germinated seeds is still not clear, because metabolism of these compounds can be affected by a number of environmental factors such as stress type, irridance, temperature, and type of ions present (Bohnert *et al.*, 1995). Germination of grains is initiated by water uptake and their successful completions are signaled by emergence of the developing root and shoot. Following uptake of water, hormone signals, probably released from the emergence, are believed to result in the synthesis of hydrolytic and other enzymes in the endosperm (Fincher, 1989). Moreover, like mature plants, germinating seeds and seedlings also can be subjected to environmental stresses. Even when they imbibe water, seeds may be exposed to elements of a hostile environment, which include high temperature of soil, salinity and varying moisture content. Failure to cope with the adversity caused by these extremes results in poor germination, seedling development, and eventually, reduced crop yields. Imposition of drought

Table 5: Effect of ABA, GA_3, NaCl and drought on germination of sorghum seeds.

Treatment	Germination
Control	99 ± 2
ABA	65 ± 4 *
GA_3	96 ± 2 ^{ns}
NaCl	35 ± 4*
Drought	Nil

* Significant difference vs. control at ($P \leq 0.05$) ^{ns} not significantly different

treatment in sorghum resulted in a very significant decrease in germination rate (nil), while germination rate under GA_3, ABA, and NaCl were 98%,65%, and 35%, respectively (Sharma *et al.*, 2004) (Table 5).

The decrease in germination rate particularly under drought and salt stress conditions may be due to the fact that seeds seemingly develop an osmotically enforced *"dormancy"* under water stress conditions. This may be an adaptive strategy of seeds to prevent germination under stressful environment, thus ensuring proper establishment of the seedlings (Gill *et al.*, 2003). However, decrease in germination rate observed under ABA treatment may be attributed to metabolic alternations. ABA may also be involved in inhibiting the seed germination by restricting the availability of energy and metabolites (Garciarrubio *et al.*, 2003). Exogenous application of ABA has been shown to inhibit embryonic germination by modulating the endogenous level of ABA (Dewar *et al.*, 1998), and when ABA synthesis is inhibited by fluridone, precocious germination occurs. During germination, another plant hormone, gibberellin (GA_3), induces embryo growth and stimulates the germination process. Therefore, in comparison to water-irrigated seeds, no substantial change in germination was observed under GA_3 treatment. GA_3 is well-documented regulator of germination and associated enzymes with generally having promotive effects (Fincher, 1989). Thus, at onset of germination, ABA and GA_3 appear to act in a fully antagonistic manner. Further, a significant reduction in tissue water content was observed in the embryos under NaCl treatment and under drought stress in the endosperm, indicating that respective tissues were under water stress. Similar observations on decrease in water level under stress conditions were made by Siddique *et al.* (2000) in wheat and Pennypacker *et al.* (1990) in alfalfa. In relation to growth, embryos were suppressed by ABA and NaCl treatments. Endosperm Fresh Weight (FW) was significantly reduced under drought stresses conditions. The reduced FW results from reduced water absorption (Prado *et al.*, 1995). Experimental evidence have shown that osmotic adjustment can reduce growth sensitivity to water stress (Cutler *et al.*, 1980) or allow growth to proceed at a slow rate under water stress by maintaining turgor. It can then be concluded that contribution of growth at lower water potential is a result of turgor maintenance, whereas, the inhibition of growth is not entirely dependent on turgor (BassiriRad and Coldwell, 1992). The fresh weight increase in distilled water was mainly due to an increase in tissue water content and is reflected in the [(FW-DW)/DW] ratios (Table, 6, 7)

Table 6: Effect of ABA, GA_3, and NaCl on fresh weight (FW), dry weight (DW) and tissue water content of embryos of sorghum.

Treatment	FW (mg embryo $^{-1}$)	DW (mg embryo $^{-1}$)	Tissue water content
Control	7.01 ± 1.10	3.50 ± 0.20	1.00 ± 0.14
ABA	0.53 ± 0.12*	0.22 ± 0.08*	1.40 ± 0.35 ns
GA_3	6.90 ± 1.20ns	3.20 ± 0.25 ns	1.15 ± 0.17 ns
NaCl	2.10 ± 0.11*	1.60 ± 0.13*	0.31 ± 0.05*

* Significant difference vs.control at ($P \leq 0.05$) ns not significantly different

Table 7: Effect of ABA, GA_3, NaCl and drought on endosperm fresh weight (FW), dry weight (DW) and tissue water content of sorghum.

Treatment	FW (mg embryo $^{-1}$)	DW (mg embryo $^{-1}$)	Tissue water content
Control	47.01 ± 2.35	41.10 ± 2.70	0.14 ± 0.05
ABA	45.03 ± 3.10 ^{ns}	39.00 ± 2.35 ^{ns}	0.15 ± 0.07 ^{ns}
GA_3	46.20 ± 2.50 ^{ns}	40.00 ± 2.11 ^{ns}	0.15 ± 0.06 ^{ns}
NaCl	46.21 ± 2.81 ^{ns}	41.54 ± 2.13 ^{ns}	0.11 ± 0.08 ^{ns}
Drought	40.01 ± 2.01*	38.12 ± 3.14 ^{ns}	0.04 ± 0.008*

* Significant difference vs.control at ($P \leq 0.05$) ^{ns} not significantly different

The results showed that imposition ABA and NaCl treatments resulted in a very significant reduction in dry matter of embryos. It is well known that as soil water availability becomes limited, plant growth usually decreases. This was previously considered to be due to turgor loss in expanded cells. However, some studies have shown that growth may be inhibited at low water potential despite complete maintenance of turgor in the growing regions as a result of osmotic adjustment. This suggests that the growth inhibition may be metabolically regulated possibly serving an adaptive role by restricting the development of transpiring leaf area in the water-stressed plants (Sharp, 1996). Acid and alkaline phosphatases activities in sorghum embryos and endosperm have been studied. In earlier research, Gill and Singh (1985) has reported that germination, growth, respiration and other related processes can be affected in seeds that are subjected to environmental stresses. Changes in anyone of these processes can affect other metabolic activities, particularly the enzymes of phosphate metabolism that plays an important role in germination and seed development (Fincher, 1989). Normally, salt and water stresses affect the physiology and biochemistry of plant cells under *in vitro* and *in vivo* conditions.

Table 8: Effect of ABA, GA_3, NaCl and drought on endosperm acid and alkaline phosphatase activities in the seeds of sorghum.

Treatment	Phosphatase activity (Units ,10^{-5} mg $^{-1}$ DW)	
	Acid	Alkaline
Control	1.61 ± 0.23	0.33 ± 0.08
ABA	6.96 ±0.45	2.32 ± 0.12
GA_3	0.17 ±0.08	2.05 ± 6.80
NaCl	17.1 ±2.30	2.01 ± 0.33
Drought	1.61 ±0.35	0.32 ± 0.09

* Significant difference vs. control at ($P \leq 0.05$) ^{ns} not significantly different

These stresses have been reported to enhance acid phosphatase activity in pea and wheat (Barrett-Lennard *et al.*, 1982). It may be due to fact that under conditions of stress, growth is restricted and delivery of phosphate is impaired, resulting in the activation of the cellular phosphatases that release soluble phosphate from its insoluble compounds inside or outside of the cells thereby modulating free phosphate uptake mechanism. A water deficit has also been shown to affect acid phosphatase in wheat (Thakur and Thakur, 1993). Embryo acid and alkaline phosphatase activities were significantly higher under salt stress. Similarly in endosperm, the acid and alkaline phosphates activities were significantly higher (Table 8).

Increase in acid or alkaline phosphatase activities under drought stress was found, indicating the cultivar and stress dependent nature of phosphatases. Olmos and Hellin (1997) observed that acid phosphatases are known to act under salt and water stress by maintaining a certain level of inorganic phosphate which can be co-transported with H+ along a gradient of proton motive force. In contrast, Szabo-Negy *et al.* (1992) and Barrettt-Lennard *et al.* (1982) reported results, indicating that phosphatase activities are independent of phosphate levels.

5. TEMPERATURE

In storage organs, as well in other non-photosynthetic tissues of higher plants, starch is the only carbohydrate reserve synthesized and stored in the amyloplasts of these tissues. In seeds of the conifer tree *Araucaria araucana* (Mol.) Koch, 61% of the dry weight is starch (Cardemil and Reinero, 1982). Starch is 31% of the dry weight of the embryo and 63% of the dry weight of the megagametophyte. The starch is digested during germination mainly by α-amylase in the embryo and by α-amylase and starch phosphorylase in the megagametophyte. Seeds germinate after 40 h of imbibition using sucrose that is synthesized from glucose, the product of starch hydrolysis (Cardemil and Reinero, 1982). The sucrose supplied by the megagametophyte is absorbed by the embryo cotyledons, acting as haustorial organs, and then transported into the embryonic tissues for seedling growth and development (Cardemil *et al.*, 1990).

The inability to germinate at higher than optimal temperatures is attributed to a condition called thermoinhibition. Tomato seeds do not germinate at high temperatures (=35°C), but do germinate after temperature is reduced to 25°C or 30°C. Thermoinhibition also occurs in other vegetable seeds, such as celery, lettuce, and spinach. In lettuce, thermotolerance differs among cultivars; this has been attributed to the degree of endo-ß-mannanase activity in the seeds (Nascimento *et al.,* 2000). Pepper seed germination was improved under supra-optimal conditions, if the seeds were first soaked in solutions of ethephon or gibberellic acid (GA_3). However, priming incurs additional costs and time considerations. A knowledge of which cultivars can germinate under supraoptimal conditions might be a more acceptable approach for transplant producers. Germination does not necessarily equate with stand establishment, but defining whether poor stand establishment is due to inhibition of germination or poor seedling growth is a logical first step.

Germination of Jalapeno M', 'Mitla', and 'TAM Veracruz' pepper variety was nearly 100% from 20 to 30°C. At 35°C, germination of 'TAM Veracruz' decreased to 48% while that of 'Jalapeno M' and 'Mitla' dropped to 82% and 72%, respectively. Germination of 'Cayenne, Large, Red Thick' and 'Ole' was below 25% at 35°C. Germination of all cultivars was equals to 1% at 40°C. The critical period for germination is appeared to be between 30 and 35°C. Germination of 'Ebano' had 95% germination at 35°C, but again germination dropped to 1% at 40°C. Cultivar differences in germination response to temperature have been reported in lettuce and spinach. Germination of 'Packman' broccoli decreased from 90% to 63% at 35°C when the germination of this cultivar was tested at a range between 5 and 40°C (Elson *et al.,* 1992).

Temperature had little affect on the time to 50% of final germination for each of the five cultivars. Only in 'TAM Veracruz' did T_{50} increase between 30 and 35°C. The T_{50} varied among cultivars from 4 to 10 days at temperatures between 20 and 35°C; the T_{50} was not calculated for 40°C, as germination percentages were equal 1% for all cultivars. Temperature and salinity interact to affect the germination of halophytes. Some species are more sensitive to changes in temperature (*Cressa cretica* and *Zygophyllum simplex*) than others (*Arthrocnemum macrostachyum* and *Suaeda fruticosa*) (Khan, 1999). Seed germination of *L. stocksii* was also influenced by temperature. The temperature cycle of 20 to 30°C to be the optimal temperature for germination and any increase or decrease in temperature inhibited germination. This inhibition progressively increased with salinity. Recruitment of *L. stocksii* in natural conditions through germination appears to take place after monsoon rains. Germination of halophyte seeds in subtropical coastal and inland salt marshes usually occurs after monsoon rains, which causes a reduction in temperature and lowering of soil salinity (Khan and Gul, 1998).

6. ALLELOCHEMICALS

Allelopathy is defined as the direct or indirect, harmful or beneficial, effects of one plant or another, including microorganisms, through the production of chemical compounds that escape into the environment. Allelochemicals are usually secondary plant products or waste products of the main metabolic pathways. The 'allelopathy' and 'competition' caused negative interference between plant species. One of the early important events during germination entails mobilization of reserve material stored in the endosperm/cotyledons. A part of these are respired and catabolised while other are translocated to growing embryonic axis. It is indicated that some toxic allelochemicals may be present in aqueous oat stubble extract, which inhibited the seed reserve mobilization and transportation of nutrients to growing embryonic axis (Gupta *et al.*, 2005).

The water soluble extract from *Aloe vera* leaves contains allelopathic chemicals promoting the initial seedling growth of four crops (lettuce, radish, turnip and rice) and re-rooting ability of rice and sweet potato seedlings. The extract did not significantly affect seed germination. Specially, appropriate extract concentrations significantly promoted the number of roots in lettuce and turnip and the root length of all the crops studied. The degree of promotion varied among the crops. In addition, all extracts significantly increased the root hair density in rice and sweet potato and the seedling height of lettuce, radish, turnip and rice (Dongzhi *et al.*, 2004). The presence of detoxification products in roots and shoots of oat (*Avena sativa*) and maize (*Zea mays*) caused inhibition of root growth in oat, timothy, crabgrass, ryegrass and lettuce. Naringin was found as a major allelopathic compound in oat straw that inhibited germination of *Chenopodium album.* Wild oat (weed) inhibited the growth of cereal crop, grain yield and seed quality in areas of intensive cereal production. Incorporation of *B. nigra* roots alone or roots and shoots in the soil reduced the emergence, plant height and dry weight per plant of *A. fatua* (wild oat). The changes in cellular processes including protein synthesis occur in allelopathic plants. Bhattacharjee *et al.* (2003) reported reduced germination, protein and insoluble carbohydrates, RNA, DNA contents as well as the activities of catalase and dehydrogeneases, when green gram seeds were treated with leaf extracts of *Casuarina equisetifolia* and *Ipomoea percaprae.* However, catalase activity in bean roots and membrane lipid peroxidation in tomato roots increases, when treated with aqueous leachates of *Callicarpa acuminata.* The oat root excretes a toxic compound that inhibits the growth of *Sinapis arvensis* and *Papaver rhoeas.* Aqueous root extract of oat up to 5% showed stimulatory effect on radicle development in mungbean at all days of germination. While stimulatory effect on plumule elongation was observed up to 2% aqueous root extract of oat application. Severe inhibition in radicle length was observed at early stages of growth as compared with that a later stage when seeds were germinated in 10% aqueous oat root extract. Length of plumule retarded by 30 and 22% at 2 DAS and 3 or 4 DAS when seeds were germinated in 10% aqueous oat root extract as compared to their controls. Oat root extract inhibited the mobilization and transportation of reserve materials from the cotyledons. This could be one of the reasons for reduction in embryo growth and also for stunted growth of embryonic axis emerged from the seeds germinated in presence of aqueous oat root extract (5 to 10%). This may be due to presence of toxic allelochemical in oat root. Oat root excreted scopoletin (fluorescent compound), a principal toxin inhibited the growth of *Sinapis arvensis* and *Papaver rhoeas.* Inhibition of *Chenopodium albus* seed germination due to presence of naringin, a major allelopathic compound found in oat straw. Aqueous stubble extract of oat was found to be more toxic as compared to that of root extract. It could be due to presence of more amount of toxic allelochemical in oat stubble.

Oat stubble extract was found to be more injurious for mobilization/ transportation of protein/amino acids as compared to that of oat root extract application during seed germination. This may be perhaps due to inhibition of protein synthesis/transport of amino acids from cotyledons to developing embryonic axes. One of the major role of storage proteins in germinating seeds is to supply amino acids for *de novo* biosynthesis of proteins both in the endosperms/cotyledons and the developing embryonic axis. Results indicate that protein mobilization particularly during initial stages of germination in response to aqueous extract of root/stubble of oat was suppressed. The observed impairment in mobilization of seeds storage proteins by aqueous extract of root/stubble of oat could have been either due to its interference with the synthesis or with the activities of protease enzymes. Initiation of protein synthesis seems not to be rate limiting for translation of either protein, and it appears that translation elongation and/or termination is a critical regulatory step in controlling storage protein accumulation in oat endosperm.

Allelopathic effects of aqueous extracts of root and stubble of oat (*Avena sativa* L.) on seedling growth and protein utilization of mungbean (*Vigna radiata* L.) during seed germination were variable depending upon the plant species. Aqueous extract of root and stubble of oat (*Avena sativa* L.) did not influence the seed germination of mung bean (*Vigna radiata* L.). Root extract upto 5% concentration, increased the radicle and plumule length of mung bean seedlings, but decreased at 10% level. Contrarily, stubble extract inhibited the growth of embryonic axes. The increasing concentrations of oat root and stubble extracts decreased the fresh and dry weights of radicle and plumule of mungbean seedlings yet increased that of cotyledons. Both aqueous extracts declined the protein content in radicle and plumule. However, it was more in cotyledons, thereby indicating the inhibition in seed reserve mobilization by root and stubble extract. The stubble extract proved more toxic than the root extract.

Laboratory and polyhouse experiments also determine the allelopathic effects of aqueous extracts of leaves of some common weeds (*Cyanodon dactylon, Cyperus rotundus, Eupatorium odoratum, Imperata cylindrica, Ipomoea sepiaria, Mikania cordata* and *Parthenium hysterophorus*) decayed for 120 and 240 h revealed that all these weeds inhibited the seed germination of tomato (*Lycopersicon esculentum* Mill) over control. Aqueous leaf extracts of *I. sepiaria* and *E. odoratum* were most inhibitory to seed germination of tomato. Leaf extracts of these weeds after 240 h decay recorded more adverse allelopathic effect on seed germination up to 7 DAS in petridishes, while up to 15 DAS in pots than weed extracts of 120 h decay. The extracts of *E. odoratum, I. Cylindrica, I. sepiaria* and *C. dactylon* completely inhibited the seed germination up to 5 DAS in soil-sand mixture in pots. *C. dactylon, I. sepiaria* and *E. odoratum* extracts reduced the shoot length of tomato seedlings. Weed extracts reduced the root length of seedlings over control and maximum inhibition was from *C. dactylon, I. sepiaria* and *E. odoratum*. It seems that water soluble extract from *Prosopis chilensis* leaves contains allelopathic chemicals, inhibitory to germination and initial seedling growth of rice. Leaf extracts of *P. chilensis* reduces the germination and early seedling growth (length and biomass) of crops or weeds support the present findings in the study. *P. juliflora* did not show its inhibitory effects on associated grasses under silvipastoral system during the early stages of tree growth. However, at later stages it exerted maximum allelopathic effect. *P. juliflora* contains germination and growth inhibitors in its leaf litter, which restricts the establishment and growth of ground cover and associated vegetation. The extract of *P. juliflora* completely inhibited the germination, decreased the root and shoot elongation, dry matter production and vigour index of parthenium. However, *P. juliflora* leaf extracts had minimum adverse effect on legumes seed germination (Sannigrahi and Chakrabortty, 2005).

Bioassay to find the allelopathic effects of aqueous leaf extracts (0, 1, 3, 5%) of *Prosopis chilensis* (Molina) Stuntz. on germination and seedling growth of rice

(*Oryza sativa* L. cv. 'Gurjari') revealed that the extracts of leaf significantly delayed and inhibited the germination and seedling growth. At 10 days after sowing, the higher concentration (5%) inhibited the germination, seedling length and dry matter as compared to control. Germination is the resumption of metabolic activity and growth by the seed tissues which starts with the imbibition of water and protrusion of embryonal axis. The initial step in germination is absorption of water by two processes *viz.*, imbibition and osmosis which caused activation of enzymes and increased metabolic activity.

The seeds imbibed with leachates/extracts of *Ocimum americanum* had delayed and inhibited germination in comparison to control. The inhibitory effect of prolonged soaking period in different concentrations of leachate/extract on seed germination might be due to imbalance in metabolism, regulated by various enzyme activities. The plasma membrane structure might have become denatured by the phytotoxins present in leachate/extract when the seeds were soaked for longer duration in leachates/extracts of different concentrations. Positive correlation was observed between increase in the concentration of leachate/extract and inhibition of germination. The allelochemicals inhibit the metabolic processes *i.e.*, cellular respiratory ability of the target plant and energy transfer responsible for ATP synthesis. The phytotoxins present in leachates/extracts of *Ocimum americanum* might have arrested the protease, α-amylase activity resulting in, inhibition of protein and starch breakdown which reduced the germination (Thapar and Singh, 2005). The leaf leachates were more phytotoxic in comparison with stem and root leachate. This might be due to synthesis and accumulation of more potential allelochemicals in leaf. The effect of allelochemicals was concentration dependent. The dry leaf extract was more inhibitory than fresh leaf leachate because the former contained more potential allelochemicals. Allelochemicals present in leachate/extract caused chlorotic and necrotic spots. The reduction in chlorophyll content was consistent with the chlorotic and necrotic spots. Allelochemicals resulted in breakdown of chlorophyll.

The different concentrations of leachates/extracts considerably caused the breakdown of chlorophyll which influenced the metabolism of carbohydrate and protein. The reduction of chlorophyll might be due to interaction of allelochemicals present in leaf leachate/extract with phosphorylation pathway or inhibition of activation of Mg^{2+} and ATPase activity or other metabolic activities. The allelochemicals reduced leaf expansion. Drooping of leaves and twigs decreases the photosynthetic area exposed to light and inhibition of photosynthetic process resulted in depletion of food reserves *i.e.*, carbohydrate and protein. Proteins play a pivotal role in biological processes. Allelochemicals inhibit rate of photosynthesis due to interference with water balance and chlorophyll contents which may result in reduction in the amount of protein. Protein serves as respiratory substrate when the carbohydrates and other compounds are inadequate. Thus, the depletion of chlorophyll and protein under the influence of higher concentration of allelochemicals might cause the death of *Parthenium*. The high content of allelochemicals in dry leaf extract of *Ocimum americanum* may be responsible for the inhibition of metabolic activities of *Parthenium* in the laboratory bioassay. *Ocimum americanum* may be used to replace *Parthenium* in nature as it has potential to inhibit seed germination and growth of target weed. Experiments were done to study the allelopathic effects of *Ocimum americanum* L. on germination and growth of *Parthenium hysterophorus*. Toxicity of *O. americanum* leachates on *P. hysterophorus* followed the order : dry leaf leachate > fresh leachate > stem leachate > root leachate.

It is reported that growth regulators stimulate and promote seed germination in a wide variety of tree crops. Treatment with growth regulators probably antagonizes the effect of growth inhibitory substances and also enhances the rate of metabolism during germination. GA_3 controls mobilization of starch which acts as a respiratory substrate, as a raw material, for synthesis of other growth regulators and increases the somatic pressure of cell leading to immediate enhancement in cell elongation (Sankhyan *et al.*, 2004).

Seeds of two blackgram (*Phaseolus mungo* L.) treated with different concentrations of aqueous leaf leachates of eight dominant weeds (*Ageratum conyzoides, Anagallis arvensis, Eclipta alba, Lippia nodiflora, Parthenium hysterophorus, Phyllanthus niruri, Pluchea lanceolata* and *Polygonum plebejum*) to assess their allelopathic effects on seed germination. Leachates of all weed spp. inhibited both the seed germination and seedling growth of both test varieties at 10% concentration. Increase in concentration of leachate was invariably associated with further decrease in germination of test cultivars, irrespective of weed species (Dongre *et al.*, 2004).

Active oxygen species (AOS) under normal conditions occur in plant cells as a product of many oxidoreduction reactions. Cells have efficient enzymatic and nonenzymatic antioxidant systems reducing the level of AOS. However, the action of different stress agents increased the AOS generation in cells. This may lead to an oxidative burst, *i.e.*, to a state, in which AOS generation will exceed the efficiency of antioxidant systems and can be a cause of oxidative damages of cellular constituents *i.e.*, lipid peroxidation, injuries of protein and strand breaks in DNA. Besides its harmful action, AOS at present are also ascribed to have a mediating role in signal transduction in cell response to a stress. Lipoxygenases (linoleate : oxygen oxidoreductases, EC 1.13.11.12, LOXs) are enzymes that catalyze hydroperoxidation of polyunsaturated fatty acids or their esters containing a *cis-cis*-1,4-pentadiene structure. During this reaction singlet oxygen is generated. In higher plants, the natural substrates for these enzymes are linolenic and linoleic acids. The physiological role of LOXs in plants is far from being clear. It has been reported that LOX activity increases in senescing and stressed tissues. One of many roles proposed for LOXs in plants is its involvement in the acclimatory response to different stress factors. The generation of AOS is one of the earliest responses of cells to various biotic and abiotic stress factors. According to the theory the formation of AOS is one of the important modes of phenolic activity. Almost any oxidation of phenolics can generate $O_2^{\bullet-}$ radical because reactive semiquinone can donate an electron to molecular oxygen. $O_2^{\bullet-}$ radical can generate additional radical species, including $^{\bullet}OH$ radical by reacting with H_2O_2. There are numerous enzymes in plant tissues, which possess the ability to oxidize phenolic compounds *i.e.*, peroxidases, laccases and tyrosinases. Besides damaging influence, AOS can act as signal molecules. The type of activity (damaging or signaling), depends on the level and kind of generated AOS. At their higher concentrations, cell death is initiated, while sublethal amounts of AOS can play a signaling role, what is probably a part of an acclimatory mechanism. Because H_2O_2 is relatively stable and diffusible through membranes, it is a perfect candidate to act as a signal molecule during stress responses. H_2O_2 and $O_2^{\bullet-}$ anion can induce different genes, in combination and separately, thereby giving more flexibility to the AOS signaling function. LOXs have a key role in signal transduction pathway in plant response to stress. Reaction products of these enzymes (fatty acid hydroperoxides) are metabolized enzymatically into compounds like thaumatin and jasmonates. Somewhat controversial is the connection of LOX activity with ethylene production. LOX-generated lipid hydroperoxides mediate the conversion of ACC to ethylene in a strictly chemical system and also enhance ethylene production in tissue. On the other hand, a rapid increase of LOX activity was observed in rye seedlings as a result of etephon or ethylene precursor – ACC (1-aminocyclopropane-1-carboxylic acid) treatment (Politycka and Bednarski, 2004).

Table 9: The effects of various Apiaceae and their essential oil components. Each was observed for inhibition of lettuce seed germination (G), root elongation (R), and root hair elongation (H). Lower case means mild, proportional.

Fennel (*Foeniculum vulgare*) GR	Cumin (*Cuminum cyminum*) GH	Anise (*Pimpinella anisum*) R
anethole fenchone GR allyl anisole camphene a-phellandrene a-pinene p-methoxyphenyl acetone	cuminaldehyde cuminyl alcohol GH p-cymene a-pinene a-phellandrene	anethole p-allylanisole p-methoxyphenyl acetone anisaldehyde R
Caraway (*Carum carvi*) G	Coriander (*Coriandrum sativum*) H	Dill (*Anethum graveolens*) G
s-carvone G carveol 2,3 butanedione furaldehyde limonene	linalool rh geraniol a-pinene p-cymene limonene	s-carvone G limonene a-phellandrene a-pinene
	Celery (*Apium graveolens*) G	
	guaiacol G palmitic acid limonene	

Lettuce seed germination assays were conducted to test the allelopathic potential of Apiaceae seeds. The effects of the individual components of the essential oils of the seeds were also assayed. The percent germination, root length, and root hair length were measured in each assay. The Apiaceae seeds: fennel, cumin, caraway, celery, dill, anise and coriander were found to inhibit germination of lettuce at various seed densities. Fennel and cumin seeds were shown to also inhibit root and root hair elongation respectively. In fennel seeds, fenchone is present in sufficient amount to inhibit germination and root elongation, but not root hair elongation of lettuce seedlings. Similarly in cumin seeds, cuminyl alcohol inhibited germination and root hair elongation, but not root elongation. Similar germination results were observed for the other Apiaceae species providing evidence that the Apiaceae family generally has allelochemicals present in seed oils which inhibit germination and root and root hair elongation of lettuce. Reduced root and root hair lengths increase the chance of desiccation in seedlings before establishment and delay growth as water uptake is required for seedling growth. All Apiaceae seeds examined inhibit seed germination of lettuce at high densities. At lower seed densities, fennel and cumin seeds inhibit root and root hair elongation, respectively. Fenchone inhibits root elongation in lettuce at low doses when germination is high with normal root hair elongation. Fenchone is the likely allelochemical in fennel seed (Table 9). Cuminyl alcohol inhibits root hair elongation in lettuce at low doses when germination is high with normal root elongation. Cuminyl alcohol is the likely allelochemical in cumin seed (Lamoureux and Koning, 1998).

REFERENCES

Andrews, D. L, D. M. MacAlpine, J. R.. Johnson, P .M. Kelley, B. G. Cobb and M. C. Drew. 1994. Differential induction of mRNAs for the glycolytic and ethanolic fermentative pathways by hypoxia and anoxia in maize seedlings. *Plant Physiol.,* **106:** 1575-1582.

Bajji, M., J. M. Kinet and S. Lutts. 2002. Osmotic and ionic effects of NaCl on germination, early seedling growth and ion content of *Atriplex halimus* (Chenopodiaceae). *Can. J. Bot.*, **80**: 297-304.

Ball, M. C. and S. M. Pidsley. 1995. Growth responses to salinity in relation to distribution of two mangrove species, *Sonneratia alba* and *S. lanceolata,* in northern Australia. *Functional Ecology,* **9**: 77–85.

Barret-Lennard, E. D, A. D. Robson and H. Greenway. 1982. Effect of phosphorous deficiency and water deficiency on phosphatse activity from wheat leaves. *J. Exp. Bot.* **33**: 682-693.

Bartels, D and Salamini, F. 2001. Desiccation tolerance in the resurrection plant *Craterostigma plantagineum.* A contribution to the study of drought tolerance at the molecular level. *Plant Physiol.,* **127:** 1346-1353.

BassiriRad, H and M. M. Coldwell. 1992. Root growth, osmotic adjustment and NO^{3-} uptake during and after a period of drought in *Artemesia tridentata. Aust. J. Plant Physiol.,* **19:** 493-500.

Basu, R. and B. Ghosh. 1991. Polyamines in various rice (*Oryza sativa*) genotypes with respect to sodium chloride salinity. *Physiol. Plant.,* **82**: 575-581.

Bhattacharjee, A., R. K. Bhakat, U. K. Kamp and R. K. Das. 2003. An investigation on allelopathic action of *Casuarina equisitifolia* and *Ipomoea pescaprae. Environmental Ecology,* **21:** 283-289.

Bliss, R.D., K.A. Platt-Alois, W.W. Thomson. 1986. The inhibitory effect of NaCl on barley seed germination. *Plant Cell Environ.,* **9**: 727-733.

Bohnert, H. J, D. E. Nelson and R. G. Jensen. 1995. Adaptations to environmental stresses. *Plant Cell* ,**7:** 1099-1111.

Bozcuk, S. 1981. Effect of kinetin and salinity on germination of tomato, barley and cotton seeds. *Ann. Bot.*, **48**: 81-84.

Cardemil, L and A. Reinero. 1982. Changes in *Araucaria araucana* (Mol.) Koch seed reserves during germination and early seedling growth. *Canadian J. Bot.,* **60:** 1629-1638.

Cardemil L, R. Lozada and M. Corteas. 1990. Sucrose uptake and anatomical studies in relation with sucrose uptake of *Araucaria araucana* cotyledons. *Plant Physiol. Biochem.,* **28:** 761-772.

Chong C. and B. B. Bible. 1995. Germination and emergence. In: M. Pessarakali, ed. Handbook of Plant and Crop Physiology, New York: Marcel Dekker, pp: 85-146.

Claes, B., R. Dekeyser, R. Villarroell, M. Van Den Bulke, G. Bauw, M. Van Montagu and A. Caplan 1990. Characterization of a rice gene showing organ-specific expression in response to salt stress and drought. *Plant Cell,* **2**: 19-27.

Close, T.J., R. D. Fenton, and F. Moonan. 1993. A view of plant dehydrins using antibodies specific to the carboxy terminal peptide. *Plant Mol. Biol.* **23**: 279-286.

Corchette, P. and H. Guerra. 1986. Effect of NaCl and polyethylene glycol on solute content and glycosidase activity during germination on lentil seeds. *Plant Cell Environ.,* **9**: 589-593, 1986.

Cutler, J. M., K. W Shahan and P. L. Steponkus. 1980. Influences of water potentials and osmotic adjustment on leaf elongation in rice. *Crop Sci.,* **20**: 314-318.

Delgado, M. J., F. Ligero and C. Lluch. 1994. Effects of salt stress on growth and nitrogen fixation by pea, faba bean, common bean and soybean plants. *Soil Biol. Biochem.,* **26**: 371-376.

Dewar, J., J. R. N. Taylor and P. Berjak. 1998 . Changes in selected plant growth regulators during germination in sorghum. *Seed Sci. Res.*, **8**: 1-8.

Dongre, P. N., A.K. Singh and K.S. Chaube. 2004. Allelopathic effects of weed leaf leachates on seed germination of blackgram (*Phaseolus mungo* L.). *Allelopathy Journal,* **14:** 65-70.

Dongzhi Lin, E., Tsuzuki, Y. Sugimoto, Yanjun Dong, M. Matsuo and H. Terao. 2004. Allelopathic effects of aqueous *Aloe vera* leaf extracts on selected crops. *Allelopathy J.,* **13**: 67-74.

Drew, M. C. 1997. Oxygen deficiency and root metabolism: injury and acclimation under hypoxia and anoxia. *Ann. Rev. Plant Physiol. Plant Mol. Biol.,* **48**: 223-250.

Duan, D., X. Liu, M. A. Khan and B. Gul 2004. Effects of salt and water stress on the germination of *chenopodium glaucum* l., seed. *Pak. J. Bot.*, **36**: 793-800.

Ehsanpour, A. A, and F. Amini 2003. Effect of salt and drought stress on acid phosphatase activities in alfalfa (*Medicago sativa* L.) explants under in vitro culture. *Afr. J. Biotechnol.,* **2**:133-135.

Elson, M. K., R. D. Morse, D. D. Wolf, and D. H. Vaughn. 1992. High temperature inhibition of seed germination and seedling emergence of broccoli. *Hort. Tech.,* **2**:417-419.

Fincher, G. B. 1989. Molecular and cellular biology association with endosperm mobilization in germination cereal grains. *Ann. Rev. Plant Physiol. Plant Mol. Biol.*, **40**: 305-346.

Garcia, A. B., J. de Almeida Engler, S. Iyer, T.Gerats, M .Van Monatgu, and A. B.Caplan 1997. Effects of osmo-protectants upon NaCl in rice. *Plant Physiol.*, **115**: 159-169.

Garciarrubio, A., J. P Legaria and A. A Covarrubias 2003. Abscisic acid inhibits germination of mature Arabidopsis seeds by limiting the availability of energy and nutrients. *Planta,* **2**: 182-187.

Gill K.S. and O.S Singh 1985. Effect of salinity on carbohydrate metabolism during paddy (*Oryza sativa*) seed germination under salt stress condition. *J. Exp. Biol.,* **23**: 384-386.

Gill P. K, A. D Sharma, P. Singh and S. S Bhullar. 2003. Changes in germination, growth and soluble sugar contents of *Sorghum bicolor* (L.) Moench seeds under various abiotic stresses. *Plant Growth Regul.,* **40**: 157-162.

Gomes Filho, E., J.T. Prisco, F.A. de Pavia Campos, J.E. Filho.1983. Effect of NaCl salinity *in vivo* and *in vitro* on ribonuclease activity of *Vigna unguiculata* cotyledons during germination. *Physiol. Plant.,* **59**: 183-188.

Guerrier, G. 1986. Tolerance for NaCl during germination of seeds, capacity to accumulating and transporting K^+ and Ca^{2+} in a salt sensitive species (tomato) and a tolerant one (cabbage). *Seed Sci. Technol.*, **14**: 15-31.

Gupta, K., V. Jain, I.S. Solanki and Tulika. 2005. Effect of aqueous extracts of root and stubble of oat (*Avena sativa* L.) on seedling growth and protein utilization in mung bean (*Vigna radiata* L.) *Alleopathy J.*, **16**: 279-278.

Hardegree, S. P., and W. E. Emmerich. 1990. Partitioning water potential and specific salt effects on seed germination of four grasses. *Ann. Bot.,* **,66**: 587–595.

Huang, J. and R. E. Redman.1995. Salt tolerance of Hordeum and Brassica species during germination and early seedling growth. *Can. J. Plant Sci.*, **75**: 815-819.

Katerji, N., J. W. van Hoorn, A. Hamdy, F. Karam and M. Mastrorilli 1994. Effect of salinity on emergence and on water stress and early seedling growth of sunflower and maize. *Agricult. Water Manag.,* **26**: 81-91.

Kennedy, R. A., T. C.Fox, D. Dybiec, and M. E.Rumpho, 1990. Biochemical adaptations to anoxia in rice and *Echinochloa* seeds. *In:* Taylorson R. B, (Ed.) Recent advances in the development and germination of seeds. New York: Plenum Press, 151-163.

Kennedy, R. A., M. E Rumpho and T. C. Fox. 1987. Germination physiology of rice and rice weeds: metabolic adaptations to anoxia. *In:* Crawford R. M. M, (Ed.) Plant life in aquatic and amphibious habitats. Oxford, UK: Blackwell Scientific, 193-203.

Khajeh-Hosseini, M., A. A. Powell and I. J. Bingham. 2003. The interaction between salinity stress and seed vigour during germination of soybean seeds. *Seed Sci. and Technol.,* **31**: 715-725.

Khan, M.A. and Y. Rizvi, 1993. Effect of salinity, temperature and growth regulators on the germination wheat embryos. *Ann. Bot.,* **72**: 97-101.

Khan, M. A. and I. A. Ungar. 1984. The effect of salinity and temperature on the germination of polymorphic seeds and growth of *Artriplex triangularis* Wild. *Am. J. Bot.* **71** : 484-489.

Khan, M.A., and I.A. Ungar. 1997. Germination responses of the subtropical annual halophyte *Zygophyllum* simplex. *Seed Sci. Technol.,* **25**: 83-91.

Khan, M.A. 1999. Comparative influence of salinity and temperature on the germination of subtropical halophytes. *In* Halophyte uses in different climates. I: ecological and ecophysiological studies. Vol. 13. (Ed.) H. Lieth, M. Moschenko, M. Lohman, H.W. Koyro, and A. Hamdy. Backhuys Publishers, Leiden. pp. 77–88.

Khan, M. A., and B. Gul. 1998. High salt tolerance in germinating dimorphic seeds of *Arthrocnemum indicum. Int. J. Plant Sci.,* **159**: 826–832.

Lamoureux, S. and R. Koning.1998. The Allelopathic Potential of Apiaceae seeds upon germination of lettuce. Poster presented at the 1998 (Biological Honor Society) Northeast District 1 Convention at Worcester.

Lin C. C. and C. H. Kao. 1996. Disturbed ammonium assimilation is associated with growth inhibition of roots in rice seedlings caused by NaCl. *Plant Growth Reg.,* **18**: 233-238.

Lin, C. C. and C. H. Kao. 1995. Levels of endogenous polyamines and NaCl inhibited growth of rice seedlings. *Plant Growth Reg.,* **17**: 15-20.

Mohammad, S. and D. N. Sen.1990. Germination behaviour of some halophytes in Indian desert. *Int. J. Exp. Biol.,* **28**: 545-549.

Nascimento, W.M., D.J. Cantliffe, and D.J. Huber. 2000. Thermotolerance in lettuce seeds: Association with ethylene and endo-ß-mannanase. *J. Amer. Soc. Hort. Sci.* **125**:518-524.

Niazi, M. L. K., K. Mehmood, and K. A. Malik. 1987. Salt tolerance studies in different cultivars of barley (*Hordeum vulgare* L.). *Pak. J. Bot.,* **19**: 17-27.

Noor, M. and M. A. Khan. 1995. Factors affecting the germination of summer and winter seeds of *Halopyrum mucronatum* under salt stress. In: M.A. Khan, I.A. Ungar, (Ed.) Biology of Salt Tolerant Plants. Department of Botany, University of Karachi, Pakistan, PP: 99-106.

Okcu, G., M. D. Kaya, and M. Atak. 2005. Effects of salt and drought stresses on germination and seedling growth of pea *(Pisum sativum L.). Turk. J. Agric. For.,* **29**: 237-242.

Olmos, E. and E . Hellin. 1997. Cytochemical localization of Atpase plasma membrane and acid phosphatase by cerium based in a salt-adapted cell line of *Pisum sativum. J. Exp. Bot.* **48**: 1529-1535.

Pennypacker B. W., K. L. Leath, W. L Stout and R. R Hill. 1990. Techniques for stimulating field drought stress in greenhouse. *Agron. J.,* **82**: 951-957.

Petruzzelli L., M. T. Melillo, T. B. Zache, B. Marano and G. Taranto. 1991. The sensitivity of germinating Triticum durum L. kenels to saline environments. *Seed Sci. Res.,* **1**:105-111.

Petruzzelli L., M. T. Melillo, T. B. Zache and G. Taranto.1991. Physiological and ultrastructural changes in isolated wheat embryos during salt and osmotic shock. *Ann. Bot.,* **69**: 25-31.

Politycka, B. and W. Bednarski 2004. Oxidative burst and lipoxygenase activity induced by hydroxycinnamic acids in cucumber roots. *Allelopathy J.,* **14**: 187-196.

Poljakoff-Mayber A., G. F. Somers, E. Werker, and J. L. Gallagher, 1994. Seeds of *Koteletzkya virginica* (Malvaceae): their structure, germination and salt tolerance. *Am. J. Bot.,* **81**: 54-59.

Prado F. E, J. A Gonzalez, M. Gallardo, M. Moris, C. Boero and A. Kortsarz. 1995. Changes in soluble carbohydrates and invertase activity in *Chenopodium quinoa* (quinoa) developed for saline stress during germination. *Curr. Top. Phytol.*, **14**:1-5.

Pujol, J. A., J. F. Calvo and L. R. Díaz. 2000. Recovery of germination from different osmotic conditions by four halophytes from southeastern Spain. *Ann Bot.*, **85**: 279-286.

Ramagopal S.1990. Inhibition of seed germination by salt and its subsequent effect on embryonic protein synthesis in barley. *J. Plant Physiol.*, **136**: 621-625.

Reinhardt D. H. and T. L. Rost. 1995. Salinity accelerates endodermal development and induces an exodermis in cotton seedling roots. Environ. Exp. Bot., **35**: 563-574.

Rogers M. E., C. L. Noble, G. M. Halloran and M. E. Nicolas. 1995. The effect of NaCl on the germination and early seedling growth of white clover (*Trifolium repens* L.) populations selected for high and low salinity tolerance. *Seed Sci. and Technol.*, **23**: 277-287.

Sadeghian, S. Y. and N. Yavari. 2004. Effect of water-deficit stress on germination and early seedling growth in sugar beet. *J. Agron. and Crop Sci.*, **190**: 138-144.

Sankhyan, H. P., R. N.Sehgal, and N. P. Bhrot. 2004. Effect of growth regulators and salt solutions on seed germination and seedling vigour of Seabuckthorn (*Hippophae tibetana* Schlecht). *Seed Res.*, **32**: 55-57.

Sannigrahi, A. K. and S. Chakrabortty, 2005. Allelopathic effects of weeds on germination and seedling growth of tomato. *Allelopathy J.*, **16**: 289-294.

Schwanz, P., C. Picon, P. Vivin, E. Dreyer, M. J. Guehi and A. Polle. 1996. Response of antioxidative systems to drought stress in pendunculate oak and maritime pine as modulated by elevated CO_2. *Plant Physiol.*, **110**: 393-402.

Sharma, A. D., M. Thakur, M. Rana and K. Singh. 2004. Effect of plant growth hormones and abiotic stresses on germination, growth and phosphatase activities in *Sorghum bicolor* (L.) Moench seeds. *African J. Biotech.*, **3**: 308-312.

Sharp R. E. 1996. Regulation of plant growth responses to low water potential. Hort. Science **31**: 36-38.

Siddique M. R. B., A. Hamid and M. Islam. 2000. Drought stress effects on water relations of wheat. *Bot. Bull. Acad. Sin.*, **41**: 35-39.

Sinha, A. and S. R. Gupta.1982 .Effect of osmotic tension and salt stress on germination of three grass species. *Plant Soil*, **69**: 13-19.

Small, J. G. C. and Y. Gutterman. 1992. Effect of sodium chloride on prevention of thermodormancy, ethylene and protein synthesis and respiration in grand rapids lettuce seeds. *Physiol. Plant.*, **84**: 35-40.

Smith S. E., and A. K. Dobrenz. 1987. Seed age and salt tolerance at germination in alfalfa. *Crop Sci.*, **27** : 1053-1056.

Szabo-Nagy A.G, G. Galiba and E. Erdei. 1992. Induction of soluble phosphatases under ionic and non-ionic osmotic stress in wheat. *J.Plant Physiol.*, **140**: 329-633.

Thakur P, and A. Thakur. 1993. Influence of triacontanol and mixtalol during plant moisture stress in *Lycopersicon esculentum* cultivars. *Plant Physiol. Biochem.* **31**: 433-439.

Thapar, R. and N. B. Singh. 2005. Effect of *Ocimum americanum* leachates on seed germination and seedling growth of *Parthenium hysterophorus* L. *Allelopathy J.*, **16**: 301-308.

Tipton J.L.1988. Drought and salinity stress effects on guayule seedling emergence. *J. Am. Soc. Hort. Sci.*, **113**: 129-133.

Ungar, I. A. 1991. Ecophysiology of vascular halophytes, CRC Press, Boca Raton, FL.

Ungar, I. A. 1996. Effect of salinity on seed germination, growth and ion accumulation of *Atriplex patula* (Chenopodiaceae). *Amer. J. Bot.*, **83**: 604-607.

Ushimaru, T., S. Kanematsu, M. Katayama, and H. Tsuji. 2001. Antioxidative enzymes in seedlings of *Nelumbo nucifera* germinated under water stress. *Physiol. Plant.*, **112**: 39-46.

Valenti G. S., L. Melane, O. Orsi and F. Riveros. 1992. Anatomical changes in *Prosopis cineraria* (L.) Druce seedlings at different levels of NaCl salinity. *Ann. Bot.*, **70**: 399-404.

Wahid A., E. Rasul and A. R. Rao. 1997. Germination responses of sensitive and tolerant sugarcane lines to sodium chloride. *Seed Sci. Technol.*, **25**: 465-470.

Wahid A., I-ul-H. Javed, I. Ali, A. Baig and E. Rasul. 1998. Short time incubation of sorghum caryopses in sodium chloride levels: changes in some pre and post germination physiological parameters. *Plant Sci*., **124**:126-131.

Wijte A. H. B. M. and J. L. Gallagher. 1976. Effect of oxygen availability and salinity on early life history stages of salt marsh plants. I. Different germination strategies of Spartina alterniflora and Phragmites australis (Poaceae). *Amer. J. Bot.,* **83**: 1337-1342.

Yan X, H. Liao, C. T. Melanie, E. B. Steve and J. P Lynch. 2001. Induction of a major leaf acid phosphatase does not confer adaptation to low phosphorous availability in common bean. *Plant Physiol.,* **125**: 1901- 1911.

Yuying S., W. Suomin and C. Yaming. 1999. The effect of saline condition on the germination and its recovery of pasture species. *Acta Prataculturae Sinic.*, **8**:54- 60.

Yokoishi, T. and S. Tanimoto. 1994. Seed germination of the halophyte *Suaeda japonica* under salt stress. *J Plant Res.,* **107**: 385-388.

6

Physiology and Biochemistry of Seed Germination and Vigour Due to Seed Management Technologies

S.Kavitha[1], M.Bhaskaran[2] and K.Vanangamudi[3]
[1]*Central Institute of Cotton Research, Coimbatore – 641 003*
[2]*National Seed Research and Training Centre, Varanasi – 221 106*
[3]*Adhiparasakthi Agricultural College, G. B. Nagar, Kalavai – 632 506*

1. INTRODUCTION

Seed is one of the major inputs in agriculture and decides the future of crops. The rationale for preplant seed treatment is to mobilize the seeds own resources and to augment them with external resources to maximum improvement in stand establishment and yield. Physiological and non-physiological treatments are needed to remove several soil and hydrological constraints. Physiological seed treatments that improve seed performance are based primarily on seed hydration and dehydration. Among several non-physiological seed treatments, coating or pelleting can indirectly improve seed germination and stand establishment. These seed management techniques are expected to help to sustain the productivity of crops even under stress situations. In this chapter, the effect of seed hardening and film coating for better germination and seedling vigour of various agricultural and horticultural crops is discussed.

2. PRESOWING INVIGOURATION TREATMENT TO ENHANCE PHYSIOLOGY AND BIOCHEMISTRY OF GERMINATION AND VIGOUR

2.1. Screening of hardening and coating treatments

To screen the best pre-sowing invigouration treatment for better germination and vigour of maize, sorghum, ragi, blackgram, cotton, tomato and marigold, the seeds were given hardening treatments with different chemicals, growth regulators and botanicals, and coated with carbendazim + imidacloprid + seed coating polymer + biofertilizer.

2.1.1. Maize

Dry seed (Control); seed hardening with 2% KH_2PO_4 + carbendazim @ 2g kg^{-1} + imidacloprid @ 1ml in 5ml of water + *Azospirillum* @ 60g kg^{-1} of seed; Seed hardening with 2% KCl + carbendazim @ 2g kg^{-1} + imidachloprid @ 1ml in 5ml of water + *Azospirillum* @ 60g kg^{-1} of seed; seed hardening with 1% prosopis leaf extract + carbendazim @ 2g kg^{-1} + imidachloprid @ 1ml in 5ml of water + *Azospirillum* @ 60g kg^{-1} of seed, Seed film coating with pink polykote @ 3g kg^{-1} seed + carbendazim @ 2g kg^{-1} + imidachloprid @ 1ml in 5ml of water + *Azospirillum* @ 60g kg^{-1} of seed; seed hardening with 2% KH_2PO_4 followed by seed film coating with pink polykote @ 3g kg^{-1} seed + carbendazim @ 2g kg^{-1} + imidachloprid @ 1ml in 5ml of water + *Azospirillum* @ 60g kg^{-1} of seed; Seed hardening with 2% KCl followed by seed film coating with pink polykote @ 3g kg^{-1} seed + carbendazim @ 2g kg^{-1} + imidachloprid @ 1ml in 5ml of water + *Azospirillum* @ 60g kg^{-1} of seed and seed hardening with 1% prosopis leaf extract followed by seed film coating with pink polykote @ 3g kg^{-1} seed + carbendazim @ 2g kg^{-1} + imidachloprid @ 1ml in 5ml of water + *Azospirillum* @ 60g kg^{-1} of seed

2.1.2. Sorghum

Dry seed (Control); seed hardening with 2% KH_2PO_4 + carbendazim @ 2g kg^{-1} + imidachloprid @ 1ml in 5ml of water + *Azospirillum* @ 40g kg^{-1} of seed; seed hardening with 2% KCl + carbendazim @ 2g kg^{-1} + imidachloprid @ 1ml in 5ml of water + *Azospirillum* @ 40g kg^{-1} of seed; seed hardening with 1% prosopis leaf extract + carbendazim @ 2g kg^{-1} + imidachloprid @ 1ml in 5ml of water + *Azospirillum* @ 40g kg^{-1} of seed; seed film coating with pink polykote @ 3g kg^{-1} seed + carbendazim @ 2g kg^{-1} + imidachloprid @ 1ml in 5ml of water + *Azospirillum* @ 40g kg^{-1} of seed; seed hardening with 2% KH_2PO_4 followed by seed film coating with pink polykote @ 3g kg^{-1} seed + carbendazim @ 2g kg^{-1} + imidachloprid @ 1ml in 5ml of water + *Azospirillum* @ 40g kg^{-1} of seed; seed hardening with 2% KCl followed by seed film coating with pink polykote @ 3g kg^{-1} seed + carbendazim @ 2g kg^{-1} + imidachloprid @ 1ml in 5ml of water + *Azospirillum* @ 40g kg^{-1} of seed and seed hardening with 1% prosopis leaf extract followed by seed film coating with pink polykote @ 3g kg^{-1} seed + carbendazim @ 2g kg^{-1} + imidachloprid @ 1ml in 5ml of water + *Azospirillum* @ 40g kg^{-1} of seed

2.1.3. Ragi

Dry seed (Control); seed hardening with 0.1 ppm brassinolide + carbendazim @ 2g kg^{-1} + imidachloprid @ 1ml in 5ml of water + *Azospirillum* @ 60g kg^{-1} of seed; seed hardening with 2% KH_2PO_4 + carbendazim @ 2g kg^{-1} + imidachloprid @ 1ml in 5ml of water + *Azospirillum* @ 60g kg^{-1} of seed; seed hardening with 1% prosopis leaf extract + carbendazim @ 2g kg^{-1} + imidachloprid @ 1ml in 5ml of water + *Azospirillum* @ 60g kg^{-1} of seed; seed film coating with pink polykote @ 3g kg^{-1} seed + carbendazim @ 2g kg^{-1} + imidachloprid @ 1ml in 5ml of water + *Azospirillum* @ 60g kg^{-1} of seed; seed hardening with 0.1 ppm brassinolide followed by seed film coating with pink polykote @ 3g kg^{-1} seed + carbendazim @ 2g kg^{-1} + imidachloprid @ 1ml in 5ml of water + *Azospirillum* @ 60g kg^{-1} of seed; seed hardening with 2% KH_2PO_4 followed by seed film coating with pink polykote @ 3g kg^{-1} seed + carbendazim @ 2g kg^{-1} + imidachloprid @ 1ml in 5ml of water + *Azospirillum* @ 60g kg^{-1} of seed and seed hardening with 1% prosopis leaf extract followed by seed film coating with pink polykote @ 3g kg^{-1} seed + carbendazim @ 2g kg^{-1} + imidachloprid @ 1ml in 5ml of water + *Azospirillum* @ 60g kg^{-1} of seed.

2.1.4. Black gram

Dry seed (Control); seed hardening with 100 ppm $ZnSO_4$ + carbendazim @ 2g kg^{-1} + imidachloprid @ 1ml in 5ml of water + *Rhizobium* @ 30g kg^{-1} of seed; seed hardening with 0.5% $FeSO_4$ + carbendazim @ 2g kg^{-1} + imidachloprid @ 1ml in 5ml of water + *Rhizobium* @ 30g kg^{-1} of seed; seed hardening with 1% prosopis leaf extract + carbendazim @ 2g kg^{-1} + imidachloprid @ 1ml in 5ml of water + *Rhizobium* @ 30g kg^{-1} of seed; seed film coating with black polykote @ 3g kg^{-1} seed + carbendazim @ 2g kg^{-1} + imidachloprid @ 1ml in 5ml of water + *Rhizobium* @ 30g kg^{-1} of seed; seed hardening with 100 ppm $ZnSO_4$ followed by seed film coating with black polykote @ 3g kg^{-1} seed + carbendazim @ 2g kg^{-1} + imidachloprid @ 1ml in 5ml of water + *Rhizobium* @ 30g kg^{-1} of seed; seed hardening with 0.5% $FeSO_4$ followed by seed film coating with black polykote @ 3g kg^{-1} seed + carbendazim @ 2g kg^{-1} + imidachloprid @ 1ml in 5ml of water + *Rhizobium* @ 30g kg^{-1} of seed and seed hardening with 1% prosopis leaf extract followed by seed film coating with black polykote @ 3g kg^{-1} seed + carbendazim @ 2g kg^{-1} + imidachloprid @ 1ml in 5ml of water + *Rhizobium* @ 30g kg^{-1} of seed.

2.1.5. Cotton

Dry seed (Control); seed hardening with 2% KCl + carbendazim @ 2g kg^{-1} + imidachloprid @ 1ml in 5ml of water + *Azospirillum* @ 80g kg^{-1} of seed; seed hardening with 100 ppm GA_3 + carbendazim @ 2g kg^{-1} + imidachloprid @ 1ml in 5ml of water + *Azospirillum* @ 80g kg^{-1} of seed; seed hardening with 1% prosopis leaf extract + carbendazim @ 2g kg^{-1} + imidachloprid @ 1ml in 5ml of water + *Azospirillum* @ 80g kg^{-1} of seed; seed film coating with pink polykote @ 3g kg^{-1} seed + carbendazim @ 2g kg^{-1} + imidachloprid @ 1ml in 5ml of water + *Azospirillum* @ 80g kg^{-1} of seed; seed hardening with 2% KCl followed by seed film coating with pink polykote @ 3g kg^{-1} seed + carbendazim @ 2g kg^{-1} + imidachloprid @ 1ml in 5ml of water + *Azospirillum* @ 80g kg^{-1} of seed; seed hardening with 100 ppm GA_3 followed by seed film coating with pink polykote @ 3g kg^{-1} seed + carbendazim @ 2g kg^{-1} + imidachloprid @ 1ml in 5ml of water + *Azospirillum* @ 80g kg^{-1} of seed and seed hardening with 1% prosopis leaf extract followed by seed film coating with pink polykote @ 3g kg^{-1} seed + carbendazim @ 2g kg^{-1} + imidachloprid @ 1ml in 5ml of water + *Azospirillum* @ 80g kg^{-1} of seed.

2.1.6. Tomato

Dry seed (Control); seed hardening with 2% KH_2PO_4 + carbendazim @ 2g kg^{-1} + imidachloprid @ 1ml in 5ml of water + *Azospirillum* @ 50g kg^{-1} of seed; seed hardening with 1% KCl + carbendazim @ 2g kg^{-1} + imidachloprid @ 1ml in 5ml of water + *Azospirillum* @ 50g kg^{-1} of seed; seed hardening with 1% prosopis leaf extract + carbendazim @ 2g kg^{-1} + imidachloprid @ 1ml in 5ml of water + *Azospirillum* @ 50g kg^{-1} of seed; seed film coating with red polykote @ 3g kg^{-1}

seed + carbendazim @ 2g kg^{-1} + imidachloprid @ 1ml in 5ml of water + *Azospirillum* @ 50g kg^{-1} of seed; seed hardening with 2% KH_2PO_4 followed by seed film coating with red polykote @ 3g kg^{-1} seed + carbendazim @ 2g kg^{-1} + imidachloprid @ 1ml in 5ml of water + *Azospirillum* @ 50g kg^{-1} of seed; seed hardening with 1% KCl followed by seed film coating with red polykote @ 3g kg^{-1} seed + carbendazim @ 2g kg^{-1} + imidachloprid @ 1ml in 5ml of water + *Azospirillum* @ 50g kg^{-1} of seed and seed hardening with 1% prosopis leaf extract followed by seed film coating with red polykote @ 3g kg^{-1} seed + carbendazim @ 2g kg^{-1} + imidachloprid @ 1ml in 5ml of water + *Azospirillum* @ 50g kg^{-1} of seed.

2.1.7. Marigold

Dry seed (Control); seed hardening with 0.75% KNO_3 + carbendazim @ 2g kg^{-1} + imidachloprid @ 1ml in 5ml of water + *Azospirillum* @ 400g kg^{-1} of seed; seed hardening with 100 ppm thiourea + carbendazim @ 2g kg^{-1} + imidachloprid @ 1ml in 5ml of water + *Azospirillum* @ 400g kg^{-1} of seed; seed hardening with 1% prosopis leaf extract + carbendazim @ 2g kg^{-1} + imidachloprid @ 1ml in 5ml of water + *Azospirillum* @ 400g kg^{-1} of seed; seed film coating with pink polykote @ 3g kg^{-1} seed + carbendazim @ 2g kg^{-1} + imidachloprid @ 1ml in 5ml of water + *Azospirillum* @ 400g kg^{-1} of seed; seed hardening with 0.75% KNO_3 followed by seed film coating with pink polykote @ 3g kg^{-1} seed + carbendazim @ 2g kg^{-1} + imidachloprid @ 1ml in 5ml of water + *Azospirillum* @ 400g kg^{-1} of seed; seed hardening with 100 ppm thiourea followed by seed film coating with pink polykote @ 3g kg^{-1} seed + carbendazim @ 2g kg^{-1} + imidachloprid @ 1ml in 5ml of water + *Azospirillum* @ 400g kg^{-1} of seed and seed hardening with 1% prosopis leaf extract followed by seed film coating with pink polykote @ 3g kg^{-1} seed + carbendazim @ 2g kg^{-1} + imidachloprid @ 1ml in 5ml of water + *Azospirillum* @ 400g kg^{-1} of seed

2.2. Development of designer seed

From the results and findings obtained from the hardening and film coating treatments, the best treatment was selected and coated using DAP and micronutrient mixture. Again this hardened and carbendazim + imidachloprid + DAP + micronutrient mixture + seed coating polymer + biofertilizer film coated seeds were evaluated under laboratory along with the best performed hardening and carbendazim + imidachloprid + seed coating polymer + biofertilizer film coating treatment and untreated seeds (control). The hardened and film coated seeds were evaluated under laboratory experiments. The treatments tried for each crop seeds are furnished herein.

2.2.1. Maize

Dry seed (Control T_1), Seed hardening with 2% KH_2PO_4 + seed film coating with pink polykote @ 3g kg^{-1} + carbendazim @ 2g kg^{-1} + imidachloprid @ 1ml in 5ml of water + *Azospirillum* @ 60g kg^{-1} of seed (T_2) and seed hardening with 2% KH_2PO_4 + seed film coating with pink polykote @ 3g kg^{-1} + carbendazim @ 2g kg^{-1} + imidachloprid @ 1ml in 5ml of water + 30g DAP + 20g micronutrient mixture + *Azospirillum* @ 60g kg^{-1} of seed (T_3) (Plate 1).

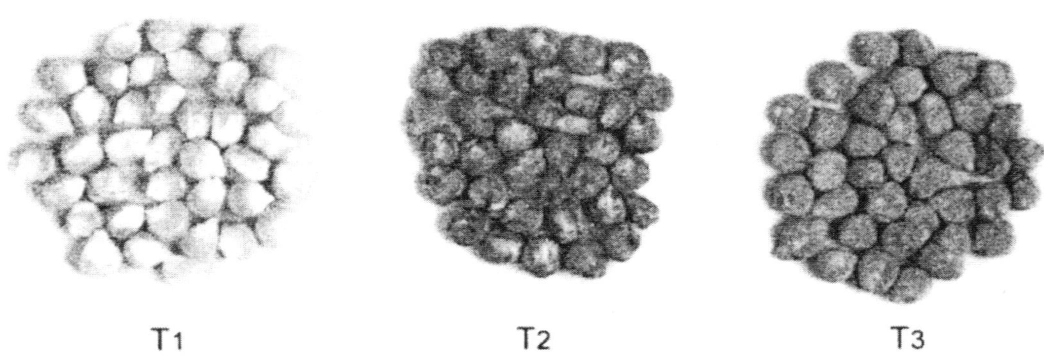

Plate 1: Hardened and pelleted seeds of maize

2.2.2. Sorghum

Dry seed (Control, (T_1)), seed hardening with 2% KH_2PO_4 + seed film coating with pink polykote @ 3g kg^{-1} + carbendazim @ 2g kg^{-1} + imidachloprid @ 1ml in 5ml of water + *Azospirillum* @ 40g kg^{-1} of seed (T_2); and seed hardening with 2% KH_2PO_4 + seed film coating with pink polykote @ 3g kg^{-1} + carbendazim @ 2g kg^{-1} + imidachloprid @ 1ml in 5ml of water + 30g DAP + 20g micronutrient mixture + *Azospirillum* @ 40g kg^{-1} of seed (T_3) (Plate 2).

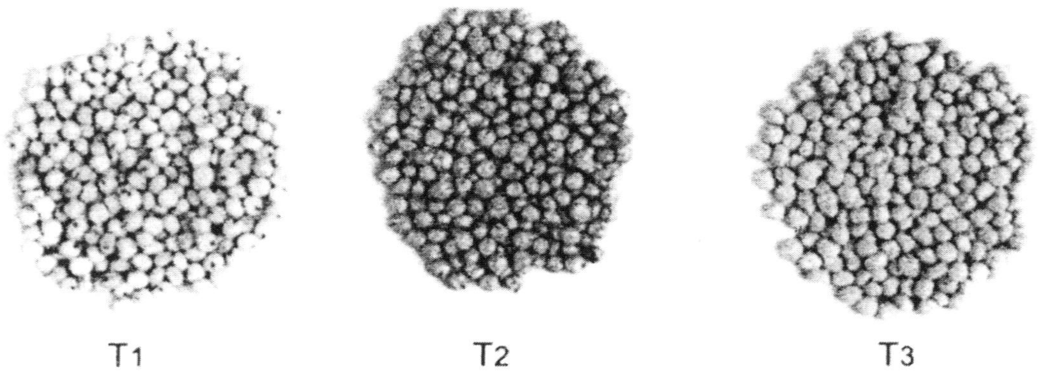

Plate 2: Hardened and film coated seeds of sorghum

2.2.3. Ragi

Dry seed (Control,(T_1)), seed hardening with 0.1 ppm brassinolide + seed film coating with pink polykote @ 3g kg^{-1} seed + carbendazim @ 2g kg^{-1} + imidachloprid @ 1ml in 5ml of water + *Azospirillum* @ 60g kg^{-1} of seed (T_2); and seed hardening with 0.1 ppm brassinolide + seed film coating with pink polykote @ 3g kg^{-1} seed + carbendazim @ 2g kg^{-1} + imidachloprid @ 1ml in 5ml of water + 30g DAP + 20g micronutrient mixture + *Azospirillum* @ 60g kg^{-1} of seed (T_3) (Plate 3).

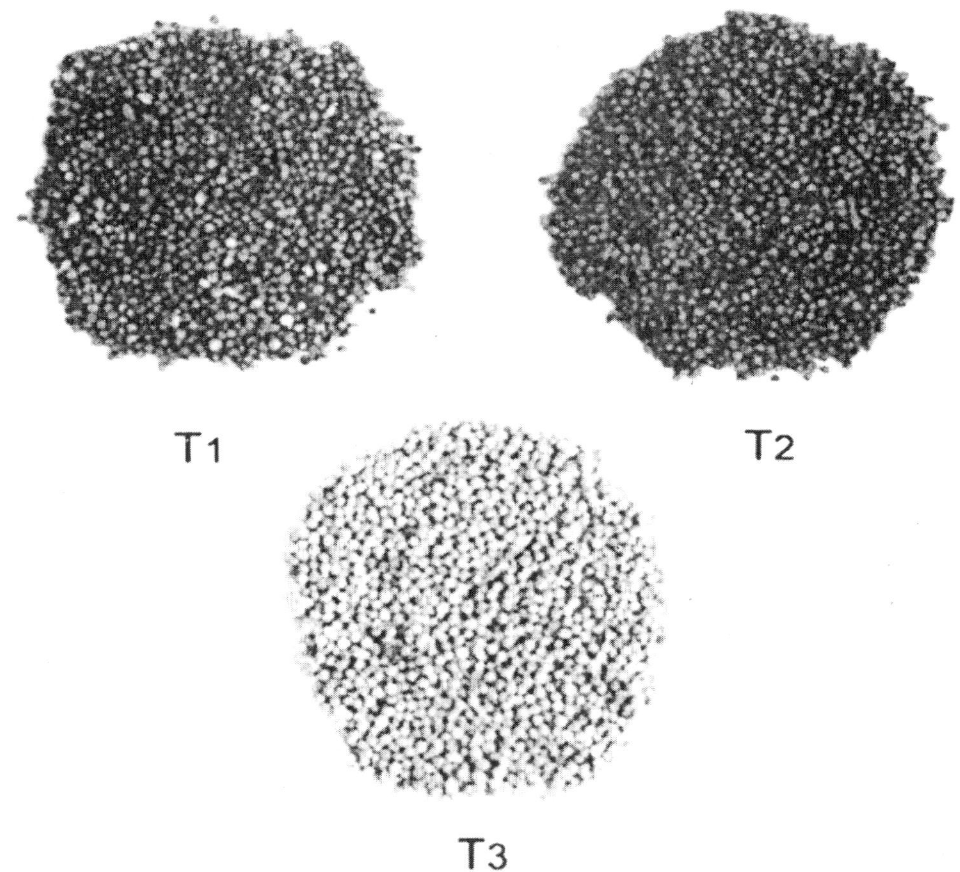

Plate 3: Hardened and film coated seeds of ragi

2.2.4. Black gram

Dry seed (Control) (T_1), Seed hardening with 100 ppm $ZnSO_4$ + Seed film coating with black polykote @ 3g kg^{-1} seed + carbendazim @ 2g kg^{-1} + imidachloprid @ 1ml in 5ml of water + *Rhizobium* @ 30g kg^{-1} of seed (T_2); and seed hardening with 100 ppm $ZnSO_4$ + Seed film coating with black polykote @ 3g kg^{-1} seed + carbendazim @ 2g kg^{-1} + imidachloprid @ 1ml in 5ml of water + 30g DAP + 20g micronutrient mixture + *Rhizobium* @ 30g kg^{-1} of seed (T_3) (Plate 4).

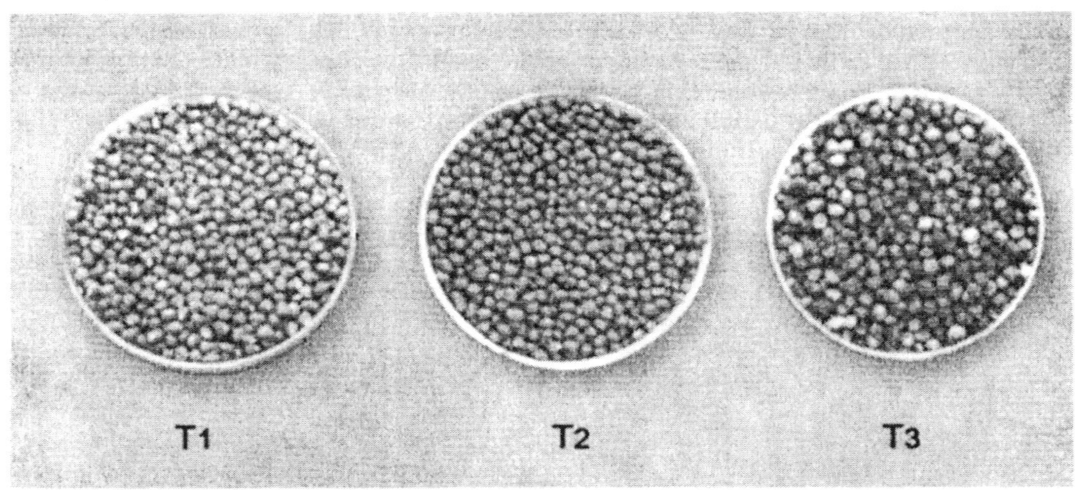

Plate 4. Hardened and film coated seeds of blackgram

2.2.5. Cotton

Dry seed (Control) (T_1), Seed hardening with 2% KCl + Seed film coating with pink polykote @ 3g kg^{-1} seed + carbendazim @ 2g kg^{-1} + imidachloprid @ 1ml in 5ml of water + *Azospirillum* @ 80g kg^{-1} of seed (T_2); and seed hardening with 2% KCl + Seed film coating with pink polykote @ 3g kg^{-1} seed + carbendazim @ 2g kg^{-1} + imidachloprid @ 1ml in 5ml of water + 30g DAP + 20g micronutrient mixture + *Azospirillum* @ 80g kg^{-1} of seed (T_3) (Plate 5).

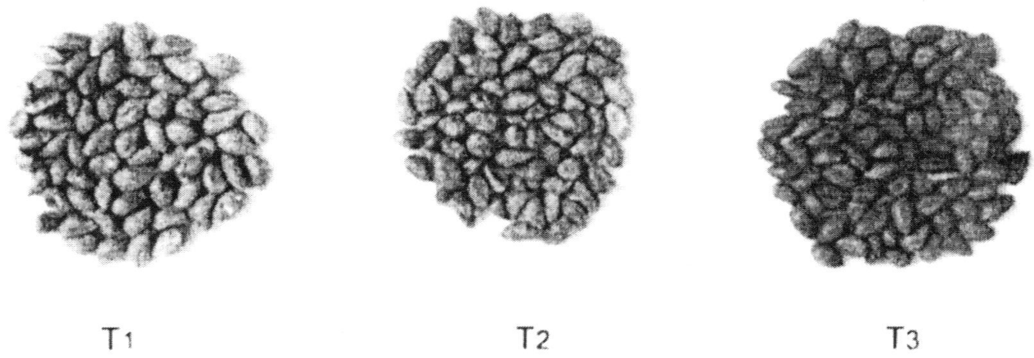

Plate 5: Hardened and film coated seeds of cotton

2.2.6. Tomato

Dry seed (Control) (T_1), seed hardening with 2% KH_2PO_4 + Seed film coating with red polykote @ 3g kg^{-1} seed + carbendazim @ 2g kg^{-1} + imidachloprid @ 1ml in 5ml of water + *Azospirillum* @ 50g kg^{-1} of seed (T_2); and seed hardening with 2% KH_2PO_4 + Seed film coating with red polykote @ 3g kg^{-1} seed + carbendazim @ 2g kg^{-1} + imidachloprid @ 1ml in 5ml of water + 30g DAP + 20g micronutrient mixture + *Azospirillum* @ 50g kg^{-1} of seed (T_3) (Plate 6).

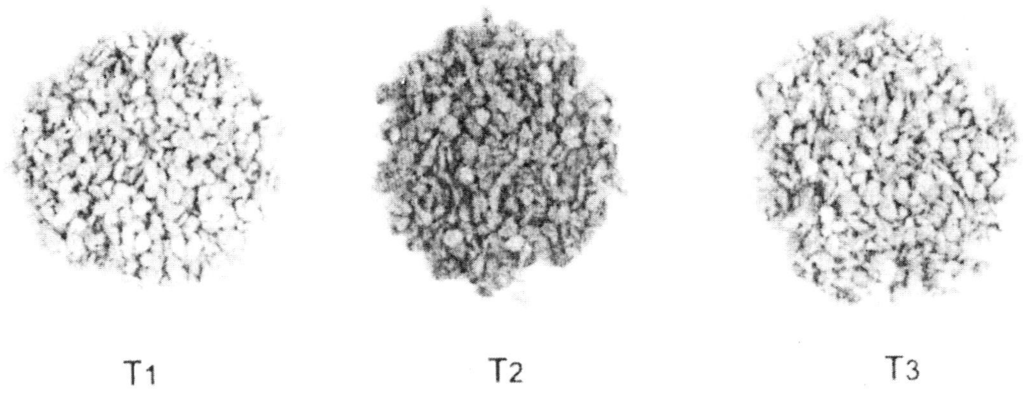

Plate 6: Hardened and pelleted seeds of tomato

2.2.7. Marigold

Dry seed (Control) (T_1), seed hardening with 0.75% KNO_3 + Seed film coating with pink polykote @ 3g kg^{-1} seed + carbendazim @ 2g kg^{-1} + imidachloprid @ 1ml in 5ml of water + *Azospirillum* @ 400g kg^{-1} of seed (T_2); and seed hardening with 0.75% KNO_3 + Seed film coating with pink polykote @ 3g kg^{-1} seed + carbendazim @ 2g kg^{-1} + imidachloprid @ 1ml in 5ml of water + 30g DAP + 20g micronutrient mixture + *Azospirillum* @ 400g kg^{-1} of seed (T_3).

The physiological and biochemical parameters like speed of germination, germination, root length, shoot length, drymatter production and vigour index and biochemical parameters such as dehydrogenase activity, α-amylase activity and total chlorophyll content of seedlings were evaluated under laboratory condition.

3. EFFECT OF HARDENING, FILMCOATING AND PELLETING ON THE PHYSIOLOGY AND BIOCHEMISTRY OF SEED GERMINATION AND VIGOUR

3.1. Speed of germination

The presoaking treatment involving wetting and drying was developed by Henckel (1964) for imparting resistance to drought conditions. The beneficial effect of seed hardening has been reported long back by many workers (Chinoy *et al.,* 1968; Dawson, 1965; Krishnasastry *et al.,* 1969; Seshidhar *et al.,* 1978). In the present study, the speed of germination was higher in 2% KH_2PO_4 hardened and carbendazim + imidachloprid + seed coating polymer + biofertilizer film coated seeds in maize, sorghum and tomato; 0.1 ppm brassinolide hardened and 2g carbendazim + 1ml imidachloprid + 3g pink polykote + 60g *Azospirillum* film coated seeds in ragi; 100 ppm ZnSO4 hardened and 2g carbendazim + 1ml imidachloprid + 3g black polykote + 30g *Rhizobium* film coated seeds in blackgram; and 2% KCl hardened and 2g carbendazim + 1ml imidachloprid + 3g pink polykote + 80g *Azospirillum* film coated seeds in cotton compared to control and hardened and carbendazim + imidachloprid + DAP + micronutrient mixture + polykote + biofertilizer film coated seeds. The quick initiation of germination in hardened and film coated seeds might be due to advancement in the physiology of germination during soaking (Kamalam, 1991; Punithavathi, 1997).

Pre-sowing hardening of crop seeds with water or solution of growth regulating compounds to induce early germination was also reported by many researchers (Joseph and Nair, 1989; Vanangamudi and Kulandaivelu, 1989). Polymer due to its hydrophilic property, helped seeds to emerge at a faster rate in laboratory condition with the available moisture. This was supported by West *et al.* (1985). The reduction in germination rate of carbendazim + imidachloprid + DAP + micronutrient mixture + polykote + biofertilizer film coated seeds might be due to the restriction imposed by the coating material, which causes slow absorption of moisture and so delayed germination.

3.2. Germination

Maize, sorghum, ragi, balckgram, cotton, tomato and marigold seeds are known to respond well to pre-sowing seed hardening, which resulted in early germination and production of vigorous seedlings. In the present investigation, all the hardening and film coating treatments showed a significant improvement in seed germination and vigour over control.

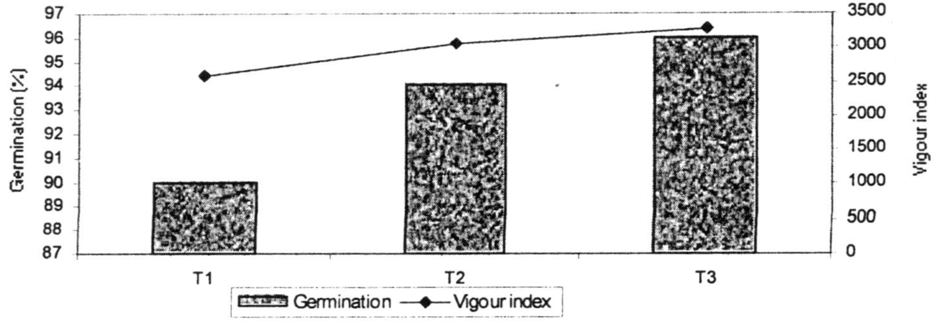

Fig. 1: Seed hardening and film coating on germination and vigour index in maize *cv.* CO 1 under laboratory condition

The maize, sorghum and tomato seed hardened with 2% KH_2PO_4 + film coated with DAP + micronutrient mixture + carbendazim + imidachloprid + polykote + *Azospirillum* recorded higher germination, which accounted for an increase of 8.0, 12.0 and 7.0 per cent, respectively over control (untreated seeds) (Fig. 1). The improvement in germination by 1% KH_2PO_4 in wheat was reported by Paul and Choudhury (1993). Encouraging results were obtained by Mehrotra *et al.* (1968), Misra and Pal (1980) in wheat and Gray *et al.* (1977) in parsnips. The 'K' element being more permeable through seed coat might have promoted the germination (Kohli *et al.*, 1992).

In the case of ragi, 0.1 ppm brassinolide hardened and 2g carbendazim + 1ml imidachloprid + 30g DAP + 20g micronutrient mixture + 3g pink polykote + 60g *Azospirillum* film coated seed recorded higher germination (98%). The improvement in germination by this treatment was 5.0 per cent over control (Fig. 2). This was supported by Fujii and Saka (2001) in rice and Misra and Reddy (1985) in wheat. Seed treatment with brassinolide might have stimulated the production of auxin and ethylene, which have positive influence on seed germination as reported by Clouse and Zurek (1991).

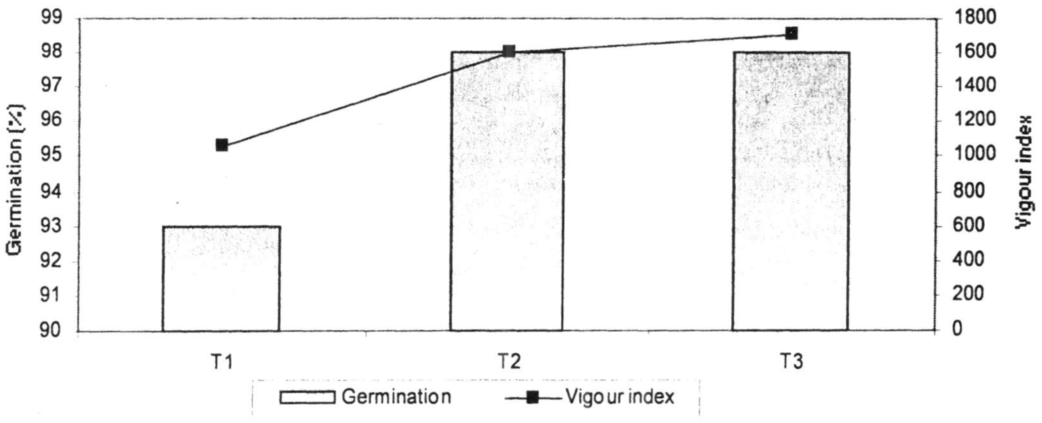

Fig.2: Seed hardening and film coating on germination and vigour index in ragi *cv.* CO 14 under laboratory condition

In pulses, the micronutrients play an important role. It was evident in the present study that the blackgram seeds hardened with 100 ppm $ZnSO_4$ and film coated with 2g carbendazim + 1ml imidachloprid + 30g DAP + 20g micronutrient mixture + 3g black polykote + 30g *Rhizobium* recorded the highest germination (Fig. 3), which was 5.0 per cent higher than control seeds. The probable reason for higher germination could due to greater hydration of colloids, higher viscosity and elasticity of protoplasm, increase in bound water content, lower water deficit, more efficient root system (May *et al.,* 1962) and increased metabolic activity (Joseph and Nair, 1989).

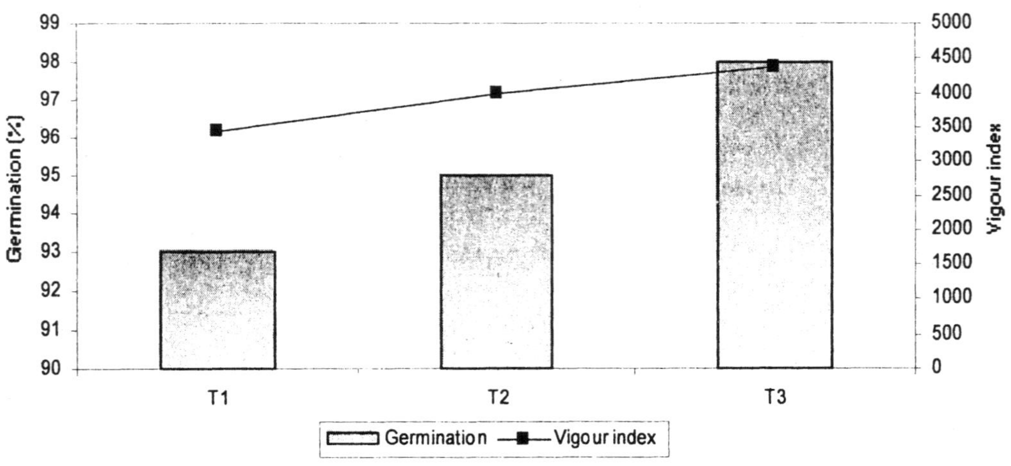

Fig.3: Seed hardening and film coating on germination and vigour index in black gram *cv.* VBN 4 under laboratory condition

In cotton, 2% KCl hardened and 2g carbendazim + 1ml imidachloprid + 30g DAP + 20g micronutrient mixture + 3g pink polykote + 80g *Azospirillum* film coated seeds gave higher germination (96%), which accounted for an increase of 6.0 per cent over control seeds. The improvement in germination could be ascribed to seed hardening as much as the early phase of germination. Pre-sowing hardening of seed is one of the methods, which result in modifying physiological and biochemical nature of seed so as to get the characters that are favourable for drought resistance (Henckel, 1964).

In the case of marigold, 0.75% KNO_3 hardened and 2g carbendazim + 1ml imidachloprid + 30g DAP + 20g micronutrient mixture + 3g pink polykote + 400g *Azospirillum* film coated seeds recorded higher percentage of germination (81%), which was 9.0 per cent higher over control seeds (Fig. 4). The superiority of KNO_3 may be ascribed to its role in making less oxygen available for the citric acid cycle and thereby enhancing the ambient oxygen level (Bewley and Black, 1982). Increased germination following treatment with KNO_3 has been documented in several species like *Carica papaya* (Nagao and Furutani, 1986), *Citrus karma* (Singh *et al.,* 1979) and soybean (Kalavathi, 1985). The promotion of germination by KNO_3 soaking has been suggested to be due to the conversion of nitrate to nitrite within the seed (Hendricks and Taylorson, 1974). Nitrite has been indicated to induce germination by enhancing pentose phosphate pathway through inhibition of catalase and increased oxidation of $NADPH_2$ (Hendricks and Taylorson, 1975; Roberts, 1973). It is also plausible that enhanced germination due to KNO_3 was the out come of quantitative and qualitative shifts in protein synthesis induced by it (Leadem, 1987).

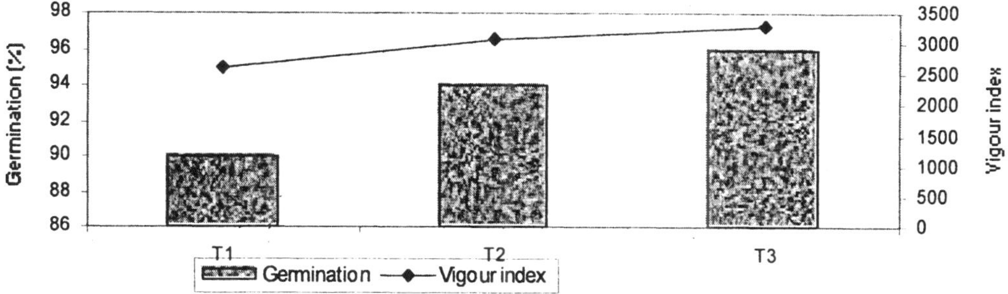

Fig.4: Seed hardening and film coating on germination and vigour index in African Marigold *cv.* MDU 1 under laboratory condition

The germination results revealed that in all the crops, seeds film coated with polykote, carbendazim, imidachloprid, nutrients, *viz.,* DAP and micronutrient mixture and biofertilizers gave higher germination over control. Similar beneficial effect of germination due to polymer coating alone and with nutrients was reported by Zinin and Imamaliev (1984) in cotton; Konstantinov and Kolarova (1988) and Min (1988) in Lucerne; Eremeeva *et al.,* (1988) in cotton; Angamuthu (1991) in minor millets and Williams *et al.* (1999).

3.2. Root and shoot length

In maize, sorghum and tomato, the 2% KH_2PO_4 hardened and carbendazim + imidachloprid + DAP + micronutrient mixture + polykote + *Azospirillum* film coated seeds improved the seedling growth (Plate 7 and 8).

Plate 7: Seedling vigour of maize due to hardening and film coating

Similar results were reported by Nagaraj (1996) in paddy and Punithavathi (1997) in ragi. Potassium present in the chemicals improve oxidative phosphorylation, utilization of sugars along with pentose phosphate cycle, formation of $NADPH_2$ and ribose-5-phosphate, synthesis of mitochondria (Kursanov *et al.,* 1965) and the activity of ATP ase enzyme (Okanenko and Bershteir, 1966). This beneficial effect of nutrients might be the reason for better seedling growth. It might also be due to enhanced metabolic activity and earliness in germination as stated by Joseph and Nair (1989).

Plate 8: Seedling vigour of hardened and film coated sorghum seed

In ragi, 0.1 ppm brassinolide hardened and 2g carbendazim + 1ml imidachloprid + 30g DAP + 20g micronutrient mixture + 3g pink polykote + 60g *Azospirillum* film coated seed recorded higher seedling growth of 28.2 and 42.3 per cent imcreased root and shoot growth, respectively over untreated control (Plate 9). This is in accordance with the findings of Hayat *et al.* (2001) in wheat. The probale reason for elite seedling growth might be due to cell elongation and cell division and enhancement in enzyme activities induced by brassinolide. In the case of blackgram, the effect of 100 ppm $ZnSO_4$ hardening and 2g carbendazim + 1ml imidachloprid + 30g DAP + 20g micronutrient mixture + 3g black polykote + 30g *Rhizobium* film coating was also much evident in growth of the seedling with 27.4 and 17.9 per cent increased root and shoot growth, respectively over control (Plate 10). Similar beneficial effect on seedling growth was reported by Vanangamudi and Kulandaivelu (1989) and Jegathambal (1996) in sorghum; Shanmugasundaram and Kannaiyan (1989) in pearl millet; Arjunan and Srinivasan (1989) in groundnut; Rangasamy *et al.* (1993) in cowpea; Bhaskaran (1995) and Nagaraj (1996) in rice; Paul and Choudhury (1993) in wheat and Thandapani and Subarayalu (1986) in cotton. The increased seedling growth observed in this treatment might be due to greater early vigour and higher percentage of germination of the seeds that had reached autotropic stage well in advance than others (Jayaraj, 1977).

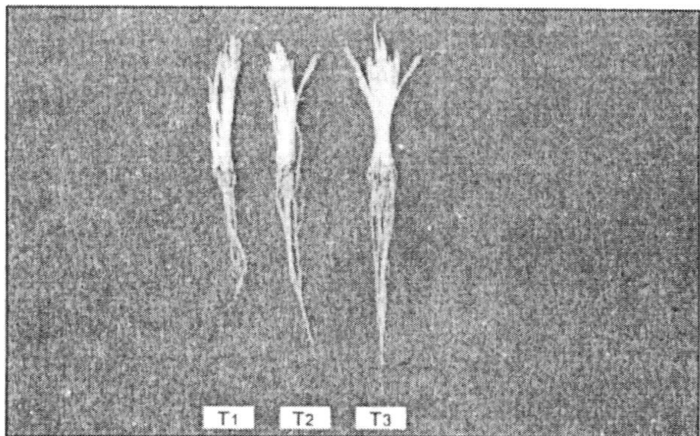

Plate 9: Seedling vigour of ragi due to hardening and film coating

Plate 10: Seedling vigour of blackgram seed hardened and film coated

In the case of cotton, 2% KCl hardened and 2g carbendazim + 1ml imidachloprid + 30g DAP + 20g micronutrient mixture + 3g pink polykote + 80g *Azospirillum* film coated seeds also showed increased seedling growth (Plate 11). The improvement in seedling growth manifested was indeed great *viz.*, 16.0 and 20.9 per cent in root and shoot length, respectively over control. The improvement in root and shoot length of seedling due to treatment might be probably due to enhanced metabolic activity, earliness in germination and seedling growth (Kamalam and Nair, 1989). In marigold, 0.75% KNO_3 hardened and 2g carbendazim + 1ml imidachloprid + 30g DAP + 20g micronutrient mixture + 3g pink polykote + 400g *Azospirillum* film coated seeds registered higher seedling growth compared to control and 0.75% KNO_3 hardened and 2g carbendazim + 1ml imidachloprid + 3g pink polykote + 80g *Azospirillum* film coated seeds.

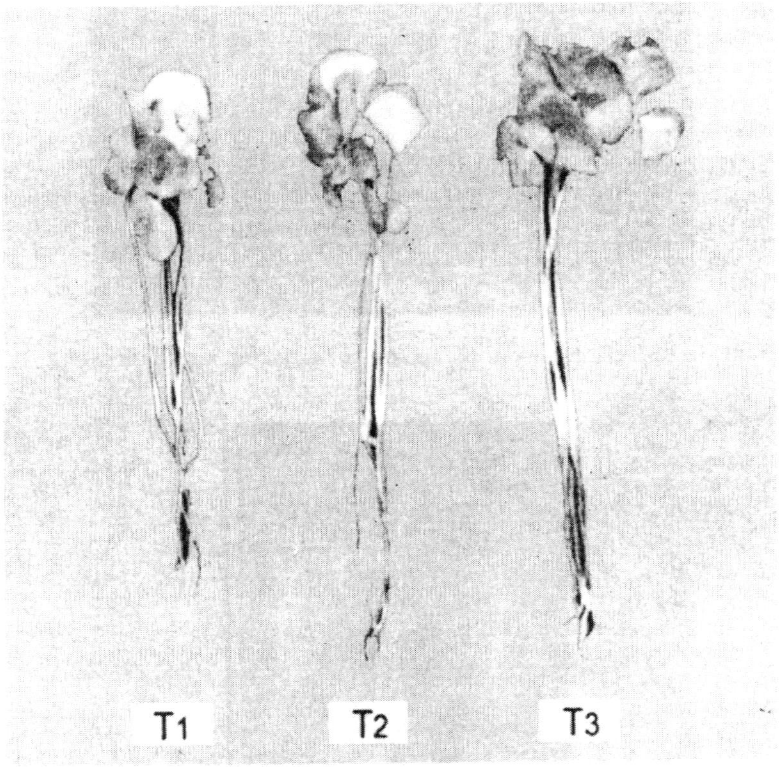

Plate 11: Seedling vigour of hardened and film coated cotton seed

In all the crops, seedling growth was increased for the seeds film coated with polykote, carbendazim, imidachloprid, DAP, micronutrient mixtures and biofertilizers. The beneficial effect of nutrients and polymer might be the reason for better seedling growth. It might also be due to enhanced metabolic activity resulted in early germination as stated by Kamalam and Nair (1989). Similar beneficial effect of seedling growth due to seed coating was reported by Saraswathi (1994), Srimathi *et al.* (2000) and Ramesh *et al.* (2001).

3.3. Drymatter production and vigour index

The drymatter production and vigour index were more in 2% KH_2PO_4 hardened and carbendazim + imidachloprid + DAP + micronutrient mixture + polykote + *Azospirillum* film coated seeds of maize, sorghum and tomato. This treatment registered the highest vigour index of 37.6, 31.4 and 53.2 per cent in maize, sorghum and tomato, respectively over untreated control seeds (Fig. 5). It might be due to higher metabolic activity, earliness in germination and seedling growth. Similar beneficial effect were also reported by Jegathambal (1996) in sorghum, Punithavathi (1997) in ragi, Kalavathi (1985) in soybean, Sanchez *et al.* (2001) in tomato, pepper and cucumber and Srimathi *et al.* (2002) in tomato.

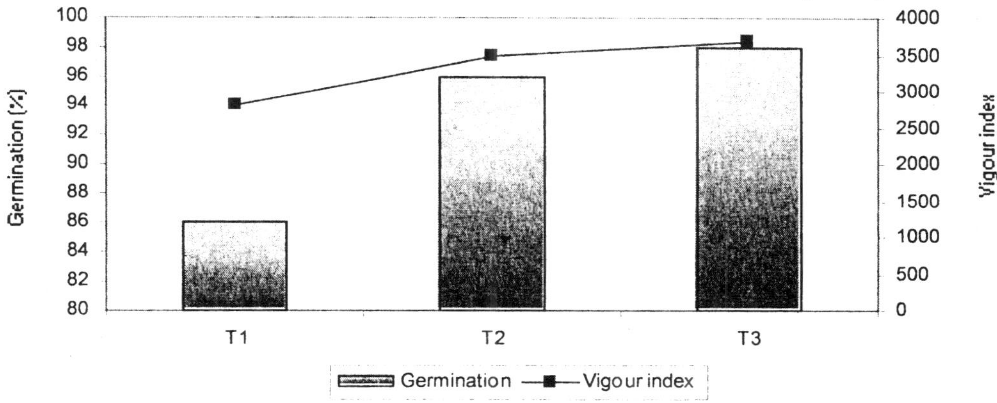

Fig.5: Seed hardening and film coating on germination and vigour index in sorghum *cv.* CO(S)28 under laboratory condition

In ragi, among the different treatments, seed hardening with 0.1ppm brassinolide and film coating with 2g carbendazim + 1ml imidachloprid + 30g DAP + 20g micronutrient mixture + 3g pink polykote + 60g *Azospirillum* improved the drymatter production and vigour index. It might be due to increase in seedling growth and germination. This was also supported by Paul and Choudhury (1993) in wheat, Fasui *et al.* (1996) in maize, and Vigneshwari (2002) in ragi.

In blackgram, 100 ppm $ZnSO_4$ hardened and 2g carbendazim + 1ml imidachloprid + 30g DAP + 20g micronutrient mixture + 3g black polykote + 30g *Rhizobium* film coated seedlings possessed 13.0 and 27.2 per cent higher drymatter production and vigour index, respectively than control seeds. Patagundi *et al.* (1997) highlighted that $ZnSO_4$, as a constituent of dehydrogenase enzyme, activator of other enzymes, and a constituent for biosynthesis of IAA and amino acid that favoured the synthesis of protein. Sulphur in $ZnSO_4$ also increased the levels of vitamins, biotins and thiamins and its coenzymes in seeds and enhanced the growth rate of seedlings. Likewise due to the active participation in seed physiology $ZnSO_4$ might have enhanced the seedling growth, drymatter production and vigour index.

In cotton, the drymatter production and computed vigour index values were higher in 2% KCl hardened and 2g carbendazim + 1ml imidachloprid + 30g DAP + 20g micronutrient mixture + 3g pink polykote + 80g *Azospirillum* film coated seeds followed by 2% KCl hardened and 2g carbendazim + 1ml imidachloprid + 3g pink polykote + 80g *Azospirillum* film coated seeds over control. In marigold, seed hardened with 0.75% KNO_3 and 2g carbendazim + 1ml imidachloprid + 30g DAP + 20g micronutrient mixture + 3g pink polykote + 400g *Azospirillum* film coated seeds performed well by recording higher drymatter production and vigour index values compared to untreated control.

Thus, the results clearly indicated the promising effect of hardening and film coating in maize, sorghum, ragi, blackgram, cotton, tomato and marigold for improving the germination, which involves early hydration and imbibition. It could also be argued that during soaking, seeds would become physiologically advanced by carrying out some of the initial steps of germination that resulted in improved germination, seedling length, drymatter accumulation and vigour index.

In this study, drymatter production and vigour index were improved by seed coating with polykote, carbendazim, imidachloprid, DAP, micronutrient mixture and biofertilizers in all the crops over uncoated control seeds. The beneficial effect of nutrients and polymer might be the reason for increased drymatter production and vigour. It might also be due to enhanced metabolic activity resulted in early germination as stated by Kamalam and Nair (1989). Similar beneficial effect of seedling growth and vigour was reported by Saraswathi (1994) and Ramesh *et al.* (2001).

Seed film coating claimed to play a vital role in modern agriculture for precision planting and to supplement nutrition through seed for uniform and vigorous seedling growth. Miller and Scooter (1979) reported the possibility of getting desired population through coating. Improved germination, seedling growth, drymatter production and vigour index by pelleting was reported in peas, soybean, cucurbits, cotton and rice by Loneragan and Mayer (1955), Iswaran and Chhonkar (1971), Sabir-Ahamed (1989), Balaji (1990) and Rathinavel and Dharmalingam (1999).

Because of *Rhizobium* seed treatment, there are possibilities for increased drymatter production. This is in consonance with the results of Rao and Sharma (1980) in soybean and blackgram and Namdeo *et al.* (1989) in chickpea. Increased vigour index by biofertilizer treatment was reported by Nirmala Devi *et al.* (1995) in chillies. Prabakaran and Srinivasan (1995) in redgram obtained increased vigour index by coir-dust application with *Rhizobium* seed inoculation, because of slow release of certain nutrients to seed and better symbiosis by coir-dust. *Azospirillum* added in the coating material stimulated the seedling growth and vigour index. Similar increase in seedling growth and vigour index due to increased seed cytokinin was reported by Khan (1980).

3.4. Dehydrogenase and α-amylase activity

Dehydrogenase activity and α-amylase activity also increased in hardened + film coated seeds. In the present investigation, activity of these enzymes was

Table 1: Effect of seed hardening and film coating on biochemical parameters of maize, sorghum and ragi under laboratory condition

Treatments	Maize			Sorghum			Ragi		
	Dehydrogenase activity (OD value)	α-amylase activity (mm)	Total chlorophyll (mg g^{-1})	Dehydrogenase activity (OD value)	α-amylase activity (mm)	Total chlorophyll (mg g^{-1})	Dehydrogenase activity (OD value)	α-amylase activity (mm)	Total chlorophyll (mg g^{-1})
T_1	0.84	12.4	1.00	0.26	4.0	1.25	0.47	0.8	1.95
T_2	0.96	14.6	1.21	0.42	5.4	1.53	0.51	1.3	2.28
T_3	0.99	15.7	1.43	0.50	6.2	1.71	0.53	1.6	2.43
Mean	0.91	14.2	1.24	0.39	5.2	1.50	0.50	1.2	2.23
CD (P=0.05)	0.02	0.16	0.06	0.017	0.16	0.02	0.02	0.16	0.04

Table 2: Effect of seed hardening and film coating on biochemical parameters of blackgram, cotton and tomato under laboratory condition

Treatments	Blackgram			Cotton			Tomato		
	Dehydrogenase activity (OD value)	α-amylase activity (mm)	Total chlorophyll (mg g^{-1})	Dehydrogenase activity (OD value)	α-amylase activity (mm)	Total chlorophyll (mg g^{-1})	Dehydrogenase activity (OD value)	α-amylase activity (mm)	Total chlorophyll (mg g^{-1})
T_1	0.316	6.8	0.81	0.903	0.3	2.03	0.093	1.2	0.70
T_2	0.318	8.7	0.89	0.932	0.6	2.32	0.110	1.6	0.78
T_3	0.319	9.2	0.92	0.959	0.6	2.47	0.112	1.7	0.81
Mean	0.318	8.2	0.87	0.931	0.5	2.27	0.105	1.5	0.76
CD (P=0.05)	0.0009	0.09	0.018	0.0004	0.09	0.01	0.0009	0.09	0.016

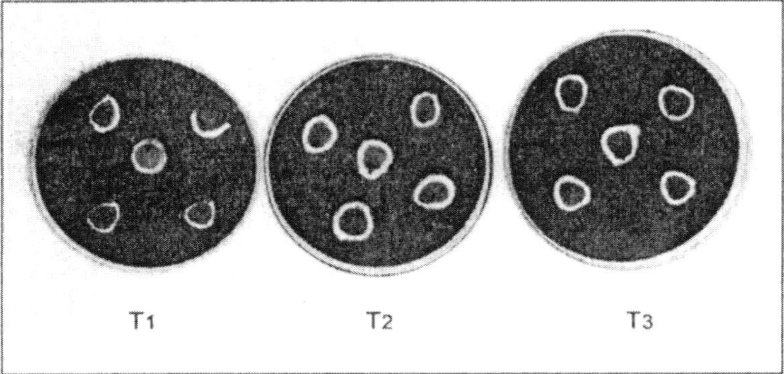

Plate 12: α-amylase activity of hardened and film coated maize seeds

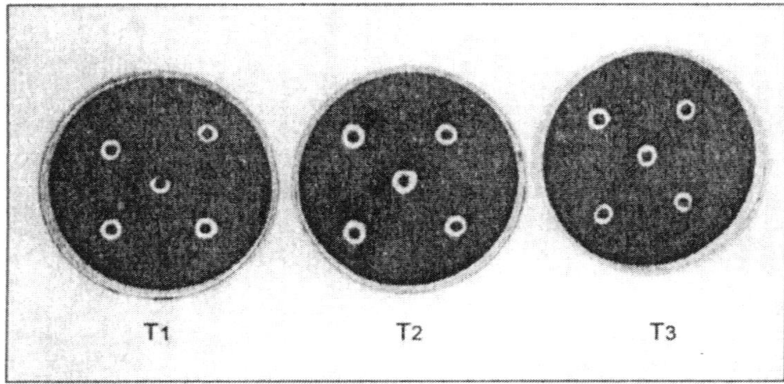

Plate 13: α-amylase activity of sorghum seed due to hardening and film coating

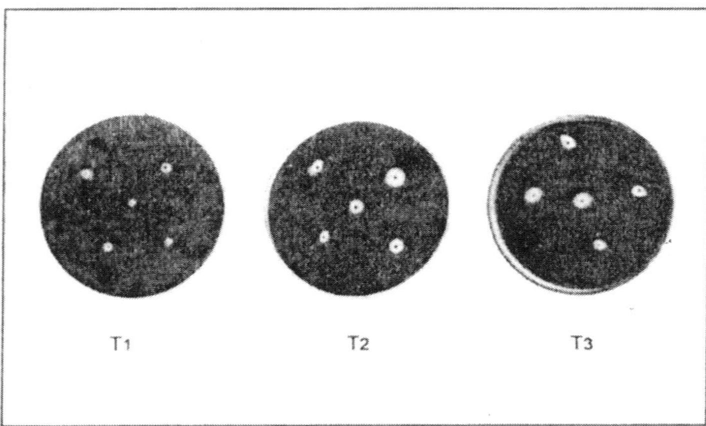

Plate 14: α-amylase activity of hardened and film coated ragi seeds

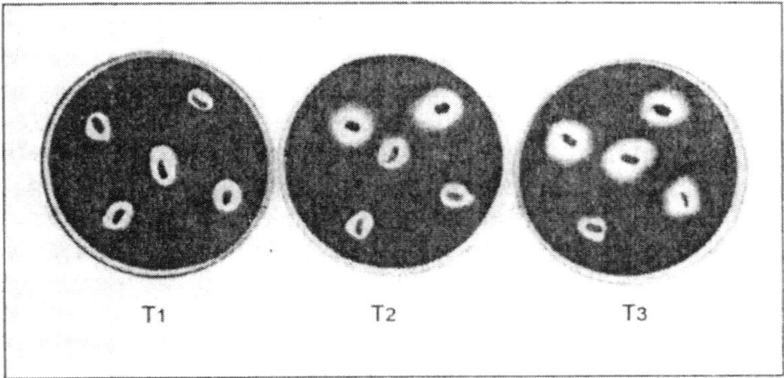

Plate 15: α- amylase activity of hardened and film coated blackgram seeds

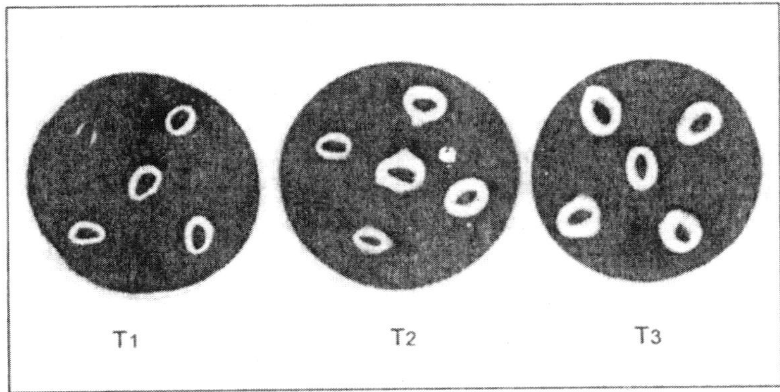

Plate 16: α-amylase activity of hardened and film coated cotton seeds

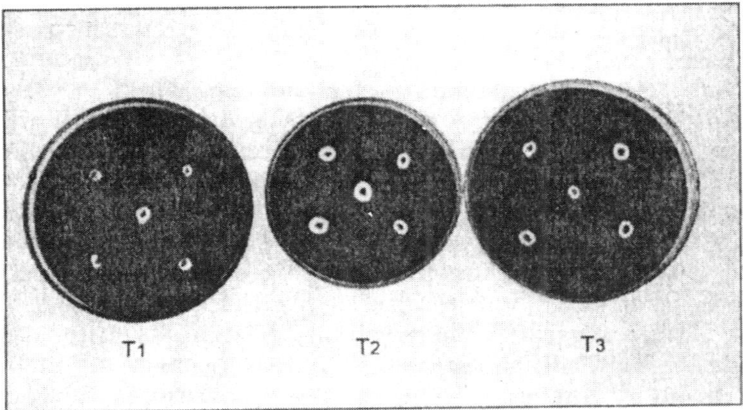

Plate 17: α- amylase activity of tomato seed due to hardening and film coating

higher in 2% KH_2PO_4 hardened and carbendazim + imidachloprid + DAP + micronutrient mixture + polykote + *Azospirillum* film coated seeds of maize, sorghum and tomato (Table 1,2; Plate 12,13,17), 0.1 ppm brassinolide hardened and 2g carbendazim + 1ml imidachloprid + 30g DAP + 20g micronutrient mixture + 3g pink polykote + 60g *Azospirillum* film coated seeds in ragi (Plate 14), 100 ppm $ZnSO_4$ hardened and 2g carbendazim + 1ml imidachloprid + 30g DAP + 20g micronutrient mixture + 3g black polykote + 30g *Rhizobium* film coated seeds in blackgram (Plate 15) and 2% KCl hardened and 2g carbendazim + 1ml imidachloprid + 30g DAP + 20g micronutrient mixture + 3g pink polykote + 80g *Azospirillum* film coated seeds in cotton (Plate 16). The parallel result was obtained by Menaka (2003) in sorghum and cumbu. The increased activity of α-amylase due to hardening in chickpea was reported by Singh and Kumar (1992).

In the process of seed hardening, the quiescent cells get hydrated and this initiates the early process of germination *i.e.*, seeds activated upto II phase of imbibition. In this II phase of imbibition, gibberellins were released from the scutellum and move through the endosperm of the aleurone layer. Then, they trigger the synthesis of hydrolytic enzymes including α-amylase and dehydrogenase. Zinc is one of the important components in micronutrient mixture, which is a constituent of several dehydrogenase enzymes and amino acids and might have attributed to the increased dehydrogenase activity (Sherin Susan John, 2003). Alpha-amylase activity was induced by growth regulator treatment in cotton (Black and Altschal, 1967). This also supported by Andoh and Kobata (2001) in wheat.

3.5. Total chlorophyll content of seedlings

The total chlorophyll content of seedlings was higher in hardened and film coated seeds. In the present study, the total chlorophyll content of seedling was higher in 2% KH_2PO_4 hardened and carbendazim + imidachloprid + DAP + micronutrient mixture + polykote + *Azospirillum* film coated seeds of maize, sorghum and tomato, 0.1 ppm brassinolide hardened and 2g carbendazim + 1ml imidachloprid + 30g DAP + 20g micronutrient mixture + 3g pink polykote + 60g *Azospirillum* film coated seeds of ragi, 100 ppm $ZnSO_4$ hardened and 2g carbendazim + 1ml imidachloprid + 30g DAP + 20g micronutrient mixture + 3g black polykote + 30g *Rhizobium* film coated seeds of blackgram and 2% KCl hardened and 2g carbendazim + 1ml imidachloprid + 30g DAP + 20g micronutrient mixture + 3g pink polykote + 80g *Azospirillum* film coated seeds of cotton when compared to control and hardened and 2g carbendazim + 1ml imidachloprid + 3g pink polykote + 80g *Azospirillum* film coated seeds. The parallel result was obtained by Vigneshwari (2002) in ragi, Menaka (2003) in sorghum and cumbu, and Afria and Narnolia (1999) in wheat. Peoples and Koch (1979) reported that potassium promoted CO_2 fixation by direct activation of RuBP carboxylase, thereby favouring synthesis of chlorophyll. Prasad and Mohammad (1987) suggested the possibility of stimulation for the production of phytochrome, which ultimately resulted in synthesis of chlorophyll by the additional supply of nutrients. The increased amount of chlorophyll by growth promoting substances might be attributed to the faster rate of synthesis of chlorophyll due to degradation effect of chlorophyllase (Sharma and Kwon, 1987).

REFERENCES

Afria, B.S. and R.E. Narnolia. 1999. Effect of cycocel and saline irrigation on physiological attributes, yield and its components in different varieties of wheat. *Indian J. Plant Physiol.,* **4**(4): 311-314.

Andoh, H. and T. Kobata. 2001. Effect of seed hardening, wetting and redrying before sowing on germination and seedling emergence of a Japanese wheat variety Norin 61 in desiccated soil. *Plant Production Science,* **4**(1): 50-55.

Angamuthu, K. 1991. Studies on seed pelleting, physiological maturity, sowing quality and storage in small millets. M.Sc.(Ag.) Thesis, Tamil Nadu Agricultural University, Coimbatore-3.

Arjunan, A. and P.S. Srinivasan. 1989. Pre-sowing seed hardening of groundnut. *Madras Agric. J.,* **76**(9): 523-524.

Balaji, D.S. 1990. Studies on the seed soil relationship in certain crops. M.Sc.(Ag.) Thesis, Tamil Nadu Agricultural University, Coimbatore-3.

Bewley, J.D. and M. Black. 1982. Physiology and bio-chemistry of seeds in relation to germination, viability, dormancy and environmental control, Vol. 2. Springer-verlag, New York. 375 p.

Bhaskaran, M. 1995. Studies on the influence of biotic and abiotic factors on seed yield, quality and storability of rice. Ph.D. Thesis, Tamil Nadu Agricultural University. Coimbatore - 3.

Black, H.S. and A.M. Altschal. 1967. Gibberellic acid induced lipase and α-amylase formation in cotton seed and their inhibition by aflatoxin (*Aspergillus flavus* metabolite). *Biochem. Biophys. Res. Commun.,* **19**: 661-664.

Chinoy, J.J., P.G. Abraham, R.B. Pandya, O.P. Saxena and I.C. Dave. 1970. Effect of pre-sowing treatment of triticum seeds with ascorbic acid and sucrose on growth, development and yield characters. *Indian J. Plant Physiol.,* **13**: 40-48.

Clouse, S.D. and D. Zurek. 1991. Molecular analysis of brassinolide action in plant growth and development. In: Brassinosteroids chemistry, Bioactivity and Application (Ed. Culter, H.G.T. Yohota and G. Adam), Am. Chem. Soc., Washington, D.C., pp. 122-140.

Dawson, M.J. 1965. Effect of seed soaking on the growth and development of crop plants finger millet. *Indian J. Pl. Physiol.,* **8**: 52-56.

Eremeeva, O.E., D. Bahaev and F.G. Mustakimova. 1988. Reaction of fine fibre cotton seeds to complex enrichement with biologically action compounds under condition of *Fusarium* wilt infection. *Seriya Biologicheskikh Nauk,* **5**: 60-64.

Fasui, H., Ma Cheng Cang, Wang XuMing, ji Guang Ming. 1996. Effect of Ca^{2+} and Zn^{2+} on seed vigour and some enzyme activities during seed germination of maize. *Pl. Physiol. comm.,* **32**(2): 100-112.

Fujii, S. and H. Saka. 2001. The promotive effect of brassinolide on lamina joint cell elongation, germination and seeding growth under low temperature stress in rice. *Plant production science,* *4*(3): 210-214.

Gray, D., R.A. Joyce and Steckel. 1977. Effect of pre-sowing treatments of seed on the germination and establishment of parsnips. *Journal of Horticultural Science*: 525-534.

Hayat, S., A. Ahmad, A. Hussain and M. Mobin. 2001. Growth of wheat seedlings raised from grains treated with 28 homobrassinolide. *Acta Physiologie Plantarum,* **23**(1): 23-30.

Henckel, P.A. 1964. Physiology of plants under drought. *Ann. Rev. Pl. Physiol.,* **15**: 363-386.

Hendricks, S.B. and R.B. Taylorson. 1974. Promotion of seed germination by nitrate, nitrite, hydroxylamine and ammonium salts. *Plant Physiol.,* **54**: 304-309.

Hendricks, S.B. and R.B. Taylorson. 1975. Breaking of seed dormancy by catalase inhibition. *Proceedings of National Academy of Science,* **72**: 306-309.

Iswaran, V. and P.K. Chhonkar. 1971. Effect of seed pelleting on nodulation and growth of pea (*Pisum sativum* L.) in acid soil. *Sci. Cult.,* **37**: 345-346.

Jayaraj, T. 1977. Study of the effect of plant protection chemicals on seed quality in sesame (*Sesamum indicum* L.) cv. KRR 2 and TMV 3. M.Sc. (Ag.) Thesis, Tamil Nadu Agricultural University, Coimbatore - 3.

Jegathambal, R. 1996. Pre-sowing seed treatment to augment productivity of sorghum cv. CO 26 under rainfed agriculture. Ph.D. Thesis, Tamil Nadu University Coimbatore-3.

Joseph, K. and N.R. Nair. 1989. Effect of seed hardening on germination and seedling vigour in paddy. *Seed Res.,* **17**(2): 188-190.

Kalavathi, D. 1985. Studies on seed viability and vigour in soybean (*Glycine max* (L.) Merril.) M.Sc.(Ag.) Thesis, Tamil Nadu Agricultural University, Coimbatore-3.

Kamalam, J. 1991. Studies on the effect of soil moisture content and seed soaking on the germination of legumes. *Legume Res.*, **14**(3): 153-154.

Kamalam, J. and N.R. Nair. 1989. Effect of seed hardening on germination and seedling vigour in paddy. *Seed Res.*, **17**(2): 188-190.

Khan, A.A. 1980. In: The physiology and biochemistry of seed dormancy and germination (Ed. A.A. Khan), North Holland Publishing Company, Amsterdam, New York, Oxford, pp. 127.

Kohli, V.K., I.K. Thakur and Y.R. Shukla. 1992. Studies on improving field emergence in the autumn crop of garden pea (*Pisum satium* L.) *Ann. Agric. Res.*, **13**(4): 394-395.

Konstantinov, G. and M. Kolarova. 1988. A study on germination of capsulated lucerne seed. *Inst. Posemefroizuodstvo.* **25**(4): 40-47.

Krishnasastry, K.S., S. Ramo Rao and K.N. Appaiah. 1969. Effect of pretreatment of seeds of *Eleusine coracana* on tolerance to high concentrations of sodium chloride and manitol solutions. *Mysore J. Agric. Sci.*, **11**: 47-49.

Kursanov, A.L., E. Byskrebentseva, I. Sveshnikova and M. Krasavina. 1965. Disorganization of energy metabolism in roots during potash metabolism. *Nauk SSSR,* **162**: 9.

Leadem, C.L. 1987. The role of plant growth regulators in the germination of forest tree seeds. *Plant Growth Regulators,* **6**: 61-93.

Lonergan, J.F. and J.F. Meyer. 1955. Lime pelleted seed for nodulation on acid soils. *J. Aus. Inst. Agrl. Sci.*, **21**: 264-265.

May, L.H., E.J. Milthorpe and F.L. Milthorpe. 1962. Pre-sowing hardening of plants to drought. An appraisal of the contributions by P.A. Genkel. *Field Crop Abstr.*, **15**: 93-98.

Mehrotra, O.N., J.P. Srivastava and S. Nath. 1968. Germination, growth and yield of wheat as influenced by pre-sowing seed treatment in phosphate solution. *Indian J. Agron.*, **13**(2): 53-56.

Menaka, C. 2003. Seed hardening and pelleting technology for maximizing the productivity of sorghum (*Sorghum bicolor* (L.) Moench.) and bajra (*Pennisetum glaucum* (L.) R.Br.) under rainfed condition. Ph.D. Thesis, Tamil Nadu Agricultural University, Coimbatore-3.

Miller, W.F. and C. Scooter. 1979. Improving emergence of pelleted seed. *Trans. Amer. Soc. Agric. Engineer.*, **16**(5): 540-549.

Min, D.H. 1988. Effects of liming and seed coating on early growth, drymatter yield and nutritive value of alfalfa (*Medicago sativa* L.) and soil properties. M.Sc. Thesis. Seoul National Univ.

Misra, N.M. and M.G.R.K. Reddy. 1985. Effect of cycocel, ascorbic acid and zinc sulphate on the yield attributes and yield of rainfed wheat. *Andhra Agric. J.*, **32**: 32-33.

Nagao, M.A. and I.C. Furutani. 1986. Improving germination of papaya seed by density separation, potassium nitrate and gibberellic acid. *Hort. Sci.*, **21**: 1439-1440.

Nagaraj, S. 1996. Seed hardening and coating studies in upland rice (*Oryza sativa* L.). M.Sc. (Ag.) Thesis, Tamil Nadu Agricultural University, Coimbatore - 3.

Namdeo, S.L., S.C. Gupta and K.S. Bangar. 1989. Effect of rhizobial strains on symbiotic traits and yield of chick pea genotypes under rainfed conditions. *Indian J. Pulses Res.*, **2**(1): 64-67.

Nirmala Devi, S., M. Abdul Vahab and Sally K. Mathew. 1995. Seedling vigour in chilli (*Capsicum annum*) as influenced by seed treatment with *Azospirillum. South Indian Hort.*, **43**(1&2): 54-56.

Okanenko, A.S. and B.I. Bershteir. 1966. Potash and sodium in the life activity of sugarbeet. *Sel' Skokbozyaistven-nykh Biologiya,* **3**: 391-405.

Patangundi, M.S., M.P. Potdar, S.T. Hundekar and S.A. Patil. 1997. Micro nutrients enhance yield of legumes. *Intensive Agriculture*, Jan-Feb, pp. 3-4.

Paul, S.R. and A.K. Choudhury. 1993. Effect of seed hardening with potassium salts of different concentrations and soaking durations on germination and seedling vigour of wheat. *Ann. Agric. Res.*, **14**(3): 357-359.

Peoples, T.R. and D.D. Koch. 1979. Role of potassium in CO_2 assimilation in *Medicago sativa* (L.). *Plant Physiol.*, **63**: 878-881.

Prabakaran, J. and K. Srinivasan. 1995. Effect of *Rhizobium* inoculation and coir dust on growth, nodulation and yield of pigeon pea (*Cajanus cajan*). *Indian J. Agron.*, **40**(3): 518-519.

Prasad, R. and G. Mohammad. 1987. Effect of gibberellic acid, nutrients and Bavistin on chlorosis and growth of *Grevileia* species. *J. Trop. For.*, **3**(11): 156-159.

Punithavathi, N. 1997. Seed hardening and pelleting studies in Ragi cv. CO 13. M.Sc.(Ag.) Thesis, Tamil Nadu Agricultural University, Coimbatore-3.

Ramesh, K., V. Thirumurugan and B. Chandrasekharan. 2001. Effect of nutrient uptake and productivity of soybean (*Glycine max*). *Indian J. of Agron.,* **46**(3): 506-510.

Rangasamy, A., S. Purushothaman and P. Devasenapathy. 1993. Seed hardening in relation to seedling quality characters of crops. *Madras agric. J.,* **80**(9): 535-537.

Rao, A.V. and R.L. Sharma. 1980. Note on effect of inoculum levels of rhizobia on symbiosis of soybean and blackgram. *Legume Res.,* **3**: 55-57.

Rathinavel, K. and C. Dharmalingam. 1999. Effect of seed pelleting on elite seedling production in cotton cv. MUC 7 (*Gossypium hirsutum* L.). *Crop Res.,* **18**(1): 137-141.

Roberts, E.H. 1973. Cytological, genetical and metabolic changes associated with loss of viability. In: Viability of seeds. Chapman and Hall Ltd., London, pp. 253-306.

Sabir-Ahamed, A. 1989. Studies on the production of quality seed and storage in soybean (*Glycine max* (L.) Merrill). M.Sc. (Ag.) Thesis, Tamil Nadu Agricultural University, Coimbatore -3.

Sanchez, J.A., B.C. Munoz and J. Fresneda. 2001. Combined effects of hardening, hydration-dehydration and heat shock treatments on the germination of tomato, pepper and cucumber. *Seed Sci. & Technol.,* **29**(3): 691-697.

Saraswathi, G. 1994. Studies on seed pelleting in relation to sowing of cotton cv. LRA 5166 (*Gossypium hirsutum* L.). M.Sc. (Ag.) Thesis, Tamil Nadu Agricultural Univeristy, Coimbatore -3.

Seshidhar, U.R., A.A. Mekhri, and K.S. Krishnasastry. 1978. Potassium content and proline accumulation following seed treatment with calcium chloride in groundnut varieties. *Indian J. Plant Physiol.,* **16**: 89-92.

Shanmugasundaram, V.S. and M. Kannaiyan. 1989. Effect of concentration of seed hardening chemicals on physiological characters of pearlmillet (*Pennisetum typhoides* stapf and Hubb). *J. Agron. & Crop Sci.,* **163**: 171-176.

Sharma, R. and E.D. Kwon. 1987. Effect of growth regulators on nodulation and some biochemical parameters in soybean. *Pl. Physiol. Biochem.,* **14**(2): 146-152.

Sherin Susan John. 2003. Seed film coating technology using polykote for maximizing the planting value, growth and productivity of maize cv. CO 1. M.Sc. (Ag.) Thesis, Tamil Nadu Agricultural University, Coimbatore -3.

Singh, A.K., G. Shanker and M. Makhya. 1979. A study of citrus seed germination as affected by some chemicals. *J. Hort. Sci.,* **8**: 194-195.

Singh, R.A. and P. Kumar. 1992. Effect of water stress and hardening on metabolic changes in chick pea (*Cicer arietinum* L.). *Indian J. Plant Physiol.,* **35**(3): 252-257.

Srimathi, P., K. Malarkodi, R. Geetha and V. Krishnasamy. 2000. Influence of pre-sowing pelleting treatment on seed yield and storability of tomato cv. PKM 1. *Orissa J. of Horti.,* **28**(2): 33-35.

Subramanyam, M.V.R. and N.M. Misra. 1980. Effect of potash seed treatment on yield and yield components of high yielding wheat under rainfed condition. *Seeds and Farms,* **6**(1): 21-24.

Thandapani, V. and M. Subarayalu. 1986. Effect of pre-sowing seed treatment with chemicals and growth regulators for drought tolerance and yield of cotton (*Gossypium hirsutum* L.) under rainfed condition. *Madras Agric. J.,* **73**(2): 668-675.

Vanangamudi, K. and R. Kulandaivelu. 1989. Pre-sowing seed treatment for dryland farming. *Seeds and Farms,* **15**(9-10): 33-34.

Vigneshwari, R. 2002. Seed hardening and pelleting in Ragi cv. CO 13 to induce drought resistance and to improve productivity under rainfed condition. M.Sc. (Ag.) Thesis, Tamil Nadu Agricultural University, Coimbatore -3.

West, S.H., S.K. Loftin, M. Wahl, C.D. Batich and C.L. Beatty. 1985. Polymers and moisture barriers to maintain seed quality. *Crop. Sci.,* **25**: 941-944.

Williams, K.D., N.W. Hopper and T.C. Wedegaertner. 1999. The germination and emergence responses of polymer coated fuzzy cotton seeds. *National Cotton Council,* : 623-625.

Zinin, T. and A. Imamaliev. 1984. The use of pelleted cotton seed. *Scientia Agricola,* (1): 11-12.

7

Physiological Seed Conditioning — An Overview

M. Djanaguiraman[1], J. Annie Sheeba[1], K.R.V. Sathya Sheela[2] and Mallika Vanangamudi[1]

[1]*Department of Crop Physiology, Tamil Nadu Agricultural University Coimbatore – 641 003*

[2]*Department of Plant Breeding and Genetics, Adhiparasakthi Agricultural College G.B.Nagar, Kalavai- 632 506*

1. INTRODUCTION

Seed technology is a critical component of agriculture. In order to utilize the genetic improvements that plant breeders and molecular biologists are continuously incorporating into crops, the plants must be propagated and established in the field. This is primarily accomplished through seeds. In many crops, the ultimate profitability of the crop can be determined by how the seeds perform to establish the plants in the field. Not only, the final number and spacing of plants important, but also the speed and uniformity of emergence can have a major impact on the crop. Many aspects of crop management are also timed according to plant growth stage. Non-uniform development of the crop leads to less than optimal results from such management practices. The efficiency of production of transplants is affected greatly by the quality of the seeds used. All of these factors have brought increasing attention to the vigour and quality of seeds. Here, some engineering or technological approaches to enhance the performance of seeds are described. The preplant seed treatment was done in order to mobilize the seed's own resources and to augment them with external resources to maximize improvement in stand establishment and yield. Mobilization of these resources is particularly critical when seeds of modern cultivars are selected for high yields under optimal growing conditions. Physiological seed treatment can improve or enhance seed performance, which include presoaking, wetting and drying, humidification, osmoconditioning, matriconditioning, pregermination, moist chilling, thermal shock, irradiation, aeration, and hormone permeation. Non-physiological seed treatments that may indirectly improve seed germination and stand establishment are seed scarification, coating or pelleting, treatment with bioactive chemicals, and inoculation or coating of seeds with beneficial microbes.

Presowing seed treatments are conducive for improving the growth and seedling establishment of different crops as they enhance the mitochondrial activity and rapid growth of the embryo which improve the growth components as well as productivity in various crops.

2. SEED PRIMING

The use of priming and related germination advancing technologies has developed widely over the last 25 years, especially for horticultural and ornamental species. Priming is now a widespread commercial tool used to help crop establishment over a range of environmental conditions, especially for carrot, celery, leek, lettuce, onion, pepper, tomato, several flower and culinary herb species and, in certain markets, for some higher-volume crops, such as grasses and sugar beet. There is usually, but not always, a statement made on the label that a seed batch has been primed. Priming exposes seeds to restricted water availability under controlled conditions to allow some physiological processes of germination to occur, but germination itself is not completed in any individual seeds.

Table 1: Influence of seed fortification on shoot length (cm), leaf area (cm^2 plant^{-1}), root volume (cc seedling^{-1}) and root dry weight (g seedling^{-1}) of jatropha

Nutrients	Shoot length (cm)	Leaf area (cm^2 plant^{-1})	Root volume (cc seedling^{-1})	Root dry weight (g seedling^{-1})
Control	24.2	76.1	0.6	0.249
$FeSO_4$ 0.5%	28.3	95.1	1.53	0.381
1.0%	31.1	132.9	2.04	0.470
1.5%	29.3	114.7	1.64	0.424
$ZnSO_4$ 0.5%	29.1	101.4	1.54	0.392
1.0%	32.1	131.7	2.11	0.466
1.5%	30.6	112.6	1.67	0.430
$MnSO_4$ 0.5%	28.7	101.4	1.50	0.399
1.0%	32.0	131.0	2.12	0.481
1.5%	30.3	109.2	1.62	0.446
CD (5%)	0.74	6.89	0.07	0.019

Primed seed usually emerges from the soil faster, and more uniformly than non primed seed of the same seed lot. Kathiravan (2004) reported that application of micronutrients *viz.*, Iron, Zinc and Manganese was able to increase the shoot length, leaf area, and root volume and root dry weight (Table 1). These differences are greatest under adverse environmental conditions in the field. In seed priming, the seeds are imbibed to a water content below that required for radicle emergence, but sufficient to allow germinative metabolism to proceed. During priming, seeds are essentially in Phase II of imbibition, but are prevented from proceeding into Phase III. After the priming treatment, the seeds can be dried, packaged, distributed and planted in the same way as untreated seeds. However, when reimbibed, the seeds retain the advancement they achieved during the priming treatment and complete germination much more rapidly and uniformly. In addition, priming can overcome some types of environmental stresses, such as high temperature inhibition of lettuce seed germination. A disadvantage of priming is that in most cases, the storage life of the seeds is shortened after the treatment and the seeds are more sensitive to poor storage conditions. However, further refinements of the priming process, particularly the handling of the seeds prior to dehydration, are able to mitigate this effect.

Biopriming, a seed treatment system that integrates the biological and physiological aspects of disease control, involves coating the seed with a bacterial biocontrol agent and incubating the seed under warm, moist conditions. Seed may be planted moist or dried for storage. Fungicide applications are one of the most common seed treatments used today. A light coating of fungicide is usually applied to the seed coat. A brightly coloured dye is sometimes added as a reminder that a fungicide has been applied. Legume seeds are frequently treated with *Rhizobium* to improve nitrogen fixation after germination. Coating and pelleting are two widely used seed treatments, particularly with small-seeded species. Coated seeds have a thin layer of material such as clay or diatomaceous earth applied to the surface of the seed to make them larger and easier to handle. However, coating does not change the overall shape of the seed. Seeds are coated to make them easier to handle, to deliver chemicals, to improve soil contact, or to inoculate seeds with microbes. To make identification easier, coated seeds are often colour-coded, so cultivars or types are not mixed at planting. A successful pelleting treatment splits upon hydration and does not pose a barrier to radicle growth or limit oxygen availability to the seeds.

Seed priming was done by soaking in variety of solutes (Table 2); including solutions of various inorganic salts, sugars and polyethylene glycol (PEG) a chemically inert, high molecular weight compound that unlike the lower molecular weight compound probably does not penetrate the cell walls. The temperature suggested during priming is between 10^{0}C – 15^{0}C. The duration of priming varies with the crop from 1 day (sorghum) to 23 day (onion). The seeds are then rinsed and redried to original water content. Heydecker *et al.* (1973) used different terms depends upon the methods adopted for priming. (i) Osmopriming – soaking the seeds in osmotic solutions, (ii) Halopriming – soaking the seeds in salt solutions, (iii) Biopriming – coating the seeds with biological agents like bacteria, (iv) Solid matripriming – This consists of mixing seeds with an organic or inorganic carrier and water for a period of time. The moisture content of the matri is brought to a level just below what is required for radicle protrusion. Seed water potential is regulated by the matri potential of the seed and during priming the water is largely held by carrier. Seed can imbibe water from the carrier till the equilibrium is reached.

Seeds in contact with an aqueous solution of the chemical begin to take up water in the normal way, but cease to do so once they have reached equilibrium with the osmotic potential of the solution. Therefore, the seed water content can be adjusted to a level that permits the seeds to go through all the essential preparatory processes of germination, but prevents cell elongation and in consequence, radicle emergence even after weeks of contact between seeds and osmotic solution. Moreover, during the waiting period at the osmotic barrier the 'slower' seeds tend to catch up with the 'faster' ones so that the subsequent germinations is more uniform, especially at lower germination temperatures, than that of untreated seeds. Favourable treatment conditions permit both the break down of food reserves and the synthesis of materials required for germination to occur, possibly to a greater extent than if the seeds had germinated immediately. This result in instantaneous growth, once the osmotic obstacle has been removed.

Seed lots vary in their response to priming, even within the same cultivar. For example, priming improves mean germination times for carrot, celery and onion seeds, but differences exist in the response among species and seed lots of the same species. Primed tomato seeds exhibited vastly different germination responses for two seed lots exposed to temperature and osmotic stress (Coolbear and McGill, 1990). Similar results were reported for pepper seed lots (Bradford *et al.,* 1990). Because of this variability, the optimum priming conditions must be determined for each seed lot to obtain optimum seed priming response.

Two environmental variables are central to the success of priming. The first determines the optimal moisture content of the seed, which is hydrated short of radicle protrusion. As the seed hydrates, chemical reactions associated with germination occur at discrete water thresholds (Bewley, 1997). The second variable is temperature, which governs the rate of these chemical reactions. Thus, the lower the internal water potential, the more water is absorbed by the seed, permitting these chemical reactions to occur at a faster rate as temperature increases.

Table 2: Osmoticants used for seed priming

Common Name	Botanical name	Osmoticum
Beet	*Beta vulgaris*	PEG, $MgSO_4$
Cabbage	*Brassica oleracea*	PEG
Carrot	*Daucus carota*	PEG
Cucumber	*Cucumis sativus*	Mannitol, NaCl
Maize	*Zea mays*	PEG, KH_2PO_4, K_2HPO_4 + KNO_3
Onion	*Allium cepa*	PEG PEG + Enriched air Glycerol, Mannitol, NaCl KNO_3 + K_2HPO_4 K_3PO_4, KNO_3
Pea	*Pisum sativum*	PEG
Peanut	*Arachis hypogea*	PEG
Pepper	*Capsicum annum*	PEG Mannitol Sodium Polypropionate KNO_3, KH_2PO_4 NaCl, $MgSO_4$
Sorghum	*Sorghum bicolor*	PEG, KNO_3, K_2HPO_4, KPO_4, KNO_3 + K_3PO_4, KNO_3 + K_2HPO_4
Soybean	*Glycine max*	PEG, Mannitol, PEG + Mannitol, PEG + GA_3 + Kinetin, Sodium Polypropionate
Sweet Corn	*Zea mays*	PEG
Tomato	*Lycopersicon sp.*	PEG, Mannitol, NaCl Sodium Polypropionate, KNO_3 K_2HPO_4, K_3PO_4, KNO_3 + K_2HPO_4 KNO_3 + K_3PO_4

Fundamental to the success of seed priming is the uptake of water that occurs in three phases in a germinating seed. Phase I is imbibition and is a result of rapid initial water uptake due to the seed's low water potential (water moves from high to low water potential). During this phase, DNA and mitochondria are repaired and proteins are synthesized using existing messenger ribonucleic acids (mRNAs). Phase II is represented by only a gradual increase in seed water content, but new physiological activities associated with germination are initiated, including the synthesis of mitochondria and proteins relying on the translation of new mRNAs.

Phases I and II represent the processes of germination (Bewley, 1997) and are the foundation of successful seed priming, where the seed is brought to a seed moisture content that is just short of radicle protrusion (Bray, 1995). The completion of germination and the start of seedling growth are signified by the resumption of radicle growth, which occurs during Phase III and is identified by another rapid increase in water uptake. This creates the turgor necessary for radicle cell elongation.

2.1. Types of priming

Four techniques are currently used commercially to accomplish seed priming. These include hydropriming, osmopriming, matripriming and pregermination.

2.1.1. Hydropriming

Soaking and misting seeds in water and redrying them before they complete germination, a process called hydropriming, is the simplest approach for hydrating seeds. This technique minimizes the use of chemicals and avoids discarding materials that may be undesirable and incompatible with the environment. One disadvantage of this method is that seeds are sometimes not evenly hydrated, which results in their failure to uniformly activate the physiological processes necessary to synchronize and improve seed performance. In addition, the rate of hydration is difficult to control and some types of seeds may actually incur imbibitional damage, if the rate of hydration is too rapid.

2.1.2. Osmopriming

Osmopriming is the process of soaking seeds in aerated osmotica of low water potential to control the amount of water they imbibe. Examples of such osmotica include polyethylene glycol (PEG), KNO_3, mannitol, KH_2PO_4, and other salts and high molecular weight (HMW) compounds (Table 3). Salts and mannitol have been used extensively as osmotica, but both are capable of being absorbed by the seeds, resulting in toxic effects in some cases. PEG is the preferred osmoticum; it is a HMW inert compound, and the large molecular size precludes it from entering the seed and creating toxic side-effects associated with the use of salts. A major disadvantage of PEG is that oxygen solubility is inversely related to its concentration. Thus, when PEG is used in priming, the solution is often aerated (Akers, 1990). During osmopriming, beneficial compounds that break dormancy or otherwise improve seed performance can be incorporated into the osmoticum so that they can be taken up by the seed. While PEG is usually considered an osmoticum, Steuter *et al.* (1981) indicated that the HMW PEG molecules regulate water potential through matri rather than osmotic forces. As a result, Taylor *et al.* (1998) suggested that PEG should be considered a matricum rather than an osmoticum.

Osmopriming has been primarily successful with small-seeded crops and less successful with large-seeded crops, such as soybean and sweet corn. Armstrong and McDonald (1992) showed that osmopriming of soybean seeds without an intervening air-drying treatment increased plumule and radicle length and weight.

2.1.3. Matripriming

Another approach to priming is the use of solid carriers with low matric potentials, a process called matripriming. Some of the ideal characteristics of the carrier are; low matric potential; negligible water solubility; high water-holding capacity; high surface-to-volume ratio; nontoxicity to the seed; and, the ability to adhere to the seed surface (Khan, 1992). Processed, natural substances with these characteristics are vermiculite and peat moss, and those available commercially include Celite and Micro-Cel, which are distomaceous silica products and Zonolite, which is vermiculite (Khan, 1992). The surface of these compounds creates matric forces that hold water, which is gradually absorbed by the seed.

Matripriming mimics the natural process of imbibition of water by the seed from soil particles. Seeds are generally mixed into the carrier at matric potentials ranging from - 0.4 to - 1.5 MPa at 15°C for 7-14 days, and brought to moisture equilibrium. The solid material is then sieved away from the primed seeds.

There is growing interest in manipulating the rhizosphere microflora to enhance the activity of certain beneficial microorganisms in order to control plant pathogens affecting seedling establishment. There are numerous reports on the application of beneficial microbes as seed coating to improve the health and the nutritional status of the growing seedlings. Among the more popular fungi that have been used to control damping-off and other plant diseases are *Trichoderma harzianum* and *Chaetomium globosum.* In several instances the application of beneficial microbes has been integrated with gel seedling and physiological seed conditioning to maximize seed performance, particularly under conditions of high disease incidence or to improve the nutritional status of the growing seedlings.

2.1.4. Seed Priming Risks

The number one risk when using primed seed is reduced seed shelf life. Depending on the species, seed lot vigor, and the temperature and humidity that the seed is being stored; a primed seed should remain viable for up to a year. If you store primed seed in hot humid conditions, it will lose viability much more quickly. Mostly in all cases however, primed seed has shorter shelf life than the non primed seed of the same seed lot. For this reason, it's best not to carry primed seed over to the next growing season.

2.2. Priming influence on germination

It is increasingly clear that priming initiates the earliest stages of germination, but not those associated with radicle growth. Generally, for priming to be successful, seed moisture content should be maintained at 40-50% on a fresh weight basis or about 90-95% of the seed moisture content necessary for germination (Gray *et al.*, 1990)

2.2.1. Proteins

Smith and Cobb (1991) reported increased protein, aldolase, and isocitrate lyase activity and found greater glucose-6-phosphate dehydrogenase and decreased alcohol dehydrogenase activity in osmoprimed sweet corn seeds (Smith and Cobb, 1992). They concluded that priming permits many of the normal events of germination to proceed. Osmopriming of peanut seeds also caused an increase in the activity of isocitrate lyase responsible for the mobilization of lipid reserves following germination (Fu *et al.*, 1988). In sweet corn seeds, osmo- and matripriming increased α-and β-amylase activity (Sung and Chang, 1993) and osmopriming led to increased protein synthesis in wheat seeds.

2.2.2. Nucleic acid synthesis

Osmopriming also increased the RNA content of leek seeds, most of this being attributed to the synthesis of ribosomal RNA (rRNA), while messenger RNA (mRNA) content appeared to remain constant. Coolbear and McGill (1990) suggested that in tomato, osmopriming improves rRNA integrity, which supports protein synthesis to permit subsequent seed germination. DNA synthesis before osmopriming was less in low than in high vigour leek seed lots, but there was little difference in DNA synthesis between the two seed lots by the seventh day of a 14-day priming treatment (Bray, 1995). Interestingly, Ashraf and Bray (1993) reported that 70-80% of DNA synthesis during osmopriming represented plastid or mitochondrial DNA. This continued into the post-osmopriming period, culminating in a significant increase in mitochondria in osmoprimed leek cells.

2.3. Mechanism of repair by priming

Priming appears to reverse the detrimental effects of seed deterioration. In sweet corn, osmo and matripriming resulted in decreased conductivity, free sugars and DNA content, while RNA content increased (Sung and Chang, 1993). Natural ageing of French bean seeds stored for up to 4 years induced membrane disruption and leakage of UV-absorbing substances, which was ameliorated by hydropriming. Lower electrical conductivity readings following hydropriming (indicating reduced membrane leakage) have also been reported for seeds of eggplant, radish and onion (Choudhuri and Basu, 1988). These beneficial effects may be due to the flushing of solutes from the seed during the priming procedure and prior to determination of leaked substances. Primed seeds often perform better in disease infested soils because of decreased electrolyte leakage and faster germination rate, which both reduce the window of opportunity for fungal attack.

Priming is also thought to increase enzyme activity and counteract the effects of lipid peroxidation. Saha *et al.* (1990) showed that matripriming caused increased amylase and dehydrogenase activity in aged soybean seeds compared to unprimed seeds. Osmopriming reverses the loss of lipid peroxidation detoxifying enzymes, such as superoxide dismutase, catalase and glutathione reductase in aged sunflower seeds, and these enzymes are present at the same activities as in unaged seeds (Bailly *et al.,* 1997).

2.4. Differences in priming technique

Smith and Cobb (1991) concluded that the osmopriming response is dependent on the duration of the treatment and the osmotic potential of the solution rather than on the specific salt used. However, other osmopriming studies have demonstrated that some crops respond better to specific priming techniques. Frett *et al.* (1991) found that, compared to the use of PEG as the osmoticum, inorganic salts on most germination blotters were more deleterious to tomato and asparagus seed germination and created membrane damage and enzymatic changes that affected subsequent growth. While salt solutions successfully osmopriming tomato and carrot seeds, the same treatments reduce percentage emergence of onion seeds, even though the emergence rate is increased (Haigh *et al.,* 1986). When salts are used for osmopriming, they are toxic to sorghum seeds, but priming seeds of three herbage grasses and tomato with salts is superior to priming in PEG. Sung and Chang (1993) showed that placing sweet corn seeds in moist vermiculite to accomplish matripriming caused earlier emergence, while osmopriming with PEG on blotters reduced subsequent field emergence.

2.5. Environmental variables affecting priming

The environment that the seed experiences during priming can influence the degree of success. The most important of these are oxygen, light and temperature.

2.5.1. Oxygen

Two studies have shown that increasing the oxygen content during osmopriming with PEG influences subsequent germination and seedling development. Bujalski and Nienow (1991) reported that using enriched air with a 75% O_2/25% N_2 ratio to aerate and suspend onion seeds in PEG solutions provided the same number of normal seedlings as untreated seeds germinated on filter paper. The high viscosity of PEG and poor oxygen permeability of the medium may explain this response. Ozbingol *et al.* (1988) found that osmopriming of tomato seeds produced the greatest response at oxygen concentrations higher than 10%. However, when water was used for hydropriming of muskmelon seeds under high (≥50%) oxygen compared to hydropriming under ambient or low (≤21%) oxygen, germination was decreased. The use of solid matrix, such as vermiculite or a finely ground shale, during matripriming allows more air to get to the seed than when an osmoticum is used. This is because the solid matrix carrier is not completely saturated to achieve the desired water potential and the seed is, therefore, in an aerobic environment.

2.5.2. Light

Seeds that require light for germination can be illuminated during priming to minimize dormancy.

2.5.3. Temperature

Lower temperatures (*e.g.* 15°C) during priming can result in enhanced seed performance. This may be because the physiological processes of germination are delayed, even as the seed absorbs increasing quantities of water. Bradford and Haigh (1994) reported a linear relationship between the median germination rate of tomato seeds after osmopriming and the accumulated hydrothermal priming time. Similarly, Ozbingol *et al.* (1988) found that the optimum temperature (27-28°C) for osmopriming with PEG at -1.0 MPa was the same as the optimum temperature for germination of tomato seeds.

Table 3: Priming technique used on various crops

Crop	Priming technique	Reference
Barley	Hydropriming	Kibite and Harker (1991)
Bean	Matripriming	Khan *et al.* (1992)
Brussels sprouts	Hydropriming	Burgess and Powell (1984)
Carrot	Osmopriming	Biniek and Babik (1993)
	Matripriming	Khan *et al.* (1992)
Cauliflower	Hydropriming	Fujikura *et al.* (1993)
Celeriac	Osmopriming	Drew and Dearman (1993)
	Hydropriming	Thomas (1983)
Celery	Matripriming	Khan *et al.*(1992)
	Osmopriming	Gray *et al.* (1991)
Guinea grass	Osmopriming	Usberti and Valio (1997)
Impatiens	Osmopriming	Finch-Savage *et. al.* (1991)
Kentucky bluegrass	Matripriming	Pill and Korengel (1997)
Lettuce	Matripriming	Taylor *et al.* (1988)
	Osmopriming	Tarquis and Bradford (1992)
Oat	Hydropriming	Hofmann *et al.* (1992)
Onion	Osmopriming	Gray *et al.* (1991)
	Hydropriming	Choudhuri and Basu (1988)
Peanut	Osmopriming	Fu *et al.* (1988)
Pepper	Osmopriming	Halpin-Ingham and Sundstrom (1992)
	Matripriming	Khan *et al.* (1992)
Petunia	Osmopriming	Finch-Savage *et al.* (1991)
Pine	Osmopriming	Huang and Zou (1989)
Radish	Hydropriming	Rudrapal and Nakamura (1988)
Soybean	Matripriming	Saha *et al.* (1990)
	Hydropriming	Tilden and West (1985)
Sugar beet	Hydropriming	Durrant and Loads (1987)
	Osmopriming	Khan *et al.* (1983)
Sweet corn	Matripriming	Chang and Sung (1998)
Tomato	Hydropriming	Penaloza and Eira (1993)
	Matripriming	Khan *et al.* (1992)
	Osmopriming	Ozbingol *et al.* (1998)
Turnip	Osmopriming	Rao *et al.* (1987)
Wheat	Hydropriming	Nath *et al.* (1991)
	Hydropriming	Kibite and Harker (1991)

2.5.4. Drying rates

Drying following priming can be performed using saturated salts, ambient or forced air. It is possible to select saturated salts that produce controlled atmospheres of specific relative humidities, in which primed seeds can be dried to desired equilibrium moisture content, usually around 10%. Rapid drying following matripriming can damage sweet corn seeds, resulting in the loss of advancement obtained by priming.

2.6. Addition of promotive substances

The addition of promotive substances to seeds during priming can further enhance seed performance. Pretreating carrot seeds with 10^{-4} M abscisic acid for 12 days produced earlier and more uniform germinating seeds. Thomas (1983) added gibberellins (GAs), ethephon and daminozide to celery and germination was enhanced. Jalapeno pepper seeds hydroprimed with ethephon and GA_3 increased in percentage germination without abnormal radicle formation (Carter and Stevens, 1998); this treatment also increased seed performance after storage of *Echinacea purpurea* seeds. Adding GA_3 to soybean seeds during osmopriming improved percentage germination and force of seedling emergence.

2.7. Advantages of matri-conditioning and differences from osmo-conditioning

1. High matric potential

In matriconditioning, seed hydration is controlled by the water retentive physical forces of the carrier matrix. In contrast, seed hydration during osmoconditioning is controlled by the osmotic forces of the dissolved substances in the carrier.

2. Good water holding capacity.

The carriers used for matriconditioning, have high water holding capacity, thus requiring a low carrier water ratio by weight for conditioning seeds. Much higher carrier: water ratio by weight is needed for a solid carrier that relies predominantly on high solute content for seed conditioning.

3. Insoluble in water

Materials found successful for matriconditioning are water insoluble. In contrast osmoconditioning depends upon ionizing and non-ionizing solutes, such as salts, alcohols, sugars, and non-penetrating organic molecules.

4. Good inoculation and delivery system.

Micro-Cel and Zonolite vermiculite have characteristics that make them ideal carriers for liquid inoculants and for multiplying beneficial microorganisms (bacteria and fungi) around the seed and for delivering them to target areas in the rhizosphere for the control of destructive soil borne insects and diseases during seedling establishment.

5. High bulk value

Because of the high bulk value of the carriers used for matriconditioning, a small amount of carrier relative to seed can be used, facilitating treatment, handling, and transport of large amounts of seeds.

6. Large available surface area

The high surface area appears to allow ready exchange and diffusion of toxic substances and inhibitors present in some seeds (*e.g.* sugar beet), thus diluting the inhibitor content of the seed and consequently improve seedling establishment.

7. Enzyme and microbe immobilization

Use of carriers as bio-catalysts and for seed conditioning requires that they can be relatively inert chemically, have high surface area, be mechanically stable, and have high water holding capacity.

2.8. Integration of seed treatments

2.8.1. Osmotic or matricconditioning with pregermination

Studies with parsnip seeds showed that osmoconditioning followed by pregermination improved the percentage seedling emergence when compared to the untreated and pregerminated seeds. Although both osmoconditioning and pregermination reduced the mean emergence time, the effect of the combined treatment, in general, was greater than either treatment alone.

2.8.2. Stratification with osmotic conditioning

Stratification or moist-chilling of seeds is a dormancy breaking treatment that restores full germinability and is generally conducted at 1-10°C for prolonged period of time. In contrast, osmoconditioning reduces germination time and invigourates the seed. However, the two physiological treatments can be combined to further improvement in germination. Other dormancy breaking treatments can also be combined with osmoconditioning.

2.8.3. Growth regulator with seed coating

Seed coating and pelletting technology have been developed to suit different needs; for seed singulation; for adding bioactive chemicals, nutrients and beneficial microbes; and for sowing seeds under dry and wet conditions. Several studies have examined the advantages of growth regulator permeation into seeds with or without seed coating.

2.8.4. Presoaking, osmotic conditioning or pregermination with bioactive chemicals

Improved seed performance has been achieved by incorporating plant growth regulators and pesticides during presoaking, osmoconditioning and other presowing treatments. Inclusion of growth regulators, such as gibberellin (GA), during prolonged dark osmotic treatment in PEG solution reduced the time to germination, improved the rate of germination, and prevented the induction of secondary dormancy in lettuce, perennial goosefoot and curly dock seeds. Pesticides and nutrients have been effectively combined with the seed hydration treatments.

2.8.5. Matricconditioning with growth regulators

Matricconditioning with the addition of ethephon has proved effective in preventing thermoinhibition at supraoptimal temperatures in lettuce and celery seeds.

3. HARDENING

Hardening (synonyms: wetting and drying, drought hardening) appears to impart drought tolerance and is particularly effective in improving percentage germination of old seeds. Seeds are hydrated at 15-25°C for various time with a limited amount of water with final water content of seeds usually ranging from 30-70% of the air-dry weight. Considerable caution is needed as the period of hydration before drying may be critical and may adversely affect seed germination. Even the uptake of a limited amount of water could be injurious to seeds such as bean, lima bean, sweet corn and soybean. These seeds are susceptible to imbibitional injury, and care must be taken to slow the entry of water into the seed, if the advantage of seed conditioning by partial seed hydration is to be realized. Henckel (1964) reported that physiological preconditioning followed by dry back to original moisture content enhanced the quality characters of seed. This technique has been widely adopted for dry land crops for early emergence; garden land for better and uniform establishment with enhanced physiological vigour status of the seed.

It is the process of hydrating the seed to initiate the pre-germinative metabolism followed by dehydration which fixes the biochemical events. This was done in order to impart resistance against stress conditions *viz*, drought and cold to the emerging seeding. The basic process involved in presowing seed hardening is increase in hydrophilic property of protoplasmic colloids namely viscosity and elasticity and increased phosphorylation activity of the mitochondria (Henckel, 1964). Reduction in solute leakage by regaining cell membrane integrity, resumption of rate of protein and RNA synthesis characterized in the first period of imbibition and shortening of the times of DNA replication in the second hydration period. Stimulation of long lived mRNA under moisture stress conditions with simultaneous protein and proline content increase after hardening treatments.

3.1. Changes associated with hardening

Physiological reorganizations

i) Significant increase in hydrophilic property of the protoplasmic colloids namely the viscosity and elasticity
ii) Quality of the proteins are changed
iii) Osmotic pressure increases
iv) These overall changes lead to increase in water holding capacity of plants

Metabolic changes

i) Increased respiration
ii) High level of synthetic reactions even during drought; leaves of hardened plants have more starch

iii) Increased phosphorylation activity of the mitochondria
iv) Higher rate of photosynthesis because of increase in the bound water
v) Higher nucleoproteins content

Anatomical and morphological changes

i) Hardened plants have a more xeromorphic morphology than unhardened ones
ii) More extensive and denser network of veins and ribs
iii) Epidermal and stomatal cells are smaller
iv) Number of stomata per unit leaf area is greater
v) Foliage area is increased
vi) Faster recovery from atmospheric drought
vii) Greater absorbing root system, as well as more number of primary roots

3.2. Hydro-hardening

Different amounts of water for different species have been new mended for hydro-hardening. Best results are claimed for seeds subjected to two or three cycles of wetting and drying, although for some species one cycle is sufficient. For spring wheat, water equivalent to 30% of the grain weight is added at each cycle with caesium, 30%, 20% and 15% are added in three successive cycles, and with leucurrum, 40%, 30% and 30%, pea and vetch seeds are given 75% water in one cycle only the water being added in small amounts at a time.

As germination and growth proceeds, the resistance to drying of the embryo decreases. At the same time the effect of treatment is claimed to be greater if the embryo is more advanced at the time of drying. Therefore,

(i) The timing of the initial imbibition period is a compromise between the above two conflicting tendencies. Besides
(ii) The temperature of the soaking and
(iii) Drying cycles and
(iv) The rate of drying may all be importance. Workers have also registered failures of drought hardening technology. Many researchers recorded inconsistencies in the responses of a different species of cultivar following treatment. This suggests that the responses of plants following wetting and drying cycles at the seed stage are very variable or that the optimal conditions have not always been established. Thus, it seems that every species, cultivar and harvest time must be investigated individually under a variety of climatic conditions to determine, if presowing treatments are a useful proposition.

Investigation on the influence of seed hardening in jatropha seedling production revealed that seeds hardened with 1% solutions of KCl, $CaCl_2$ adopting equal volume of seed to solution ratio had 14% and 11% higher survival, respectively, than untreated seeds. This might be due to enhanced oxidative phosphorylation, utilization of sugars during pentose phosphate cycle, formation of $NADPH_2$ and ribose-5-phosphate (Kursanov *et al.,* 1965). Apart from that, hardening with $CaCl_2$ and KCl also increased shoot length, leaf area, root volume, dry weight of the root and chlorophyll content in jatropha (Table 4). Fujii and Saka (2001) in rice reported similar results by exogenous application of chemicals. Kamalam and Nair (1989) expressed that potassium applied in the form of KCl enhanced the cell elongation, cell division as well as enzyme activities which increased the metabolic activities, speed of germination and total germination. KCl also proved its effectiveness in total chlorophyll content of leaf of jatropha (Kathiravan, 2004).

Table 4: Influence of presowing seed hardening treatment on shoot length (cm), leaf area (cm^2 $plant^{-1}$), root volume (cc $seedling^{-1}$), dry weight of root (g $seedling^{-1}$) and total chlorophyll content (mg g^{-1}) of jatropha

Nutrients	Shoot length (cm)	Leaf area (cm^2 $plant^{-1}$)	Root volume (cc $seedling^{-1}$)	Dry weight of root (g $seedling^{-1}$)	Chlorophyll content (mg g^{-1})
Dry seed	24.4	89.8	0.63	0.355	3.25
Water soaking	26.9	109.5	1.22	0.422	4.12
$CaCl_2$ 0.5%	29.5	116.5	1.77	0.491	4.52
1.0%	32.6	127.6	2.35	0.650	5.41
1.5%	31.2	120.6	1.92	0.510	4.59
KCl 0.5%	33.4	128.4	2.40	0.699	5.05
2.0%	31.2	118.6	1.85	0.554	4.45
3.0%	30.8	113.2	1.72	0.511	4.32
CD (5%)	0.41	4.77	0.107	0.027	0.0277

3.3. Humidification

Humidification is achieved by exposure of seeds to a water vapour saturated (100% RH) or nearly saturated atmosphere. A relationship between the relative humidity and seed moisture content has been developed for various species. In most species, moisture uptake is rapid for the first 2-3 days and then decreased. The disadvantages and difficulties in humidifying seeds at high RH, include (1) difficulty in maintaining constant temperature conditions and preventing water condensation on the seeds; (2) impossibility of keeping a uniform moisture content during the treatment; (3) extreme sensitivity of the seed hydration process to differences in carbondioxide and oxygen concentrations in confined atmospheres; (4) the danger of seed deterioration during protracted period of seed hydration; and (5) substantially less hydration of seeds in water vapour equilibrated atmosphere than in seeds imbibed in contact with liquid phase, especially, the part of the seed that can retain "free" water which can be utilized by the seed during germination. It has become increasingly apparent that osmoconditioning is particularly suited to improve the performance of slow emerging small vegetable seeds under cold and/or wet conditions and where the paramount consideration is increasing the biomass of the vegetative portion of the plant rather than seed or fruit yield.

4. SEED INFUSION

It is a method of impregnation of seeds with bioactive chemicals through organic solvents instead of water. This technique of infusion helps to avoid the damage caused by the seed due to water soaking. Hence, highly suitable to seeds that suffer with soaking/seed coat injury, simultaneously, it also provides protective, regulatory and selective functions of the chemicals to improve the performance of seed with the help of organic solvents. The depth and amount of chemicals penetrated into the seed vary with (i) duration of treatment (amount of chemical reaching the embryo increased with increase in time), (ii) concentration of solutes (higher the concentration more will be the penetration), and (iii) type of chemical (compared to IAA, GA_3 reaches the embryo lesser in quantity in a given time due to higher molecular weight of GA_3).

5. FORTIFICATION

It is a process of enriching the seeds with bioactive chemicals for improving the germination and seeding vigour and to overcome dormancy. Seed fortification helps to improve germination and seed vigour due to the supplemental nutrients provided. The bioactive chemicals used for fortification may help to overcome the effects of chemical dormancy. Fortification of seeds with ferrous sulphate helps in increasing seed storability by bringing about significant reduction in lipid peroxidation. Seed fortification is impregnation of required substances for invigorating the seed for improved production where the seeds are soaked either in water or dilute solutions of bioactive chemicals such as micronutrients, growth regulators, vitamins and seed protectants. The seeds are soaked in solutions in equal volume for a duration of 6-24 hours depending on the crop to increase the moisture content of the seed to 20-25% just enough for endogenous impregnation of the chemicals by exogenous application. The various nutrients used in fortification are (i) nitrogen (constituents of amino acids, nucleotides and also acts as a coenzymes), (ii) phosphorus (constituents of phospholipids, nucleotides and also sugar phosphate), (iii) potassium (required for activation of enzymatic process), (iv) sulphur (constituent of amino acid like cystine, methionine, cystenine), (v) magnesium (constituent of chlorophyll), (vi) iron (integral part of protein and constituent of phytochromes and thereby, it involves in photosynthesis and respiratory action of the system, It also forms a consistent of enzymes such as catalase , peroxidase and dehydrogenase enzyme) and (vii) zinc (constituent of dehydrogenase, alcohol dehydrogenase and lactic dehydrogenase enzyme).

The seeds were soaked in micronutrient solutions of $FeSO_4$, $ZnSO_4$, and MnSO4 with three different concentrations (0.5, 1.0 and 1.5%) and soaking durations (3, 6 and 9 hour) along with untreated control was evaluated in jatropha. The observations on 60 DAS revealed that both the nutrients and durations of soaking significantly improved the shoot length, leaf area, root volume and dry weight over control. Among different concentrations, seeds fortified with 1% solution each of $FeSO_4$, $ZnSO_4$ and $MnSO_4$ recorded the highest survival percentage (81, 82 and 80) than control (66%). The lesser (0.5%) and higher (1.5%) concentration was sub and supra-optimal, respectively, for seed fortification, while 1% irrespective of nutrients was optimum (Kathiravan, 2004).

Presoaking sunflower seeds in 0.5% sodium molybdate, 1% $MnSO_4$ or 2% $ZnSO_4$ increased seed and oil yields of sunflower (Muthuvel *et al.,* 1983). Vijayakumar *et al.,* (1988) also reported improved germination in bhendi by seed soaking in 0.2 or 0.5% $FeSO_4$. Seeds of cowpea and black gram fortified with $ZnSO_4$ + $MnSO_4$ + $NaMO_4$ registered the maximum field emergence, rate of germination, root length, shoot length, dry matter production and vigour index (Vijaya, 1996). Anbazhagi (1997) found that soaking the groundnut kernels in solutions of KH_2PO_4 (0.5, 0.75 and 1%), $CaCl_2$ (0.5 and 1%) and groundnut micronutrient mixture (0.25 and 0.5%) increased the germination potential, speed of germination, seedling length, dry matter production and vigour. Bioactive chemicals may be of (i) seed protectants (fungicide, bactericide), (ii) herbicide antidotes (to prevent the ill effects of herbicide, herbicides antidotes are infused) (iii) growth regulator (GA_3, IAA and kinetin are incorporated to nullify the endogenous deficit of the growth substances), (iv) nutrients (micro and macro elements) and (v) substances to resist adverse stress conditions (substances like fusicoccin alleviate the adverse effect in soil).

6. FILM COATING

The primary way of enhancing seed performance is through the addition of chemicals to protect the seed from pathogens. These are generally fungicides, although insecticides and other protectants can also be applied in this way. Until recently, these chemicals were applied mainly as slurry that coated the seeds and then was dried. Disadvantages of this method are that it is difficult to get a uniform coating of the material on the seed and much of the material can be rubbed off the seed during transport and handling. In recent years, film-coating methods have been developed to overcome these problems. In film coating, the chemicals are applied in a polymer that is sprayed on the seeds as they fall though a specialized machine. The polymer is rapidly dried, so that the seeds emerge with a complete, dry polymeric coating. This coating does not rub off the seeds, greatly reducing worker exposure and waste. In addition, colors can be added to the polymers to code different varieties or seed lots. Large-scale units capable of treating large quantities of seeds per hour have been developed, making it feasible to treat high-volume agronomic crops by this method as well.

6.1. Coatings to modify germination properties

Several studies have been published on ideas to use treatment or film coating techniques to manipulate seed imbibitions characteristics. Aksenova *et al.* (1994) reported that treatment with non-ionic surfactants can stimulate the germination of soybeans and wheat under water stress. Alternatively, efforts have been made to delay germination or artificially impose 'exogenous coat dormancy' by film coating with water resistant polymers. Pellet types have been developed using hydrophobic materials to allow seeds to germinate under wet conditions, where germination might otherwise be erratic, such as for onion. Conversely, water-attracting material can be used to aid imbibitions, or pellet coatings may be designed to disintegrate or split after imbibitions to expose the seed.

Some novel applications have also been developed using the film coating method. For example, artificial polymers have been developed that exhibit temperature-sensitive permeability to water. These polymers are permeable to water at warm temperatures, but not at cool temperatures. Modifying the composition of the polymer can set the temperature at which the permeability changes. Seeds coated with these polymers will not imbibe water if the temperature is below the set point of the polymer, potentially protecting the seed from imbibitional chilling injury. The coatings are also being used to delay germination after planting, such as for timing the emergence of male parent lines at different times for hybrid seed production. A starch based biopolymer is also being used in film coating to slow water uptake and alleviate chilling injury.

The picture below shows (Fig. 1) bare lettuce seed at the far left, next is film coated seed (green), then encrusted seed (pink) and finally lettuce seed with a full seed pellet around it (white). Film coating puts a very thin, colored film around the seed, increasing the weight of the seed by about 5%, without changing the shape of the seed. This is in contrast with seed encrustment, which can increase seed weight from 50% to 200, and a full pellet which can increase the weight of a seed by as much as 3500%, (35:1) and completely change the shape of the seed to a uniform sphere.

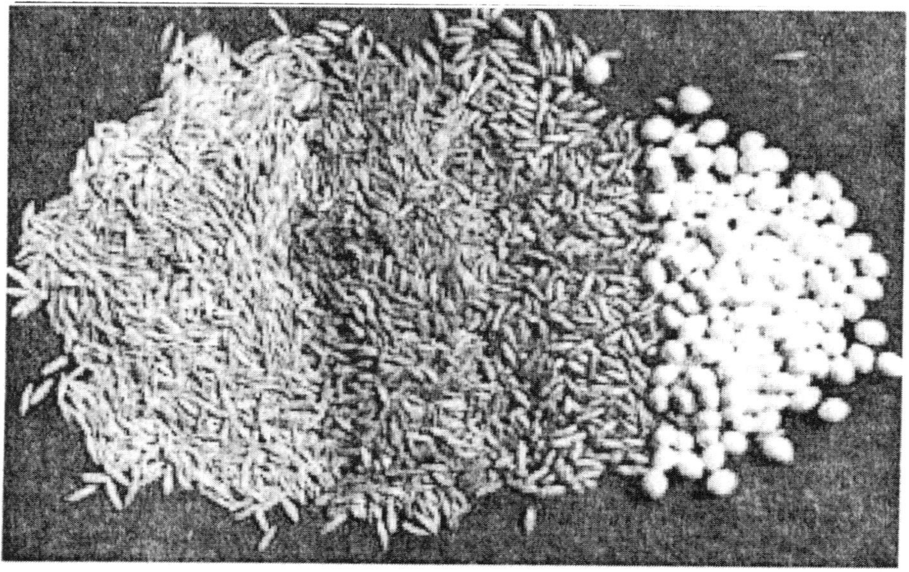

Fig. 1: Film coating compared to other seed coats and pellets

In its most simple form, the film coat material consists of a polymer (large molecules in repeating chains), a plasticizer (helps make the coat more elastic and/or less tacky after drying), and a colorant.

6.2. Advantage of film coat

- Fungicides and pesticides are completely enclosed around the seed by the non-toxic film coating material. The pesticides stay directly on the seed for maximum affect and workers handling the seed are not exposed to any toxic chemicals.
- The film coating material can help the seed flow better in the planter box, depending on the type of film coating material used.

- The colour makes the seed easier to pick out of the soil when checking for seed spacing during planting.

6.3. Use of film coating

In film coating, treatment formulations are premixed in aqueous suspension with a binder, adjuvants, pigments and opacifiers, and sprayed onto a moving seed mass. The binder forms a film as the mixture dries, so that the additives applied are stuck firmly to the seed surface. Typical binders are polymers of various types, such as derivatised starches and celluloses (*e.g.* with hydroxypropy or methyl-substitution, polyvinyl acetate, polyvinyl alcohol or poly-vinylpyrrolidone).

7. HYDRATION - DEHYDRATION

It is the process of soaking the low and medium vigour seeds in water with or without added chemicals usually for short durations to raise the seed moisture content to 25-30% and bring back the seed to safe limits for dry storage. It is done to improve the vigour of medium and low vigour seeds by bringing about certain preventive, corrective and restoration action. The hydration and dehydration treatments include soaking – drying , dipping - drying , spraying – drying, stepwise hydration - drying, moisture equilibration – drying, moisture equilibration soaking – drying, moist sand conditioning–drying, *etc.* The choice of the treatment depends upon the characteristics of seed and initial vigour status of the seeds.

8. STEEPING AND OTHER WET ENHANCEMENT TECHNIQUES

It has long been known that steeping and drying, with or without one or more wetting treatments can improve seed germination performance. A washing or steeping treatment is now routinely conducted on a large scale to remove residual amounts of inhibitors in sugar beet, and thus raise field establishment levels in affected seed bulks. Steeping is also used to introduce chemical treatments to seeds. At the simplest level, there is long-standing practice in some quarters for farmers to sow seed that has been steeped to a day to begin the germination process.

9. PREGERMINATION

Pregermination is characterized by seed hydration to the point of radicle protrusion. The result is faster, more uniform germination and almost 100% seedling establishment. The seed is permitted to complete germination as evidenced by radicle protrusion, but not to proceed beyond cell division. Seeds that complete germination will float and are separated from those that do not by using liquid density gradients. Because of rapid seed deterioration, pregerminated seeds should be used within 1 or 2 weeks. Recent research has focused on the imposition of desiccation tolerance to extend the shelf-life of pregerminated seed. To accomplish this, studies have focused on redrying techniques and germinated seeds are now available with a longer shelf-life (3–4 months). Pregerminated seeds offer significant advantages to those interested in optimum and uniform seed performance. Speed and uniformity of seedling emergence are attributes of pregerminated seeds, since they have already completed germination and have been selected to be at the same morphological stage of development. Methods of germinating and selecting the seeds were developed, but as they generally were not tolerant of dehydration after radicle emergence, they had to be planted wet.

10. SEED PELLETING

The idea behind seed pelleting is to take a seed that is comparatively light in density and oddly shaped and turn it into a seed with a pellet around it that makes that seed heavier, more uniform, and as perfectly round as possible. A planter will be able to plant that seed/pellet in the most accurate and precise way that is mechanically possible. Creating a pellet is both a science and an art. Most of the science comes when selecting the powders and adhesive used to create the pellet around the seed. The materials should form a pellet structure that is porous enough to provide immediate oxygen to the germinating seed, and absorb water in a way that is most beneficial to the seed. It is necessary sometimes to have a mixture of 4 or more powders in order to achieve the desired characteristics and also end up with the desired pellet density. The adhesive is also very important. It needs to be strong enough to hold the powder together, without making the pellet so hard that it slows the germination process. Putting together the right combination of adhesives and powders to create a pellet that actually splits open much like an egg cracks open is in need at present. This splitting action is said to lead to even faster oxygen availability and thus less stress on the germinating seed. The seeds of many horticultural species are either small or irregularly shaped, making it difficult to singulate them for planting, even if the seeds can be dropped individually to control spacing, it is difficult with small seeds to control the depth of planting. Seed pelleting with clay or other materials give them a uniform shape and size and increasing their size and density to allow more precise placement in the soil. Small and valuable seeds, such as petunias or geraniums, are often pelleted to increase their size and facilitate their use in automatic planters in greenhouses. Modern types of pellets split upon imbibition, freeing the seed from any impediment by the coating material.

11. PHYSIOLOGICAL AND BIOCHEMICAL ASPECTS OF CONDITIONING

11.1. Enhancement in germination potential

Physiological and biochemical changes occur during osmoconditioning that allow seeds to develop a high germination potential (radicle thrust) or the ability to remove the seed coat restraint. This might be the basis for an improvement in the rate of germination and emergence often seen in osmoconditioned seeds. The improved germination potential can be demonstrated readily by germinating the conditioned and untreated seeds in solutions of PEG with a progressively lower osmotic potential to counter the radicle thrust.

The ability of osmoconditioned seeds, particularly those conditioned in the presence of ethephon and/or kinetin; to generate high germination potential appears to be the basis for the alleviation of a wide variety of stresses imposed by supraoptimal temperatures, salinity, and drought in lettuce and perhaps other seeds. The development of a high radicle thrust during osmoconditioning may be related to activation and/or synthesis of numerous enzymes, increased metabolic activity, and the production of osmotically active substances.

11.2. Role of gibberellin, ethylene and cytokinin

Aminoethoxyvinyl glycine (AVG), a specific inhibitor of the conversion of S-adenosyl methionine to 1-aminocyclopropane-1-carboxylic acid (ACC) in the ethylene biosynthetic pathway, failed to inhibit the germination advancement effect of osmoconditioning in the light or in the presence of GA in the dark, indicating that ethylene biosynthesis was not involved in the germination advancement effect of seed osmoconditioning. These results support the suggestion that the action of GA is directly related to germination while the role of ethylene, cytokinin, and ABA is secondary, involving the removal or imposition of stress on germination.

12. CONCLUSION

Seed priming has been successfully applied to many crops where rapid, uniform, and complete emergence is desirable attributes. It is clear that the technology of seed priming continues to evolve. In some of the more complex systems, such as osmopriming, osmotica (e. g. salts, PEG) are used to control hydration. These viscous substances, especially PEG, must be aerated, removed from the seed prior to drying and properly disposed off. An extremely complex but increasingly popular priming technique is pregermination, where seeds are hydrated to the point of radicle protrusion. Separation of the seeds with protruded radicles assures 100% performance.

It seems that priming initiates the earliest physiological events associated with seed germination. However, priming also appears to improve seed performance through its amelioration of deteriorated seeds. Most evidence suggests that priming allows time for the seed to 'repair' damage from deteriorative events associated with mitochondrial dysfunction, enzyme inactivation, membrane perturbations, and genetic damage incurred during seed storage and aging. Thus, it may be concluded that priming has two roles in improving seed performance. The first is the advancement of events leading to germination. The second is the repair of seed damage that allows more efficient germination.

Glossary of terms used in this chapter

Aeration, oxygenation: A germination promoting treatment applied during seed hydration by bubbling air or oxygen.

Biopriming: A combined treatment involving seed bacterization and seed moisturizing.

Conditioning: Enhancement of physiological and biochemical events (those associated with rapidity and synchrony of germination, repair and improved germination potential) in seeds during controlled hydration and suspension of germination by low water potential of the imbibing medium (liquid or moist solid). Seed is hydrated until water potential of the seed and the imbibing medium become equal (reach equilibrium water potential). Conditioning is usually conducted at 15 to 20°C.

Hardening, wetting and drying, drought hardening: Soaking of seeds with a limited amount of water followed by drying and sometimes, repeating the wetting-drying cycle.

Humidification: Exposure of seeds to a water vapor saturated (100% RH) or nearly saturated atmosphere. Equilibrium moisture content of seeds at specific RH varies with crop species.

Irradiation: Exposure of seeds in a hydrated state to light in order to break seed dormancy or to improve the rate of germination.

Matriconditioning: Enhancement of physiological and biochemical events in seeds during suspension of germination by low matric potential and negligible osmotic potential of the imbibing medium.

Moisturizing: A treatment to partially hydrate seeds on a material with hydrophilic surfaces in order to prevent germination and to invigorate the seed.

Osmoconditioning: Enhancement of physiological and biochemical events in seeds during suspension of germination by low osmotic potential and negligible matric potential of the imbibing medium. Salts or non-penetrating organic solutes in liquid medium or matrix solution are used to establish equilibrium water potential between seed and the osmotic medium needed for conditioning.

Permeation with hormones: soaking of seeds for a few minutes to several hours in aqueous or non-aqueous (organic solvents) hormone solutions. Response of seed to permeated hormones is elicited during subsequent hydration.

Pregermination: Seed soak under optimal hydration and temperature conditions to a point when the radicle is just visible, usually 1 to 2 millimeter in length.

Presoaking: Pregermination water soak of seeds in unlimited amount of water at low to moderate temperatures.

Priming, osmopriming, osmotic priming: These terms are used interchangeably with osmoconditioning.

Solid matrix priming: The term has been used to refer to seed conditioning by osmotic and/or matric component of the solid matrix water potential.

Stratification: A dormancy breaking seed treatment generally conducted at 1-10°C during prolonged seed soak in water.

Thermal shock: A dormancy breaking seed treatment achieved in a hydrated state by a brief shift to high temperature (*e.g.* 1 h at 40°C) followed by a return to ambient temperature (20-25°C) condition.

REFERENCES

Akers, S.W. 1990. Seed response to priming in aerated solutions. *Search*, **19**:8-17.

Aksenova, L.A., M. V. Dunaeva, E. A. Zak, F. Yu Osipov and N. L. Klyachko. 1994. The effect of Tween 80 on seed germination in winter wheat cultivars differing in drought resistance. *Russian J. Plant Physiol.*, **41**:557-559.

Armstrong, H. and M. B. McDonald. 1992. Effects of osmoconditioning on water uptake and electrical conductivity in soybean seeds. *Seed Science and Tech.*, **20**:391-400.

Anbazhagi, G. 1997. Nutrient management on seed yield, quality and storability of bunch groundnut (*Arachis hypogaea*) cv. CO 2. M.Sc. (Ag.) Thesis, TNAU, Coimbatore, India.

Ashraf, M. and C. M. Bray. 1993. DNA synthesis in osmoprimed leek (*Allium porrum*, L.) seeds and evidence for repair and replication. *Seed Science Res.*, **3**:15-23.

Bailly, C., A. Benamar, F. Corbineau, and D. Come. 1997. Changes in superoxide dismutase, catalase and glutathione reductase activities in sunflower seeds during accelerated ageing and subsequent priming, *In* Basic and Applied Aspects of Seed Biology (eds, R. H. Ellis, M. Black, A. J. Murdoch and T. D. Hong), Kluwer Academic Publishers, Boston, pp. 665-672.

Bewley, J.D. 1997. Seed germination and dormancy. *Plant Cell.* **9**:1055-1066.

Biniek, A. and I. Babik. 1993. The influence of osmoconditioning in polyethylene glycol (PEG 6000) on the germination and emergence of carrot and parsley seeds. *Acta Horticulturae.* **371**:77-81.

Bradford, K. J. and A. M. Haigh. 1994. Relationship between accumulated hydrothermal time during seed priming and subsequent seed germination rates. *Seed Science Res.,* **4:** 63-69.

Bradford, K.J., J. J. Steiner, and S. E. Trawatha. 1990. Seed priming influence and emergence of pepper seed lots. *Crop Sci.* **30**:718-721.

Bray, C.M. 1995. Biochemical processes during the osmoconditioning of seeds *In* Seed development and Germination (eds. J. Kigel and G. Galili). Marced Dekker, NY. pp. 767-789.

Bujalski, W. and A. W. Nienow. 1991. Large scale osmotic priming of onion seeds a comparison of different strategies for oxygenation. *Scientia Horticulturae.* **45:**13-24.

Burgess, R. W. and A. A. Powell. 1984. Evidence for repair processes in the invigoration of seeds by hydration. *Ann. Bot.*, **53:**753-757.

Carter, A. K. and R. Stevens. 1998. Using ethephon and GA_3 to overcome thermoinihibition in 'Jalapeno M' pepper seed. *Hort. Sci.* **33**:1026-1027.

Chang, S.M. and J. M. Sung. 1998. Deteriorative changes in primed sweet corn seeds during storage. *Seed Science and Tech.*, **26**:613-626.

Choudhuri, N. and R. N. Basu. 1988. Maintenance of seed vigour and viability of onion (*Allium cepa* L.) *Seed Science and Tech.* **16**:51-61.

Coolbear, P. and C. R. McGill. 1990. Effects of a low temperature presowing treatment on the germination of tomato seed under temperature and osmotic stress. *Scientia Horticulturae,* **44**:53-54.

Drew, R.L.K. and J. Dearman. 1993. Effect of osmotic priming on germination characteristics of celeriac (*Apium graveales* L. var. rapaccum). *Seed Science and Tech.,* **21**:411-415.

Durrant, M.J. and A. H. Loads. 1987. Experiments to determine the optimum advancement for the treatment for sugar beet seed. *Seed Science and Tech.*, **15:**185-196.

Finch-Savage, W.E., D. Gray and G. M. Dickson. 1991. The combined effects of osmotic priming with plant growth regulators and fungicide soaks on the seed quality of five bedding plant species. *Seed Science and Tech.*, **19:**495-503.

Frett, J.J., W. G. Pill, and D. C. Morneau. 1991. A comparison of priming agents for tomato and asparagus seeds. *Hort Sci.*, **26**:1158-1159.

Fu, J.R., X. H. Lu, R.Z. Chen, B.Z. Zhang, Z.S. Liu, Z.S. Ki, and C.Y. Cai. 1988. Osmoconditioning of peanut (*Arachis hypogaea* L.) seeds with PEG to improve vigour and some biochemical activities. *Seed Science and Tech.*, **16**:197-212.

Fujikura, Y., H.L. Kraak, A.S. Basra, and C.M. Karssen. 1993. Hydropriming, a simple and in expensive priming method. *Seed Science and Tech.*, **21**:639-642.

Fujji, S. and H. Saka. 2001. The promotive effect of brassinolide on lamina joint cell elongation, germination and seedling growth under low temperature stress in rice. *Plant Production Sci.*, **4:** 210-214.

Gray, D., J.R.A. Steckel and L.J. Hands. 1990. Response of vegetable seeds to controlled hydration. *Ann. Bot.*, **66:**227-235.

Gray, D., R. L. K. Drew, W. Bujalski and A.W. Nienow. 1991. Comparison of polyethylene glycol polymers, betaine, and L-proline for priming vegetable seed. *Seed Science and Tech.*, **19:**581-590.

Haigh, A.M., E.W.R. Barlow, F.L. Milthorpe and P.J. Sinclair. 1986. Field emergence of tomato carrot and onion seeds primed in an aerated salt solution. *J. American Soc. Horticultural Sci.*, **111**:660-665.

Halpin-Ingham, B. and F.J. Sundstrom. 1992. Pepper seed water content, germination response and respiration following priming treatments. *Seed Science and Tech.*, **20**:589-596.

Henckel, P.A. 1964. Physiological response of plants under drought. *Ann. Rev. Pl. Physiol.*, **15** : 363-386.

Heydecker, W., J. Higgiris, and R. L. Gullivar. 1973. Accelerated germination by osmotic seed treatment. *Nature*, **246**: 42 -44.

Hofmann, P., C. Aschermann-Koch, and A. M. Steiner. 1992. Presowing treatment for improving seed quality in cereals. II. Field emergence. *Seed Science and Tech.*, **20**:441-446.

Huang, Y.G. and Q. Zou. 1989. Effect of osmoconditioning and drying on germination of *Pinus sylvestris* var. Mongloica and *Larix gmeliniii* seeds. *Seed Science and Tech.*, **17**:235-242.

Kamalam, J. and N. R. Nair 1989. Effect of seed hardening on germination and seedling vigour in paddy. *Seed Res.*, **17:** 188-190.

Kathiravan, M. 2004. Seed production, processing, testing and storage techniques in jatropha. Ph.D. thesis, TNAU, Coimbatore, India.

Khan, A. A. 1992. Preplant physiological seed conditioning. *Hort. Rev.*, **14:** p.131-181.

Khan, A.A., N. H. Peck, A. G. Taylor, and C. Sammmy. 1983. Osmoconditioning of beet seeds to improve emergence and yield in cold soil. *Agronomy J.*, **75**:788-794.

Kibite, S. and K.N. Harker. 1991. Effects of seed hydration on agronomic performance of wheat, barley and oats in central Alberta. *Canadian J. Plant Sci.*, **71**:515-518.

Kursanov, A. L., E. Byskrebentseva, I. Sveshnikova and M. Krasavina. 1965. Disorganization of energy metabolism in roots during potash metabolism. Nauk *SSSR*, **162**:9.

Muthuvel, P., B. Habeebullerh and A. Chamy. 1983. Effect of presoaking of sunflower seeds in micronutrients on seed yield and oil yield. *Madras Agri. J.*, **70:** 375-376.

Nath, S., P. Coolbear, and J. G. Hampton. 1991. Hydration-dehydration treatments to protect or repair stored "Karamu" wheat seeds. *Crop Sci.*, **31**:822-826.

Ozbingol, N., F. Corbineau and D. Come. 1988. Responses of tomato seeds to osmoconditioning as related to temperature and oxygen. *Seed Science Res.*, **8**:377-384.

Penaloza, A.P.S. and M. T. S. Eira. 1993. Hydration-dehydration treatments on tomato seeds (*Lycopersicon esculentum* Mill). *Seed Science and Tech.*, **21**:309-316.

Pill, W.G. and T.K. Korengel. 1997. Seed priming advances the germination of Kentucky bluegrass (*Poa pratensis* L.). *J. Turf grass Manage.*, **2**:27-43.

Rao, S.C., S.W. Akers and Ahring, R.M. 1987. Priming brassica seed to improve emergence under different temperatures and soil moisture conditions. *Crop Sci.*, **27**:1050-1053.

Rudrapal, D. and S. Nakamura. 1988. Use of halogens in controlling eggplant and radish seed deterioration. *Seed Science and Tech.*, **16**:115-122.

Saha, R., A. K. Mandal and R. N. Basu. 1990. Physiology of seed invigoration treatments in soybean (*Glycine max* L.). *Seed Science and Tech.*, **18**:269-276.

Smith, P.T. and B.G. Cobb. 1991. Physiological and enzymatic activity of pepper seeds (*Capsicum annum*) during priming. *Physiol. Plant.* **82**:433-439.

Smith, P.T. and B.G. Cobb. 1992. Physiological and enzymatic characteristics of primed germinated pepper seeds (*Capsicum annum* L.). *Seed Science and Tech.*, **26**:503-513.

Steuter, A. A., A. Mozafar and J. R. Goodin. 1981. Water potential of aqueous polyethylene glycol. *Plant Physiol.*, **67**: 64-67.

Sung, F.J.M. and Y.H. Chang. 1993. Biochemical activities associated with priming of sweet corn seeds to improve vigour. *Seed Science and Tech.*, **21**:97-105.

Tarquis, A.M. and K.J. Bradford. 1992. Prehydration and priming treatments that advance germination also increase the rate of deterioration of lettuce seed. *J. Exp. Bot.*, **43**:307-317.

Taylor, A.G., D.E. Klein and T.H. Whitlow. 1988. SMP: Solid matrix priming of seeds. *Science Horticulturae*, **37**:1-11.

Thomas, T.H. 1983. Stimulation of celeriac and celery seed germination by growth regulator seed soaks. *Seed Science and Tech.* **11**:301-305.

Tilden, R.L. and S.H. West. 1985. Reversal of the effects of ageing in soybean seeds. *Plant Physiology*, **77**:584-586.

Usberti, R. and I.F.M. Valio 1997. Osmoconditioning effects on germination of guinea grass (*Pannicum maximum*) seeds. *Seed Science and Tech.* **25**:303-310.

Vijaya, J. 1996. Standardisation of presowing seed management techniques for pulse. M.Sc. (Ag.) thesis, TNAU, Coimbatore, India.

Vijayakumar, A., C. Dharmalingam and S.Sambandamurthi. 1988. Effect of presowing treatment on seed yield on quality in Bhendi. *South India Hort.*, **36**:118-120.

Seed Dormancy

8

S. Kavitha[1], K. Natarajan[2] and R. Renuka[2]
[1]*Central Institute for Cotton Research, Coimbatore - 641 003*
[2]*Adhiparasakthi Agricultural College, G.B.Nagar, Kalavai- 632 506*

I. INTRODUCTION

The seed phase is the most important stage in the higher plant life cycle with respect to survival as a species. The seed is well equipped to sustain extended periods of unfavourable conditions. The embryo, containing the full genetic makeup, is protected by one of several layers of other tissue, such as endosperm, perisperm, seed coat, and fruit tissues. These structures protect the embryo from physical damage but also play a role in spreading the seeds after abscission. Often, the surrounding layers are important elements in the regulation of dormancy and germination.

2. DEFINITIONS

Over the years many attempts have been made to define dormancy. In general, different types of dormancy have been related to the different conditions that induce dormancy, *e.g.*, thermodormancy and photodormancy, which are induced by elevated temperatures and high light intensity, respectively. Other classifications are based on the timing of dormancy, *e.g.*, primary of innate dormancy and secondary of induced dormancy. The former type is imposed during seed development and the latter essentially post harvest. However, all these conditions are descriptive and not based on differences in mechanism. Here, we will use a strictly phenomenological definition of dormancy, the absence of germination, growth of a viable seed under otherwise optimal conditions.

It is often interesting to speculate on how any particular physiological phenomenon evolved; for instance, in respect to what selective pressures did dormancy arise? It is possible that it came about as a mechanism allowing the seed, or equivalent structure, to be retained for longer periods on the parent plant, thus extending the period of seed development and maturation. The necessity to inhibit the germination of such a seed while it continued to accumulate reserves could have resulted in the origin of a dormancy system as a suppression of vivipary. The extension of this phase into the period subsequent to shedding and reimbibition would then carry the selective advantages of aiding temporal and spatial dispersion of the propagules.

After seed matures, germination follows, but these two processes are normally separated both by time and space. The interval between the two events may vary from a couple of hours to many years and from a few centimeters to thousands of kilometers. According to Pollock and Toole (1961) the function of the seed is to carry its embryo plant through the hazards of time and space to a time and space where the new plant can grow, flower, and in its turn produce seeds. Delayed germination is not accidental it represents the physiological mechanisms that keep the seed in a nongerminating state. Since young plants are vulnerable to the hazards of drought and extremes of heat and cold it is advantageous for the seed to remain in an inactive condition until it reaches a time and place appropriate for germination. In the nongrowing condition the seed moisture content is low and the cell protoplasm is protected from damage, owing to its low metabolic rate. The seed thus survives on its own nutrient reserves for a long period.

3. BIOCHEMICAL MECHANISM OF DORMANCY

The term "Seed dormancy" has been used to describe two inactive conditions: one resulting from unfavourable environmental conditions and the other due to internally imposed germination blocks. Seed germination may be delayed, for example, by inadequate water supply or unfavourable temperatures. In some seeds, germination may be prevented by the presence of blocking mechanisms within the seed, which must be removed before germination can occur. The terms "rest", "resting" or "rest periods" have been used to describe seeds or buds that are inactive because of such internal blocks. It is necessary to distinguish seed dormancy resulting from an unfavourable environment from that due to the presence of inherent blocks.

Seeds possess remarkably complex and effective protective mechanisms to ensure their long survival. Such prolonged survival of the embryo within the seed ensures dispersal, distribution, or spread of the plant species over long distances. Many seeds have the ability to adapt to seasonal rhythms *i.e.* they have a natural time clock or a delayed-action mechanism that ensures that the seeds will remain dormant until another growing season rolls around, which may be long enough to permit another generation of seeds to mature (Boswell, 1961).

The length of dormancy and the nature of delaying or resting mechanisms differ greatly among various species and cultivars of crops. Annual cycles of dormancy undergone by the seeds of some weed species was reported by Karseen and Hilhorst, (1992). Citing the work of Karseen, (1982), the author schematically represented various changes in primary dormancy, its release, germination, quiescence, and secondary dormancy. Primarily dormant seeds, either dry or imbibed, require exposure to a certain temperature for a certain period to relieve dormancy. Only then will seeds become sensitive to factors that stimulate germination. However, germination will occur only when the complete set of required factors is present. If not, germination will be inhibited. The state being called quiescence, when quiescence is extended for longer periods, the seed may enter a new state of dormancy, called secondary dormancy, which can be broken under suitable temperature conditions. The cycle appears to depend on seasonal fluctuations of the soil temperature (Bouwmeister, 1982).

Bewley and Black (1982) distinguished the term quiescence from dormancy; the former refers to nongermination of seed due to either lack of water or suitable environmental conditions, where the response of the seed is due to its internal germination blocks. The block may disappear slowly from the dry seed (after – ripening). The dormancy can be terminated under the influence of some factor not required for germination itself but only to prime the seed subsequently to respond to the conditions that support germination. Thus, a dormant seed may be promoted to germinate by a discontinuity of the conditions, one condition, being necessary to prime or potentiate seeds to germinate and another adequate for germination itself. Seeds of many plant species, for example, require chilling for several weeks at 4°C for germination. These seeds normally do not germinate at low temperatures or do so very sluggishly, but germinate readily when they are transferred to warmer conditions, *i.e.*, in order to germinate, these dormant seeds must experience a discontinuity in conditions a low temperature to break dormancy and a higher one to support germination. Similarly, some seeds germinate only after they have been exposed to light, that to only for a few seconds. The dormancy of these seeds is broken by another environmental factor light, but seeds do germinate perfectly under dark conditions. Chilling and light are not factors strictly required continually during germination but only to trigger or potentiate the process (Bewley and Black, 1982). Seed dormancy varies with external conditions like temperatures and with time both in the developing and in the mature seed.

The term dormancy or restricting cannot be applied rigidly to describe the response of certain seeds. Lettuce, for example, germinates promptly in total darkness if it is planted in moist soil at 13.9°C, since it does not contain a germination block. If it is planted at 28.9° C however, it remains inactive. After a few days at 28.9°C, if the temperature drops to 13.9°C the seed still does not germinate. Exposure of the seed to high temperature induces the formation of a germination block that did not exist previously. This block can be removed by exposing the lettuce seed to red light. It is, therefore, necessary to list carefully the conditions under which seed germination is attempted. The variety of the seed crop and its previous history are also important.

Germination blocks are not absolute but rather relative. The growth of some seeds, *e.g.*, sour cherry (*Prunus cerasus*), is not completely stopped even in a blocked seed at low temperature. Root and shoot cell division continues and the whole embryonic axis grows slightly at a time when the seed does not germinate even under the most favourable conditions. Something obviously does occur during after-ripening to permit the subsequent germination, indicating that the blocks are only relative. A blocked seed is like a motionless automobile with its motor running at idling speed with its gears disengaged (Pollock and Toole, 1961).

It is common knowledge that all viable seeds when planted do not germinate; the proportion of the seeds that do germinate varies with the conditions of germination, because all seeds are not genetically identical. The expression of a seed's heredity in the form of germination depends on the environment during seed formation, maturity and the variable germination blocks. The genetic and environmental factors interact to produce greater variability in the magnitude of germination of different kinds of seeds.

Germination blocks act through a number of different physiological mechanisms; some may be simple and well understood while others are complex and unknown. The presence of a hard seed coat impermeable to water may completely block germination in some seeds such as water lily, cotton, okra, clover and some beans. During ripening and drying processes, the seed coats of these species become impermeable to water, with the seed moisture content reaching a low level. Some germination blocks may be localized to affect the embryo. The blocked apple seeds germinate only after undergoing a period of after – ripening *i.e.*, remains moist at a temperature around 4.4°C. The imbibed seeds chilled for 2-3 months germinate promptly, producing normal seedlings.

Germination blocks can be removed by a natural combination of time and exposure to the elements. A hard seed coat may be softened in the soil by alternate freezing and thawing or wetting and drying. Microorganisms using seed coat as a source of organic carbon may rot it off in the soil and a low soil temperature may meet the chilling requirement for seed germination. Drying for a period of time constitute after – ripening in the dry condition, which induces favorable changes in the seed coat to permit germination. Mechanical scarification consisting of blowing seeds against abrasive points or rubbing them over an abrasive surface may be employed to artificially remove the hard seed coats. Hot water or acids also are used to scarify the seed coats chemically (*e.g.*, drop seed grass and cotton).

The leaching action of water and light initiates changes in seeds, permitting their germination. Some desert plant seeds require a large amount of water to remove the inhibitors blocking germination. The conditions against which blocks protect the seed often are the same one that serves to remove the blocks, which is closely related to the seasonal changes that seeds encounter (Pollock and Toole, 1961).

Both germination blocks and after ripening are reversible after ripened seeds are ready to germinate. However, blocking may again be induced in such seeds if they are exposed to conditions such as oxygen deficiency, an excess of water, or a temperature too high for germination. After ripening is a major factor in the longevity of some weed seeds. The carefully after ripened seeds in a refrigerator may revert to a blocked conditions if planted at higher temperatures (above 21.1°C), resulting in a second dormancy (*e.g.*, seeds of pine, rose, apple, cocklebur).

Germination blocks may be physical or chemical, the former being caused by structural changes in seed coats surrounding the embryo. Chemical blocks are of two types:

1. Those inhibiting chemicals in tissues surrounding the seed and
2. Chemical inhibitors within the embryo itself

Many seeds may have more than one type of block (clover). Physical blocks are associated with the structure of the seed coats and other tissues surrounding the embryo. In addition to acting as germination blocks, these tissues also protect the embryo from mechanical damage and from attack by microorganisms. Some blocks act by preventing the entry of water or exchange of oxygen and carbon dioxide through the seed coats. Most seeds need an abundant supply of oxygen for enhanced metabolism during germination. A restricted supply of oxygen may induce changes in seed metabolism imposing germination blocks.

Chemical blocks may be present in tissues surrounding the embryo or in the outermost covering of the ovary of fruit wall. Failure of seeds to germinate in the fruit may be attributed to such chemical blocks. Inhibitors are also known to be present in the seed coats and other membranes around the embryo. Hundreds of inhibitors of seed germination have been isolated and chemically characterized. Most of these are nonspecific, blocking germination in many kinds of seeds besides those from plants in which they are found. Blocks in seed coats and membranes are simpler and better understood than those present within the embryo. Although most growth phenomena are chemical processes, it may be misleading to suggest that all embryo blocks are chemical (Pollock and Toole, 1961). The embryo blocks are not necessarily caused by growth-inhibiting chemicals and may result from deficiency of some essential compound. After ripening may permit accumulation of the missing compound to a level permitting germination (*e.g., loblolly pine*) as given in Fig. 1

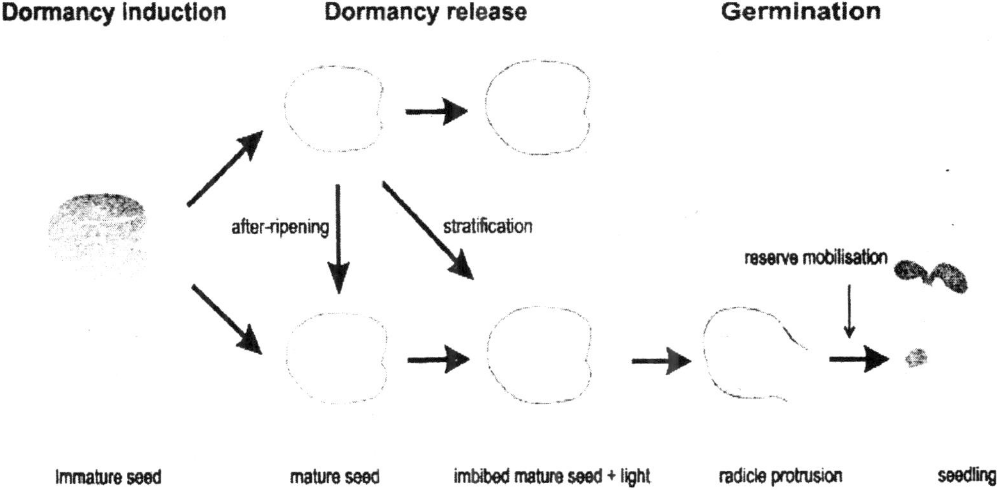

Fig. 1: Release of dormancy by after ripening and stratification

It is rather difficult to isolate and identify an inhibitor from an embryo causing germination blocks. After isolation, the active compound must be measured outside of the plant using some bioassays. Cyanide is a powerful inhibitor controlling the germination of many seeds. It occurs as a glycoside in the embryo, having no growth inhibiting properties. Free cyanide is formed only when the cells are damaged. However, evidence is not available to show that cyanide actually functions as a germination inhibitor. Vigorous test are to be made to decide whether an extracted compound functions in the seed as an inhibitor. The true inhibitor should change in quantity parallel to changes in the physiological condition of the seed (Pollock and Toole, 1961). If the seed is strongly blocked, a germination inhibitor should be present in high concentration. As the block is removed, the concentration of the inhibitor should decline to a minimum, leading to prompt germination. If after ripening is reversed by high temperature or lack of oxygen, the concentration of the inhibitor should increase. It was thus concluded that though growth inhibiting chemicals may very well control germination, they are not the only possible mechanisms. Changes in cellular organization and processes of respiration, photosynthesis and protein synthesis occurring in microsomes and mitochondria may also account for certain mechanism controlling germination, which are yet to be investigated fully.

The mechanism of stimulation of germination by light also controls several other growth responses of crop plants, including flowering. Temperature may influence one type of chemical reaction in plants differently than others. The low temperature requirement of some plant seed for germination is not fully understood. Some physical and/or chemical blocks may be interdependent. Artificially applied chemicals and growth regulators have been known to interact with light and temperature. Since interactions exist and germination is controlled by such interacting blocks, germination appears to proceed by a number of alternate pathways. These pathways may be closed by blocks imposed and removed by various environmental conditions (light and temperature) as well as by chemical treatments. The activity of cellular reactions catalyzed by enzymes is regulated by many factors, including temperature, which may reflect alternate pathways of germination.

4. CONCEPTS OF DORMANCY

4.1. Types of dormancy

Because mechanisms of dormancy are still to be elucidated, the classification of the different types of dormancy is entirely based on its expression under various conditions. For example, dormancy induced by elevated temperatures is usually indicated as thermodormancy. However, it is not known whether this type of dormancy differs from, *e.g.*, photodormancy in a mechanical sense. Thus, it was proposed to distinguish only two types of dormancy: primary and secondary dormancy (Karseen, 1982 and Hilhorst, 1995). This classification is based on timing. Primary dormancy (exogenous) is associated with development, whereas secondary dormancy (endogenous) occurs essentially after dispersal and is basic to seasonal dormancy cycling in soil seed banks.

4.2. Phenomenology of dormancy

The basic characteristic of dormancy is simply the absence of germination. Unfortunately, this is yet to be the only way to quantify dormancy. A non-germinating seed may be quiescent or dormant. A quiescent seed will germinate under the right set of conditions, whereas a dormant seed will not. Seeds with true dormancy require another set of conditions to break their dormancy. The requirements for dormancy relief may be different from those of germination. For example, many seeds require cold temperatures to release dormancy-so-called prechilling. For subsequent germination higher temperatures are generally required. Thus, the dormancy breaking treatment shifts the seeds from the dormant to the quiescent state without inducing germination.

4.3. Regulation and mechanisms

4.3.1. Primary dormancy

During the first third to first half of seed development most embryos become germinable when taken from the fruit and seed. This marks the completion of the histo differentiation phase. Some species produce ripe seeds containing immature embryos, *e.g.*, *Hordleum* and celery. In species like these completion of embryo development preludes germination *sensu stricto*. However, the majority of higher plant species produce seed with fully developed embryos of which precocious germination is suppressed during development. At the end of seed development and maturation seeds may be dormant or nondormant (quiescent). As believed hitherto, plant species with a long history of domestication generally have lower levels of seed dormancy than wild species.

The clearest form of primary dormancy is displayed by seeds with hard and/or thick seed coats. The coats prevent germination of embryo growth of the often nondormant embryo. Germination may be prevented by the mechanical properties of the coat. Only factors from outside may be able to reduce this restraint, *e.g.*, microbial attack, extreme temperature (*e.g.*, fire) or drought or passage through an animal's digestive tract. Coats may also have properties that prevent uptake of water or diffusion of oxygen and thus prevent embryo growth (Miller, 1956).

Primary dormancy and its regulation have been reviewed by Hilhorst, (1995). ABA appears to play a central role in the regulation of dormancy during development and acts on two distinct processes: (i) the prevention of precocious germination and (ii) the imposition of primary dormancy. Evidence for these functions of ABA has been obtained using plants in which ABA levels were manipulated, either chemically or genetically. Absence of ABA during seed development may lead to vivipary. This had been shown clearly for the VP_1 mutants in maize. These mutants were identified by screening for vivipary. Most of them (*e.g.*, vp_2, vp_2, and vp_7) contained reduced embryonic ABA contents, which were approximately one-tenth of the corresponding wild type. In contrast, the vp_1, mutant contains normal levels of ABA but appears to be less sensitive to ABA, in that more exogenous ABA is required to inhibit germination and seedling growth (Robichaud and Sussex). Interestingly, viviparous seedling of the vp_1 mutant develops into normal plants with normal response to water stress. This indicates that vp_1 gene expression has a temporal specificity. Vivipary occurs normally in mangroves (*Rhizophora* spp.). In these recalcitrant species very high levels of ABA are required to suppress viviparous germination, which is also indicative of ABA insensitivity (Sussex, 1975). In wheat the preharvest sprouting cultivars appear to have normal ABA content but are also less sensitive to exogenous ABA (Walker - Simmons, 1988). Chemical reduction of plant ABA levels may also induce vivipary. Embryo from plants treated with fluridone grew readily on culture, whereas embryos from untreated plants did not (LePage – Degivry and Garello, 1992). In *Arabidopsis,* the process of suppressing precocious germination appears to be extremely sensitive to ABA. Single mutation resulting in progeny that is either ABA deficient or showed a reduced response to exogenous ABA did not display viviparous growth. Only the ***aba1, abi3,*** double mutant, which is both deficient for and insensitive to ABA, appeared to be viviparous.

From these examples it is clear that ABA indeed plays a role in the suppression of precocious germination. However, the response to ABA is stage specific and may be modified by the environment. Important factors are temperature and water. Preharvest sprouting normally occurs at elevated temperature and humidity. In addition, the osmotic environment of the embryo and the seed will play a role in the suppression of growth. This may particularly be the case for seeds developing and maturing in a permanently moist environment, *e.g.*, within fleshy fruits. Evidently, in such an environment maturation drying dose not occur although the water content of the seeds drop significantly during the second half of development. For tomato it has been shown that the osmotic environment of the embryo and the seed will play a role in the suppression of growth. This may particularly be the case for seeds developing and maturing in a permanently moist environment, *e.g.*, within fleshy fruits. In tomato it has been reported that the osmotic potential of the fruit tissues effectively suppresses precocious germination (Miyamoto and Everson, 1958). Studies with the ABA deficient sit mutant, which shows vivipary in overripe fruits, suggested that the mutant seeds are less sensitive to osmotic stress than the corresponding wild type (Groot and Karssen, 1992 and Hilhorst and Downie, 1996).

For the prevention of precocious germination the permanent presence of ABA seems to be a prerequisite. However, in the case of primary dormancy. ABA needs to be present only during a limited period of development to effectuate a persistent dormancy. A comparative study of wild type of sit^{w} tomato (cv. Moneymaker) seed development showed that only in the wild type did a transient rise in ABA content occur midway through development. However, in the mature seeds of both genotypes ABA levels were barely detectable, yet only the wild type showed dormancy (Groot *et al.,* 1991). On the other hand, in seeds of the cultivar Caruso, ABA content increased halfway development but hardly decreased upon maturation. Remarkably, this cultivar hardly ever shows dormancy (Miyamoto and Everson,1958). Therefore, correlations between ABA content of seeds and germination are not always evident and this raises the question whether ABA is the major regulator of dormancy. An equally important factor is the sensitivity to ABA. In several studies it has been shown that the sensitivity to ABA decreases towards the end of maturation which is observed in case of alfalfa and maize (Neill *et al.,* 1987). The conclusion seems justified that, in general, the absence of ABA during seed development results in the absence of dormancy in the mature seed but the presence of ABA during development has a variable influence on the acquisition of primary dormancy.

Several conditions are known that are conducive to the release of dormancy from ripe seeds, apart from the previously mentioned methods to relieve hard-coated dormancy. In many species dormancy is reduced by a period of dry storage, particularly at elevated temperatures. This effect may refer to a possible hot and dry climatic organization of the particular species in which the hot season is unfavorable for growth but is used for dormancy itself. Many other species lose dormancy upon imbibition at low temperatures so called prechilling. This may be indicative of a more temperature origin where the cool winter temperatures thus prepare the seeds in the soil for the subsequent growth season (Esashi *et al.,* 1977). To date, there is no consensus on mechanisms underlying the termination of primary dormancy. Several suggestions have been that during both dry after-ripening and chilling, degradation of ABA occurs. This has been shown convincingly for lettuce and sunflower, mature lettuce seeds contain relatively large amounts of ABA. Upon imbibition ABA leaches out and germinability increases (Dulson *et al.,* 1988). However, for leaching out of ABA low temperatures are not a prerequisite. Dry storage of mature sunflower seeds for several weeks also reduced the ABA content and increased germination (Thomas *et al.,* 1979). It has also been shown that phenolic compounds leach out of imbibing seeds. These compounds are also known to inhibit germination. Clearly, the evidence for ABA maintaining dormancy in mature seeds is at best circumstantial.

Evidence is accumulating, that these dormancy breaking treatments are associated with an increase in sensitivity to the germination promoting gibberellins (GAs). A short drying treatment induced germinability and increased responsiveness to GAs in sunflower embryos (Thomas *et al.,* 1979). In *A. thaliana* GA sensitivity increased upon dry storage. This also occurred in the GA-deficient mutants of this species, indicating that GA synthesis was not involved in dormancy relief (Karssen and Lacka, 1986). In lettuce seeds it was shown that red light irradiation or application of GAs caused a decreased in ABA content that appeared to be metabolization rather than leaching (Toyomasu *et al.,* 1994). Red light also caused an increase in the level of GAs in these seeds. Chilling of *A. thaliana* seeds caused an increase in the levels of GAs but a correlation between GA content and germination could not be found (Derkx *et al.,* 1994). It may be concluded that the breaking of primary dormancy is associated with an increase in GA sensitivity and possibly GA content. Concomitantly, the levels of ABA or other inhibitors may decrease, either by leaching or by metabolic inactivation.

4.3.2. Secondary dormancy

Seeds that have terminated primary dormancy are quiescent and are ready to germinate. A specific set of conditions has to be present to ensure germination, *e.g.,* a certain temperature, light, nitrate, and oxygen. When these conditions, are not met, sometimes even when only one factor of a multitude is missing, seeds may become dormant again and nonresponsive to germination stimulants. This state of dormancy may be denoted as secondary dormancy. The similarity with primary dormancy is the comparable effectiveness of dry storage of prechilling on dormancy release.

The induction of secondary dormancy is often stimulated by elevated temperature. In fact, this may be one of the reasons why chilling is particularly effective in dormancy breaking. Primary dormancy may be terminated by higher temperatures, up to 20-25°C, but secondary dormancy is induced simultaneously. This shows that temperature is a major regulating factor in dormancy changes (Karssen, 1982). In soil seed banks cycling of seeds through successive phases of dormancy and quiescence occurs in a seasonal or annual fashion (Miyamoto *et al.,* 1961). There are strong indications that the temperature changes or trends during the year are perceived by the seeds in the seed bank to "sense" the right season for growth (Karssen, 1982). Thus, summer annuals require low temperatures for dormancy breaking, whereas winter annuals require the warmer summer and fall temperatures for this purpose. Temperature has dual role in seedling emergence. First, the timing of dormancy cycles over the year is determined by the long-term changes in the field temperature (T_1). Second, the germination of nondormant seeds is under temperature control (T_2). Thus, optimum temperatures for dormancy release and germination are not necessarily similar.

The termination and induction of dormancy of seeds in the seed bank are characterized by a respective widening or narrowing of the germination temperature range (Karssen, 1982). If during dormancy breaking this "temperature window" overlaps with the actual field temperature, the temperature requirements for germination are met and germination can commence if other required environmental cues are present.

Apart from water and oxygen, light and nitrate may be considered the most important naturally occurring factors required for seed germination (Roberts and Smith, 1977). Interactions between temperature, light and nitrate are common among a large group of weeds of the temperature zones (Vincent and Roberts, 1977 and Hilhorst *et al.,* 1986). Simultaneously with a widening of the germination temperature window upon relieve of dormancy, the responsiveness (sensitivity) to these environmental factors increases and it decreases again when dormancy is (re) induced (Bouwmeiester, 1990). A detailed dose-response analysis has been made of the changes in responsiveness to light and nitrate during dormancy cycles of *Sisymbrium officinale* (hedge mustard) and *A. thaliana*, both by lab and field experiments. One important conclusion was that under field conditions light and nitrate are the limiting factors for germination of these species, thereby overriding all other possible germination modifying factor (Hilhorst *et al.,* 1996). The dose-response studies also showed that the changes in responsiveness to light (through phytochrome) and nitrate could be well explained by the receptor occupancy theory with the assumption that phytochrome and nitrate receptors were associated with membranes. Through its action on membrane properties temperature could thus modify the activity of the nitrate and phytochrome receptors. There are good indications that membranes are the primary targets in the biological perception of temperature (Vigh *et al.,* 1985). They could perhaps function as long-term memory to store temperature information. This has lead to a model for dormancy cycling that integrates the effects of temperature, light and nitrate on dormancy and germination (Hilhorst *et al.,* 1996 and Vigh *et al.,* 1985).

4.4. Implications for agricultural practice

Agriculture has to deal with seed dormancy in two respects. First, dormancy is generally an unwanted characteristic in crops. Second, dormancy is the major regulating factor for the emergence of weeds in crop fields.

4.4.1. Dormancy in crops, a relation with vigor?

In many crop species dormancy has disappeared during domestication of their wild progenitors. However, in many species dormancy may reestablish in variable ways when germination and seedling growth occur under stress (Field) conditions. This is most likely a result of the fact that dormancy has both genetic and environmental components. A well-known effect of temperature stress can be found in lettuce seed germination. Lettuce seeds often fail to germinate at elevated temperatures. This so-called thermodormancy is a practical problem often encountered in Mediterranean and subtropical regions. It may be argued that the lettuce varieties that are able to germinate at relatively high temperature are the less dormant ones because these have widest germination temperature windows. Similarly, the resistance to water stress may be considered an expression of the level of dormancy. Germination of the nondormant ABA deficient mutant of tomato is considerably less affected by osmotic stress than the corresponding wild type. These examples make clear that the definition of dormancy as given previously needs to be extended. Apparently, dormancy can only be defined in relation with the environment. Another observation that necessitates a re-evaluation of the definition of dormancy can be made in seeds with a coat-imposed dormancy. For example, peach seeds require a prolonged cold period to break their primary dormancy. When the seed coat of the dormant seed is removed the embryo is capable of growth and the radicle will extend. Thus, germination *sensu stricto* occurs and it may be argued that by removing the coat, dormancy was terminated. However, when observing growth of the embryo into a seedling extreme dwarf growth is apparent (Pollock, 1962). It was shown that in this species reserve food mobilization did not take place. In other words, dormancy is also associated with seedling growth rate and with formation of abnormal plants. Clearly, germination percentage is not a good measure of dormancy. Indeed, germination rate is a much more sensitive tool to measure dormancy. This was realized when the so-called germination resistance test more dormant seeds took longer to germinate than the nondormant ones despite the fact that in both cases germination percentage could be 100%. Thus, this was the first attempt to extend the definition of dormancy as proposed previously. The germination resistance test measures the time between sowing and germination and is very sensitive to minor changes in test conditions. For example, both wild-type and the ABA-deficient tomato mutant germinate to almost 100% in water or in –0.5 MPa PEG. Judged by these observations it might be concluded that both genotypes are nondormant. However, when observing germination in time substantial differences between genotypes become apparent and it must be concluded that the wild type is more dormant than the mutant. In a study of a large number of barley seed lots good correlations were found between the germination – resistance test, seed size, and seedling dry weight. Seedling dry weight is often used as a measure of vigour. Thus, this study suggested an overlap of factors that determine either dormancy or vigor. Vigour is often defined as "those seed properties which determine the potential for rapid uniform emergence and development of normal seedlings under a wide range of field conditions". Clearly, this definition contains references to properties associated with dormancy "rapid emergence" refers to germination resistance and "a wide range of field conditions" refers to stress resistance.

Both dormancy and vigour are routinely improved by various seed enhancement methods. Interestingly, the same treatment that reduced dormancy *e.g.* osmotic priming may also enhance seedling vigour. The concept of dormancy and vigour similarities may be extended even further. Dormancy of many seeds can be broken by application of GAs. It is a powerful method and effective in many cases when other treatments are not effective. Application of GAs to relieve thermodormancy in lettuce seeds indeed made seeds germinable at higher temperatures (Tran and Cavanagh, 1984). However, another effect of GA treatment was observed when seedlings were grown into plants. The treated seeds produced plants that showed bolting and early flowering. Similar results were obtained when osmotic priming was performed at 2°C or when other growth regulators, such as kinetin and thioureas, were used. It may be argued that these treatments represent strong means of dormancy breaking. We may speculate that conditions that relieve seed dormancy may have a long-term effect on further growth and flowering time. Cold treatment, *i.e.*, vernalization, is often used to obtain earlier flowering time. It may be hypothesize that dormancy breaking treatments influence "developmental time", which stretches beyond the germination process.

The term developmental time or "biological time" has been discussed in relation to seed germination and dormancy (Berry and Bewley, 1992). The term was introduced as a consequence of the so-called hydrothermal time concept. For a dormant seed, biological time is proceeding at a slow rate compared to clock time. Dormancy breaking speeds up the developmental clock marking the rate of progress towards germination. This concept may be extended to post germinative growth and flowering. This implies that the occurrence of dormancy is not restricted to the seed only. Evidently, this idea needs further exploration.

4.4.2. Dormant or Dead?

If a seed fails to germinate under any conditions it is either (deeply) dormant or dead. How can they be distinguished? Most tests are based on determining germination. Tests that do may also fail. For example, the tetrazolium vitality stain is based on the activity of a group of enzymes that reduce tetrazolium compounds to insoluble red formazans. It may be hypothesized that activity of these enzymes can be low or absent in deeply dormant seeds. The conductivity test, which is widely used as a measure for seed deterioration, will show increased values when dead seeds are tested. However, in lettuce seeds induction of thermodormancy is also accompanied by an increased leakage of amino acids (Hendricks and Taylorson, 1979). Cell membranes appear to be involved both in seed deterioration by aging and in changes in dormancy. During deterioration, membranes are damaged, leading to loss of cell components and ultimately death. Because dormancy induction is readily reversible, and the alterations of seeds membrane properties must also be reversible. Thus, a decrease in vigour caused by deterioration does not seem to overlap with dormancy induction in this case. Although, evidence is scarce, we should not preclude the possibility of at least a partial overlap. It is known that seed deterioration can be partially reversed by priming treatments. As shown for artificially aged cauliflower seeds (Fujikura and Karssen, 1992). Furthermore, accelerated aging of tomato seeds at high moisture content and temperature may induce dormancy in part of the seeds.

5. STRUCTURAL FACTORS INFLUENCING DORMANCY

Two types of dormancy mechanisms are known; (1) coat-imposed dormancy and (2) embryo dormancy. The mechanism(s) of dormancy should be able to explain the nature of the block(s) within the seed preventing germination under apparently favourable conditions and how they operate. A full appreciation of the mechanism of dormancy eventually should also consider which environmental factors cause dormancy and what is required to break dormancy. The basic dichotomy of embryo and coat-imposed dormancy is still retained, and in some species both types may exist simultaneously or consecutively (Bewley and Black, 1982).

5.1. Coat – imposed dormancy

In many plant species seed dormancy is imposed by the structures surrounding the embryo (seed coat), which may include glumes, palea and lemmae (grasses), the pericarp, testa, perisperm, and endosperm. The embryo in these cases when isolated experimentally from the enclosing structures is not dormant. Certain physical or chemical treatments of the seed coat also permit the embryo to germinate, *e.g.* abrasion (scarification), perforation, or treatment with mineral acids.

The following are the possible mechanisms of prevention of embryo germination by the covering structures

1. Interference with water uptake
2. Interference with gaseous exchange
3. Presence of chemical inhibitors
4. Seed coats as barrier for release of inhibitors from the embryo
5. Modification of light reaching the embryo
6. Seed coats exerting a mechanical restraint.

One or the combined action of two or three of these effects may be responsible for the maintenance of dormancy in seeds.

Impermeability of the seed coat to water can be an overriding cause of dormancy and delayed germination in most hard seeds of leguminous crops as well as in some members of the Cannaceae; Chenopodiaceae, Convolvulaceae, Convalaiaceae, Geraniaceae, Graminae, Liliaceae, Malvaceae, and Solanaceae. In most hard-coated leguminous and other seeds, germination may be delayed owing to inadequate uptake of water by the embryo, though these seeds may be generally described as dormant.

In legume seeds, the testa is the main barrier to the entry of water and the outermost region of the testa, particularly the waxy cuticle, is the major impermeable layer, which can be washed out with ethanol to confer water permeability. In the species of Trifoliceae, the inner regions of the testa may be responsible for restricting the entry of water. The testa of leguminous seeds is composed of hilum, strophiole, and micropyle; the latter often is occluded in impermeable seeds of *Phaseolus lunatus* and other plant species. The hilum is also impervious to water and under moist conditions the fissure in the hilum is closed but tends to open only in a low relative humidity. This device plays an important role in the drying of the seed, including its testa, resulting in the establishment of its impermeability to water. The strophiole also does not conduct water unless the seed has experienced certain conditions, at which point it becomes the major site for the entry of water.

It has been observed that the removal of the covering tissues allows the embryo to germinate, even in seeds in which the seed coats do not prevent entry of water. Clearly, the coat that imposes dormancy, when removed, allows the embryo to come in contact with oxygen. The covering tissues in such cases probably hinder exchanges of gases, *i.e.*, intake of oxygen and escape of carbon dioxide from the embryo. Some seeds possess mucilaginous coats (*e.g.*, *Sinapis arvensis*), and the mucilage of the testa in such seeds may be responsible, at least partially, for the impermeability of the gaseous exchange. The testa of *Blepharis persica* with abundant mucilage, when exposed to excessive amounts of water, swells greatly to occupy completely the spaces between multicellular hairs protruding from the coat's surface. This forms an effective barrier against oxygen, leading to delayed germination. An atmosphere of oxygen overcomes the inhibitory effect of the over wetted muscilage, allowing germination.

The presence of phenolics in the hull has been implicated in the coat-imposed dormancy with oxygen consumption (barley). Barley testa has enzymes like peroxidase, cytochrome oxidase, and phosphoglycerate dehydrogenase which could account for some oxygen consumption. In rice (*Oryza sativa*), dormancy is also imposed by the hull. The naked caryopsis is known to germinate readily, and therefore intact rice grains (with hull present) can be made to germinate by subjecting them to an atmosphere of oxygen or by puncturing the glumes over the embryo. The imbibed hull takes up oxygen because of its peroxidase activity. The oxygen consumption of dormant grain hulls is about double that of nondormant seeds, which coincides with the twofold activity of peroxidase in the dormant hulls. Also, the oxygen uptake of the intact dormant grain is more than 50% higher than that of the nonformant grain, apparently due to the hull effect. The coat-imposed dormancy in rice and other cereals is caused by the consumption of oxygen by the peroxidases present in the hull limiting the amount of oxygen reaching the embryo(Bewley and Black, 1982). The finding that peroxidase activity declines during the natural loss of dormancy (after ripening) appears to be consistent with this concept.

As a consequence of the effect of the seed coats in limiting the amount of oxygen available to the embryo, the respiration needed for germination would also be affected, enforcing the embryo to remain dormant. The isolated embryos of some seeds are, however, capable of germinating under extremely low oxygen, tensions and even under nitrogen. The deleterious effects of low oxygen levels, because of coat impedance) may also be attributed to factors other than reduced respiration, such as the production of inhibitors or the failure to oxidize inhibitors already present.

The seeds of many plant species contain germination inhibitors, known to be present in the coat. Limited evidence is available. However, to support the contention that these inhibitors present in the covering tissue might actually account for the coat's imposition and maintenance of dormancy. More vigorous tests of a correlation between the magnitude of dormancy and level of inhibitor are needed to prove that the inhibitor is an important component of the dormancy regulating system.

Table 1: Inhibitors present in the coat of some seeds

Species	**Location of inhibitor**	**Nature of inhibitor**
Acer negundo	Pericarp	Possibly ABA
Aegilops ovata	Hull	A moneooxylignanolide
Avena sativa	Hull	Unknown
Beta vulgaris	Pericarp	Varoius phenolic acids, possibly ABA, cis4-cyclohaxene-1, 2-dicarboxymide, high concentration of inorganic ions
Betula pubescens	Pericarp	Unknown
Bouteloua curtipendula	Glumes, lemma, palea	Possibly coumarin and its derivatives
Compotonia peregrain	Testa	Possible ABA
Corylys avellana	Testa	ABA
Elaeagnus angustifolia	*Pericarp, testa*	*Possibly coumarin*
E. umbellate	Pericarp, testa	Possibly coumarin
Fraxinus Americana	Pericarp	ABA
F. ornus	Pericarp	ABA
Hordeum vulgare	Hull	Coumarin, phenolic acids, scopoletin
Iris spp.	Endosperm	Unknown
Oryza sativa	Hull	Probably ABA
Prunus persica	Testa	ABA
Rosa canina	Pericarp, testa	ABA
Sinapsis arvensis	Testa	Unknown
Triticum spp.	Pericarp, testa	Catechin, catechin tannins, several unknown
Zilla macroptera	Pericarp	Unknown

Not all named chemicals were rigorously characterized (Bewley and Black, 1982)

The inhibitors present in the outer coats of rice and wheat have been known to decrease as dormancy is lost during after-ripening process (Kermode,1995). Wheat testa contains catechins (tannins), which are precursors of the red pigment phlobaphene (Balboa – Zavala and Dennis, 1977). The red-coated wheats are, therefore, thought to have high levels of inhibitory catechins and their derivatives, whereas the white wheats have low levels of these compounds. The white wheats are also highly susceptible to preharvest sprouting as compared to the red wheats. There appears to be persuasive evidence to support that dormancy in some wheat cultivars may actually be controlled by the inhibitors present in their seed coats (Table 1).

Seed coats may also impair the outward diffusion of inhibitors present in the embryo, forming an impermeable barrier. Thus the embryo is not able to germinate because of its retention of high levels of inhibitors. The isolated embryos, when wetted with water, lose their inhibitors and regain germination capacity. Dormancy in wild oats (*Avena fatua*) for example, is normally imposed and maintained by the hull (lemma and Palea). It is lost when naked caryopsis are held on the wet substratum, ensuring rapid germination. In contrast, if they are held under high humidity, and caryopsis is able to take up the same quantity of water, dormancy is maintained. In the latter conditions, since the hull is removed, constraints due to either mechanical or gaseous exchange cannot operate. The inhibitor present in the caryopsis, however, cannot diffuse out through the hull. The absence of surrounding water thus prevents the outward diffusion of inhibitor from the naked caryopsis. The hull in these seeds appears to form a barrier against the escape of the inhibitor.

Many light-sensitive seeds possess coat-induced dormancy; naked embryos germinate irrespective of the light/dark conditions, with only a few showing true embryo dormancy. Intact light-requiring seeds can be stimulated to germinate when a certain ratio of P_{fr} (active) and P_r (inactive) forms of phytochrome (P) is established within the embryo by the combined action of red and far-red lights; the ratio needed depends upon the plant species. Since the light must pass through the surrounding tissue (coat) enclosing the embryo, they could act as a filter, altering the proportion of P_{fr} and P_r radiation reaching the embryo. Thus, in addition to coat-induced dormancy, the embryo of light – sensitive seeds may also suffer from an inadequate light environment, resulting in dormancy. The coat thickness and pigmentation, as affected by the environment during the development of seed on the mother plant, modify the coat to act as a light filter. Seeds with thick, dark coats are comparatively less responsive to light than those with thin, light-coated seeds (*e.g. Chenopodium album*).

In addition to the foregoing mechanisms, the coat might also act by exerting a mechanical restraint. The coats of many seeds and nuts (indehiscent fruits) consist of hard, tough tissues offering mechanical resistance to the growth of embryo. The endosperm of lettuce, for example, imposes dormancy by acting as a mechanical barrier against germination unless hydrolases (cellulose and pectin hydrolyzing enzymes) are synthesized newly to weaken the mechanical barrier of these polymers; various methods of breaking exogenous dormancy were described by Copeland and McDonald (1995).

6. EMBRYO DORMANCY

The failure of an isolated (viable and mature) embryo to germinate constitutes a case of embryo dormancy. Naked embryos, when placed on a wet substratum, remain dormant under the conditions suitable for germination. This dormancy may be due to either a deficiency in the axis or some metabolic blocks within cotyledons. The embryos of the freshly harvested seeds of *Xanthium pennsylvanicum* are dormant, while one-year seeds of the same species are not, indicating existence of some metabolic deficiency in the cotyledons and in the axes of the dormant embryos (Wareing and Saunders, 1971).

Seeds of the plant species from Rosaceae (woody trees) suffer greatly from embryo dormancy. In apple (*Malus sylvestris*), the intensity and a magnitude of embryo dormancy may depend upon variety, provenance, year of harvest, and other factors.

The control of embryo dormancy involves both cotyledons and germination inhibitors. In many plant species the cotyledons may be responsible for inhibiting the growth of the axis in dormant embryos. The dormancy of such seeds is broken off by removing one or both the cotyledons from the isolated dormant embryos. The embryonic axis of the dormant isolated embryo can be to germinate by excising its scutellum (modified cotyledon). In apple, the degree of dormancy appears to be a function of the amount of cotyledon left attached to the embryo, suggesting that the dormancy of the axis in the intact embryo, in these cases, is maintained by some inhibitory action of the cotyledon(s). In both epicotyl dormancy and physiological dwarfism, the cotyledons exert a similar inhibitory influence over the axis. The cotyledons also seem to be inducing secondary dormancy. When isolated embryos of nonormant applied seeds, for example, are held under unfavourable conditions for germination, a secondary dormancy is induced, which does not occur in embryos from which portions of cotyledons have been removed.

The physiological and biochemical basis of cotyledon action in embryo dormancy is not fully understood. The presence of inhibitors such as abscisic acid in cotyledons is thought to be involved. These chemical inhibitors are leached out of cotyledons resting on the wet substratum, inducing germination. Enlargement and greening are prevented in cotyledons wetted with a solution of abscisic acid. This observation suggests that abscisic acid-type inhibitors present in cotyledons may be controlling the growth of the axis.

Experimental evidence exists to show that

1. Inhibitors are found in embryos of many plant species possessing embryo dormancy.
2. Leaching out of the inhibitors promotes germination of the isolated, dormant embryos.
3. Embryo dormancy can be induced by treating nondormant embryos with a known inhibitor.
4. Treatments that break dormancy in some can cause a drop in inhibitor level in the embryo.

Abscisic acid (ABA) has been found to be the most predominant inhibitor in several species. The extent of embryo dormancy can be correlated with the concentration of ABA present. Prolonged washing (leaching) can induce germination of the dormant embryos, probably by leaching out of the inhibitors (ABA) from the dormant embryos.

Embryo dormancy may also be related to its morphological immaturity. Such embryos do not germinate when isolated, but may require a period of further development to be able to germinate. Immature embryos are characterized by their small size and poor differentiation, needing further growth and development before being ready for germination. The developmental period may in some cases by followed by a period of embryo dormancy broken by chilling. Developing of immature embryos within the seeds of *Heracleum sphondylium* best occurs at low temperatures (2-3°C) over a period of 12-15 weeks. Light may also support embryo maturation in some species (alpine), where in continuous light or several long daily photoperiods encourage embryo growth and its subsequent emergence from the seed (Bewlwy and Black, 1982).

Environmental conditions during seed development and maturation such as day length experienced by the mother plant, position of the seed on the mother plant, and age of the mother plant at the time of flower induction influence seed dormancy. Plants differ in their response to photoperiod and their expression of seed dormancy (17-19). For example, longer days seem to promote the germination of *Polypogon mospeliensis* (Table 2). The data presented in Table 3 show that seeds from primary umbels of three cultivars to celery (*Apium graveolens*) are heavier, more mature, and more dormant than those produced elsewhere on the plant (Bewley, 1955).

Table 2: Influence of day length under greenhouse and outdoor conditions on *Polypagon mospeliensis* during growth and seed maturation and germinability

Day length during growth of mother plant and seed maturation (h)	Seeds from greenhouse plants germination(%)		Seeds from outdoor plants, germination (%)	
	3 days	7 days	3 days	7 days
9.0	0.0	0.0	1.5	10.0
11.0	0.0	0.0	9.5	17.0
12.0	17.5	19.0	64.0	86.5
13.5	38.0	44.5	98.5	98.5
15.0	60.5	66.0	97.0	98.5
18.0	90.5	91.0	98.0	99.0
Control (natural day length)	93.0	96.0	91.5	93.0

Boeswinkel and Bouman (1995)

Kermode (1995) described the regulatory mechanisms in the transition from seed development to germination. Embryo development occurs under the controlling influence and protection of the maternal environment. The interactions between the embryo and the surrounding tissues (seed environment), which are in initimate contact with the maternal sporophyte (environment), modulate and control the course of embryogeny. Important cues in regulating embryogenesis may be generated both internally by the embryo and externally from the surrounding seed tissues and other plant parts.

Table 3: Seed position of Umbel, weight (g) and germination (%) after 21 days at 18°C in light in three celery cultivars

Cultivar	Umbel position	Mean seed weight[a] (mg)	Germination
Green snap	Primary	0.590	51
	Secondary	0.440	85
	Tertiary	0.386	94
	Quaternary	0.382	80
		(0.069)[a]	(9.8)
Lathom blanching	Primary	0.474	50
	Secondary	0.438	72
	Tertiary	0.380	94
	Quaternary	0.348	82
		(0.069)	(9.2)
Ely white	Primary	0.590	59
	Secondary	0.468	62
	Tertiary	0.490	80
	Quaternary	0.520	87
		(0.086)	(7.3)

[a]LSD at 5% in parenthesis Thomas *et al.* (1979)

6.1. Combination of impermeable seed coat and dormant embryo

Some kinds of seed with dormant embryos do not respond to stratification treatments because their coats are impermeable to water, and stratification is not effective unless seed are fully imbibed. In nature, such seed are exposed to warm summer temperature (they mostly mature in spring or early summer) in contact with soil. The warm temperature stimulates attack on seed coverings by fungi, which "Softens" the coats, making them permeable to water. Then as the seed imbibe water in the fall, the low winter temperature causes the "Stratification effect" with the result that the seed germinate the following spring.

Nurserymen handling seed with hard seed coats and dormant embryos usually "soften" the seed by treatment with concentrated sulphuric acid for 60-80 minutes, or by 3-4 months storage in moist peat or peat/soil at 25° to 30°C.

6.2. Epicotyl dormancy

In some species, the site of dormancy is rather specific in the embryo it resides in the epicotyl. Seed with this type of dormancy will produce a root at favourable warm temperature, but the shoot or bud will not develop unless the rooted seedlings are exposed to low temperature (stratified) for one to several months. In nature, such seed do not produce growing seedlings until the second summer after maturity, thus, they are referred to as 2-year seeds. They mature and are dispersed to the soil in spring. During the first summer the root develops, then the epicotyl is "after-ripened" during the winter, and finally during the second summer the shoot begins to grow and develop. A list of species having epicotyl dormancy is given in Table 4.

6.3. Double dormancy

A few kinds of seed have been identified in which both root and epicotyl are dormant. Such seed require a low temperature to break dormancy of root, followed by a warm temperature favourable for root growth and develoment, followed by a low temperature period to break dormancy in epicotyl, followed by a warm temperature favourable for growth and development of epicotyl. In nature such seed may not complete germination for a full two years. They are the real 2-year seed. Table 5 deals a few of the kinds of seed that have double dormancy along with recommendations for breaking dormancy.

6.4. Physiological dwarfs

Some kinds of seed that have embryo dormancy can be "forced" to germinate at room temperature by removal of all the seed coverings. The seedlings that do development very sluggishly to produce slowly growing dwarf seedlings are dwarfed. Physiological seedlings may not initiate "normal" growth for months or even years, unless the dwarf seedlings are exposed to low temperature. In some cases, physiological dwarfism has been overcome by treatment with chilling or gibberellins, while it has also been demonstrated that removal of the cotyledons permits normal growth in a few species. Presumably, removal of the cotyledons also removes some inhibitor. Khan (1977) has shown that the cytokinetins can also overcome physiological dwarifism, and explained this action as the basis of cytokinetin – inhibitor antagonism.

7. Involvement of hormones and growth regulators

7.1. Germinator inhibitors

The basic mechanism of dormancy in seed (except hard seed) is inhibition of some basic step in the germination process by a specific chemical inhibitor or group of inhibitors.

Specific inhibitors that have been implicated are coumarin and its derivations and relatives, xanthatin, etc. Structural formulas of some inhibitors (phenolic compounds) are shown below.

Table7.2: Germination stimulators

KNO_3	General stimulant under temperature of 20°C or less.
Thiourea	Stimulates dark germination of some seed kinds.
Ethylene	Effective on some grass seed.
Ethylene	Breaks dormancy in buds, potato tubers and some seed, notably, peanuts.
Gibberellin	Effective to over come the osmotic inhibition.
Cytokinin	Over comes physiological dwarfism.
Na or Ca Hypochlorite	Effective in breaking dormancy in rice.
Indole acetic acid (IAA)	Stimulates some seed, inhibits other, perhaps depends on concentration.

A most powerful inhibitor has been isolated the identified as abscisic acid (or abscisin 11 or dormin). It was identified and isolated from cotton by Addicoft, 1963.

There are of course many other chemicals that inhibit germination that may not occur naturally. Nacl and mannitol (through osmotic pressure) 2, 4-D, other herbicides, etc.

It is hypothesized that seed dormancy is regulated by the interaction of growth promoters and inhibitors, the latter in particular controlling the onset and maintenance of dormancy. ABA is one of the most important naturally occurring inhibitors including dormancy. It is specifically involved with the onset and maintenance of dormancy. The kinetics of production and disappearance of the inhibitor development of dormant and nondormant plant species have established the role of ABA in dormancy. ABA, which appears in grains during their development, is though to be retained into maturity by grains during their development, is thought to be retained into maturity by dormant varieties but not by nondormant ones. Balboa Zavala and Dennis (1977) mentioned the germination ability of isolated embryos of apple seeds during seed development and found that neither free nor bound ABA leaves in the embryonic axis and cotyledons showed any consistent correlation with the acquisition of dormancy. The germination capacity of the embryos dropped to zero before an increase in ABA concentration occurred. Such studies question of the universal role of ABA in the induction of dormancy. The dormancy may be induced by an increase in the embryos sensitivity to the inhibitors. On this basis, the total and extractable inhibitor need not change at all, yet it could play a role in the inception of dormancy. Attempts to establish the role of inhibitors and other hormones in the induction of dormancy by correlating total extractable growth substance with the plants physiological behaviour have proved to be rewarding. The action of inhibitors is counteracted by promotive hormones such as gibberellins and cytokinins. It is, therefore, possible that dormancy in some plant species may begin when the levels of these promotive hormones in the developing seeds decrease. The onset of dormancy in some plant species (*Avena fatua* and *A. Ludoviciana*) has been prevented by treating mother plants with gibberellic acid (GA) during seed development. The treatment presumably obviates dormancy by artificially elevating the level of promoting hormones (GA) in the maturing caryopsis.

The mechanisms of dormancy and its release involve the action of inhibitor(s) and the growth-promoting hormones. These include gibberellins, cytokinins, ethylene and abscisic acid, which occur commonly in plants and probably have specific physiological and biochemical effects on induction, generally adopted is to monitor the changes in the levels of extractable hormone in relation to continuance or termination of dormancy.

According to the model given in Fig. 2, ßGlu I (glucanase I), which is transcriptionally down-regulated by abscisic acid (ABA), contributes to endosperm wall weakening and breaking of coat-imposed dormancy. ABA accumulated during seed maturation is sufficient to delay ßGlu I induction and endosperm rupture. After ripening decreases ABA levels and possibly sensitivity to ABA allowing the induction of ßGlu I during imbibitions; which helps modulate coat-imposed dormancy. After-ripening promoted testa rupture of seeds imbibed in the light and is also correlated with loss of gibberellin (GA) requirement for dark-germination. The model does not exclude the possibility that other factors contribute to after ripening and modulation of coat-imposed dormancy, or that ABA has additional functions, *e.g.* in photodormancy (Metzger and Meins, 2000).

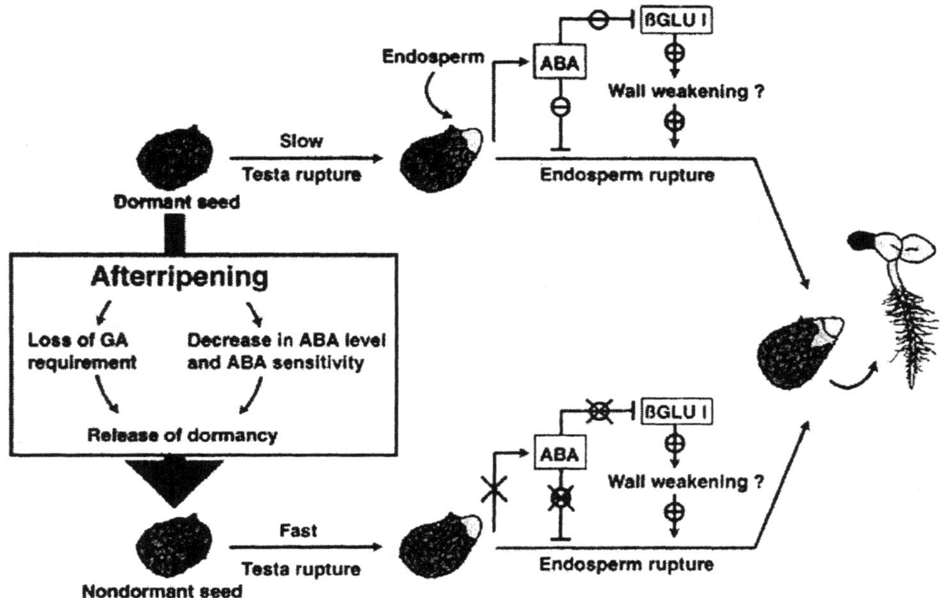

Fig. 2: A speculative model relating after ripening, class I ß-1,3-glucanase (ßGlu I) and endosperm rupture of tobacco (Metzger and Meins, 2000).

Bewley and Black (1982) have cautioned about the pitfalls of this approach, leading to erroneous interpretations of the results. Further, they set out the following criteria, which should be met before acceptable conclusions can be made about the role of hormones in seed dormancy.

7.2.1. The time course of germination

The stage of seed germination at which hormone contents are determined should be defined (prior to embryo growth), excluding germinated seeds. The change in hormonal level should occur during termination of dormancy, leading to seed germination.

7.2.2. Controls

Dormant seeds which have not been subjected to dormancy breaking treatment must be included.

7.2.3. Extraction and separation

Methods of extraction and separation of hormones should be reliable and reproducible. Free and bound hormones should be examined using internal standards.

7.2.4. Identification and measurement

Physiochemical methods or bioassays used must be specific, sensitive and reproducible.

7.2.5. Occurrence of hormones in the seed

It may be more useful to know the hormonal status of the individual seed parts, such as radicle, seed coat, cotyledons, etc., rather than using whole seeds.

7.2.6. Concentration of hormone in the seed

Measured levels of extracted hormone should be transformed into concentration within the seed or its part.

7.2.7. Turnover of hormones

The rates of production, utilization and degradation of a hormone are more meaningful than its concentration at any one time.

7.2.8. Sensitivity to endogenous hormone

Changes in sensitivity of the seed tissues to its own endogenous hormone, along with the changes in amounts of hormone, will be more meaningful.

According to the generally accepted hormonal theory of seed dormancy, the latter is imposed and maintained by the inhibitor(s), which may decline during disappearance of dormancy. Promotes are responsible for the release of dormancy. Thus, light, chilling and other factors may break seed dormancy by causing a drop in the inhibitor level and a rise in the level of promoters.

Investigators have tried to elucidate the mechanisms of seed dormancy by studying the application of growth substance to seeds to release them from dormancy and promote germination. Since all active growth regulators (gibberellins, cytockinis, and ethylene) are found endogenously in seeds, it is argued that if these chemicals are active when supplied to the speed, the endogenously present hormones should act in a similar manner. It has been suggested that seed dormancy is controlled by the balance of promoters and inhibitors (Baskin and Baskin, 1985 and Bewley and Black, 1994).

Gibberellins have been found to release dormancy in seeds of several plant species. They also accelerate germination of non-dormant seeds and even among dormant seeds, both types (*i.e.* seeds with coat imposed dormancy (*e.g.* Barley, Lettuce) and embryo-imposed dormancy (wild oats) are promoted. The effective physiological concentrations vary from 10^{-5} to 10^{-3} M and a mixture of GA_4 and GA_7 (gibberellic acid) has been found to be effective.

Miller (1956) investigated the effects of cytokinins on the germination of lettuce seeds. The dormancy of lettuce seeds could be overcome by supplying kinetin in darkness. However, kinetin was found to be effective only in presence of low light, which in themselves were ineffective to promote germinations in complete darkness, kinetin and other cytokinins and break the dormancy of only a small percentage of lettuce seeds, with abnormal emergence of embryo and cotyledons growing first. Cytokinins act by counteracting the effect of inhibitors, especially ABA when applied simultaneously. The dormancy breaking effect of gibberellic acid is prevented by ABA. Whose inhibition is overcome by kinetin. Cytokinins are thus though to have a "permissive" role, allowing a second promoter (GS) to act. Cytokinins also reverse the action of ABA on isolated lettuce embryos in the absence of gibberellins, which are required only by the intact seed.

Ethylene effectively breaks the dormancy of many seeds when supplied either directly as a gas or in the form of ethylene – releasing substances like Ethrel. The effective concentrations range from 0.1 to 200 µL/L. Robichaud and Sussex. (1986) examined 43 species of weed seeds and found hat 32 species shown no response to ethylene. Several responsive species demonstrated interactions between light and ethylene.

The precise roles played by ABA and the osmotic potential within the developing seed had its own simulations (Karseen, 1995). It is difficult to conclude that control over maturation and germinability rests universally with these "Cues". Important advances have resulted from a movement away from traditional approaches to a study of regulatory factors in seed development. The pathways of ABA biosynthesis and degradation and developmental changes in metabolism both in embryo and maternal seed tissues are being elucidated. The factors involved in the partitioning of ABA; particularly in the control of ABA movement from the material tissues (seed coat) to the embryo, are also becoming understood.

The relationship between the regulation by ABA and a low osmotic potential is complex. ABA and osmotic potential in some cases may act synergistically, whereas in other instances they seem to operate through separate pathways, eliciting different metabolic effects .Desiccation, whether natural or imposed, plays an important role in switching sees from a developmental mode to one that is essential to promote germination. Premature drying may permanently redirect metabolism from a developmental to a germination mode, affecting a normal temporally regulated "switch". It is not known, how, at the molecular level, drying leads to suppression of protein required for the germination and subsequent growth of the seedling. Kermode (1995) has described the regulatory mechanisms involved in the transition from seed development to germination Bewley (1955) reviewed some fundamental physiological aspects of seed development to germination. According to Kermode (1995) these basic studies will complement and provide the foundation for molecular studies to elucidate mechanisms regulating the fine – tuned temporal changeover in gene expression from a developmental to germinative mode. A process, which probably involves a complex temporal sequence of cues generated both internally from the embryo and externally from the seed environment, remains to be unreavled.

The study of hormone mutants has strongly contributed to our understanding of the control of dormancy and germination (Derkx and Karssen, 1994), although these studies have been restricted only to a few species. Conclusion drawn from these studies might differ for seeds with embryo dormancy and those with other forms of dormancy. ABA appears to play an important role during seed development in the induction of dormancy and desiccation tolerance. GAs is not required for normal seed development and their role appears to be restricted to the control of fruit growth. Both classes of hormones are directly involved in the regulation of dormancy in the nature seed. Dormancy is actually controlled at the level of availability or activity of the receptor of P_{fr} Nitrate might be a cofactor at that level. An active P_{fr} receptor complex is required to start the signal transduction pathway leading to germination. GA synthesis and sensitivity to Gas are essential parts of this pathway. Seed germination depends on the presence of endogenous of exogenously applied GA and may be inhibited by the presence of ABA.

7.2.8. Respiration

The metabolism and growth associated with germination require an active synthesis of ATP. The early work reported that a sharp rise in respiratory rate occurred during stratification of sour cherry (***Pruns cerasus***) seeds prior to the enhancement of the capacity for growth of his embryonic axis. Thus a simple hypothesis was proposed, that dormancy was imposed by a block in energy metabolism and was broken by increased availability of phosphate acceptors such as glucose, presumably via hydrolysis of stored forms. The observation that dormancy could be alleviated in many species by elevated oxygen concentrations seemed to support this hypothesis. Again, increased respiration was the key to dormancy breakage.

However, it became apparent that many chemical inhibitors or respiration could also promote germination and this paradox is still with us.

Many of the inhibitors used were those known to block the terminal cytochrome oxidase (EC 1, 9, 3, 1) of the mitochondrial electron transport chain. As a consequence this led to the proposal that the alternative suspiratary (cyanide – insensitive) path way was involved in the termination of dormancy. However, others inhibitors of respiration that did not act directly on the electron transport chain could proposal that the pentose phosphate pathway was involved in some crucial step. Attempts have been made to examine these metabolic events in vivo and in some cases to draw together the two hypotheses but so far the situation remains, unresolved.

Detailed arguments have been presented elsewhere, but in essence it was suggested that the pentose phosphate pathway employed cyanide – insensitive oxidase system and in the dormant seed this oxidase is in competition with the conventional electron transport system. An inhibition of the conventional system or oxidation of HADPH stimulates the pentose phosphate system some intermediate is made available that allows breakage of dormancy.

The cyanide - type stimulation of germination is mediated by an inhibition of catalase (EC 1.11.1.6) activity, which would normally hold hydrogen peroxide (H_2O_2) at a depressed level (the H_2O_2 pressumably coming from glyoxysonal – Oxidation of fatty acids). The resultant sparing of the H_2O_2 allows peroxidase (EC 01.11.1.7 activity through a linked quione redox system to oxidize any available NADPH. In this way the pool of NADP is kept at a high concentration and this would allow the pentose phosohate pathway to proceed rapidly. There is unfortunately no ascribed function for the pentose phosphate pathway in the metabolism of dormancy breakage. An obvious function would be to provide the biosynthetically useful reducing power of NDPH, but the scheme outlined above seems to preclude this. None of the various intermediates of the pentose phosphate has yet been shown to stimulate germination when applied exogenously.

A number of reports have suggested that the activity of the pentose phosphate pathway is higher in dormant compared to dormant seed wild oat. However, cold-soil inhibition of wild oat does indeed increase the activities of two enzymes belonging to this pathway, glucose 6-phophate dehydrogenase (EC 1.1.1.49) and 6-phospho-gluconate dehydrogenase (EC 1.1.1.44, this treatment does not terminate dormancy. Conversely, warm (25°C) soil incubation does break dormancy but does not give significant changes in these two dehydrogenase activities.

In two species which respond positively to cold stratification, sour cherry and hazel, there is some evidence that the pathway may be of some importance. The pentose phosphate pathway operates at increased rates in the stratified sour cherry seed. The C_6:C_1 ratio method was used which is claimed to distinguish between the pathway and normal glycolysis, although may criticisms can be raised over its validity. Using partially purified extracts of stratified hazel seed, increased specific activities of both dehydrogenases, the increases occurring concurrently with the release from dormancy.

These types of observation are not ubiquitous, and it was unable to show any association between dormancy breakage of Xanthium seeds and the pentose phosphate pathway. The C_6:C_1 ratio technique and also measured dehydrogenase activities and there was no evidence linking ethylene stimulated dormancy breakage of spergula arvensis with pentose phosphate pathway activity. There results tended to suggest a decreased participation during the crucial phase. These authors did demonstrate an increased respiratory rate 60 hours after soaking, but attributed this to the heightened activity of the germinating seeds.

The extensive and thorough investigation into the dormancy of Xanthium seeds has done using inhibitions of both the conventional and the alternative respiratory systems they argue that alternative respiration is necessary for germination and that secondary dormancy arises from its inactivation. They further claim that ethylene exerts its effect through a stimulation of the alternative system.

When genetically homogenous lines of wild oat were investigated a similar situation was revealed. Sodium azide, which blocks conventional respiration, stimulates germination of dormant seed, whereas salicylhydroxamic acid, an inhibitor of alternative respiration, completely inhibits the azide – stimulated germination. However, this latter inhibitor has no effect when applied to genetically non-dormant lines of wild oat. Although alternative respiration is necessary for the azide – stimulated germination of dormant fresh seed, it does not have a role in the germination of after-ripened seed, nor in genetically non-dormant varieties.

In two species with a stratification requirement, pear (*Pyrus communis*) and sugar pine, it has been suggested that alternative respiration is not of great importance. Pear seeds show a progressive linear increase in respiratory capacity during stratification, but inhibitor studies suggest that alternative respiration remains at a constant level which is only a small part of the total. Hence it was concluded that the conventional cytochrome oxidase – mediated pathway gave the increase in the respiratory capacity.

When dormant sugar pine seeds are stratified they undergo a rapid increase in respiration rate and in ATP content, both then plateau for some time, until after 60 days at 5°C there is a second significant rise. Those seeds held at 25°C show a similar early increase during the first four days, but this is followed by a decline in both respiration and ATP concentration with no subsequent secondary increase. During the plateau phase of the stratified seed, between days 5 and 60, the conventional cytochrome oxidase (cyanide – sensitive) pathway accounts for about 80% of the respiratory activity, whereas in those seeds held at 25°C it constitutes only 60% of the total activity. One conclusion drawn this work is that high temperature imbibition and incubation is actually detrimental to the metabolism of the seed.

Thus the only clear statement that can be made about respiratory metabolism during dormancy is that those treatments leading ultimately to the termination of dormancy give enhanced respiratory activity. The roles of he pentose phosphate pathway and cyanide – insensitive alternative respiration remain enigmatic. However, as this is a field in which well established techniques are available it is to be hoped that some understanding will soon be obtained.

8. BREAKING OF DORMANCY

8.1. Natural breakdown of impermeable seeds

There have been no comprehensive investigations of mechanisms of natural breakdown of hard seeds even though most artificial treatments are probably based on speculative observation of the effects of natural agents on the seed, when impermeable seeds become permeable naturally, it is usually assumed that the seed coat has been damage in some way, for instance, by fire, by abrasion due to wind and water movement over the ground, by microbiological attack in the soil, by high and fluctuating temperatures, or by passage through the digestive tracts of birds and animals.

Simple observation of affected seeds without comprehensive anatomical investigation has often led to false conclusions being drawn. Thus as fire or heat produces a network of cracks, for example, in Acacia, *Albizzia julibrissin Coronilla varia Rhus ovata,* it is usually considered that water enters the hard coat at these points. However, sections through cracked seeds coats reveal that the cracks frequently do not penetrate to the base the Malpighian cells and thus water entry through then us unlikely. It is now well established in *Acacia, Albizzia, Abutilon theophrasti, Crotalaria, Gossypium* and in many papilionoid legumes that heat causes a rupture of the seed coat at the lens or chalazal discontinuity and that is the first site of water entry.

Fluctuating temperatures in the field have been shown to render papilionoid legumes permeable by causing a fracture at the lens tissue. Seed of *Acacia karoo* and *Acacia bidwillii* had damaged lens tissue and were permeable, giving to 80% germination. It is assumed that the action of soil bacteria is responsible for such damage. Germinating seeds of *Albizzia julibrissin* in soil containing various fungi that *Rhizoctonia* sp. in particular was able to raise germination from 8.7% in controls to 37.1% in 90 days. Scanning electron microscope (SEM) photographs showed the fungi and altered the seed surface in random areas and have not necessarily attacked specialized sites such as the hilum or lens.

The role of birds and animals in distributing seed and enhancing its germination has been known for many years. Once again, there is little information concerning the effects on the seed coat of passage through the digestive tract though it is usually supposed that in birds at least, the grinding action of grift in the gizzard is equivalent to mechanical scarification. Also, stomach aids are though to corrode the coat and allow fluid penetration 16 species of seeds fed to captive birds, 8 had marked improvement in germination after passage through the bird's intestines. Weight losses of up to 20% were recorded as compared with controls and this was attributed to seed coat thinning. In a SEM study, the thickness of the seed coat was important in protecting the seed from damage and that smaller seeds were more likely to be degraded due to longer retention in the gizzard. The major damage suffered by the seed appeared to be mechanical scarification and random seed coat thinning though the author did not examine the area of hilum - lens - chalazal discontinuity.

More detailed information is available on the role played by animals in dispersing and germinating seed. Once again there is little direct knowledge of the effects of such digestion on the seed coat (Table 4). It was speculated that because germination enhancement was relatively low in most cases (frequently less than 40%), the seed probably suffered relatively litter damage in the months of larger animals, such as elephants (*Loxodonta africana*) and cattle with impala (*Aepyceros melampus*) indicated that they destroyed most of the seed consumed.

Seed held for considerable periods of time in the intestines of animals, for sample, six days in the case of cattle would be subject to possible enzymic action and degradation. Consequently, it is likely that coat damage and chemical degradation occur and that both are responsible for enhanced germination observed after animal ingestion.

Table 4: Per cent germination after ingestion by birds and animals

Species	Dispersal agent	% germination		
		Control	After passage	% Bruchid damage
Acacia giraffae	Eland[a]	45%-49%	60%-66%	-
	Elephant[b]	92%	78%	-
Acacia albida	Cattle	85% at 90 days	85% at 90 days	-
Acacia raddiana	Dorcas gazelle[c]	1%	13% ± 4%	-
	Dorcase gazelle	4%	21%	
Acaci tortilis	Impala	-	8%	72%
subsp. spirocarpa	Thompsons gazelle[e]	-	12%	26%
	Dikdik	-	11%	31%
	Elephant[b]	-	19%	45%
	Dorcas gazelle[c]	-	13.2%	22%
Acacia cyclopis	Various	9%	30%	-
Amaranthus palmeri		95%	11%	-
Rhus glabra	Starlings[g]	0%	20%	-
Abutilon theophrasti		1%	10%	-

8.2. Artificial treatments

The methods used to render 'hard' seeds permeable have been described by porter. Essentially, most methods are concerned with damaging the seed coat, or specific area of the seed coat and thus facilitating water entry. The availability of water can initiate the process of germination; it is likely that practices which weaken or damage the coat will also bring about other changes, for instance, increased sensitivity to light and temperature, permeability to gases and removal of inhibitors. Similarly, it is often believed that the mechanical strength of the coat of surrounding covers such as pericarp, glumes and lemmas acts as a restraint to germination and radicle penetration, or prevents adequate expansion and development of the embryo.

8.2.1. Effects of treatments on the seed coat

Except in few instances, there has been little critical evaluation of how the treatment acts on the coat. Most experiments have been concerned with enhancing percentage germination rather then with understanding the basic mechanisms involved, consequently our knowledge of effects is much poorer than our knowledge of results. Additionally, most studies have been made on unsorted material containing both permeable and non-permeable seeds. A more realistic assessment of treatment effectiveness would be obtained by working with samples containing only known impermeable seeds obtained by soaking beforehand. Seed coat treatments are found to enhance the germination percentage (Fig. 3).

Fig. 3: Germination of seeds after seed coat treatment (left); no germination of control (right)

Wet treatments

Seed coat treatments can be divided into 'wet' and 'dry' treatments.

Wet

Chemical action	:	Sulphuric acid, alcohols, acetone, oxidizing agents like KNO_3 and ethylene.
Thermal action	:	Hot or boiling water, liquefied gases

Dry

Mechanical action	:	Manual and mechanical scarification, impaction, high pressure
Thermal action	:	Dry heating, using radiant or electromagnetic waves, field temperature fluctuations

In general, alcohols, acetone and organic solvents have been relatively ineffective or have caused high seed mortality. Water entry was facilitated by penetration of alcohol through cracks in the seed coat or through the pleurogram of mimosodi seeds. Results of other studies have been inconclusive. More aggressive reagents such as methanol, corrode random holes through the coat and sulphuric acid is also believed to behave similarly, for example, in *Coronilla varia* although the counter palisade tissue of the hilum of lupins, however, can be hydrolyzed, presumably permitting water entry. In recent years, sulphuric acid has falled into disuse, mainly because of the dangers of handling it but also because it offers few advantages over other techniques such as soaking in hot water reported that *Lupinus cosentinii* required upto 7 hours soaking to become permeable, and as much as 12 hours may be required in other species with very thick coats. SEM examination that sulphuric acid exposed the lumens of the Malpighian cells in *Coronilla varia* and considered that such damage allowed the seeds to become permeable. However, they did not determine the actual depth of impermeability. Freezing treatments with liquid air, liquid nitrogen and liquid oxygen have been successful in reducing impermeable seeds in several legumes but not in theirs, whereas freezing in ice has no effect speculated that differential expansion and contraction of the seed coat in both boiling water and liquid nitrogen caused cracks in the palisade layer to the base of the Malpighian cells, thus permitting water entry.

Hot and boiling water have long been known as effective agents for reducing impermeability. It is usually considered that boiling softens the seed coat, allowing random water uptake, whereas in reality, boiling, like dry heat or microwave heating, has little effect on the properties of the coat and renders the seed permeable through action on the lens. Sufficient evidence is now available to indicate that the lens and the chalazal discontinuity are widespread in the plant kingdom and these features perform the specific role of providing predetermined points of weakness which rupture under treatment.

Dry treatments

Microscopic examination of scarified seeds shows that, as expected, the seed coat is damaged to varying degrees by the process. Any mechanical impact might be expected to separate the highly stressed cells at the lens of papilionoid legumes but this is less effective with other legumes. The role of alternating temperature in breaking impermeability has been known now well. Field and laboratory trials showed that impermeable seeds which become permeable naturally do so at the lens area. The critical factor which determined the rate of breakdown was the maximum daily temperature to which the seeds had been exposed. There is almost no knowledge of the part played by temperature fluctuation in reducing impermeability of other types of seeds.

The use of various radiation and electrical processes in enhancing germination but such studies have been characterized by an almost complete lack of a basic understanding of the physiological and structural effects of the treatments. The situation has been partly remedied with microwave energy and its effect on Acacia seeds. Radio and microwave frequency treatments produce only a rapid heating effect whereas others consider that non-thermal effects may be present. Adverse effects observed include molecular breakdown of enzyme structure due to resonance and partial protein denaturation and inactivation of key enzymes. The effectiveness of high frequency heating on papilionoid legumes was established and demonstrated with serial photography that the lens was the first site of water entry. Treated seed retained viability and a high level of germination after up to 14 years of controlled storage. However, the treatment, though simple and inexpensive, is not effective on all impermeable seeds.

9. THE BIOLOGICAL IMPLICATIONS OF DORMANCY

The understanding of the complexities of the mechanisms of seed dormancy is of great biological significance. An understanding of the phenomenon will lead to control of seed germination, which is of enormous practical importance (*e.g.*, control of weeds). Seed germination is a critical stage in the life of each crop plant, affecting both the vigor and stand of the crop and the final crop yield. Understanding the natural mechanism(s) controlling germination would provide a scientific basis for raising better crops. Knowledge of dormancy mechanisms would enable us to cause or prevent germination at will and possibly control the storage life of seeds. An understanding of the mechanisms of germination blocks in seeds could also contribute to a larger understanding and control of growth and development in other organisms.

Dormancy is a device for optimizing the distribution of germination in time or space, having great ecological importance. Extension of germination over an extended period of time occurs because seeds of many species show variability in depth of dormancy, the population exhibiting sporadic release from dormancy and irregular germination. The temporal dispersal thus enhances the spread and survival of the species. Seasonal environmental changes (light and temperature) ultimately regulate seed dormancy and germination, controlling the seed dispersal.

Four basic patterns of germination have been recognized in relation to its temporal distribution.

1. Quasi – simultaneous, when germination of all the seeds occurs over a relatively after period.
2. Continuous, in which seeds germinate over an extended period of time.
3. Intermittent, irregular germination over long periods of time (multimodal distribution), and
4. Periodic multimodal germination showing more regular periodicity.

These patterns result from the dormancy characteristics and from their interplay with various environmental factors like temperature. Dormancy is also important for preventing vivipary and precocious germination, *i.e.*, germination while the seed is on the mother plant, which is important in certain crop species like cereals.

REFERENCES

Amen, R.D. 1968. A model of seed dormancy. *Bot. Rev.,* **34**: 1.

Balboa – Zavala, O. and F.Q.Dennis. 1977. Abscisic acid and apple seed dormancy. *J. Am. Soc. Hort. Sci.,* **102**: 633.

Barton, L.V. 1965. Seed dormancy : General survey of dormancy types in seeds and dormancy imposed by external agents. *Encyclopedia of Plant Physiology,* 15/2 (W. Ruhland, ed.), Springer – Verlag, Berlin, 1965, p. 909.

Barton, L.V. 1965a. **In:** Encyclopedia of Plant Physiology, XV (2) (W. Ruhland, ed.), pp. 727-745. Springer – Verlag, New York.

Baskin, J.M. and C.C. Baskin. 1985. The annual dormancy cycle in buried weed seeds. *A Continuum. Bioscience,* **35**: 492-498.

Berry, T. and J.D.Bewley. 1992. A role for the surrounding fruit tissues in preventing the germination of tomato (*Lycopersicon esculentum*) seeds. A consideration of the osmotic environment and abscisic acid. *Plant Physiol.,* **100**: 951-957.

Bewley, J.D. 1955. Challenges in seed physiology, challenges for the second decade (J.S. Burns, ed.). Proceedings of the Tenth Ann. Seed Technology Conf., Iowa State University Press. Ames I. p. 51.

Bewley, J.D. and M.Black. 1982. Physiology and Biochemistry of Seeds in Relation to Germination, Springer – Verlag, Berlin.

Bewley, J.D. and M.Black. 1994. "Seeds. Physiology of Development and Germination". 2nd ed. Plenum, New York.

Bianco, J., G. Garello and M.T LePage.Degivry. 1994. Release of dormancy in sunflower embryos of dry storage. Involvement of gibberellins and abscisic acid. *Seed Sci. Res.,* **4**: 57-62.

Boeswinkel, F.D.F. and Bouman. 1995. The seed structure and Function. **In:** "Seed Development and Germination" (J. Kigel and Q. Galili, Eds.), pp. 1-24, Dekker, New York.

Boswell, V.R. 1961. Flowering habit and production of seeds. Seeds : The Yearbook of Agriculture, U.S. Department of Agriculture, Washington, DC. p. 57.

Bouwmeiester, H.J. 1990. The effect of environmental conditions on the seasonal dormancy pattern and germination of weed seeds. Thesis, Wageningen Agricultural University.

Bouwmeister, H.J. and C.M.Karssen. 1992. The dual role of temperature in the regulation of the seasonal changes in dormancy and germination of seeds of *Polygonum persicaria* L. *Oecologia,* **90**: 88-84.

Bouwmeister, H.J. 1982. The effect of environmental conditions on the seasonal dormancy and germination of weed seeds. Ph.D. Thesis, Agricultural University, Wageninggen, The Netherlands.

Bradford, K.J. 1995. Water relations in seed germination. **In:** "Seed Development and Germination" (J. Kigel and Q. Galili, Eds.) pp. 351-396, Dekker, New York.

Copeland, L.Q. and M.B.McDonald. 1995. Principles of Seed Science and Technology, 3rd ed., Chapman and Hall, New York.

Derkx, M.P.M. and C.M. Karssen.1994. Are seasonal dormancy patterns in *Arabidopsis thaliana* regulated by changes in seed sensitivity to light nitrate and gibberellin. *Ann. Bot.,* **73**: 129-136.

Derkx, M.P.M., E. Vermeer and C.M.Karssen.1994. Gibberellins in seeds of *Arabidopsis thaliana.* Biological activities, identification and effects of light and chilling on endogenous levels. *Plant Growth Regul.,* **15**: 223-234.

Dulson, J., J.D.Bewley and R.N. Johnston 1988. Abscisic acid is an endogenous inhibitor in the regulation of mannase production by isolated lettuce (*Lactuca sativa* cv. Grand Rapids) endosperms. *Plant Physiol.,* **87**: 660-665.

Esashi, Y., H..Katoh, Y.Haba and N.Goto. 1977. Dormancy and impotency of cocklebur seeds. *Plant Phyisol.,* **59**: 122.

Fujikura, Y. and C.M.Karssen. 1992. Effects of controlled deterioration and osmopriming on protein synthesis of cauliflower seeds during early germination. *Seed Sci. Rev.,* **2**: 23-31.

Gordon, A.Q. 1973. The rate of germination. **In:** "Seed Ecology" (W. Heydecker, Ed.), pp. 391-410. Butterworths, London.

Groot, S.P.C. and C.M.Karssen, 1992. Dormancy and germination of abscisic acid deficient tomato seeds : Studies with the sitiens mutant. *Plant Physiol.,* **99**: 952-958.

Groot, S.P.C., I.I.Van Yperen and C.M.Karssen. 1991. Strongly reduced levels of endogenous abscisic acid in developing seeds of tomato mutants sitiens do not influence *in vivo* accumulation of dry matter and proteins. *Physiol. Plant.,* **81**: 73-78.

Hendricks, S.B. and R.B. Taylorson. 1979. Dependence of thermal responses of seeds on membrane transitions. *Proc. Natl. Acad. Sci., USA,* **76**: 778-781.

Hilhorst, H.W.M. 1995. A critical update on seed dormancy. I. Primary dormancy. *Seed Sci. Res.,* **5**: 61-73.

Hilhorst, H.W.M. and B.Downie. 1996. Primary dormancy on tomato (*Lycopersicon esculentum* cv. Moneymarker) : Studies with the sixteen's mutant. *J. Exp. Bot.,* **47**: 89-97.

Hilhorst, H.W.M., A.I.Smitt and C.M.Karssen. (1986). Gibberellin biosynthesis and sensitivity mediated stimulation of seeds germination of *Sisymbrium officinale* by red light and nitrate. *Physiol. Plant.,* **67**: 285-290.

Hilhorst, H.W.M., M.P.M. Derx and C.M.Karssen 1996. An integrating model for seed dormancy cycling. Characterizing of reversible sensitivity. **In:** "Plant Dormancy : Physiology, Biochemistry and Molecular Biology" (Q.A. Lang, ed.) pp. 341-360. CAB International, Wallingford, UK.

Karseen, C.M. 1982. Seasonal patterns of dormancy in weeds. The Physiology and Biochemistry of Seed Development Dormancy and Germination. (A.A. Khan ed.), Elsevier Biomedical Press, Amsterdam, p. 243.

Karseen, C.M. 1995. Hormonal regulation of seed development, dormancy and germination studied by genetic control, Seed Development and Germination (J. Kigel and G. Galili, ed.), Marcel Dekker, New York, p. 333.

Karseen, C.M. and H.W.M.Hilhorst. 1992. Effect of chemical environment on seed germination, Seeds : The Ecology of Regeneration in Plant Communities (M. Fenner, ed.). C.A.B. International, Wallingford, UK, p. 327.

Karssen, C.M. 1982. Seasonal patterns of dormancy in weed seeds. In "The Physiology and Biochemistry of Seed Development. Dormancy and Germination" (A.A. Khan, Ed.). pp. 243-270, Elsevier, Amsterdam.

Karssen, C.M. and E. Lacka. 1986. A revision of the hormone balance theory of seed dormancy. Studies on gibberellin and / or abscisic acid – deficient mutants of *Arabidopsis thaliana.* In "Plant Growth Substances 1985" (M. Bopp, Ed.) pp. 315-323. Springer – Verlag, Berlin, New York.

Kermode, A.R. 1990. Regulatory mechanisms involved in the transition from seed development to germination. *Crit. Rev. Plant Sci.,* **9**: 155.

Kermode, A.R. 1995. Regulatory mechanisms in the transition from seed development to germination ; Interactions between the embryo and the seed environment, Seed Development and Germination (J. Kigel and G. Galili, eds.), Marcel Dekker, New York, p. 273.

Khan, A.A. 1977. **In:** The Physiology and Biochemistry of Seed Dormancy and Germination (A.A. Khan, ed.) , pp. 29-50. North Holland Publishing Co., Amsterdam.

Koornneef, M., M.Hanhart, H.W.M Hilhorst and C.M.Karssen. 1989. *In vivo* inhibition of seed development and reserve protein accumulation in recombinants of abscisic acid biosynthesis and responsiveness mutants in *Arabidopsis thaliana. Plant Physiol.,* **90**: 463-469.

LePage – Degivry, M.T. and G. Garello. 1992. *In situ* abscissic acid synthesis. A requirement for induction of embryo dormancy in *Helianthus annus. Plant Physiol.,* **98**: 1386-1390.

McDonald, M.M. 1980. Vigour test subcommittee report. *Assoc. Off. Seed Anal. Newslett.,* **54**: 137-140.

Metzger, G. L. and F. Meins. 2000. Sense transformation reveals a novel role for class I ß-1,3-glucanase in tobacco seed germination. *Plant J.,* 23: 215-221.

Miller, C.O. 1956. Effects of cytokinins on germination of lettuce seeds. *Plant Physiol.,* **31**: 318.

Mitchell, B., C.Armstrong, M.Black and J.Chapman. 1980. Physiological aspects of sprouting and spoilage in developing *Triticum aestivum* L. (wheat) grains. **In:** "Seed Productio n" (P.D. Hebblethwaite, Ed.,) pp. 339-356, Butterworth, London.

Miyamoto, T. and E.H. Everson. 1958. Biochemical and physiological studies of wheat seed pigmentation. *Agron. J.,* **50**: 733.

Miyamoto, T., N.E. Tobert and E.H. Everson. 1961. Germination inhibitors related to dormancy in wheat seeds. *Plant Physiol.,* **36**: 739.

Neill, S.J., R.Horgan and A.D.Parry. 1986. The carotenoid and abscisic acid content of viviparous kernels and seedlings of *Zea mays* L. *Planta.,* **169**: 87-96.

Neill, S.J., R.Horgan and A.F.Rees. 1987. Seed Development and vivipary in *Zea mays* L. *Planta.,* **171**: 358-364.

Nelson, S.O. 1965. **In** : 'Electromagnetic Radiation in Agriculture, pp. 60-68, p.166-165, American Society of Agricultural Engineers, St. Joseph, Michigan.

Oishi, M.Y. and J.D.Bewley. 1990. Distinction between the responses of developing maize kernels to fluridone and desiccation in relation to germinability, amylase activity and abscisic acid content. *Plant Physiol.,* **94**: 592-598.

Pollock, B.M. 1962. Temperature control of physiological dwarfing in peach seedlings. *Plant Physiol.,* **37**: 190-197.

Pollock, B.M. and V.K.Toole. 1961. After ripening, rest period and dormancy. Seeds : The Year book of Agriculture, U.S. Department of Agriculture, Washington, DC, 1961. p. 106.

Roberts, E.H. 1969. **In:** 'Society for Experimental Biology Symposia XIII'. pp. 161-192. Cambridge University Press, Cambridge.

Roberts, E.H. 1973. **In:** 'Seed Ecology' (W. Heydecker, ed.), pp. 189-218. Butterworths, London.

Roberts, E.H. and R.D.Smith. 1977. Dormancy and the pentose phosphate pathway. **In:** "The physiology and Biochemistry of Seed Dormancy and Germination" (A.A. Khan, Ed.). pp. 385-411, Elsevier, Amsterdam.

Roberts, E.H. and R.D.Smith. 1977. In 'The Physiology and Biochemistry of Seed Germination' (A.A. Khan, ed.), pp. 385-412, Elsevier / North Holland Biomedical Press, Amsterdam.

Robichaud, C. and I.M. Sussex. 1986. The response of viviparous – 1 and wild type embryos of *Zea mays* to culture in the presence of abscisic acid. *J. Plant Physiol.,* **126**: 235-242.

Sussex, I.M. 1975. Growth and metabolism of the embryo and attached seedlings of the *Viviparous mangrove, Rhizophora mangale. Am. J. Bot.,* **62**: 948-953.

Thomas, T.H. 1977. Cytokinins, Cytokinin – active compounds and seed germination. The Physiology and Biochemistry of Seed Dormancy and Germination (A.A. Khan, ed.). North Holland, Publishing Co., Amsterdam, p. 111.

Thomas, T.H., N.L.Biddington and D.F.O'Toole, 1979. Relationship between position on the parent plant and dormancy characteristics of seeds of three cultivars of celery (*Apium gravelens*). *Physiol. Plant.,* **45**: 492.

Thornber, S. 1903. Some practical suggestions concerning seed germination. Timely Hints for Farmers No. 50, Univ of Arizona Agric. Expt. Station, Arizona.

Toyomasu, T., H.Yamane, N.Murofushi and Y.Inoue. 1994. Effects of exogenously applied gibberellin and red light on the endogenous levels of abscisic acid in photoblastic lettuce seeds. *Plant Cell Physiol.,* **35**: 127-129.

Tran, V.N. and Cavanagh, A.K. (1984). Structural aspects of dormancy. **In:** 'Seed Physiology' (Murray, I. and David, R. ed.), Vol. 2. p.35.

Vigh, L., Horvath, I., Hasselt, P.R. and Kuiper, P.J. (1985). Effect of frost hardening on lipid and fatty acid composition of chloroplast thylakoid membranes in two wheat varieties of contrasting hardiness. *Plant Physiol.,* **79**: 756-759.

Vincent, E.M. and E.H.Roberts. 1977. The interactions of light, nitrate and alternating temperatures in promoting the germination of dormant seeds of common weed species. *Seed Sci.& Technol.,* **6**: 659-670.

Vleeshouwers, I.M., H.J. Bouwmeester, and C.M. Karssen. 1996. Redefining seed dormancy. An attempt to integrate physiology and ecology. *J. Ecol.,* **83**: 1031-1037.

Walker – Simmons. M. 1988. Enhancement of ABA responsiveness in wheat embryos by high temperature. *Plant Cell Environ.,* **11**: 769-775.

Wareing, P.F. and P.F.Saunders. 1971. Hormones and dormancy. *Ann. Review. Plant Physiol.,* **22**: 261.

9

Ecological Aspects of Germination

K. Sivasubramaniam[1] and E. Sujatha[2]

[1]*Seed Science and Technology Unit, Department of Plant Breeding and Genetics Agricultural College and Research Institute, Madurai – 625 104*
Department of Soil and Crops, Adhiparasakthi Agricultural College, G.B.Nagar Kalavai – 632 506

1. INTRODUCTION

Spread of germination over a period of time can have survival value and can protect the species from eradication as a result of adverse conditions that follow germination. Survival of plant species thus, depends on dormancy, a distinctive phase which requires the perception and interpretation of environmental influences to permit germination (Villers, 1975). The inter-related factors of temperature, light, seed impermeability, balance of growth regulators and enzyme mediated metabolic processes combine to impose dormancy on seed and then allow germination at favourable times. Ecological aspects do determine the germination behaviour of seeds, the micro-climate prevailing at the immediate vicinity of the seed will act as the determining factors. Germination immediately upon shedding may in fact accelerate the depletion of the species and prevent its colonization, whereas spacing the germination over time helps in conquering the space compared to other species.

2. ENVIRONMENTAL CONTROL OF GERMINATION

Environment plays a key role in seed survival. Seeds are naturally in tune with environmental cues to enhance their survival ability. Both pre harvest and post harvest environment (upon sowing) affect the germination potential of seeds. Seeds upon maturation do undergo an increase in inhibitors that prevent precocious germination or pre harvest sprouting – in commercial parlance, is necessary for further storage on the mother plant or in soil to space the time and distance from the source of production. This helps in preventing competition for nutrients with the mother plant system and among siblings. The impact of environment starts prior to harvest and continues till the seed germinates and establishes. Systematic planning and management during production and post harvest helps in providing seed a favourable environment in agriculture. This is not the case in wild.

The environment plays a major role in seed germination in two ways viz., either during the seed formation or its location on the mother plant which determines subsequent germination behaviour. The differences in degree of dormancy in *Avena fatua* it was opined, cannot be due to complete shut down of the metabolism, but could be due to specific metabolic block. Further, environment factors may act on specific sites of metabolic sequence in the seed and it is likely that an environmental factor controls germination by general depression or activation of entire metabolic sequence in seed.

Post harvest short duration dormancy has been known to practicing farmers cultivating wheat, barley and oats and is generally due to a special type of dormancy that disappears during dry storage under ambient conditions (Harrington, 1923; Cóme and Corbineau, 1984). Freshly harvested seeds germinate at temperature between 0-10°C, but dormancy is due to the inability to germinate at high temperature which upon storage leads to improvement in germination at wide range of temperatures. Studies in wheat and barley have shown that the dormancy is due to the presence of inhibitors like phenolic compounds that absorb the oxygen during oxidation by polyphenoloxydase in pericarp and seed coat or the adjoining glumes. This barrier to oxygen is especially strong at higher temperatures as seen in barley where dormant seeds absorb more oxygen than non-dormant ones. Though there is no qualitative change in phenolic compounds or polyphenoloxydases the oxidation gets delayed (Lenoir, 1983; Corbineau *et al.,* 1984). In rice, it was found that seeds harvested during kharif had a higher degree of dormancy (8-62 days) than those for rabi harvest (2-40days) in Kerala (Sukumara Dev,1982) and opined that it varied with varieties and within a variety for different seasons. Generally longer duration varieties exhibited longer dormancy and shorter with less dormancy. The phenomenon of cardinal temperature was used to express the right temperature for germination. Among the 22 tobacco cultivars studied, the cardinal temperature ranged between 15-34°C and the average optimum temperature for all cultivars was 24.6°C and at 36°C most of the cultivars exhibited thermo-dormancy in contrast to variety Speight G140 that gave 90% germination at 36°C (Mahopatra and Suggs, 1989). Such studies would help in selecting cultivars suitable or tailored to soil temperature prevailing during different seasons.

Seeds are able to sense their environment for temperature cues. Even brief exposure of 6h to an unfavorable temperature in peas led to prolonged effect on phospholipids metabolism even when seeds were transferred to favourable temperature regimes. Mitochondrial structure failed to develop and few golgi bodies were noticed with lack of endoplasmic reticulum. Lipid bodies located near the plasma membrane disappeared or fused more slowly with membrane thereby affecting its semi-permeability (Hodson *et al.,* 1987).

Photoperiod can also affect seedling recruitment in plants sensitive to light. In *Chenopodium album,* seeds exposed to long days had reduced germination than those exposed to short days (Table 1). In *Ononis sicula,* seeds exposed to long day had permeable seed coat, while short day exposed plants had seeds with reduced seed coat permeability. Fruits of *Cucumis sativus* exposed to red light had seeds that produced 92% dark germination and fruits exposed to far-red light or dark had germination of only 28% proving the effect of mother plant environment in subsequent seed germination. Many plant show marked response to light stimulated emergence (Table 2).

Table 1: Effect of mother plant environment on seed germination in *Chenopodium album*

Culture condition of mother plant		Long day	Short day	Short day with interrupted night
2 weeks after harvest	Light	10	92	24
	Dark	0	77	12
4 months after harvest	Light	24	100	100
	Dark	3	100	96

Table 2: List of species showing marked response to light (Mayer and Poljakoff-Mayber, 1982)

Species whose seeds are stimulated by light		Species with seeds whose germination is retarded by light	
1.	*Arceuthobium oxicedri*	*1.*	*Bromus sp.*
2.	*Daucus carota*	*2.*	*Datura stramonium*
3.	*Elatine alsinastrum*	*3.*	*Lycopersicum esculentum*
4.	*Ficus aurea*	*4.*	*Liliaceae various species*
5.	*Ficus elastica*	*5.*	*Nigella sp.*
6.	*Gloxinia hybrida*	*6.*	*Phacelia sp.*
7.	*Graminae many species*	*7.*	*Primula spectabilis*
8.	*Gesneriaceae many species*		
9.	*Lactuca sativa*		
10.	*Lobelia cardinalis*		
11.	*Lobelia inflate*		
12.	*Loranthus europaeus*		
13.	*Lythrum ringens*		
14.	*Mimulus ringens*		
15.	*Nicotiana tabacum*		
16.	*Nicotiana affinis*		
17.	*Oenothera biennis*		
18.	*Primula obconica*		
19.	*Phoradendron flavescens*		
20.	*Raymondia pirenaica*		
21.	*Rumex crispus*		
22.	*Verbascum thapsus*		

2.1. Effect of pre-harvest environment in subsequent germination

Boyce *et al.* (1976) reported that in *Festuca arundinacea* seeds maturing at 15°C had greater dormancy than those maturing at 26.6°C. Germination of orchard grass, tall fescue and meadow fescue was positively correlated with harvest time (Saiga and Araki, 1976). In rice, high temperatures noticed 10-15 days after flowering induced dormancy (Ikehashi, 1975). The degree of dormancy increased when the germination temperature shifted from 20°C to 30°C, indicating imposition of secondary dormancy. Differences in dormancy breaking among rice varieties were noticed by Chandrasekariah *et al.* (1976) and Mitra *et al.* (1975), but were related to the prevalent relative humidity and temperature prior to seed maturity. In *Brassica japonica* and B. *cernua*, seasonal periodicity was noticed in degree of dormancy (Tokumasu, 1977).

Baskin and Baskin (1976) found high summer temperature requirements for after ripening of the winter annuals, *Stellaria media*, *Valerianella umbilicata* and *Phacelia purshii*. Low winter temperatures induced secondary dormancy in *Phacelia dubia*, (winter annual) (Baskin and Baskin, 1978). Temperature mediated dormancy in seed occurred in 11 tulip cultivars. Cold stratification overcome dormancy but subsequent temperatures above 15°C reimposed dormancy (Fortanier and van Brenk, 1975).

The dormancy process was studied in detail in two European tree species, European white birch (*Betula_pubescens*) and sycamore maple (*Acer pseudoplatanua*). In an experimentally induced dormant birch, the leaves were removed and the buds were exposed to long days. The result was the breaking of dormancy by the buds. In another experimentally induced dormant plant, the buds were exposed to long days and the leaves to short days. The result was that the buds did not break dormancy. This suggests that the dormant state is controlled by the leaves. This could attribute the reason why trees in temperate climates shed leaves prior to onset of winter. In order to break dormancy in spring, native species of trees and shrubs require exposure to an extended cold period. The length of the period and the degree of cooling has not been determined for all species. However, among cultivated plants, peach trees require 400 or more hours of exposure to temperatures below 7°C, and blueberry plants require 800 hours of exposure at this level in order to break dormancy. After the period of cold exposure, plants require increasing temperature and increasing photoperiod to resume growth. Most deciduous tree species require a minimum of 300 hours at temperature near 25°C before new growth begins.

The dormancy requirement of plant species has a profound influence on geographic distribution. If such a plant is not exposed to the required minimum cold period it may not break dormancy in spring or if it does break dormancy, growth may be weak and the plant devoid of flowers. Thus a species may be limited in its southward distribution by the length and severity of winter temperatures. This may be the reason why apple trees do not flourish in Florida as it is not cold long enough to fulfil the dormancy requirements of apple trees.

2.2. Effect of post harvest environment in seed germination

Generally all seeds will possess some degree of resting period known as quiescence. This temporary phenomenon lasts until the seed encounters favourable temperature, light, moisture and air. Some seeds that have a prolonged resting period may be termed to as dormant.

The classes of seed dormancy are physiological (PD), morphological (MD), morphophysiological (MPD), physical (PY), and combinational (PY+PD) (Baskin and Baskin, 2003).

Physiological dormancy is caused by a physiological inhibiting mechanism in the embryo, and it is the most common class of seed dormancy on earth. PD may be broken during exposure to cool (0.05-10°C) moist conditions such as occur in winter (called cold stratification) or warm (≥15°C) conditions such as occur in summer, depending on the species. In some species, dormancy-break during summer may occur under warm, dry conditions (after ripening), but in other species warm, moist conditions (called warm stratification) are required. Information on the phenological (ecological) life cycle of a species and seasonal temperature changes in the habitat will help one decide if seeds should be given warm and/or cold treatments to break dormancy.

Physical dormancy (PY) due to an impermeable seed or fruit coat is the second most common class of seed dormancy on earth. PY occurs in some members of 15 plant families, including the *Anacardiaceae*, *Bixaceae*, *Cannaceae*, *Cistaceae*, *Cochlospermaceae*, *Convolvulaceae*, *Curcurbitaceae*, *Dipterocarpaceae*, *Fabaceae*, *Geraniaceae*, *Malvaceae*, *Bombacaceae*, *Sterculiaceae*, *Tiliaceae*, *Nelumbonaceae*, *Rhamnaceae*, *Sapindaceae*, and *Sarcolaenaceae*. Impermeable seeds (or fruits) in each of these families have a distinct "water plug" that must be dislodged or moved before imbibition can occur. In nature, seeds with PY may germinate after exposure to low winter temperatures (*Melilotus spp.*), drying (*Geranium carolinianum*), high daily temperature fluctuations (winter annual *Trifolium spp.*), or fire (*Iliamna corei*) (Baskin and Baskin, 2003).

The seed coat of charlock seeds shows no resistance to water uptake yet offers a resistance to oxygen diffusion. Naked embryos could germinate at low oxygen tensions. The calculated partial pressure of oxygen at the surface of the embryo in intact seeds is low but apparently sufficient for germination. The low germination was due to the effect of seed coat to the establishment of low internal oxygen concentrations that lead to the production of growth inhibiting substances in the embryonic tissue. The seed coat therefore acts both as a diffusion barrier for oxygen and as a deposit for endogenous inhibitors. In later stages this oxygen diffusion barrier may also control germination by impeding the oxidation and subsequent destruction of endogenous inhibitors.

3. GEOGRAPHICAL DIVERSITY - AN ADAPTIVE STRATEGY

Species vary in their ability to interact with environment and their divergence in dormancy characteristics is a direct result of adaptation to the environment.

Heuchera parviflora var. *parviflora* grows primarily in southeastern North America and *Heuchera cylindrica* var. *cylindrica* in northwestern North America. Seeds of *H. parviflora* were collected in Kentucky in November 1995 and 1996 upon maturity but prior to dispersal stage of the life cycle and those of *H. cylindrica* were collected in Washington in August 1996 to study and compare the germination patterns of these two intracontinental disjunct species of *Heuchera*. During 2–12 weeks of incubation over a range of thermoperiods (15/6, 20/10, 25/15, 30/15, 35/20°C (12 h : 12 h)) that stimulated habitat temperatures, fresh *H. parviflora* seeds germinated to 62–87% at 20/10 - 30/15°C in light. Cold stratification at 5°C effectively broke conditional dormancy in *H. parviflora* seeds: 76–96% germination took place during 2 weeks of incubation at 15/6 – 35/20°C in light following 12 weeks of cold stratification in light. In contrast, fresh H. *cylindrica* seeds germinated to 69–94% at 15/6 – 25/15°C in light during 2–12 weeks of incubation in light, but cold stratification at 5°C did not improve germination. Peak germination of *H. parviflora* seeds sown in November 1996 in an ambient-temperature greenhouse occurred in early March 1997, whereas that of *H. cylindrica* seeds sown in September 1996 occurred in late October 1996. Adaptive divergence of these species apparently has occurred with respect to their seed germination ecology. Their dormancy and germination characteristics are similar to many other species growing in the same area (Hidayati and Walck, 2002).

Germination timing of *Arabidopsis thaliana* displays strong plasticity to geographic location and seasonal conditions experienced by seeds. Selection on germination timing varied across geographic location, indicating that germination timing can be a critical stage in the establishment of genotypes in new locations. The rate of evolution of germination timing may therefore strongly influence the rate at which species can expand their range (Donohue et *al.,* 2005).

4. WEED SEED BANK

The adage "one year seeding; seven years weeding" is appropriate when we consider the presence of seeds in soil. Competition from weeds can result in yield reduction from 27-71% depending upon type and intensity of weeds and their time of occurrence. Weed seed exist in soil. A reserve of viable / non-viable seeds in a soil is called soil seed bank. The dormant status of weed seed in soil seed bank and their sporadic germination often determines the population dynamics of weeds in a given eco-system.

Baskin and Baskin (1998) recorded that of the 155 families, 71 were represented by 2 or more genera (*Apiaceae, Aisteraceae*). Among these, the seeds possess myriad types of dormancy. Twenty nine families viz., *Amaryllidaceae, Anonaceae, Apiaceae, Berberidaceae, Caprifiliaceae, Dilleriaceae, Fumariaceae, Grossulariaicerae, Haemodoraceae, Hydrophyllaceae, Illiciaceae, Iridaceae, Tiliaceae* and *Magnoliaceae* found in soil seed banks are found to possess rudimentary embryo and 29 families have morphological or morpho-physiological dormancy, 10 with physical dormancy and 110 families with physiological dormancy thus underlying the nature of regeneration capabilities of soil seed bank.

Presence or deposit of such seeds can occur due to falling of seeds from plants, a phenomenon known as seed rain, seed spills that occur during normal agricultural activities involving production and transport, seed dispersal and seed accumulation by carnivores – animals that consume seeds (Fig. 1). Similarly withdrawal from the seed bank can occur by:

- Germination, loss of viability,
- Predation by animals, birds and insects,
- Moment of seed into deeper layers of soil,
- Wash-off by forces of erosion and
- Other causes.

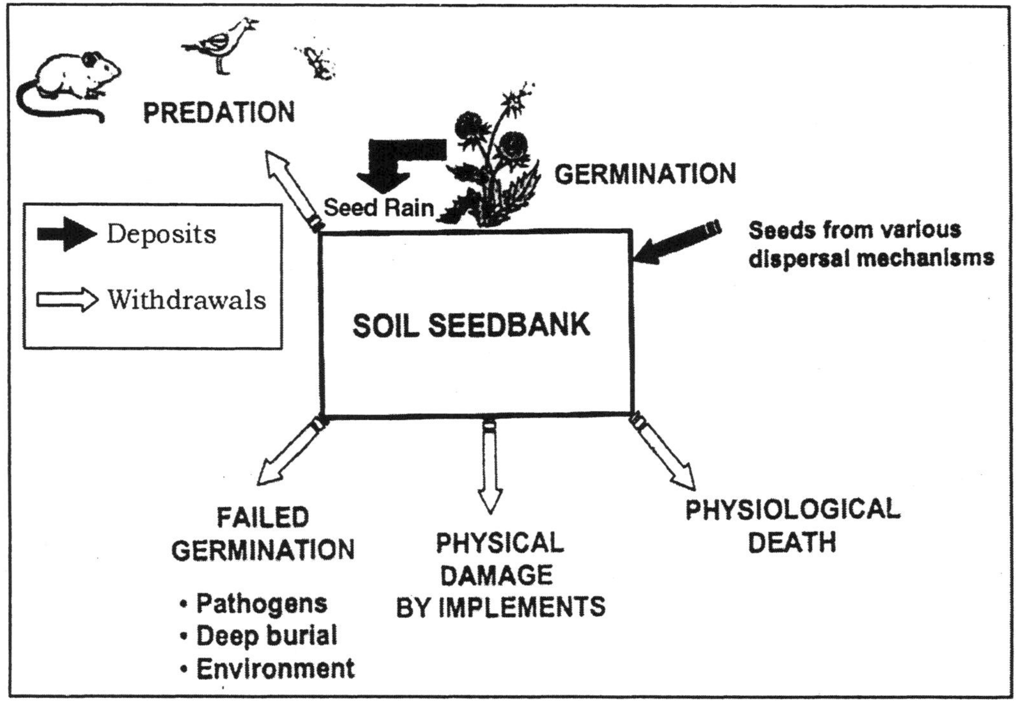

Fig. 1: Seed bank characteristics

Such inputs directly affect the composition, density and genetic reserve of soil seed banks.

5. SEED RAIN AND SEED SHADOW

Seed rain is the presence to seed in the vicinity of the tree base while seed shadow is presence of seed away from the seed base (Fig. 2). The area of seed shadow will depend on the dispersal mechanism and dispersal agents.

Howe and Primack found higher densities of bird dispersed *Casearia corymbosa* seeds under than away from two fruiting trees in Costa Rica rainforest. Eight weeks later upto three times as many seedlings survived 10-30m away from the crown edges as under the trees probably due to localized rodent herbivory. Jensen (2003) report mortality of *Andira inermis* seeds by weevils killed 60% of the seeds under parent trees, 30% under feeding roots and 10% of the single seeds scattered by bats. Dispersal of North American *Juniperus monosperma* and Galapagos, *Bursera graveolens* respectively for only a few metres results in dramatic enhancement of survival to the seedling stage. Chinese elm (*Ulmus parvifolia*) seedlings directly under parents suffered 580 times the mortality from leaf beetles (*Pyrrahalta luteoloa*) as those a few metres away on suburban lawns.

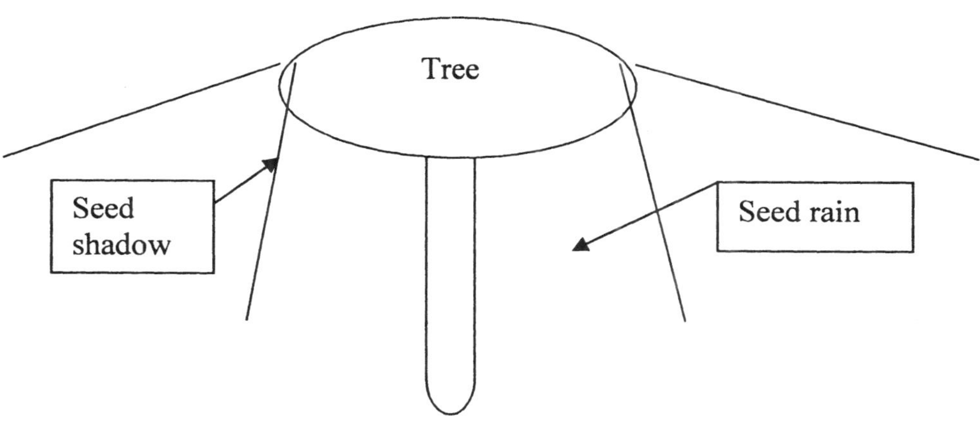

Fig. 2: Concept of seed rain and seed shadow

Thompson and Grime (1979) opined that most of the soil seed bank consists of buried seeds and a few over the surface or buried in litter / humus. Seeds can fall into soil cracks, covered by sediment during flood or have particles blown over them by wind. Initial dispersal of seed over the soil will be horizontal and upon burying vertical. Seed size is important character in the moment of seed in soil. Small seeds are trapped by small and large seeds by large particles.

6. TYPES OF SOIL SEED BANKS

Seed banks can be transient or persistent. Transient seed banks posses seeds that germinate within a year of dispersal. Persistent seed banks remain in soil for more than a year. This persistent component of seed bank represents a reserve of genetic potential accumulated over time. Seed bank input is determined by seed rain especially from local dominant species. Such local dispersal may be passive means shattering of pods, capsules or siliqua and from distant sources from dispersal modes namely wind, rain, animals etc.

Thompson and Grime (1979) further subdivided the transient types into two: Type 1 and Type 2 and persistent into Type 3 and Type 4. In Type 1 and Type 2, depending on season (autumn, winter) seeds germinate. In Type 3, while a few seeds germinate soon after dispersal, a large reserve of viable seeds still remains ungerminated. Germination of Type 3 and 4 will be enhanced if they are brought to surface.

Weeds vary considerably with respect to the longevity of their seed dispersal depending on species, depth of seed burial, soil type and level of disturbance. Baker *et al.* (1996) described it as short term persistent and long term persistent. Seeds in short term persistent live for at least 1 year but less then five years, while in long term persistent live for 5 years or more. Agronomists and biologists in general are more interested in the persistent soil seed bank than transient because of their genetic diversity and ability to colonize an ecosystem over time. Annual agricultural weeds predominate in soils found beneath ancient human dewellings. Viable lambs quarters (*Chenopodium album*) and corn spurry (*Spergula arvensis*) were discovered in soil associated with habitations known to be 1700 years old. The persistent seed bank changes very little from year to year as long as the cropping system / pattern in a field remain unchanged.

Knowledge of weed seed banks and their persistence is important for developing longterm weed management strategies. Kropac (1966) and Robers (1981) proved that more than 80,000 weed seeds per metre square can be found in top 15-20cm of soil. Weed seeds generally found at the top 5cm germinate and perish, if they do not possess sufficient reserves for germination and establishment. Physiological mechanisms may prevent germination of buried seeds and allow persistence in soil. Till or churning of soil layers during ploughing can result in germination of buried seeds. Weed seed banks decline exponentially over the years and the rate of decline progresses with cultivation frequency. Robers (1981) calculated that the number of seedlings emerging from a disturbed soil declines exponentially at a constant rate for each species between 45-62% per year.

Study of seed-bank seeds of sweet fern (*Comptonia peregrina* (L.) Coult.), an actinorhizal nitrogen-fixing shrub (Dow and Schwintzer,1999), revealed that seeds extracted from the soil contained approximately 2000 seeds m^{-2} with the greatest concentration at a depth of 60-80 mm. The seeds were 4.0-5.5 mm long, enclosed by a pitted, woody pericarp, and 8% contained embryos. Many seedlings emerged in disturbed plots (vegetation removed and upper soil mixed) in May and June 1997, but none appeared after 24 July. Seeds collected in May and June germinated readily in a growth chamber (30-45% germination) whereas only 2-5% of July and August collected seeds germinated indicating induction of secondary dormancy. August collected seeds showed strong germination after 15 days of moist chilling at 4°C indicating relief of secondary dormancy by chilling. Temperature fluctuation with an amplitude of 10°C strongly stimulated germination. Presence of annual secondary dormancy cycles and stimulation by strong temperature fluctuation assures that seed-bank seeds germinate under conditions that allow the seedlings to become established.

Another pre-requisite for longevity of seeds in the soil is that they do not germinate while buried. Though germination is possible only when seeds are non dormant, the germination of buried seeds are prevented by environmental factors, including darkness, CO_2/O_2 ratios, volatile metabolites and dampened temperature fluctuations. Pons (1991) revealed that small seeds require less oxygen compared to large seeds and hence they survive well in soil seed banks. Sahoo *et al.* (1994) opined that in many species, the length of time seeds remain viable and ungerminated in soil increases with an increase in depth of burial.

6.1. Population dynamics of soil seed banks

Several scientists have attempted to gather data to use in the equations regarding soil seed banks. While conducting research on 200 seeds of

Tragopogon pratensis found that at the end of 12-13 months, 1-13 germinated seeds, 184-199 dead, 1-4 in enforced dormancy and 0-1 seeds in induced dormancy and found that after 12 months, 71% of *Acacia derlandieri* and *Leucaena pulverulenta* had impermeable seed coats.

The persistent seed banks are found in many kinds of habitat ranging from tropics to the arctic tundra and deserts to marshes. Mathematical model of soil seed banks was propounded by Schafer and Chilcote (1969)

$$S = Pe\underline{x} + Pend + D\,g + D\,n$$

S (100%)	: Total number of seeds buried in the soil at a point of time.
Pex	: Percentage of persistent seeds that are prevented from germination by exogenous or environmental factors.
Pend	: Percentage of persistent seeds that are prevented from germination by endogenous factors and will not germinate even if they are exhumed and placed in endo or exogenous dormancy relieving treatments.
Dg	: Percentage of seeds germinated *in situ.*
Dn	: Percentage of seeds dead at the time of burial.

Roberts (1972) modified the above formula as follows:

S = Pinn + Pend + Penf + Dgd + Dge + Dni + Dne + Dnp

S	: Total buried in the soil
Pend	: Endogenous dormancy
Pinn	: Innate dormancy
Penf	: Enforced dormancy
Dgd	: *In situ* germination deep in the soil
Dge	: *In situ* germination on soil surface
Dni	: Part of the seed loose viability before being taken into soil bank
Dne	: Dead as result of physiological ageing
Dnp	: Killed by predation and pathogens

Seeds buried in soil for many years accumulate cytological that can cause a shift in genetic material (Levin, 1990). The death of seed in a seed bank can be due to the following reasons (1) food reserves are exhausted and thus cells have no energy supply with which to make repairs. (2) food reserves are altered chemically therefore, they are no longer usable by the embryo, (3) many lesions gradually accumulate in the DNA that cell division and subsequent growth be impossible and (4) membrane become non functional.

Soil organisms that interact with buried seeds are (1) animals that predate or eat seeds (2) and bacteria that attack seeds, and (3) organisms that do not use seeds as a food source. Weed seed population is also decided by fire and flooding (Fig 3). The survival of seeds in risky environment by a persistent seed bank, where if a species fails to produce seeds in one year, the persistent seed bank ensures that the species can persist at the site without immigration. Seed banks can have various effects on population dynamics, especially of annual species. The effects include a speeding up of the oscillary return to equilibrium (1) or a slowing down (0) exponential return to equilibrium. The seed bank has a greater demographic effect in a poor than in a good year for population growth and the actual number of individuals of species present at a population site. The variation is measured by the presence of a seed bank. In fact genetic variations stored in the bank may decrease the genetic changes and thus decrease the rate of evolution.

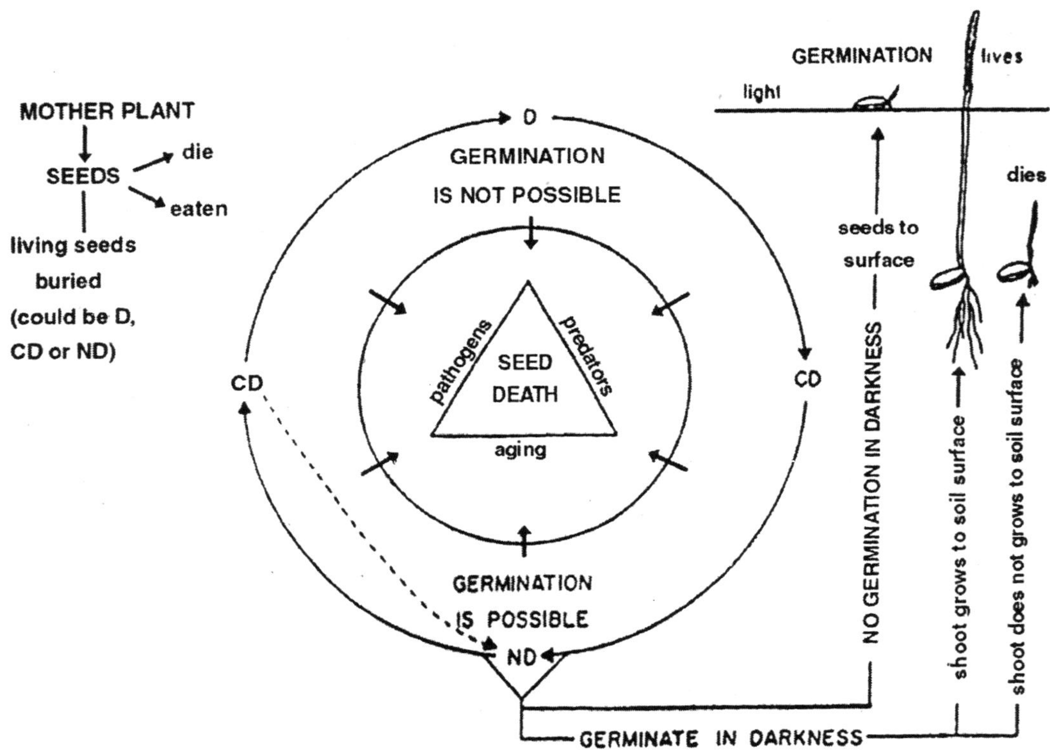

Fig. 3: Seed bank dynamics

The germination ecology (Jensen, 2003), including primary and secondary dormancy, mortality of seeds, and seed-bank type, of 20 fen grassland species from Northern Germany was investigated using a combination of burial and germination experiments. To analyze primary dormancy and effects of after-ripening (dry storage for 28 days) on freshly matured seeds, germination was measured at two fluctuating temperature regimes (15/25°C and 5/15°C) in the light and in darkness. Temporal changes in dormancy were investigated by burying seeds at 8–10 cm depth in nylon bags, exhuming samples at bimonthly intervals over a period of two years, and analyzing germination at the above-mentioned temperature and light treatments. Additional seed samples were retrieved five years after burial to determine seed longevity and seed-bank type. Freshly matured seeds of all species except *Bromus racemosus* showed primary dormancy. Dry after-ripening significantly increased germination in *Parnassia palustris* and *Triglochin palustre.*

Mortality of buried seeds of *Bromus racemosus*, *Sanguisorba officinalis*, and *Succisa pratensis* reached 100% within 12 months (transient seed banks). All other species germinated both in the first and in the second growth period after burial and (with the exception of *Briza media*) showed annual changes in dormancy. For *Silene flos-cuculi* and *Juncus filiformis*, dormancy cycles were detected only in the dark treatments. Most species had a lower percentage germination in darkness than in the light, and the greatest suppression of germination in darkness was found in the small-seeded species (*Juncus filiformis* and *Parnassia palustris*). The retrieval of seed samples after five years revealed that most of the fen grassland species examined have short-term, persistent seed banks and thus, are buffered against years of poor seed production and/or seedling survival. In addition, a large proportion of the species maintain long-term, persistent seed banks from which re-establishment is possible if management practices and site conditions of degenerated fen grasslands become appropriate following restoration measures.

Table 3: Longevity of some of the weed spp.

Species	Period (years)	Extent of germination (%)
1. *Verbascum blattaria*	100	42
2. *Verbascum thapsus*	35	1
3. *Rumex crispus*	80	2
4. *Oenothera biennis*	80	10
5. *Rumex crispus*	70	8
6. *Amaranthus retropexus*	40	1
7. *Brassica nigra*	50	1
8. *Polygonum hydropiper*	50	1
9. *Portulaca oleraceae*	40	1
10. *Setaria glacua*	30	1
11. *Stellaria media*	30	1
12. *Centaurea maculosa*	>10	1

The regeneration of plant communities from seed depends on seeds being in the right place at the right time. They must also be in the right physiological state to germinate and established seedlings when there is good chance for regeneration. In some species these requirements are small and the seeds germinate as soon as they are shed. In others, seeds may remain for long periods in soil seed bank with intermittent germination of a part of the population. Some of these seeds are macrobiotic (long lived) (Table 3, 4). If the seeds are to survive in the soil, viability must be maintained as long as germination is arrested by dormancy or queiscence. Dormancy is a failure to germinate under given moisture, and suitable temperature for radicle emergence and seedling growth. If these minimum requirements for germination are lacking, the seed is described as queiscent since metabolism often be arrested or retarded. The adage 'one year seeding, seven years weeding' is an epitomy of weed seed longivity in soil. The composition and density of weed seeds in the soil vary greatly but are closely linked to the cropping history of the land. The density of seed bank ranges from 0 in newly developed soils and 4,100 to 137,700 seed per square metre in cultivated soil. Seed density is influenced by past farming practices may vary from field to field and between areas within fields. But seed banks do have a number of similarities. Generally they are made up of numerous species, although several species comprise 70 to 90% of the total seed bank.

Table 4: Fecundity of some weeds species and their ability to survive in soil

Weed species	No. of seeds produced per plant	Length of seed survival in undisturbed soil (years)
Canada thistle (*Cirsium arvense* (L))	680/srem	21
Common Lambs quarters (*Chenopodium album* L.)	72,450	39
Purslane (*Portulaca olracea* L)	52,300	30
Common sunflower (*Helianthus annuus*)	7,200	1
Seven foxtail (*Setaria viridis* (L.) Beauv)	34,000	39
Pigweed (*Amaranthus retroflexus* L.)		10
Wild oats (*Avena fatua* L.)	117,400	1
Shallow foxtail (*Setaria glauca* (L.) Beauv.)	250 6,420	30

The 120 year Dr.Beals seed burial experiment carriedout in moist, well aerated sand, was excavated for the 15th time during April 2000 and studied by Telewski and Zeevart (2002) and reported that 23 seeds of *Verbascum blattaria* and two seeds of *Verbascum* sp. germinated and produced normal plants. Upon cold treatment (6 weeks), one seed of *Malva rotundifolia* germinated and produced a normal plant. All crossed and selfed flowers of V. *blattaria* produced viable seeds. While *Verbascum* sp. failed to produce seeds after artificial self pollination (Table 5.). Seeds produced by selfing in *Malva rotundifolia* required cold treatment before germinating. It was ample proved that even after remaining buried for 120yrs, seeds of *V. blattaria* and *M. rotundifolia* can produce viable-propagating progenies.

Table 5: Dr. Beals seed burial experiment – data of weed seed viability over 120 years

Species	Habit	1884	1889	1894	1899	1904	1909	1914	1920	1930	1940	1950	1960	1970	1980	2000
Years in burial		5	10	15	20	25	30	35	40	50	60	70	80	90	100	120
Agrostemma githago	Annual	0	0	0	0	0	0	0	0	0	0	0	0	0	0	0
Amaranthus retroflexus	Annual	42	Few	2	8	Few	Few	Few	2	0	0	0	0	0	0	0
Ambrosia artemisifolia	Annual	0	0	0	0	0	0	0	4	0	0	0	0	0	0	0
Anthemis cotula	Annual	52	Few	10	0	Few	Few	0	0	0	0	0	0	0	0	0
Brassica nigra	Annual	0	Few	8	18	Few	Few	Few	38	8	0	0	0	0	0	0
Bromus secalinus	Annual	0	0	0	0	0	0	0	0	0	0	0	0	0	0	0
Capsella bursa-pastoris	Annual	100	Few	42	42	Few	Few	Few	0	0	0	0	0	0	0	0
Erechities hieracifolia	Annual	0	0	0	0	0	0	0	0	0	0	0	0	0	0	0
Euphorbia maculata	Annual	0	0	0	0	0	0	0	0	0	0	0	0	0	0	0
Lepidium virginicum	Annual	94	Few	34	58	Few	Few	Few	2	0	0	0	0	0	0	0
Malva rotundifolia	*Annual or Biennial*	2	0	0	6	0	0	0	0	0	0	0	0	0	2	2
Plantago major	Annual	0	0	2	0	0	0	10	0	0	0	0	0	0	0	0
Polygonum hydropiper	Annual	6	Few	8	2	Few	0	0	0	4	0	0	0	0	0	0
Protulacca oleracea	Annual	38	Few	18	14	Few	0	0	2	0	0	0	0	0	0	0
Setaria glauca	Annual	68	Few	Few	0	Few	Few	0	0	0	0	0	0	0	0	0
Stellaria media	Annual	72	Few	6	6	Few	Few	0	0	0	0	0	0	0	0	0
Trifolium repens	Annual or Biennial	4	0	0	0	0	0	0	0	0	0	0	0	0	0	0
Verbascum thapsus	Annual	84	-	38	32	0	0	Few	0	0	0	0	0	0	2	0
Oenothera biennis	Annual	82	Few	12	28	Few	Few	0	38	38	24	14	10	0	0	0
Rumex crispus	Annual	90	-	14	16	Few	Few	Few	18	52	4	8	2	0	0	0
Thuja occidentalis	Annual	0	0	0	0	0	0	0	0	0	0	0	0	0	0	0
Verbascum blattaria	Annual	-	-	-	-	-	-	-	-	62	68	74	70	20	42	46
Verbascum sp.	Annual	-	-	-	-	-	-	-	-	-	-	-	-	-	-	4

6.2. Seed bank and agriculture

Lack of soil disturbance is also important for longevity of buried seeds of many species. The number of buried, viable seeds for individual species, as well as the total for all species decline in cultivated than in non-disturbed soil. However frequency, and thoroughness of cultivation, all temperature at time of cultivation in cycles of species and depth of disturbance (Dessaint *et al.,* 1997) each play a role in determining what proportions of the seeds in the seed bank germinates. Further, the cropping systems or rotation of crops, season of year, kind of ploughing, soil type can influence the rate of decline in the number of buried weed seed, associated with increased disturbance. Jordan *et al.,* (1995) showed that crop rotation greatly decreased seeds of *Setaria viridis,* but it had little effect on *Abutilon theophrasti.* Soil disturbances may enhance *in situ* germination and/or result in seeds being brought to the surface, where they are exposed to light, improved oxygen levels and so on that promote germination. In addition to ploughing, seeds could be exposed to favourable germination conditions by drainage of wetlands or digging by animals. Earthworms are responsible for bringing many seeds to the soil surface via casts, which may accumulate on the soil surface at rates of 2-4 to 15-20 mm per year or 750 per year (Chinnusamy *et al.,* 2001).

Table 6: Cropping system and weed seed bank density

Crop history	Soil depth (cm)	Average no. of seeds collected (seed/m^2)	Total number of most frequently encountered species		
			Species	Number	% of total
Vegetables	0-15	4,100	76	9	89
Potatoes	0-20	16,000	80	6	78
Barley-corn-sugarbeets	0-25	1,37,700	8	2	86
Corn-soybeans-corn	0-18	10,200	25	4	85
Corn-field beans – sugarbeets	0-15	20,400	19	3	85
Potatoes - wheat	0-30	51,000	23	--	90

The size and composition of the weed seed bank after 5 year of maize *(Zea mays* L.) continuous cropping under four crop management systems was evaluated. The management systems included a conventional system (ploughing, full rate fertilization, broadcast pre-emergence herbicides), an organic system (ploughing, organic fertilization, physical weed control), a reduced input system (rotary harrowing, low-rate fertilization, physical post-emergence herbicides) and a strip cultivation system (non-tillage, banded fertilization and herbicides at sowing, broadcast top dress fertilization and post-emergence herbicides). Seed bank sampling was performed after maize harvest by taking 270 soil cores per management system at depths of 0-10, 10-20 and 20-30 cm. Weed seed bank size was largest under the organic system (1,00,761 seeds m^{-2}) and smallest under the conventional system (27, 601 seeds m^{-2}). Differences among maize management systems depended mainly upon weed control efficacy rather than upon tillage effects. Four of the eight major weed species showed differential responses to tillage and weed control methods carried out within maize management systems (Table 6).

6.3. Contamination of crop seed

The size and composition of the weed seed population in the soil, at the time a crop is sown, is determined by seed production in previous crops, previous cultural practices and introduction by various natural and agricultural agencies. Agrestals that have poorly developed abiotic methods of dispersal and rely almost entirely on humans for their dissemination. This is achieved largely through seed contamination. Of particular importance is the introduction of weed seeds as contaminants of the crops grown. Salisbury has suggested that the contamination of crop seed has been the most important single agency for the introduction of alien species. Furthermore, he attributes the diminution of several formerly characteristic cornfield weeds to improvements in seed cleaning technology.

Seed production of residual weed population needs to be taken into account when estimating the long-term impact especially in low-input agronomic practices that has been promoted by many sustainable agriculture enthusiasts. Studies were carried out to measure the effects and interactions of crop, weed control, tillage practice and nutrient source on the seed production of the dominant residual weed species *Echinochloa crusgalli* (L.) Beauc on a Sainte-Rosalie clay and *Chenopodium album* L. on a Duravin clay loam in a maize-soyabean rotation at two sites. Seed productjon per unit area was estimated in each experimental unit. Weed seed production was greater under mechanical weed control compared with chemical weed control. In 1997, *E. crusgalli* seed production reached over 3,26,000 seeds m^2 in mechanical weed control treatments, but averaged less than 500 seeds per m^2 in the chemical weed control treatments. *Chenopodium album* produced in the range of 766,000 and 73,000 seeds per m^2 in mechanical and chemical weed control treatments respectively. Very few or no weed seeds were produced in soybean under chemical control. Tillage intensity and nutrient source did not affect seed production of either weed species, with the exception that *E. crusgalli* produced more seeds in chisel than in mouldboard plough tillage in soyabean. Weed control method had more impact on seed production than tillage intensity and nutrient source in a maize soybean rotation.

Generally all mechanized field operations result in the movement of some seeds, but the farmer is concerned with those operations that introduce weed seeds or pose weeds to germinate and establish. Concern has been expressed

Table 7: Land requirements for various crops (Tunwar and Singh, 1988)

Land requirement criteria	**Crops**
Previous season Land free of volunteer plants in the previous season	**Cereals** Rice variety and hybrid, wheat variety and hybrid
	Millets Maize - open pollinated variety, hybrids, synthetics, composities, sorghum variety and hybrid, bajra variety and hybrid, italian millet, little millet, kodo millet, common millet
	Pulses Blackgram, greengram, bengalgram, cowpea, horsegram, lablab, lentil, redgram
	Oilseeds Linseed, niger, safflower, sesame, soybean, castor variety and hybrid
	Fibres Cotton variety and hybrid, Jute
	Forages Berseem, clover, lucerne, forage sorghum, buffel grass, dharaf grass, guinea grass, sudan grass, napier, oats, marvel grass, setaria grass, stylo
	Vegetables Peas, french bean, guar, yam, ashgourd, bittergourd variety and hybrid, bottlegourd variety and hybrid, chow-chow, cucumber variety and hybrid, longmelon, muskelon variety and hybrid, pumpkin variety and hybrid, ridgegourd variety and hybrid, snakegourd, spongegourd, watermelon, wintersquash variety and hybrid, brinjal variety and hybrid, chillies, bhendi, tomato, amaranthus, asparagus, celery, lettuce, fenugreek, cabbage, cauliflower, knol-khol, garlic, onion (aggregate), onion variety and hybrid, carrot variety and hybrid, beet, radish
Previous one year Land in which the crop grown in previous year was of same kind / variety and of equivalent or higher class and eligible for certification	Sunflower variety and hybrid
Previous two years Land in which crops grown in previous two seasons were of same kind / variety and of equivalent or higher class and are eligible for certification	Groundnut
Previous five years (Foundation stage alone) Land must not have been used for cultivation of same crop for past five years	Buffel grass (*Cenchrus ciliaris*); bird wood grass (*Cenchrus setigerus*), daraf grass, dinanath grass, marvel grass, setaria grass, stylo

that replacing the ~r-binder with the combine harvester would enhance weed problems. The delay of harvest associated with the combine compared with the reaper, has provided better opportunities for seed dispersal among some species, particularly those that shed seed later than the time at which would have been carried out. Soil weed seed bank and its expression of population dynamics are often harmful in weed control measures but the seed banks are important in the long term survival of individual species as in all plant communities. However, not all the species in a community are represented in the seed bank (Milberg and Hanson, 1994).

The weed seeds are unlikely to escape seed cleaning, evolutionary selection may favour early flowering forms and hence precocious seed setting and shedding.

7. SOIL SEED BANK AND SEED CERTIFICATION

Emergence of seeds from soil seed banks (previous crop seed) is viewed seriously in seed production. Seeds emerging from previous crop are known as volunteers and may led to reduction in seed standards and thus seed fields face rejection by certification authorities. The Minimum Seed Certification Standards indicate that the land selected for seed production must be free from volunteers and for this the previous crop requirements are to be followed scrupulously. A whole season or a year will have to be considered in case of persistent crop seeds like sunflower or groundnut and sometimes even five years as in forage crops (Table 7).

The release of genetically modified, herbicide tolerant canola (*Brassica napus* L.) genotypes in western Canada has increased interest in the persistence of volunteer canola. Gulden (2003) found that in Canadian farms, average seed bank additions during harvest were approximately 20 times the normal seeding rate of canola. High secondary seed dormancy potential which was genotype specific, prolonged seed bank persistence in fields, irrespective of tillage system. To prevent the spread of 'super weed' canola, it was concluded i) that on some farms, seed bank additions may be lowered by preventing the spillage by adopting more diligent harvest practices and ii) seed bank persistence of *B. napus* may be reduced by growing low dormancy genotypes and avoiding seed burial for one year by following crop rotation.

8. GENETICS OF DORMANCY

Genetics of dormancy has been researched widely especially in commercial wheat and barley cultivars that are resistant to preharvest sprouting. Indications are that the dormancy is controlled by many genes that also interact over each other (epistases).

Dormancy is a genetically complex trait controlled by polygenes with effects modified by the genetic background and environmental factors (Johnson 1935; Anderson *et al.* 1993). Quantitative Trait Loci were determined for seed dormancy in Arabidopsis (*Arabidopsis thaliana*), barley (*Hordeum vulgare*), sorghum (*Sorghum bicolor*), rice (*Oryza sativa*) (Xing-You Gu *et al.*, 2004), and wheat (*Triticum aestivum*; Anderson *et al.* 1993; Oberthur *et al.* 1995; Lin *et al.*

1998; Lijavetzky *et al.* 2000; Mares *et al.* 2002; Alonso-Blanco *et al.* 2003). Six dormancy QTLs, designated as *qSD^S-4, -6, -7, -1, -7, -2, -8,* and *-12,* were identified in wild rice. The locus *qSD^S-7-1* was tightly linked to the red pericarp color gene *Rc.* Two, three and four orders of epistases were detected with four, six, and six QTL, respectively. The higher order epistases indicated the presence of genetically complex network in the regulation of variation for seed dormancy in natural populations. Quantitative trait loci analysis can also be used a tool prerequisite to clone and characterize genes that directly regulate seed dormancy and germination and to facilitate marker-assisted selection for resistance to pre harvest sprouting in breeding programs.

In another study, two mutants of Arabidopsis thaliana were found with reduced dormancy among freshly harvested seeds and designated as RDO1 and RDO2, that are specifically affected in dormancy determined by the embryo (Leon-Kloosterziel et al., 1996). The rdo1 and rdo2 mutants showed normal levels of abscisic acid and the same sensitivity to abscisic acid, ethylene, auxin and cytokinin as the wild type. The rdo2 mutant had a reduced sensitivity to the gibberellin biosynthesis inhibitor tetcyclacis. Double-mutant analysis suggested that the RDO1 and RDO2 genes are involved in separate pathways leading to the development of dormancy and it is assumed RDO2 gene controls a step in the induction of dormancy that is most likely induced by abscisic acid and is expressed as an increase of the gibberellin requirement for germination.

9. CONCLUSION

With colonization of most plants and eons of breeding has resulted in reduced dormancy in cultivated crops. Research is still underway to understand the genetic aspects of genes responsible for dormancy and the ability to switch them on or off. Future crops may be able to sense the environment and behave accordingly. GM seeds of future may not germinate if the temperature/moisture is sub-optimal. Thus, making resowing redundant. Till then weed seeds in soil seed banks make a dynamic phenomenon that are being constantly reworked by man's interference with habitat. Further, such seed banks, during habitat disturbance, that is routine as in agriculture and rare as in natural disturbances like erosion, results in the death of plants, ensures continuation of a species in a site. Also seed banks are receiving attention as a means of restoration of various kinds of plant communities. Seed banks are also important for re-vegetation of lands that have been severely disturbed by mining activities or natural calamities (Vivian-Smith and Handel, 1996). The fear of escape of herbicide resistant gene into weeds and the persistent nature of herbicide resistant canola (rape seed mustard) is a pointer towards future risks. But the knowledge on the behavioural aspects of seed bank in a particular ecosystem can also decide the type of management to be adopted to arrest the disturbing interference of weeds in crop growth and production of seed or grain.

REFERENCES

Alonso-Blanco, C., I. Bentsink, C. J. Hanhart, H. B. E. Vries and M. Koornneef. 2003 . Analysis of natural allelic variation at seed dormancy loci of *Arabidopsis thaliana.*. *Genetics,* **164**:711-729.

Anderson, J. A., M. E. Sorrells and S. D. Tanksley. 1993. RFLP analysis of genomic regions associated with resistance to pre-harvest sprouting in wheat. *Crop Science,* . **33**:453-459.

Baker, J. P., I. P. Poschlod, R.J. Strykstra, R.M. Bakker and K. Thompson.1996. Seed banks and seed dispersal : Important topics in restoration ecology. *Acta Botanica, Netherlands* **45**:461-190.

Baskin, J.M. and C.C. Baskin. 1976. Effect of photoperiod on germination of *Cyprus inflexus* seeds. *Botanical Gazette,* **137**(3) : 269-273.

Baskin, J.M. and C. C. Baskin.1978. Temperature requirements for after ripening of seeds of winter annual induced into secondary dormancy by low winter temperatures. *Bulletin of the Torrey Botanical Club,* **105**(2):104-107.

Baskin, J.M. and C. C. Baskin.1998. **In:** Seeds : Ecology, Biogeography and Evolution of Dormancy and Germination. Academic Press Ltd. London. Pp : 133-179.

Baskin, C. C. and J. M. Baskin. 2003. How to get the most information on dormancy-breaking and germination requirements from the fewest seeds. In: E. Guerrant, K. Havens, and M. Maunder (eds.). Strategies for Survival. Island Press, Covelo, CA.

Boyce, K.G., D.F. Cole and D.O. Chilcote.1976. Effect of temperature and dormancy on germination of tall fescue. *Crop Science* , **16**(1): 15-18.

Chandrasekariah, S. R., T. Govindappa and G. N. Kulkurni.1976. Pattern of dormancy breaking in seeds of few paddy varieties. *Current Science,* **45**(12):463-464.

Chinnusamy, C., O. S.Kandasamy and N. Sankaran. 2001. Role of weed seeds in quality seed production. In. Recent Techniques and Participatory Approaches in Quality Seed Production. (eds) Vanangamudi,K. et. al. Tamil Nadu Agricultural University,Coimbatore. Pp :147-156.

Côme, D. and Corbineau, F. 1984. La dormance des semences des céréales son élimination. I. Principles caractéristiques. *Compres Rendus de i Academie d' Agriculture de France,* **70** : 141-147.

Corbineau, F., Lenoir, C., Leeat, S. and Come, D. 1984. La dormance des semences des céréales son élimination. II. Mecanismes mis en jeu. *Compres Rendus de i Academie d' Agriculture de France,* **70** : 148-155.

Dessaint, F., R. Chadoeuf and G. Barralis. 1997. Nine years soil seed bank and weed vegetation relationships in an arable field without weed control. *J. Applied Ecology,* **34**: 123-130.

Donohue, K., Dorn, L., Griffith, C., Kim, E. Aguilera, A., Polisetty, C. R. and Schmitt, J. 2005. The evolutionary ecology of seed germination of *Arabidopsis thaliana*: variable natural selection on germination *timing.* Evolution: Vol. 59, No. 4, pp. 758–770.

Dow, M. A. and C . R Schwintzer. 1999. Seed germination, seedling emergence, and seed bank ecology of sweet fern (*Comptonia peregrina* (L.) Coult.). *Can. J. Bot.,* **77**(9): 1378–1386

Fortainer, E.J. and G. van Brenk.1975. Dormancy of seeds as compared with dormancy of bulbs in tulips. *Acta Horticulturae,* **47**: 331-338.

Gulden, R.H. 2003. Secondary seed dormancy and the seedbank ecology of *Brassica napus* L. in Western Canada. Ph. D. Thesis, University of Saskatchewan.

Harrington, G. 1923. Forcing the germination of freshly harvested wheat and other cereals. *Journal of Agricultural Research,* **23** : 79-100.

Hidayati, S. N. and J. L.Walck, 2002. Contrasting seed germination patterns for intra-continental disjuncts of *Heuchera* (Saxifragaceae) in North America: the eastern temperate *H. parviflora* and western montane *H. cylindrica* *Can. J. Bot.*, **80**(11): 1185–1192.

Hodson, M.J., Di Nola L., A. M. Mayer.1987. The effect of changing temperatures during imbibition on ultrastructure in germinating pea embryonic radicles. *Journal of Experimental Botany,* **38** : 525-534.

Ikehashi, H. 1975. Dormancy formation and subsequent changes of germination habits in rice seeds. *JARQ,* **9**(11): 8-12.

Jensen, K. 2003. Dormancy patterns, germination ecology, and seed-bank types of twenty temperate fen grassland species. *Wetlands,* **24** (1): 152–166.

Johnson, L. P. V. 1935. The inheritance of delayed germination in hybrids of *Avena fatua* and *A. sativa.. Canadian Journal of Research,* **13**:367-387.

Jordan,A., D.A. Mortensen, D.M.Prenzlow and K.C. Cox. .1995. Stimulation analysis of crop rotation effects on weed seed banks. *American Journal of Botany,* **82** : 390-398.

Kropac,Z. 1966. Estimation of weed seeds in arable soil. *Pedobiologia,* **6** : 105-128.

Lenoir,C. 1983. Recherches sur les mecanismes de la dormance des semences d' Orge (*Hordeum vulgare* L.) et de son élimination. Thesis 3éme cycle. University of Paris 6.

Leon-Kloosterziel, K.M., G. A. Van de Bunt, J. A. Zeevaart and M. 1996. Arabidopsis mutants with a reduced seed dormancy. *Plant Physiology,* **110**(1):233-240.

Levin, D. A. 1990. The seed bank as source of genetic novelty in plants. *American Naturalist,* **135** : 563-572.

Lijavetzky, D., M. C. Martinez, F. Carrari and H. E. Hopp. 2000. QTL analysis and mapping of pre-harvest sprouting resistance in sorghum. *Euphytica,* **112**:125-135.

Lin, S. Y., T. Sasaki and M. Yano. 1998 . Mapping quantitative trait loci controlling seed dormancy and heading date in rice, *Oryza sativa* L., using backcross inbred lines. *Theoretical and Applied Genetics,* :997-1003.

Mahopatra, S. C. and C. W. Suggs. 1989. Cultivar differences in tobacco seed response to germination temperature. *Seed Sci, & Technol.,***17** : 639 - 647.

Mares, D., K. Mrva, M. K. Tan and P. Sharp. 2002. Dormancy in white-grained wheat: progress towards identification of genes and molecular markers. *Euphytica,* **126**:47-53.

Mayer, A.M. and A. Poljakoff-Mayber.1982. Ecology of seed germination. In: The Germination of Seeds. (ed) Mayer, A. M. and A. Poljakoff-Mayber. Pergamon Press, New York. : Pp. 181.

Milberg, P. and M. L. Hanson. 1994. Soil seed bank and species turnover in lime stone grass land. *Journal of Vegetable Science,* **5** : 35-42.

Mitra, A.K., D.K. Mukherji and P. Mukerjee.1975. Inheritence of dormancy in rice. *Sabrao Journal,* **7**(2): 197-200.

Oberthur, L., T. K. Blake, W. E. Dyer and S. E. Ullrich. 1995. Genetic analysis of seed dormancy in barley (*Hordeum vulgare* L.). *Journal of Quantitative Trait Loci,* **1**:5.

Pons, T. L. 1991. Dormancy, germination and mortality of seeds in a chalk-grassland flora. *Journal of Ecology,* **79**: 765-780.

Robers, H. A. 1981. Seed Banks in Soil. *Advances in Applied Biology,* **5** : 1-55.

Roberts, E. H. 1972. Dormancy. In: Viability of Seeds (ed) Roberts, E. H. Chapman and Hall, London. Pp : 348.

Sahoo,U.K., R. S. Tripathi, H. N. Pondey and J. Misra. 1994. Population dynamics of buried weed seeds as influenced by shifting and terrace cultivation in the humid tropics of India. *Weed Research,* **34** : 157-165.

Saiga, S. and H. Araki. 1976. Speeding-up of germination in cool season grasses by breaking the dormancy of fresh seeds. I. Germination of orchard grass, tall fescue and meadow fescue in seed development process. *Research Bulletin of the Hokkaido National Agricultural Experiment Station,.* **16** : 35-42.

Schafer, D. E. and D. O. Chilcote. 1969. Factors influencing persistence and depletion in buried seed populations. 1. A model for analysis of parameters of buried seed persistence and depletion. *Crop Science*, **9** : 417-419.

Sukumara Dev, V. P. 1982. Post – harvest dormancy of important high yielding paddy varieties cultivated in Kerala. *Seed Research*, **10**(2):167-171.

Telewski, F.W. and J.A.D. Zeevaart. 2002. The 120 –yr period for Dr.Beal's seed viability experiment. *American Journal of Botany*, **89**(3): 1285-1288.

Tokumasu, S.1977. Seasonal periodicity of the loss of dormancy of imbibed seeds in Brassica vegetables. *Scientific Horticulture*, **6**(2): 101-106.

Thompson, K. and J. P. Grime. 1979. Seasonal variation in the seed banks of herbaceous species in ten contrasting habitats. *J. Ecol.*, **67**:893-921.

Tunwar, N. S. and S. V. Singh. 1988. Indian Minimum Seed Certification Standards. The Central Seed Certification Board, Department of Agriculture and Co-operation. Ministry of Agriculture, Govt. of India, New Delhi : 1-352.

Villiers, T. A. 1975. Dormancy and the survival of plants. *Institute of Biology Studies in Biology*, **57** : 62.

Vivian-Smith, G. and S. N. Handel. 1996. Fresh water wetland restoration of an abandoned sands mine: Seed bank recruitment dynamics and plant colonization. *Wetlands*, **16** : 185-186.

Xing-You Gu, F.K. Shahryar, M. E. Foley. 2004. Multiple loci and epistases control genetic variation for seed dormancy in weedy rice (*Oryza sativa*). *Genetics*, **166**:1503-1516.

10

Physiological Mechanism of Seed Deterioration

M. Djanaguiraman[1], J. Annie Sheeba[1], Mallika Vanangamudi[1] and G. Thilagavathi[2]
[1]*Department of Crop Physiology, Tamil Nadu Agricultural University Coimbatore – 641 003*
[2]*Department of Plant Protection, Adhiparasakthi Agricultural College G. B. Nagar Kalavai – 632 506*

1. INTRODUCTION

Orthodox seeds are characterized by their ability to tolerate desiccation and to retain their viability for a long time in the dry state. However, these seeds age during storage and eventually lose their ability to germinate. Several comprehensive reviews have identified free radical-mediated lipid peroxidation, enzyme inactivation or protein degradation, disruption of cellular membranes, and damage to genetic (nucleic acids) integrity as major causes of seed deterioration (Walters, 1998; McDonald, 1999). During the last 20 years, considerable research has been conducted to understand better the physiology of seed ageing, since the primary processes and their interactions involved in seed ageing are not yet fully understood (McDonald, 1999). The mechanisms of seed deterioration during storage have been the subject of much discussion and controversy (Walters, 1998); multiple mechanisms are likely to be involved in seed deterioration and in its prevention in dry seeds (Fig. 1). Free radical generation and lipid peroxidation, resulting in membrane damage as well as the generation of toxic by-products, is a viable hypothesis (Fig. 2). Oxidative damage to DNA and proteins is almost certainly involved in seed deterioration. Failure of repair mechanisms upon rehydration is also likely to contribute to loss of viability. Regardless of the specific mechanism of deterioration, at present all that can be done to delay the inevitable loss of seed viability is to control the storage conditions to maintain low moisture content and temperature.

Lipid peroxidation can be defined as the oxidative deterioration of lipids containing any number of carbon–carbon double bonds. Seeds usually exhibit their maximum germination potential soon after harvest and as storage period increases; they loose vigour and viability and eventually die. Physiological changes accompanying physiological ageing are oxidation of unsaturated lipids which leads to free radical formation. Lipid peroxidation could be a direct cause of seed deterioration and was first proposed by Kaloyereas (1958). Free radical is a molecule or atom with an unpaired electron. By virtue of their unpaired electron, free radicals are usually unstable and are quite reactive. These free radicals abstract an electron from stable compounds that in turn are transformed into a new free radical. Therefore, one radical can initiate a self propagating chain reaction of destructive process by removing electrons from stable compounds, and forming many free radicals that are Reactive Oxygen Species (ROS). Such reactions tend to proceed until the free radical is deactivated by a chain reaction breaking antioxidant molecule. The most studied free radical chain reaction in living systems is lipid peroxidation which is mediated by oxygen free radicals, and is believed to be an important cause for cell membrane destruction and cell damage. Lipid peroxidation has considerable potential to damage membranes and may be important in the deterioration of stored seeds and reduced longevity of seeds under natural conditions.

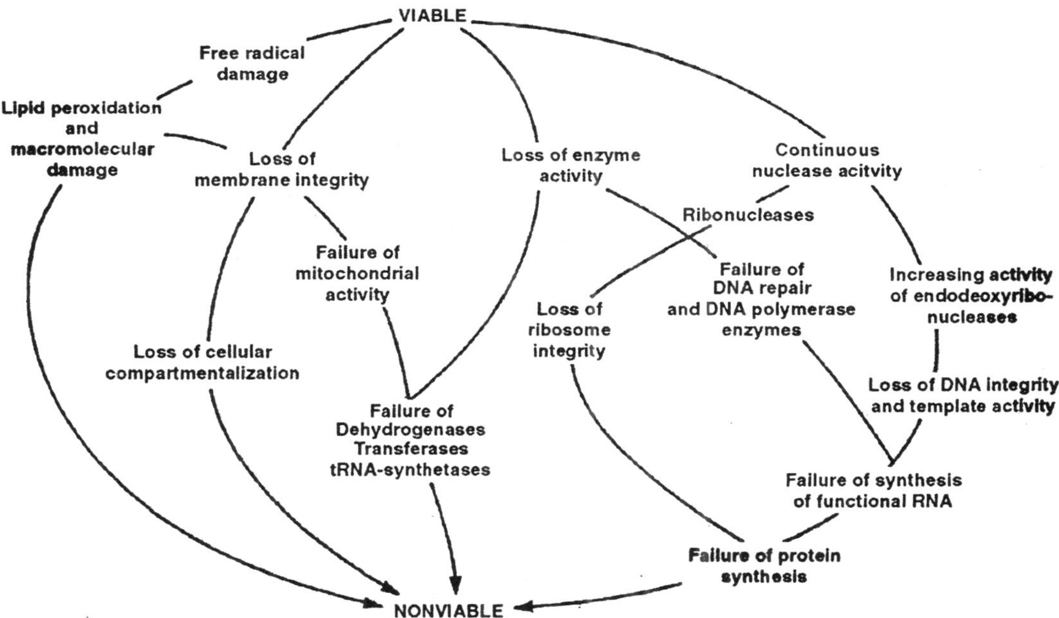

Fig. 1: A scheme to illustrate the variety of causes that may be involved in the loss of viability in stored seeds. Osborne (1980).

Most vascular plants can withstand little desiccation except during the later stages of seed development and the early stages of germination, when embryos are able to survive considerable desiccation up to the point of radical emergence. This desiccation tolerance progressively lessens as the seed germinates. There is increasing evidence that the development of intolerance is correlated with increased free radical attack and damage during water loss (Hendry *et al.,* 1992). Desiccation in tissues generally exhibit extensive peroxidative damage to cellular membranes, resulting in loss of semi permeability, changes in their micro viscosity and lipid phase properties (Senaratna *et al.,* 1987), increases in free fatty acids (Senaratna *et al.,* 1987), and accumulation of thiobarbituric acid reactive substances associated with lipid peroxidation (Hendry *et al.,* 1992). As suggested by the extensive literature (Halliwell, 1987; Bowler *et al.,* 1992; Winston, 1992), the origin of such peroxidative damage occurring under stress conditions is likely to be the formation of transient but highly reactive, partially reduced or activated forms of O_2.

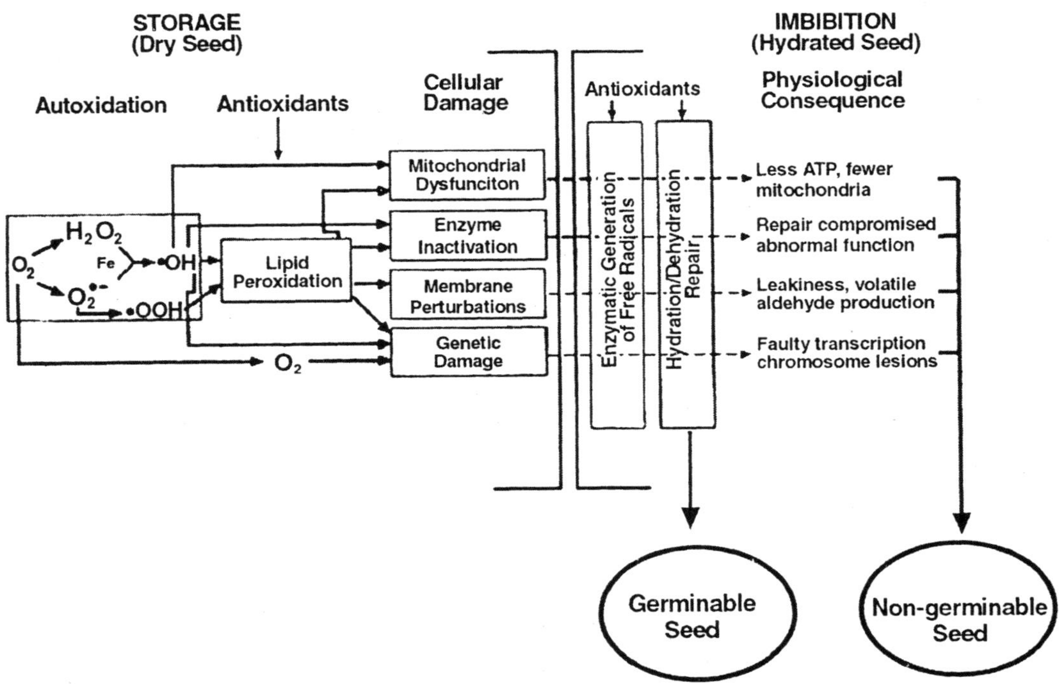

Fig. 2: Possible roles of free radicals and lipid peroxidation in seed deterioration (McDonald, 1999)

2. POSSIBLE CAUSES OF SEED DETERIORATION

An understanding of the fundamental factors that induce aging is essential in a study of seed deterioration. It is fairly certain that seed deterioration occurs from a combination of several causes.

2.1 Degradation of functional structures

As seed deterioration progresses, cellular membranes lose their selective permeability, permitting the cytoplasmic metabolites to leach into the intercellular spaces. Membrane degradation occurs from both hydrolysis of phospholipids by phospholipase and phospholipid autoxidation; therefore, in the strictest sense, it might be considered as a result of aging rather than as a cause. Mitochondrial degradation and functional changes appear to play a major role in seed deterioration. Two important aspects of mitochondiral deterioration are, an increase in ATPase and a decline in oxidative phosphorylation ability necessary to complete its respiratory function (Weinbach *et al.*, 1963). Since ATPase catalyzes the breakdown of ATP and ADP, it depletes energy available in the mitochondria. Levels of ATP are lower in aged seeds of a number of crops.

2.2. Inability of ribosomes to dissociate

Associated with the degradation of function in deteriorating seeds is the dissociation of ribosomes. Evidence indicates that the dissociation of polyribosomes must occur before the attachment of performed mRNA occurs, leading to protein synthesis in germinating seedlings (App *et al.*, 1971). In nonviable seeds, the ribosomes fail to dissociate and protein synthesis is retarded. Such declines in protein synthesis are a measurable symptom of aging.

2.3. Enzyme degradation and inactivation

Decreased activity of enzymes such as catalase, dehydrogenase and glutamic acid decarboxylase in deteriorating seeds is well documented. The general decrease in enzyme activity in the seed lowers its respiratory potential, which in turn lowers both the energy (ATP) and food supply to the germinating seed. Several changes in the enzyme macromolecular structure may contribute to their lowered effectiveness (Walter, 1963). They may undergo compositional changes by losing or gaining certain functional groups, by oxidation of sulfhydryl groups, or by conversion of amino acids within the protein structure. The enzymes may also undergo configurational changes such as (1) partial folding or unfolding of ultrastructure, (2) condensation to form polymers, or (3) degradation to subunits.

2.4. Formation and activation of hydrolytic enzymes

As seed moisture content approaches levels necessary for germination, hydrolytic enzymes are activated. If the seed moisture content remains high or reaches higher levels, normal germination may occur; however, if moisture levels for germination are not attained, the seed deteriorates because of energy expenditure or accumulation of breakdown products. The increase in free fatty acids, a symptom of deterioration caused by activation of lipase enzymes in oil-containing seeds has been an example for this. A related group of enzymes, the phospholipases, hydrolyzes the phospholipids destroys the membrane structure of the seed. Activation of phosphatase enzymes converts ATP to ADP, resulting in an energy loss accompanied by increased phosphate acidity. Other hydrolytic enzymes activated by high moisture levels are amylases and proteolases. This kind of deterioration is rapid and is important only over short periods at moisture levels around 20% and above. Below 20% moisture, other kinds of deterioration predominate.

2.5. Genetic degradation

A great deal of indirect evidence supports the view that seed deterioration is associated with random somatic mutations that impair the cellular functions of vital seed tissues. The increase in chromosomal aberrations in deteriorating seeds as a result of somatic mutation has been observed in many species. Two general observations have been made on barley seeds (1) the properties of chromosomes change gradually with age, and (2) mutations in seeds are highly correlated with seed age and decline in viability in aging seeds. The following observations have been made in support of the genetic theory (1) extracts from aged seeds retard the germination of fresh seeds, (2) mutations in seeds are highly correlated with seed age and decline in viability, (3) spontaneous mutations arise in aging seeds and become evident by chromosome breakage in the presplit phase, (4) differences in the shoot and root tips in aged seed is closely parallel to those in irradiated seeds.

2.6. Depletion of food reserves

Depletion of food reserves is one of the oldest theories on deterioration. In fact, most seeds contain enough food materials to last thousands of years. Even 1000 to 2000 year old wheat seeds that have been found in ancient tombs still retain most of their food reserves; thus, even nonviable seeds contain enough food reserves for seedling growth and development.

2.7. Starvation of meristematic cells

The meristematic cells, starve even though only a few cells away from abundant reserves of energy may die from lack of food or from injury. It was noted that respiration may deplete the tissues involved in the transfer of nutrition from reserve storage areas and thus prevent them from reaching the embryo.

2.8. Accumulation of toxic compounds

Under low-moisture storage, the reduced respiration and enzyme activity may be responsible for accumulation of toxic substances that reduce seed viability. When aged wheat embryos were transplanted onto young endosperms and young embryos onto aged endosperms, a progressive decline in germination and vigor of both transplants has been observed, strongly indicating a gradual accumulation of toxic metabolites. It has been suggested that the presence of abscisic acid, a germination inhibitor, in several seeds supports this theory as a probable cause of aging as well as phenolic compounds and polyamines.

2.9. Lipid Peroxidation Model of seed deterioration (Wilson and McDonald, 1986)

Oil rancidity as a cause of declining seed germinability was first proposed in 1958 by Kaloyereas. He found that oil from deteriorated pine seeds which performed poorly in a germination test would oxidize the iodine ion indicating that seed lipids has reacted with oxygen. Kaloyereas (1958) demonstrated improved laboratory germination by treating seeds before storage with starch phosphate an antioxidant molecule. This study indicated for the first time that lipid peroxidation could be a direct cause of seed storage deterioration.

Initiation

$RH + O_2$ ---> Free radical (R^+)

ROOH--> Free radicals (ROO^+)

Propagation

$R+O_2$--->RO_2

$RO_2 + RH$---------------------------------------> R+ROOH (Hydro peroxide)

Lipid peroxidation may be caused by spontaneous oxidation of fatty açid hydrocarbon chains highly reactive free radical intermediates called hydroperoxides and wide variety of secondary products form hydroperoxides decomposition (Fig.3).

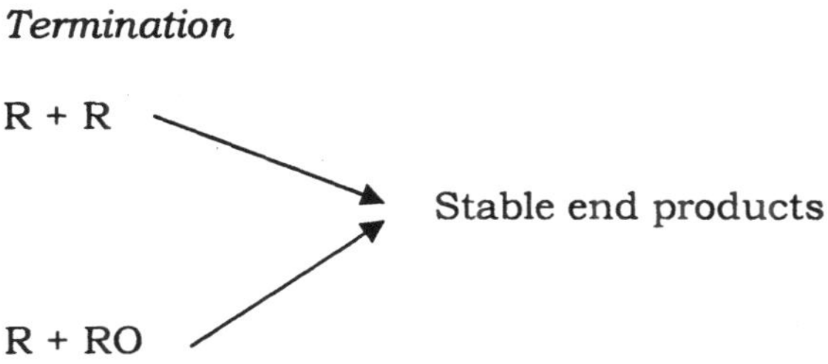

Fig. 3: Auto oxidation process

i) Auto Oxidation (non-enzymic) – By oxygen attack – a chain of reaction is initiated which creates additional reaction cycles by removal of Hydrogen atom from α – Methylene group.

ii) Chemical catalysis (non-enzymic) by cytochromes, when hydroperoxides may be partially degraded and may undergo further peroxidation cycles.

iii) Oxidation catalyzed by lipoxygenase (lipoxidase), this enzyme greatly accelerates the rate of reaction.

3. Oxygen, free radial process and seed longevity

Storage temperature, moisture content of seeds and oxygen pressure are the factors suspected very much as a cause for the loss of seed viability by increasing the rate of respiration. It has been pointed out that deleterious effects of oxygen on peas extend to condition (60°C at 12% moisture content) in which it is unlikely that conventional respiration is taking place. Furthermore, in this species it was shown that the rate of respiration was unaffected by oxygen concentration over the range 1.4 to 21% of atmospheric pressure and yet variation of oxygen level over this range did affect the rate of loss of viability. When seeds are relatively dry, it has been shown that oxygen tends to decrease viability period perhaps not by increasing the respiratory rate but by involving in free radical process. In contrast, unless oxygen is present, the viability of the lettuce may be reduced to a few days when they are fully hydrated probably because free radicals are quenched by the excess water. The role played by oxygen .in ageing, disease, death and decay is brought together in the free radical theory of ageing first formulated with mammalian systems.

3.1 Oxygen nomenclature and definition

Trioxygen	O_3	(Ozone) (high altitudes)
Monooxygen (atomic)	O	transient form

Dioxygen O_2 abundant

If dioxygen,

i) Loses one electron to form the di–oxygenyl radical

$$O_2 \text{ -------------------> } e^- + O_2^+$$

ii) Gains one electron to form the superoxide radical

$$O_2 + e^- \text{ ---------------> } O_2^-$$

iii) Gain 2 electrons to from the peroxide anion

$$O_2 + 2e^- \text{ ----------------> } O_2^{2-}$$

iv) Addition of protons to oxygen leads to hydrogen peroxide and water formation

$$O_2 + 2e^- + 2H^+ \text{ ----------------->} H_2O_2$$

$$O_2 + 4e^- + 4H+ \text{ ------------------>} 2H_2O$$

Then

$$O_2^- + H_2O_2 \text{ ------------------------>} O_2 + HO + HO^-$$

(Syperoxide radical) (Hydroxyl radical)

i) O, O_3, O_2, O_2^+, O_2^- and HO are free radicals

ii) O_2^{2+} and H_2O_2 are not free radicals

3.2. Oxygen reactivity

3.2.1. Electronic configuration

Atmospheric oxygen in its ground-state is distinctive among the gaseous elements because it is a biradical, or in other words it has two unpaired electrons. This feature makes oxygen paramagnetic; it also makes oxygen very unlikely to participate in reactions with organic molecules unless it is "activated". The requirement for activation occurs because the two unpaired electrons in oxygen have parallel spins. According to Pauli's exclusion principle, this precludes reactions with a divalent reductant, unless this reductant also has two unpaired electrons with parallel spin opposite to that of the oxygen, which is a very rare occurrence. Hence, oxygen is usually non-reactive to organic molecules which have paired electrons with opposite spins. This spin restriction means that the most common mechanisms of oxygen reduction in biochemical reactions are those involving transfer of only a single electron (monovalent reduction).

Activation of oxygen may occur by two different mechanisms: absorption of sufficient energy to reverse the spin on one of the unpaired electrons, or monovalent reduction. The biradical form of oxygen is in a triplet ground state because the electrons have parallel spins. If triplet oxygen absorbs sufficient energy to reverse the spin of one of its unpaired electrons, it will form the singlet state, in which the two electrons have opposite spins (Fig. 4). This activation overcomes the spin restriction and singlet oxygen can consequently participate in reactions involving the simultaneous transfer of two electrons (divalent reduction). Since paired electrons are common in organic molecules, singlet oxygen is much more reactive towards organic molecules than its triplet counterpart.

Triplet Oxygen (Ground state)	•O-O•
Singlet Oxygen	O-O:
Superoxide	•O-O:
Perhydroxy radical	•O-O:H
Hydrogen peroxide	H: O-O :H
Hydroxy radical	H: O•
Hydroxyl ion	H: O:

Fig.4: Nomenclature of the various forms of oxygen

The second mechanism of activation is by the stepwise monovalent reduction of oxygen to form superoxide (O^{2-}), hydrogen peroxide (H_2O_2), hydroxyl radical (OH) and finally water. Superoxide can act either as an oxidant or as a reductant; it can oxidize, sulphur, ascorbic acid or NADPH; it can reduce cytochrome C and metal ions. A dismutation reaction leading to the formation of hydrogen peroxide and oxygen can occur spontaneously or is catalysed by the enzyme superoxide dismutase. In its protonated form superoxide forms the perhydroxyl radical (.OOH) which is a powerful oxidant (Gebicki and Bielski, 1981), but its biological relevance is probably minor because of its low concentration at physiological pH.

The univalent reduction of superoxide produces hydrogen peroxide which is not a free radical because all of its electrons are paired. Very often the reduction products of oxygen are referred to by biologists as oxygen free radicals which is a misnomer because in chemistry a free radical is defined as an atom or molecule with an unpaired electron. It is more appropriate to refer to the intermediate reduction products of oxygen as activated and not as free radicals because triplet oxygen (ground state) is a radical and hydrogen peroxide is not. Hydrogen peroxide is noteworthy because it readily permeates membranes and it is therefore not compartmentalised in the cell. The well-known reactivity of hydrogen peroxide is not due to its reactivity *per se,* but requires the presence of a metal reductant to form the highly reactive hydroxyl radical which is the strongest oxidizing agent known and reacts with organic molecules at diffusion-limited rates. Fenton described in the late nineteenth century (Fenton, 1894; 1899) the oxidising

Potential of hydrogen peroxide mixed with ferrous salts. Forty years later, Haber and Weiss (1934) identified the hydroxyl radical as the oxidising species in these reactions:

(1) $Fe^{2+} + H_2O_2 \rightarrow Fe^{3+} + \cdot OH + OH^-$

In biological systems the availability of ferrous ions limits the rate of reaction, but the recycling of iron from the ferric to the ferrous form by a reducing agent can maintain an ongoing Fenton reaction leading to the generation of hydroxyl radicals. One suitable reducing agent is superoxide which participates in the overall reaction 2 as two half reactions shown in reactions 3 and 4:

$$(2)\ \bullet O_2^{-} + H_2O_2 \rightarrow O_2 + \bullet OH + OH^{-}$$

$$(3)\ Fe^{2+} + H_2O_2 \rightarrow Fe^{3+} + \bullet OH + OH^{-}$$

$$(4)\ \bullet O_2^{-} + Fe^{3+} \rightarrow O_2 + Fe^{2+}$$

Therefore, in the presence of trace amounts of iron, the reaction of superoxide and hydrogen peroxide will form the destructive hydroxyl radical and initiate the oxidation of organic substrates. Metals other than iron may also participate in these electron transfer reactions by cycling between oxidised and reduced states. Several reactive oxygen species (ROS) are known. Among them, the most frequently studied are given below:

3.3. Activated oxygen species

3.3.1. Superoxide radical ($O_2^{\cdot-}$)

This ROS is formed when oxygen takes up one electron and as leaks in the mitochondrial electron transport but its formation is easily increased when exogenous components (redox cycling compounds) are present. Its first production site is the internal mitochondrial membrane (NADH ubiquinone reductase and ubiquinone cytochrome *C* reductase). This species is reduced and forms hydrogen peroxide (H_2O_2). Superoxide radical can be directly toxic; it has a limited reactivity with lipids, raising questions about its true toxicity. Thus, its action is frequently considered to result from a secondary production of the far-more reactive ·OH species by the iron-catalyzed Haber-Weiss reaction.

3.3.2. Hydrogen peroxide (H_2O_2)

Hydrogen peroxide is mainly produced by enzymatic reactions. These enzymes are located in microsomes, peroxysomes and mitochondria. Even in normoxia conditions, the hydrogen peroxide production is relatively important and leads to a constant cellular concentration between 10^{-9} and 10^{-7} M. The superoxide dismutase is able to produce H_2O_2 by dismutation of $O_2^{\cdot-}$, thus contributing to the lowering of oxidative reactions. The natural combination of dismutase and catalase contributes to remove H_2O_2 and thus has a true cellular antioxidant activity. H_2O_2 is also able to diffuse easily through cellular membranes.

3.3.3. Hydroxyl radical (OH)

In the presence of Fe^{2+}, H_2O_2 produces the very active species -OH by the Fenton reaction [$Fe^{2+} + H_2O_2 \rightarrow Fe^{3+} + OH + OH^{-}$]. This iron-catalyzed decomposition of oxygen peroxide is considered the most prevalent reaction in biological systems and the source of various deleterious lipid peroxidation products.

3.4. Origins of ROS and their damaging effects

Any circumstance in which cellular redox homeostasis is disrupted can lead to oxidative stress or the generation of ROS. Production of ROS is one of the main causes for injury and death that accompany. ROS are produced in both unstressed and stressed cells, and in various locations (Halliwell and Gutteridge, 1989). They are generated endogenously during certain developmental transitions such as seed maturation and as a result of normal, unstressed, photosynthetic and respiratory metabolism. An initial oxyradical product, the superoxide radical ($O_2^{\cdot-}$), upon further reaction within the cell, can form more ROS such as hydroxyl radicals and singlet oxygen (Table 1, Fig. 5 and 6). Superoxide is a charged molecule and cannot cross biological membranes. Subcellular compartmentation of defense mechanisms is, therefore, crucial for efficient removal of superoxide anions at their sites of generation throughout the

Table 1: Mechanisms for production and scavenging of ROI in cells. AOX, alternative oxidase; APX, Ascorbate peroxidase; CAT, catalase; GPX, glutathione peroxidase; SOD, superoxide dismutase.

Production of ROI	**Localisation**	**Primary ROI**
Photosynthesis ET and PSI or II	Chl	O_2^-
Respiration ET	Mit	O_2^-
Glycolate Oxidase	Per	H_2O_2
Excited Chlorophyll	Chl	O_2^1
NADPH oxidase	PM	O_2^-
Fatty acid β oxidation	Per	H_2O_2
Oxalate oxidase	Apo	H_2O_2
Xanthine oxidase	Per	H_2O_2, O_2^-
Peroxidases, Mn^{2+} and NADPH	CW	O_2^-
Amine Oxidase	Apo	H_2O_2
Scavenging of ROI		
SOD	Chl,Cyt,Mit,Per,Apo	O_2^-
APX	Chl,Cyt,Mit,Per,Apo	H_2O_2
CAT	Per	H_2O_2
GPX	Cyt	H_2O_2,ROOH
Peroxidases	CW,Cyt,Vae	H_2O_2
Thioredoxin peroxidase	Chl,Cyt,Mit	H_2O_2
Ascorbic acid	Chl,Cyt,Mit,Per,Apo	H_2O_2, O_2^-
Glutathione	Chl,Cyt,Mit,Per,Apo	O_2^1
α - Tocopherol	Membranes	ROOH ,O_2^1
Carotenoids	Chl	O_2^1
AOX	Chl,Mit	O_2^-

cell. Hydrogen peroxide, on the other hand, which is formed as a result of SOD action, is capable of diffusing across membranes and is thought to fulfil a signaling function in defense responses (Mullineaux *et al.,* 2000). In the presence of metal ions such as Fe or Cu(II), hydroxyl radicals are formed very rapidly. Hydroxyl radicals can cause damage to all classes of biologically important macromolecules, especially nucleic acids. Hydroxyl radicals can also modify proteins so as to make them more susceptible to proteolytic attack.

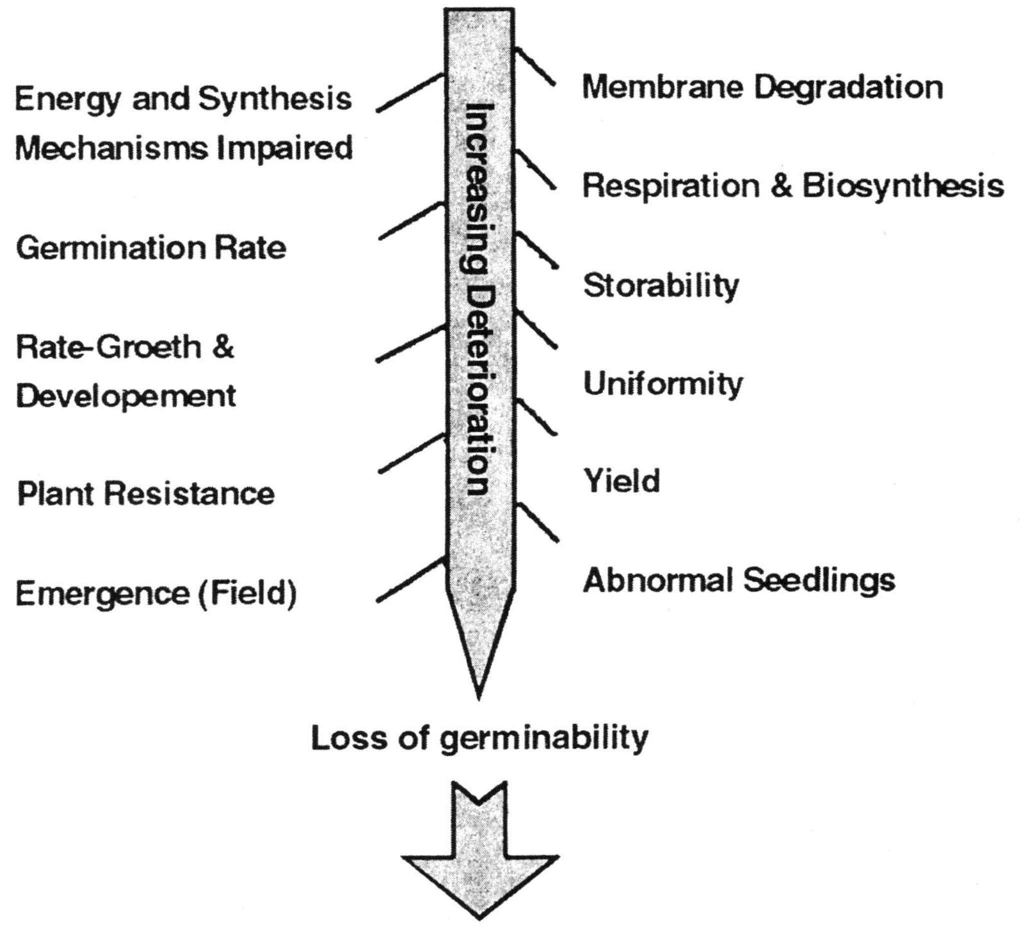

Fig. 5: Events governing loss of germinability

Abbreviations:

Apo, apoplast; Chl, chloroplast; CW, Cell wall; Cyt, cytosol; ET, electron transport; Mit, mitochondria ; O_2^1 ,Singlet oxygen Per, peroxisome; PM, plasmamembrane; PS, photosystem ; Vac, Vacuole.

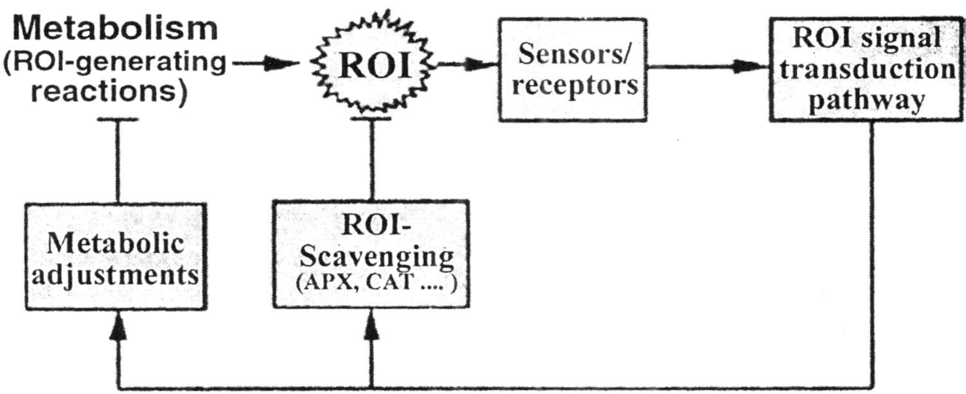

Fig 6: The basic ROI cycle. This cycle modulates the cellular levels of ROI during normal metabolism. Some of the key ROI scavenging enzymes of plants, ascorbate peroxidase (APX), and catalase (CAT) are indicated (Halliwell and Gutteridge, 1989).

3.5. The Chemistry of Lipid Peroxidation

The chemistry underlying lipid peroxidation is reasonably well established, although the precise involvement of free radicals in the initiation of peroxidation is less certain. The process of lipid peroxidation is initiated by a free radical (with its unpaired electron) reacting with neighbouring unsaturated fatty acid. The precise identity of the initiating species is however, unclear. It may be the hydroxyl radical itself. However, lipid peroxidation can proceed without the apparent or detectable presence of the hydroxyl radical in the presence of Fe^{2+} or Cu^{+} which may have been formed *in vivo* by preceeding activated oxygen–transitional metal reaction. The chemistry is complex, not least because the composition of membranes with their associated metal containing proteins is considerably more complex than the *in vitro* reactions between pure lipids and .HO. The subject is discussed in more detail by Gutteridge and Halliwell, (1995).

The attack by the highly reactive hydroxide radical or other highly reactive radical, results in the formation of a lipid (Carbon–centered) radical from the lipid in membrane, which undergoes random rearrangement and attack by molecular oxygen. The oxygen molecule is incorporated into the lipid to form the first of a series of lipid peroxy radicals, whose further random rearrangement is propagated until another hydrogen is abstracted from, say, a neighbouring lipid (resulting in a further chain of hydrogen abstractions and infinitum) or from a fat–soluble chain–breaking antioxidant such as a tocopherol (so terminating the lipid peroxidation).

3.6. Biological effects of lipid peroxidation

Harman (1956) hypothesized that ageing was a consequence of free radical attack on cellular constituents and connective tissue. Free radicals were formed by ionizing radiation, splitting of oxygen by enzymes or transition metals and by normal metabolic processes. The role played by oxygen in its several forms as the molecular source of damaging radical reactions that cause ageing, disease death and decay, has been discussed elsewhere. Lipid peroxidation as a major source of free radicals in tissue was proposed by Harman (1956). Biomembranes represent a key site of direct injury from lipid peroxidation. Membranes possess an inherently larger surface area and usually more unsaturated than storage lipids. Evidence exists that lipid peroxidation leads to an increase in membrane permeability. Mitochondrial inner membranes are composed of a higher proportion of unsaturated acyl chains than other membranes (Moreau *et al.,* 1974). Increased mitochondrial membrane permeability can lead to a breakdown of the proton gradient necessary to maintain respiratory coupling. Thus, a primary consequence of lipid peroxidation in tissues can decrease respiratory competence as well as general decline in membrane integrity. Co-oxidation of associated cellular components by free radical transfer from oxidizing lipids is another possible mechanism of damage. Normally stable molecules are incorporated into the authoritative chain reacting by abstracting of hydrogen atoms. In addition, free radicals from peroxidising lipids cause proteins to polymerize.

While free radicals possess a potential for short distance damage due to their short life time, more stable chemical species formed during lipid peroxidation provide a mechanism for long distance detrimental effects. Aldehydes formed by hydroperoxide breakdown, especially 4- hydroxyalkenals, produce a wide variety of cytotoxic effects. These decreased the mitotic index in liver cells and have been shown to inhibit tubulin, the main functional protein of microtubules which is necessary for mitotic spindle formation. Respiratory stimulation, induced by addition of ADP or dinitrophenol to isolated mitochondria, was decreased by trans-4-hydroxy-non-2-enal. The ADP/O ratio remained unaffected but the acceptor control ration was decreased by 40% at a concentration of 0.1 mm of the aldehyde using hepatoma mitochondria. In addition, aldehyde products of lipid peroxidation strongly inhibited protein synthesis, protein and lipoprotein secretion, and DNA synthesis at low concentration (10^{-3} M) *in vitro* (Dianzani, 1979) and protein synthesis *in vivo* in plants (Dhindsa, 1982). At the organ level, some exogenously applied secondary products of lipid peroxidation have shown toxicity to germination seeds. Lipid peroxidation, therefore, is potentially damaging to organisms in three ways, by destruction of membrane lipids, by co-oxidative assault by free radicals, and by the formation of cytotoxic aldehydes. Most of these possible mechanisms have been studied using exogenously applied substances *in vitro* systems. Determination of the relative importance of lipid peroxidation as a cause of seed deterioration, however requires investigation using intact seeds.

3.7. Lipid peroxidation and respiration

In order to understand the possible consequences of lipid peroxidation in the seed, it is important to examine the interaction between peroxidation and respiration. On one hand lipid peroxidation provides a parsimonious explanation for many structural and functional manifestation of storage deterioration. Not so obvious is the possibility that mitochondria can cause or affect the out come of peroxidation synergistically damaging the respiration integrity of the seed.

The mitochondrial inner membrane represents a logical site of attack by oxygen. It seems to require a high degree of fluidity and is typically composed of a high proportion of unsaturated fatty acids. Ageing has been shown to reduce respiratory competence in corn, barley, soybeans and wheat. Membrane oxidation could interfere directly with respiration in at least three ways.

1. Viscosity or polarity changes of the inner membrane
2. Interferences with mitochondrial assembly during imbibitions
3. Co-oxidation of critical enzymes and co-factors.

In order to function properly, biomembranes must exist in liquid – crystalline state. Lipid peroxidation can be alter or interrupt this liquid crystalline phase by introduction of polar oxygenated moieties on the hydrophobic acyl chains of phospholipids molecules or simply by selective destruction of the unsaturated acid's necessary fluidity. Lipid peroxidation has been shown to increase membrane permeability. Mitochondria are among the organelles to exhibit age related ultra structural damage and this damage can be explained by deterioration of membranes (Roberts, 1972). In conclusion, an expected effect of lipid peroxidation is decreased membrane fluidity. Aside from a general deleterious effect on respiration, dry seeds are particularly prone to this damage as a consequence of the extensive reorganization required to assemble the electron transport system during imbibition. A convenient method exists for disruption the membrane assembly in seeds, imbibitional chilling injury.

Leopold and Musgrave (1979) utilizing intact soybean axis and sectioned cotyledons, demonstrated that the cold treatment during first hours of imbibition greatly decreased respiration *viz.,* the cytochrome pathway. The alternate pathway of respiration was not damaged. The authors suggested that the lowering of respiratory activity and the shift in respiratory pathways may play a major role in decline of germinability and vigour. Dupont (1971) has found lipoxygenase activity which is apparently endogenous to mitochondria was unmasked by destabilization of membrane structure by phospholipids alteration. This activity resulted in damage of fatty acids which might be a mechanism of volatile aldehyde formation and accumulation of malondialdehyde in mitochondria from cauliflower. The unmasking agent may be endogenous phospholipase *A* also found in mitochondrial membranes. Whether these mechanisms could occur in dry seeds is unknown, but the unimbibed state might alter membrane structure sufficiently to unmask lipoxygenase activity. Mitochondrial cytochromes have been found to catalyse the oxidation of both fatty acids and especially fatty acid hydroperoxides. This reaction may be significant as *in vivo* oxidation mechanisms. The implication is that cytochrome molecules in an attack could be randomly initiated at some of the poly unsaturated fatty acids located in the plasma membrane. These sites would enlarge *via* the free radical chain reactions into polar bridges across the hydrophobic barrier of the membrane leading to an increase in permeability associated with ageing. Two approaches were used to demonstrate this association of lipid peroxidation with deterioration of seeds (i) Changes of phospholipids content were monitored and (ii) An effort was made to detect malondialdehyde as secondary product of lipid peroxidation.

3.8. Deterioration and Volatile compounds

Fielding and Goldsworthy (1982) were the first to describe the release of volatile compounds on heating dry wheat seeds in nitrogen as a sensitive indication of their physiological age and potential vigour. Of particular interest was peak 6 which had similar retention time to one of the volatile compounds obtained by heating pure wheat germ oil and was possibly one of its peroxidation products. It should be noted that peak 6 was approximately twice the size in the older seed samples and might therefore provide an indication of the seed physiological age.

When the levels of peak 6 in nitrogen are compared for the two naturally aged seed lots, it was seen to be about three times the size in the older sample. It should be emphasized that at the time of these experiments both seed lots, was about three times the size in the older sample. It should be emphasized that at the time of these experiments both seed lots had a high germination and growth rate and was difficult to distinguish in a normal germination test. This method could therefore form the basis of a very sensitive technique for the detection of physiological age in seed, which is considerably more rapid than tests based on germination. Volatiles identified by gas chromatography were hexanal, pentanal and butanal, when expressed as a percentage of the total volatiles produced were significantly correlated with seed germination, seed vigour and hydroperoxide levels in the six batches of soybeans investigated. The authors provided the evidence that the breakdown of lipid hydroperoxides is the source of these volatiles and the lipid peroxidation may be an important determinant in seed ageing. They also reported that, decline in PUFA with decreasing vigour in soybean and cabbage seeds. The results indicate a definite relationship between lipid peroxidation and seed vigour.

A mechanism of seed deterioration in relation to the volatile compounds in dry seeds reported that "some of the seed volatiles, which were mainly composed of methanol, acetaldehyde, ethanol and acetone, caused the loss of seed germinability during storage". In general, the deleterious effects of the volatiles increased with increasing RH and temperature during storage. Acetaldehyde had strongest deleterious effect regardless of RH and temperature, while ethanol caused seed deterioration only at high RH. Acetone was slightly deleterious to some species, while methanol was almost inert in most seeds.

Seeds constituents related to viability, such as protein, DNA and RNA, are all long lived because of the low metabolic activity of dry seeds and are maintained throughout the long term storage period until the next generation. Amino groups in protein, DNA and RNA would be attacked by volatile aldehydes, as a result the amino groups in these macro molecules would be modified which would there by cause ageing of dehydrated seeds under long term storage. In a dry living seed, having an integrated cell structure, many potentially reactive substances would be separated from each other and thus some non enzymatic and/or enzymatic reaction causing seed deterioration would not occur.

However, since the seed evolved volatile compounds with low boiling points have high mobility they would easily react with many substances without being restricted to even dehydrated seeds. Based on these facts, it is suggested that endogenous deterioration which often occurs under low RH and/or temperature throughout long-term storage. It should be noted that acetaldehyde, the most toxic endogenous volatile was very abundant in dry seeds which could modify the amino basis in protein and DNA. ADH (Alcohol dehydrogenase) in dry seeds can transform acetaldehyde to relatively non-poisonous ethanol, which perhaps is a detoxification pathway in dry seeds. On the other hand, it is also possible that ethanol accumulated within seeds is converted back to acetaldehyde *via* ADH. Therefore, acetaldehyde content in seeds would also depend upon the inter conversion between ethanol and acetaldehyde *via* ADH. Very few volatiles were detected from pea seeds whereas, lettuce and carrot seeds evolved large amount of various aldehydes.

Kalpana and Madhava Rao (1994) reported that free radical induced lipid peroxidation did not seed to play a role in the rapid ageing of pigeon pea seeds. Superoxide dismutase activity tended to increase, while catalase and peroxidase activities were affected to varying degrees exhibiting a declining trend with rapid 'ageing' indicating the role of hydrogen peroxide (produced by the increased activity of SOD and decreased activity of catalase and peroxidase) in the rapid ageing process of pigeon peaseeds. The accumulation of reducing sugars beyond 2 days ageing; their correlation with loss of seed viability indicated the possibility for the occurrence of Amadori and Mailard reactions in rapidly aged seeds, consequently leading to seed deterioration. Amadori reaction involves a simple non-enzymic attack by reducing sugars on amino groups (mostly proteins) to form fructosyl derivatives or glycated proteins. Maillard reaction represents subsequent complex interactions between glycated Amadori products to form polymeric brown coloured compounds. These reactions have been reported to cause protein inactivation and DNA damage. Thus, the accumulation of reducing sugars in pigeon pea seeds during accelerated ageing could lead to the occurrence of Amadori and Maillard reactions which may cause protein inactivation and nucleic acid damage, thereby threatening the viability of the seeds. The Maillard product theory of seed ageing is conceptually attractive, since the reaction occurs at moisture contents 6-15% at which seeds deteriorates more rapidly. The non-enzymatic modifications of proteins through Amadori and Maillard reactions play an important role in the loss of seed viability during storage. In the present study, the contribution of sugar hydrolysis and lipid peroxidation to Amadori and Maillard reactions, and to seed deterioration was investigated in mung bean (*Vigna radiata* Wilczek). The contents of glucose and lipid peroxidation products in seed axes increased significantly during storage. The accumulation of Amadori products in seed axes was correlated to the lipid peroxidation, whereas the accumulation of Maillard products was closely correlated to sugar hydrolysis. The rate of accumulation of Maillard products was not well correlated to the content of Amadori products in both seed axes and protein/glucose model system, reflecting the complex nature of Amadori and Maillard reactions. The content of Amadori products in seed axes increased during the early stages of seed ageing, whereas the content of

Maillard products increased steadily during the entire period of storage. The accumulation of Maillard products in seed axes was associated with the decline of seed vigour. These data suggest that, during seed ageing, sugar hydrolysis and lipid peroxidation are coupled with non-enzymatic protein modification through Amadori and Maillard reactions (Murthy and Sun, 2000).

Murthy *et al.* (2003) found that two primary biochemical reactions in seed ageing (lipid peroxidation and non-enzymatic protein glycosylation with reducing sugars) have been studied under different seed water contents and storage temperatures, and the role of the glassy state in retarding biochemical deterioration examined. The viability loss of *Vigna radiata* seeds during storage is associated with Maillard reactions; however, the contribution of primary biochemical reactions varies under different storage conditions. Biochemical deterioration and viability loss are greatly retarded in seeds stored below a high critical temperature (approximately 40°C above glass transition temperature). This high critical temperature corresponds to the cross-over temperature (T_0) of glass transition where molecular dynamics changes from a solid like system to a normal liquid system. The data show that seed ageing slows down significantly, even before seed tissue enters into the glassy state.

4. METABOLIC DEFENSE MECHANISMS AGAINST ROS

Plant cells respond defensively to oxidative stress by removing the ROS and maintaining antioxidant defense compounds at levels that reflect ambient environmental conditions (Scandalios *et al.,* 1997). Metabolic containment mechanisms for ROS involving antioxidant genes and associated processes are likely to have predated or co-evolved with the appearance of aerobiosis and represent fundamental adaptations of aerobic systems to an oxygen dependent metabolism. The mechanisms that act to adjust antioxidant levels to afford protection include changes in antioxidant gene expression.

ROS themselves play a role in intracellular redox sensing, activating antioxidant resistance mechanisms, among other adaptive processes. Functional roles of these responses include the protection of redox sensitive enzymatic processes, the preservation of membrane integrity, and the protection of DNA and proteins (Scandalios, 1997). Redox-sensitive regulatory enzymes such as fructose-1,6-bisphosphatase can be protected from oxidation or inactivation by the action of antioxidants such as glutathione. Under unstressed conditions, the formation and removal of O_2 are in balance. The defense system, when presented with increased ROS formation under stress conditions, can be overwhelmed when it is unable to remove the toxic molecular species with increased enzymatic or non-enzymatic antioxidant processes. ROS are produced at high levels in peroxisomes. Hydrogen peroxide is produced in the peroxisomal respiratory pathway by flavin oxidase. Fatty acid beta oxidation and glycolate oxidase action are other sources of hydrogen peroxide production in the peroxisome. Developmental transitions such as seed maturation, in which peroxisomes play an important role, also involve oxidative stress (Leprince *et al.* 1990; Walters, 1998).

Ascorbic acid, glutathione, and α-tocopherol have each been shown to act as antioxidants in the detoxification of ROS. These compounds have central and interrelated roles, acting both non-enzymatically and as substrates in enzyme-catalyzed detoxification reactions (Chaudiere and Ferrari-Iliou 1999). Metabolic cycles located within the aqueous phase of the peroxisome, chloroplast, cytosol, and the mitochondrion successively oxidize and re-reduce glutathione and ascorbate, using NAD (P) H as the ultimate electron donor. Ascorbate, reduced glutathione (GSH), ascorbate peroxidase (APX), glutathione reductase (GR), superoxide dismutase (SOD), and monodehydroascorbate reductase (MDHAR) are involved in several contexts in antioxidant regeneration throughout the plant cell.

4.1. Superoxide dismutase

Superoxide dismutase (SOD) was first thought to be a copper storage protein. Subsequently, the enzyme was identified by a number of names, erythrocuprein, indophenol oxidase, and tetrazolium oxidase until its catalytic function was discovered. SOD is now known to catalyse the dismutation of superoxide to hydrogen peroxide and oxygen: Therefore, the activity of this enzyme determines the relative proportions of the two constituents of the Haber-Weiss reaction that generates hydroxyl radicals. Since SOD is present in all aerobic organisms and most (if not all) subcellular compartments that generate activated oxygen, it has been assumed that SOD has a central role in the defence against oxidative stress (Scandalias, 1997). There are three distinct types of SOD classified on the basis of the metal cofactor: the copper/zinc (Cu/Zn - SOD), the manganese (Mn-SOD) and the iron (Fe-SOD) isozymes (Bannister *et al.*, 1991). These isozymes can be separated by native polyacrylamide gel electrophoresis, their activity detected by negative staining and identified on the basis of their sensitivity to KCN and H_2O_2. The Mn-SOD is resistant to both inhibitors, whereas the Cu/Zn-SOD is sensitive to both inhibitors; Fe-SOD is resistant to KCN, and sensitive to H_2O_2. The subcellular distribution of these isozymes is also distinctive. The Mn-SOD is found in the mitochondria of eukaryotic cells; some Cu/Zn-SOD isozymes are found in the cytosol, others in the chloroplasts of higher plants. The Fe-SOD isozymes are often not detected in plants, but when detected, Fe-SOD is usually associated with the chloroplast compartment (Bowler *et al.*, 1992). The prokaryotic Mn-SOD and Fe-SOD, and the eukaryotic Cu/Zn-SOD enzymes are dimers, whereas the Mn-SOD of mitochondria is tetramers (Scandalias, 1997). Peroxisomes and glyoxysomes shown to contain both Cu/Zn- and Mn-SOD activity but there are no reports of extracellular SOD enzymes in plants. All forms of the SOD are nuclear-encoded and are targeted to their respective subcellular compartments by an amino terminal targeting sequence. Several forms of SOD have been cloned from a variety of plants (Bowler *et al.*, 1992).

4.2. Catalase

Catalase is a heme-containing enzyme that catalyses the dismutation of hydrogen peroxide into water and oxygen. The enzyme is found in all aerobic eukaryotes and is important in the removal of hydrogen peroxide generated in peroxisomes (microbodies) by oxidases involved in ß-oxidation of fatty acids, the glyoxylate cycle (photorespiration) and purine catabolism. Catalase was one of the first enzymes to be isolated in a highly purified state. All forms of the enzyme are tetramers in excess of 220,000 molecular weight. Multiple forms of catalase have been described in many plants. Maize has three isoforms termed cat-1, cat-2 and cat-3 that are on separate chromosomes and are differentially expressed and independently regulated (Scandalias, 1997). Cat-1 and cat-2 are localised in peroxisomes and the cytosol, whereas cat-3 is mitochondrial.

4.3. Ascorbate and APX : Multiple roles in antioxidant defenses

Ascorbate has been found in the chloroplast, cytosol, vacuole and extra-cellular compartments of the cell. About 20-40% of the ascorbate in the mesophyll leaf cell is in the chloroplast. The chloroplast contains all the enzymes to regenerate reduced ascorbate from its oxidised products. Foyer and Halliwell (1976) proposed that hydrogen peroxide was dissipated in the chloroplast by the coupling of ascorbate and glutathione redox cycling. Illuminated chloroplasts produce superoxide and hydrogen peroxide from sites on the thylakoids, most commonly PSI. Superoxide is converted into hydrogen peroxide by either spontaneous dismutation or by the SOD enzyme. Hydrogen peroxide is scavenged by ascorbate and the enzyme ascorbate peroxidase (Asada, 1994). The monodehydroascorbate has two routes of regeneration, one via monodehydroascorbate reductase, the other via dehydroascorbate reductase and glutathione. The terminal electron donor is NADPH. This pathway serves two functions. One is the detoxification of hydrogen peroxide that might otherwise participate is Fenton reactions, and the second in the oxidation of NADPH. The latter function is an apparently energy consuming, wasteful process analogous to photorespiration. It might at first seem more logical that the chloroplast contain catalase because it would allow the dissipation of hydrogen peroxide without "wasting" NADPH. However, it should be realised that conditions favouring electron transfer from PSI to oxygen generally cause a high redox potential, *i.e.* high NADPH/NADP ratio. By reducing this redox potential through the Halliwell-Asada pathway, the tendency of PSI to reduce oxygen is minimised.

Although ascorbate metabolism has been studied in most detail in the chloroplast, it is likely that all enzymes for its regeneration also exist in the cytosol of both photosynthetic and non-photosynthetic cells. For example different isozymes for ascorbate peroxidase have been shown to exist in the cytosol and chloroplast compartments. The cell wall is also an important site of ascorbate metabolism because it contains mM concentrations of ascorbate. Here, ascorbate may have a role in cell wall biosynthesis. Since the enzymes to recycle oxidised forms of ascorbate are not present in the cell wall, it has been proposed that the plasmalemma may have an ascorbate translocator to shuttle oxidised and reduced forms between the cytosol and cell wall (Foyer, 1993).

APX exists as a multigene family in Arabidopsis. APX 1 and 2 are both cytosolic enzymes. A membrane associated APX has been described in the peroxisome, and also in the chloroplast (Mullen and Trelease, 2000). The chloroplast contains two distinct APX enzymes, one which is free in the stroma and one which is associated with the thylakoids.

In addition to its role in the scavenging cycle, ascorbate acts as a reductant in the regeneration of α-tocopherol and in the zeaxanthin cycle (Foyer, 1993). A third role of ascorbate is at the thylakoid surface within the chloroplast where it acts as reductant in the APX-mediated scavenging of H_2O_2. Ascorbate is reduced to the monodehydroascorbate radical as a result of this thylakoid-associated process (Grace *et al.,* 1995). Ascorbic acid is regenerated by a light-dependent process at the thylakoid that can utilize ferredoxin as the source of reductant. In cases where fully oxidized dehydroascorbic acid (DHA) is produced, reduced glutathione (GSH) is the source of reductant for the regeneration of ascorbic acid. Since APX appears to play such an essential role in the scavenging process, the processes within the chloroplast associated with oxygen uptake and APX function and the reduction of molecular oxygen have been named "Mehler-ascorbate peroxidase photorespiration" (Mullineaux *et al.,* 2000).

4.4. Glutathione

Glutathione (GSH) is a tripeptide (Glu-Cys-Gly) whose antioxidant function is facilitated by the sulphydryl group of cysteine. On oxidation, the sulphur forms a thiyl radical that reacts with a second oxidised glutathione forming a disulphide bond (GSSG). Some legumes contain homoglutathione (hGSH) that is a homologous tripeptide of Glu-Cys-Ala. GSH is found in most tissues, cells and subcellular compartments of higher plants. Its levels declines with tissue age and vary among growth environments; for example, glutathione levels are higher in the light than in the dark. At the subcellular level, GSH concentration is highest in the chloroplast, averaging between 1 and 4 mM, but significant quantities also accumulate in the cytosol. GSH exists predominantly in the reduced form with estimates varying from 70%.

GSH can function as an antioxidant in many ways. It can react chemically with singlet oxygen, superoxide and hydroxyl radicals and therefore function directly as a free radical scavenger. GSH may stabilise membrane structure by removing acyl peroxides formed by lipid peroxidation reactions. GSH is the reducing agent that recycles ascorbic acid from its oxidised to its reduced form by the enzyme dehydroascorbate reductase. GSH can also reduce dehydroascorbate by a non-enzymatic mechanism at pH > 7 and GSH concentrations greater than 1 mM. This may be a significant pathway in chloroplasts whose stromal pH in the light is about 8 and GSH concentrations may be as high as 5 mM (Foyer and Halliwell, 1976).

There are alternative functions for GSH in cellular metabolism independent of its antioxidant properties. It may have a significant role in the transport of reduced sulphur from leaves to sink tissues such as the root. The synthesis and degradation of GSH occurs continuously through the glutamyl cycle that has been well characterised in animals and at least portions have been confirmed in plants. The first step in GSH synthesis is the combination of glutamate and cysteine to form glutamylcysteine by the enzyme glutamylcysteine synthetase. The subsequent step involves the addition of glycine by the enzyme glutathione synthetase. In the legumes that accumulate hGSH, this second step involves the addition of alanine by the enzyme homoglutathione synthetase.

The reduction of GSSG to GSH is catalysed by the enzyme glutathione reductase (GR) which has been purified from a number of plant tissues (Smith *et al.*, 1989), and cDNA for GR has been cloned from pea (Creissen *et al.*, 1991). Animals contain a selenium containing enzyme, glutathione peroxidase that reduces hydrogen peroxide forming GSSG and thereby serves as an alternative means of detoxifying activated oxygen.

Upon the imposition of oxidative stress, the existing pool of reduced glutathione (GSH) is converted to oxidized glutathione (GSSG) and glutathione biosynthesis is stimulated. The rate-limiting step for glutathione synthesis is thought to be gamma-glutamylcysteine synthetase, which is feedback regulated by GSH and is controlled primarily by the level of available L-cysteine. Increasing glutathione biosynthetic capacity has been shown to enhance resistance to oxidative stress (Arisi *et al.*, 1998; Zhu *et al.*, 1999). Glutathione reductase (GR) activities increase as the glutathione pool increases through a multi-level control mechanism, which includes coordinate activation of genes encoding glutathione biosynthetic enzymes and GR (Xiang and Oliver, 1998).

Glutathione acts as a redox sensor of environmental cues and forms part of the multiple regulatory circuitry coordinating defense gene expression. There are two known genes encoding GR (Madamanchi *et al.*, 1992). The redox state of the GSH/GSSG couple may act as a direct link between environmental cues and crucial molecular adaptive responses of plant cells (Baginsky *et al.*, 1999). Glutathione has been reported to regulate rates of cell division (Sanchez-Fernandez *et al.*, 1997) and the induction of antioxidant defenses, as exemplified by the induction of Cu-Zn SOD (Herouart *et al.*, 1993). Glutathione has been suggested as an intermediary in a redox sensing signaling pathway in plants involving the ROS-mediated oxidation of membrane lipids to oxylipins as the initial step (Ball *et al.*, 2001). It is also thought to act specifically as a regulator of the transcription of chloroplast genes (Baginsky *et al.*, 1999). Previous biochemical data demonstrated the existence of multiple isoforms of GR, which have been assigned to various organelles, including the peroxisome, the chloroplast and the mitochondrion (Creissen *et al.*, 1991).

The ROS scavenging mechanism in living cells by ascorbate and glutathione in chloroplast is depicted below (Fig.7)

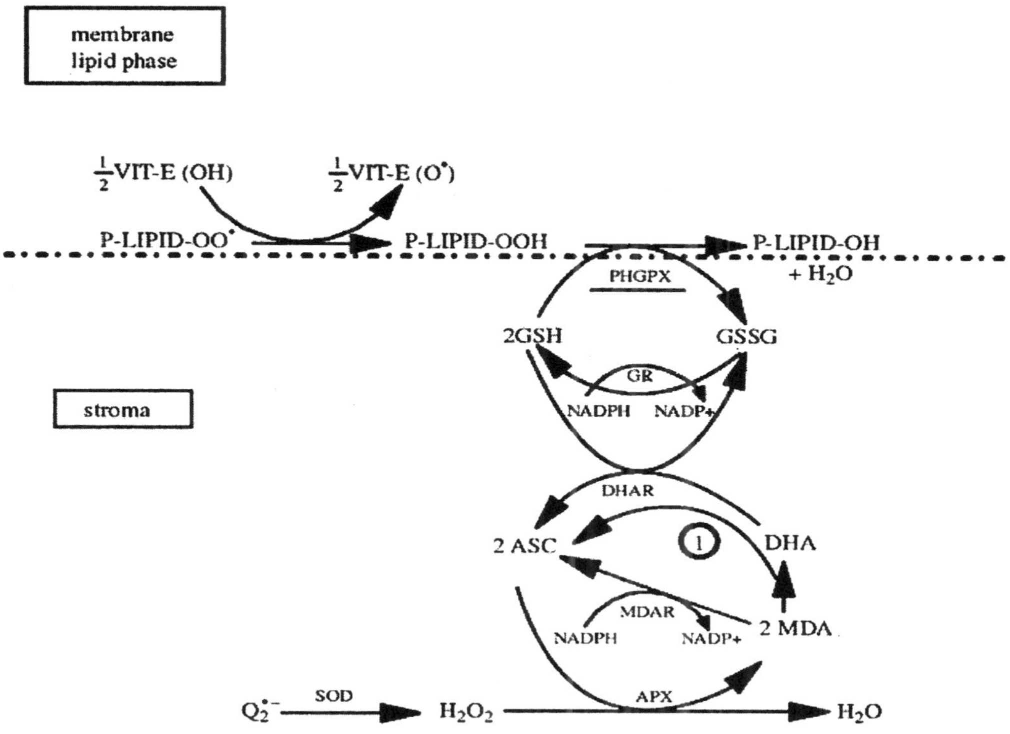

Fig 7: The scavenging of active oxygen species in the chloroplast in both the lipid membrane phase and the stroma, linked to redox cycles for ascorbate and glutathione and the oxidation of α-tocopherol (Vitamin E).

Abbreviations are as follows: P-LIPID -OO, phospholipid peroxy radical; P-LIPID-OOH, phospholipid peroxide; P-LIPID-OH, phospholipid alcohol; VIT-E(OH), α-tocopherol (vitamin E); VIT-E (O*), α-chromanoxyl radical; PHGPX, phospholipid hydroperoxide-dependent glutathione peroxidase; GSH, reduced glutathione; GSSG, glutathione disulfide; GR, glutathione reductase; DHAR, dehydroascorbate reductase; ASC, ascorbic acid; DHA, dehydroascorbate; MDA, monodehydroascorbate free radical; MDAR, monodehydroascorbate free radical reductase; APX, ascorbate reductase; SOD, superoxide dismutase; O_2-, superoxide ion. Reaction 1 is the non-enzyme-catalyzed spontaneous dismutation of two MDA molecules to one ASC and one DHA, respectively. (Mullineaux *et al.*, 2000).

5. FACTORS INFLUENCING RATE OF SEED DETERIORATION

5.1. Internal factors

The physical condition and physiological state of seeds greatly influence their life-span. Seeds that have been broken, cracked, or even bruised deteriorate more rapidly than undamaged seeds. Even in the absence of physical symptoms, seeds may be physiologically impaired and become susceptible to rapid deterioration. Several kinds of environmental stresses during seed development, and prior to physiological maturity, can impair the longevity of seeds for example, deficiency of minerals (N, K and Ca), water, and temperature extremes. Initial germination capacity also influences the life-span of seeds, although it may only be a symptom of physical or physiological injury. Some seed are genetically and chemically equipped for longer life. Fatty seeds of peanuts, flax, and certain pines are examples of short-lived seeds, partly because they are less hygroscopic, and available water is most likely to be loosely held. Seeds of other species may be chemically very similar but have vastly different storability due to differences in genetic potential.

5.2. External factors

5.2.1. Relative humidity and temperature

Relative humidity and temperature are the principle external factors that influence seed longevity. The effects of relative humidity and temperature are highly interdependent. Most crop seeds lose viability rapidly at relative humidities approaching 80% and temperatures of 25° to 30°C, but can be kept ten years or longer at relative humidities of 50% or less and a temperature of 5°C or lower. According to Harrington (1973) the sum of the per cent relative humidity plus the temperature in degrees Fahrenheit should not exceed 100 for safe storage. It has been suggested that the relative humidity should be no higher than 60% for seeds at 21°C and no higher than 70% for seeds at 4° to 10°C; however, at 5°C and 45 to 50% relative humidity, many crop seeds can be safely stored for ten years or longer.

5.2.2. Gaseous composition of storage environment

Seed storage in an atmosphere different from that of air (20% oxygen, 0.03% carbon dioxide, and about 80% inert nitrogen gas) can be controlled in a gas tight room or container, where the desired atmosphere can be maintained. An atmosphere of higher oxygen concentration than that of air hastens deterioration in seeds, while an atmosphere of carbon dioxide is reported to increase the storage life. An atmosphere of nitrogen gas was found to delay the onset of deterioration of wheat seed; however, once deterioration started, it proceeded as rapidly as in air.

5.2.3. Seed moisture relationships

5.2.3.1. Types of seed moisture

The way in which water interacts with plant substances is more complex. Water has been classified as bound, adsorbed and free water depending on the way it is held by the plant substances. The bound water is tightly held to ionic groups such as amino or carboxyl groups and exists as a monolayer around macromolecules of the seed. Adsorbed water is considered to exist in multilayers loosely held by bonding to hydroxyl and amide groups above the monolayer of bound water. Free water is considered as capillary or solution water held only by capillary forces to the seed tissues.

5.2.3.2. Moisture equilibrium

The hygroscopic nature of seeds allows them to adjust their moisture content to be in equilibrium with any given relative humidity. Equilibrium is attained when the seed has no further tendency to absorb or lose moisture. Hygroscopic equilibrium curves, also called absorption isotherms, are graphic expressions of the relationship between the moisture content of seeds and their ambient relative humidity at constant temperatures. They are established by measuring the absorption or desorption at successive relative humidities and can be used to predict seed moisture content at any given relative humidity. The hygroscopic equilibrium curve is a sigmoid like curve, with three rather distinct phases representing different stages of water absorption or desorption. Phase one represents very tightly held water that may actually be a part of the chemical structure of the seed. This kind of water cannot be removed without destruction of the seed tissue. This phase may also include some water held as discrete molecules in bonding interactions with the seed tissue molecules. Phase two of the moisture equilibrium curve represents water that is more loosely held than that of phase one. For most seeds, this portion of the moisture equilibrium is represented by a straight-line relationship between relative humidity and moisture content. Water represented by the upper portion of phase two is easily removed by drying; however, the lower portion representing very strong bonding is difficult to remove. Water in the upper portion of phase two contributes significantly to seed deterioration during storage. Water in phase three represents water loosely held by very weak bonding and free water in the intercellular and inter-tissue spaces. It is easily eliminated during drying; but if not eliminated, it contributes to rapid seed deterioration.

6. IMPLICATION FOR SEED STORAGE

6.1. Controlled atmosphere and seed storage

Seeds of most species may be safely stored for several years by careful control of temperature and relative humidity. Although such storage conditions are too costly for most agricultural seed lots, they may be extremely valuable for preserving germplasm and certain high value seed stocks.

6.2. Sealed storage

In recent years there has been considerable interest in packaging seeds in moisture resistant or *hermetically sealed* containers for storage and marketing. The purpose of such containers is to keep sealed at safe storage moisture levels. Ordinary paper and cloth containers were least effective, while various laminate and polyethylene materials were somewhat effective. Metal cans were completely effective in maintaining seed moisture at the initial five per cent level. Such completely moisture proof containers hermetically seal the seed and are effective for long term seed storage up to ten years or more. The effectiveness of other material was directly associated with their ability to resist moisture.

6.3. Humidity control and containerised seed storage

Absolute humidity control in seed storage areas requires considerable investment in specially constructed rooms and dehumidifying equipment. Saturated salt solutions or acid solutions may be used. Sulfuric acid is a common acid desiccant and is diluted with water to provide increasing relative humidity levels. Saturated salt solutions are less hazardous to both seed and personnel; however, different salts must be used to attain various humidity levels. Complete saturation is attained when a given quantity of water dissolves its full capacity of salt crystals. To maintain complete saturation, some additional salt should be added so that the relative humidity is held constant. Care must be used to keep the seed separate from the liquid desiccant (Table 2).

Table 2: Relative humidity in equilibrium with different concentrations of aqueous acid solutions at various temperatures

Acid	Temperature °F	Acid by weight, per cent			
		20	**40**	**60**	**80**
H_2SO_4 (Sulfuric acid)	0	87.3	55.7	15.0	3.14
	50	87.4	56.6	15.8	3.88
	68	87.7	56.7	16.3	4.76
	86	87.5	56.6	17.0	5.75
	104	87.6	57.5	17.8	6.88
	112	88.8	58.2	18.8	8.2
		Acid by weight, per cent			
		20	30	40	50
HNO_3 (Nitric acid)	0	89.2	78.4	65.3	45.7
	50	86.7	77.0	63.0	45.6
	68	86.6	75.2	63.0	
	86	86.6	74.9	61.5	
	104	85.9	74.1	60.5	
	112	86.5	74.6		
	140	86.9	75.6		
		Acid by weight, per cent			
		10	20	30	40
HCl (Hydrochloric acid)	0	83.5	56.0	27.4	8.9
	50	83.5			
	68	83.2			
	86	84.2			

7. CONCLUSION

Lipid peroxidation has been continuously suspected and investigated as a cause of seed deterioration on long term stored dry seeds. The studies on aldehydes and free radicals have been used to prove the involvement of lipid peroxidation in seed deterioration. Impairment of DNA synthesize, selective protein synthesis, decrease in enzyme activity have been reported in both long term stored seeds and rapidly aged seeds. Lipid peroxidation has been reported to cause biological implications including i) decrease in membrane integrity ii) co-oxidation of associated cellular components, iii) decrease *in vivo* and *in vitro* protein synthesis iv) decrease in respiration due to damage of mitochondria. Exclusive studies with rapid seed ageing (which is physiologically different from natural ageing) have indicated that seed deterioration after rapid seed ageing, i) biochemically "mimics" seed germination, ii) suffers from quantitative variation in protein synthesis and enzyme activity, iii) impairment in DNA synthesis, iv) prevention of development of new mitochondria, v) damage due to accumulation of cytotoxic hydrogen peroxide and vi) accumulation of reducing sugars (Maillard Reaction) which inactivate proteins and DNA. It seems that both long term stored seeds and rapidly aged seeds share in common the impairment in protein synthesis, DNA impairment and decrease in respiration. However rapidly aged seeds might be subjected to a different type of damage due to the accumulation of hydrogen peroxide and biochemical activities which mimics "seed germination", but culminate in loss of viability.

It is thus conceivable that the total damage to the seed as described above is mainly due to the involvement of free radicals, might be in "Lipid peroxidation" or "Maillard reactions". Hence, apart from the conventional prophylactic treatments which guard the seeds against insects and pests, identification and standardization of effective treatments in the line of "Halogenations" will go a long way in quenching the free radicals and increasing the storage life of seeds.

REFERENCES

App, A. A., M. G. Bulis and W. J. McCarthy. 1971. Dissociation of ribosomes and seed germination. *Plant Physiol.*, **47**:81-86.

Arisi, A.M., G. Cornic, L. Jouanin, and C.H. Foyer. 1998. Overexpression of iron superoxide dismutase in transformed poplar modifies the regulation of photosynthesis at low CO2 partial pressures or following exposure to the prooxidant herbicide methyl viologen. *Plant Physiol.* **117**: 565–74.

Asada, K. 1994. Production and action of active oxygen species in photosynthetic tissue: *In* Causes of Photooxidative Stress and Amelioration of Defense Systems in Plants. (C.H. Foyer and P.M. Mullineaux, (eds), pp. 77–104. Boca Raton, CRC Press.

Baginsky, S., K. Tiller, T. Pfannschmidt, and G. Link. 1999. PTK, the chloroplast RNA polymerase-associated protein kinase from mustard (*Sinapis alba*), mediates redox control of plastid *in vitro* transcription. *Plant Mol. Biol.* **39** :1013–23.

Ball, L., O. Richard, U. Bechtold, C. Penkett, H. Reynolds, B. Kular, G. Creissen, S. Karpinski, W. Schuch, and P. Mullineaux. 2001. Changes in Global Gene Expression in Response to Excess Excitation Energy in *Arabidopsis thaliana. J. Exp. Bot.*, **52**: 1245-1251.

Bannister, W.H., J.V. Bannister, D. Barra, J. Bond, and F. Bossa. 1991. Evolutionary aspects of superoxide dismutase: The copper/zinc enzyme. *Free Radic. Res. Commun.*, **12-13**: 349–61.

Bowler, C., M. Van Montagu, and D. Inze. 1992. Superoxide dismutase and stress tolerance. *Annu. Rev. Plant Physiol. Plant Mol. Biol.*, **43**: 83-116.

Chaudiere, J., and R. Ferrari-Iliou. 1999. Intracellular antioxidants: from chemical to biochemical mechanisms. *Food Chem. Toxicol.* **37**: 9-10. 949–62.

Creissen, G., E.A. Edwards, C. Enard, A. Wellburn, and P. Mullineaux. 1991. Molecular characterization of glutathione reductase cDNAs from pea (*Pisum sativum* L). *Plant J.* **2**: 129–131.

Dhindsa, R. 1982. Inhibition of protein synthesis by products of lipid peroxidation. *Phytochemistry,* **21**: 309-314.

Dianzani, M. V. 1979. Biological activity of methyglyoxal and related aldehydes. *In.* Submolecular Biology and Cancer. CIBA Foundation Series 67, pp. 245-270. Excerpta Medical Amsterdam.

Dupont, J. 1971. Lipoxygenase mediated cleavage of fatty acids in plant mitochondria. *Physiol. Plant.,* **52**: 225-232.

Fenton, H. J .H. 1899. Oxidation of certain organic acids in the presence of ferrous salts. *Proc. Chem. Soc.,* **25**:224.

Fenton, H. J. H. 1894. Oxidation of tartaric acid in the presence of iron. *J. Chem. Soc.,* **65**:899.

Fielding, J.L. and A. Goldsworthy. 1982.The evolution of volatiles in relation to ageing in dry wheat seed. *Seed Science and Technology,* **10**: 277-282.

Foyer, C.H. 1993. Ascorbic acid. *In:* Antioxidants in higher plants (Eds). R.G. Alscher and J.L. Hess, pp. 31–58. CRC Press, Boca Raton, Fl.

Foyer, C.H., and B. Halliwell. 1976. The presence of glutathione and glutathione reductase in chloroplasts: A proposed role in ascorbic acid metabolism. *Planta,* **133**: 21–25.

Gebicki, J. M. and B. H. J. Bielski. 1981. Comparison of the capacities of the perhydroxyl and superoxide radicals to initiate chain oxidation of linoleic acid. *J. Am. Chem. Soc.,* **103**: 7020-7022.

Grace, S., R. Pace, and T. Wydrzynski. 1995. Formation and decay of monodehydroascorbate radicals in illuminated thylakoids as determined by EPR spectroscopy. *Biochim.. Biophys. Acta.* **1229**: 155–165.

Gutteridge, J.M.C. and B. Halliwell. 1995. Free Radicals in Biology and Medicine. Clarendon Press, Oxford.

Haber, F. and J. Weiss. 1934. The catalytic decomposition of hydrogen peroxide by iron salts. *Proc. Royal Soc. A.,* **147**:332.

Halliwell, B. 1987. Oxidative damage, lipid peroxidation and antioxidant protection in chloroplasts. *Chem. Phys. Lipids,* **44**: 327-340

Halliwell, B. and J.M.C. Gutteridge. 1989. Free Radicals in Biology and Medicine: Oxford, Clarendon Press.

Harman, D. 1956. Ageing: a theory based on free radical and radiation chemistry. *J Gerontal* **11**: 298-300.

Harrington, J.F. 1973. Biochemical basis of seed longevity. *Seed Sci. & Technol.,* **1**: 453-461.

Hendry, G. A. F., W. E. Finch-Savage, N. M. ThorpeAtherkon, S. H. Buckland, K. A. Nilsson, and W. E. Seel. 1992. Free radical processes and loss of seed viability during desiccation in the recalcitrant species *Quercus robur* L. *New Phytol,* **122**: 273-299.

Herouart, D., M. Van Montagu, and D. Inze. 1993. Redox-activated expression of the cytosolic copper/zinc superoxide dismutase gene in Nicotiana: *Proc. Natl. Acad. Sci.* USA. **90**: 3108–3112.

Kaloyereas, S.A. 1958. Rancidity as a factor in the loss of viability of pine and other seeds. *J. Amer. Oil Chem. Soc.,* **35**: 176-179.

Kalpana, R. and K.V.Madhava Rao, 1994. Absence of the role of lipid peroxidation during accelerated ageing of seeds of pigeon pea cultivars. *Seed Science and Technology,* **22**: 253-260.

Leopold, A.C. and M.E. Musgrave. 1979. Respiratory changes with chilling injury of soybeans. *Plant Physiology,* **64**: 702-705.

Leprince, O., P.C. Thorpe, R. Deltour, N.M. Atherton, and G.A.F. Hendry. 1990. The role of free radicals and radical processing systems in loss of desiccation tolerance in germinating maize. *New Phytol.,* **116**: 573–580.

Madamanchi, N., R. Alscher, K. Hatzios, and C. Cramer. 1992. Acquired resistance to herbicides in pea cultivars by exposure to sulfur dioxide. *Pest. Biochem. Physiol.,* **48**: 31–40.

McDonald, M. B. 1999. Seed deterioration: physiology, repair and assessment. *Seed Science and Technology,* **27**: 177-237.

Moreau, F., J. Dupont, and C. Lance. 1974. Phospholipid and fatty acid composition of outer and inner membranes of plant mitochondria. *Biochem. Biophys. Acta.,* **345**: 294-304.

Mullen, R.T, and R.N. Trelease. 2000. The sorting signals for peroxisomal membrane-bound ascorbate peroxidase are within its C-terminal tail. *J. Biol. Chem.* **275**: 16337–44.

Mullineaux, P., L. Ball, C. Escobar, B. Karpinska, G. Creissen, and S. Karpinski. 2000. Are diverse signalling pathways integrated in the regulation of arabidopsis antioxidant defence gene expression in response to excess excitation energy?. *Trans. R. Soc. Lond. B. Biol. Sci.* **355**:1531–40.

Murthy, N. U.M., P.P. Kumar, and W.Q. Sun. 2003. Mechanisms of seed ageing under different storage conditions for *Vigna radiata* (L.) Wilczek: Lipid peroxidation, sugar hydrolysis, Maillard reactions and their relationship to glass state transition. *J. Exp. Bot.,* **54**: 1057-1067.

Murthy, N.U.M. and W.Q. Sun. 2000. Protein modification by Amadori and Maillard reactions during seed storage: roles of sugar hydrolysis and lipid peroxidation. *J. Exp. Bot.,* **51**: 1221-1228.

Osborne, D.J., 1980. Senescence in seeds. In KV Thimann [ed.], Senescence in plants, CRC Press, Boca Raton, Florida. pp. 13-37.

Roberts, E. H. (1972). Cytological genetical and metabolic changes associated with loss of viability. *In.* Viability of seeds. (Ed.) E.H. Roberts, Chapman and Hall Ltd. London.

Sanchez-Fernandez, R., M. Fricker, L.B. Corben, N.S. White, N. Sheard, C.J. Leaver, and M. Van Montagu. 1997. Cell proliferation and hair tip growth in the Arabidopsis root are under mechanistically different forms of redox control. *Proc. Natl. Acad. Sci.* USA., **94**: 2745–2750.

Scandalios, J. G., L. Guan and A. N. Polidoros 1997. Catalases in plants: gene structure, properties, regulation and expression. Oxidative stress and the molecular biology of antioxidant defenses. J. G. Scandalios. Cold Spring Harbor, NY, Cold Spring Harbor Laboratory Press: 343-406.

Scandalios, J.G. 1997. Molecular Genetics of Superoxide Dismutases in Plants. *In* Oxidative Stress and the Molecular Biology of Antioxidative Defenses. J.G. Scandalios, (Ed.) pp. 527–568. Plainview, Cold Spring Harbor.

Senaratna, T, B. D. McKersie, and A. Borochov. 1987. Desiccation and free radical mediated changes in plant membranes. *J. Exp. Bot.,* **38**: 2005-2014.

Smith, I.K., T.L. Vierheller, and C.A. Thorne. 1989. Properties and functions of glutathione reductase in plants. *Physiol. Plant.,* **77**:449-456.

Walter. H. 1963. Chemical reactivity of a macromolecule as a function of its age. *Biochemica et Biophysica Acta.,* **69:** 410-411.

Walters, C. 1998. Understanding the mechanisms and kinetics of seed ageing. *Seed Sci. Res.,* **8**: 223–244.

Weinbach, E. C., H. Sheffield and J. Garbus. 1963. Restoration of oxidative phosphorylation and morphological integrity of swollen, uncoupled rat liver mitochondria. *Proc. Nat. Acad. Sci.,* **50**: 561-568.

Wilson, D.O. and M.B. McDonald. 1986. The lipid peroxidation model of seed ageing. *Seed Science and Technology,* **14**: 269-300.

Winston, G. W. 1992. Physicochemical basis for free radical formation in cells: production and defenses. *In:* R. G Alscher, J. R. Cumming, (eds) *Stress Responses in Plants: Adaptation and Acclimation mechanisms.* Wiley-Liss, New York, pp 57-86.

Xiang, C. and D. Oliver. 1998. Glutathione metabolic genes coordinately respond to heavy metals and jasmonic acid in Arabidopsis. *Plant Cell,* **10**: 1539–1550.

Zhu, Y. L., E. A. H. Philon-Smits, A. S. Tarun, S. U. Weber, L. Jouanin, and N. Terry. 1999. Cadmium tolerance and accumulation in Indian mustard is enhanced by overexpressing gamma-glutamylcysteine synthetase. *Plant Physiol.,* **121**: 1169–1178.

11

Seed Vigour

K. Natarajan[1], T. Saravanan[1] and A.P. Sivamurugan[2]
[1]*Department of Plant Breeding and Genetics*
[2]*Department of Soil and Crops*
Adhiparasakthi Agricultural College, G. B. Nagar, Kalavai 632 506

1. INTRODUCTION

In agricultural practice, the achievement of specified plant population and synchronous emergence is often necessary for both the optimum yield and uniform maturity (ISTA, 1995). Demands have therefore, been made on the seed quality of all seeds in trade, despite high levels of germination (Matthews, 1980). The physiological quality of a seed lot is routinely evaluated by the standard germination test, which states that germination is the emergence and development from the seed embryo of those essential structures which are indicative of the ability to produce a normal plant under favourable conditions (AOSA, 1991). The germination test, however, does not correspond to the performance of the seed lots when they are sown under field conditions and does not provide complete evaluation of seed lot deterioration or quality (McDonald, 1980a) and also does not differentiate strong and weak seedlings. This had led the seed growers and technologists to seek for methods, which offer more reliable information about the physiological potential of a seed lot. The recognition of performance differences among high germinating seed lots is generally known as seed vigour (ISTA, 1995). Relatively large differences in seed vigour at high viability percentage cannot be discriminated by the standard germination test (Roberts, 1986). Consequently, a great interest has arisen in vigour tests which are used as an attempt to identify possible differences in physiological quality among seed lots that present a similar germination potential. Vigour tests are intended to supplement the germination test and are designed to provide additional information about the physiological quality of a seed lot i. e. information such as storage and planting potential (Ferguson,1993). Thus, different vigour tests have been developed and are continuously being improved.

2. SEED GERMINATION AND VIGOUR

The laboratory germination of seeds describes the percentage of seeds that are capable of producing a normal seedling under optimum conditions in the laboratory. In India, minimum seed certification standards for germination in each crop are specified, which cannot be sold commercially. This ensures that emergence failure does not result from the inability of seeds to germinate. Differences in the emergence of germinable seeds can however occur, particularly in poor soil conditions, due to what has come to be termed seed vigour (Perry, 1970; Matthews, 1980). The poor emergence can result from sowing low-vigour seeds, even though the germination in the laboratory is perfectly acceptable. Thus, seed vigour is not equivalent of seed germinability. So, seed vigour indicates the ability of seed/seed lot to germinate and establish seedlings even under sub optimal or poor field conditions. As a result, seed lots of a genotype with similar germination percentages may differ significantly in terms of their vigour levels.

3. CONCEPT OF SEED VIGOUR

Seed vigour is a multiple concept. This concept has evolved through attempts of several research workers throughout the world. The origin of the term 'seed vigour' is found in the book written by Fedriech Nobbe (Nobbe, 1876) on 'Handbuch der Samen Kunde'. Later this term was mostly used in agronomic sense rather than physiological. Then vigour is defined as tantamount to the property of seeds which promotes a satisfactory performance, its antithesis has often been referred to as 'weakness' (De Tempe, 1961) and in this double sense it has become widely accepted. German pathologist Hitlner called the "successful seedlings" of cereals from the brickgravel test as 'Trieb kraft'. The term 'Trieb kraft' means both "shoot strength" and "driving force". The concept of 'Triebkraft' however, assumed a life of its own; it was translated into 'French' and 'English' as 'Vigueur' and 'Vigour', respectively. The original connotation was that of 'seedling vigour', the ability which enables a seedling upon germination to grow rapidly (Heydecker, 1972). Further, he raised the objections, a cultivar may have bred for dwarf growth; and the ultimate size of the plants may have little to do with their potential yield, either individually, where the product consists of fruits or seeds and not of the plant itself or collectively, per unit area of ground, where close plant spacing can make up for small plant size (Heydecker, 1972). According to Delouche and Caldwell (1960), seed vigour is, in a negative sense, generally thought of as "something" not adequately measured or reflected by the standard germination test. On the positive side, no such precise definition is possible. Isely (1957) suggested that two ideas on concepts of vigour; i) rapidity of growth and ii) non-susceptibility to unfavourable growing conditions, then he concluded; vigour is "sum total of all seed attributes, which favour stand establishment under unfavourable field conditions" which in turn is primarily a function of the seed and soil microorganism interrelationship and presented a graphical representation of the relationships among seed germinability, vigour and deterioration (Fig.1 and 2). Ader (1965) stated that vigour as the percentage of seeds able to produce normal seedlings under suboptimal conditions. Whereas, Neeb (1970) defined the vigour as "the totality of properties contributing to the defence against and successful resistance to biotic and abiotic hazards during germination under suboptimal conditions.

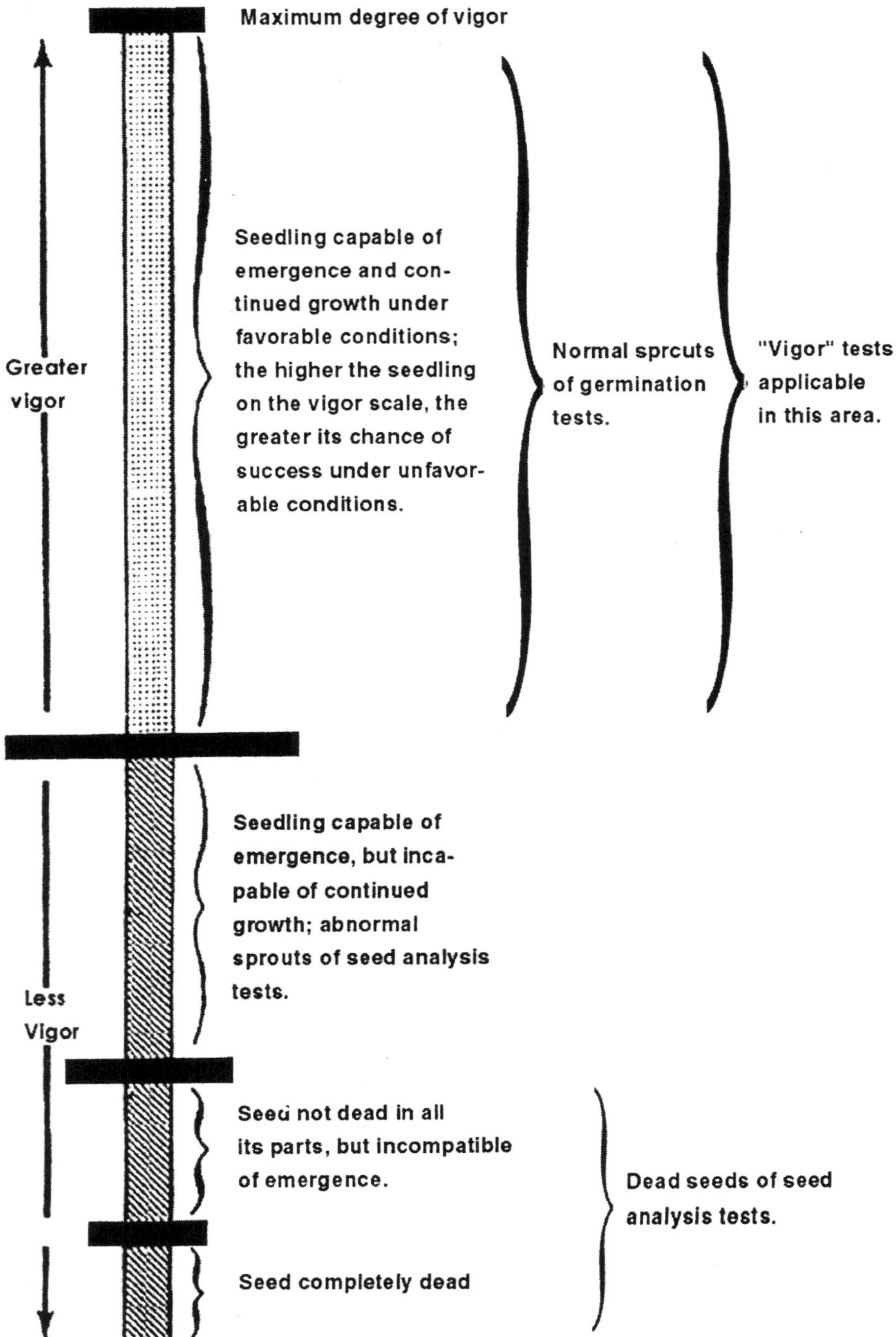

Fig. 1: Schematic representation of the relationship between germination and vigour (Isley, 1957)

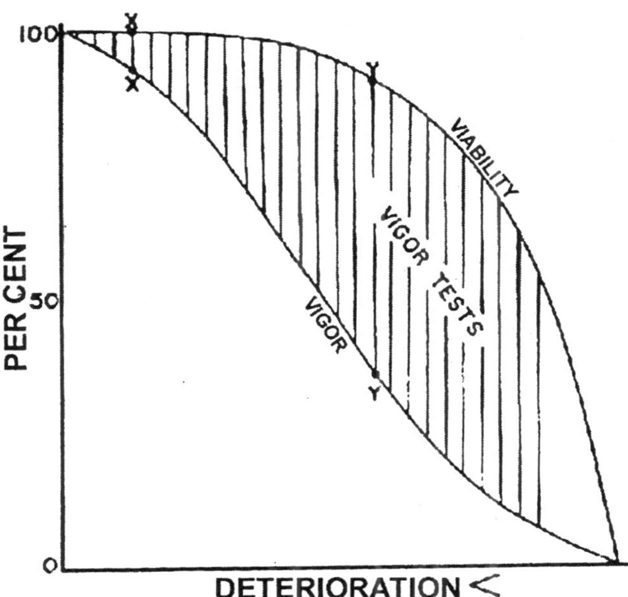

Fig.2: Relationships among seed vigour, viability (germinability) and deterioration and the area for application of vigour tests. The X and Y points on the viability and vigour curves illustrate the increasing "gap" between germinability and vigour with increasing deterioration

Vigour is viewed in more positive sense as the "ability of the seeds to produce vigourous seedlings (Nutile, 1964), as the "condition of active good health and natural robustness in seeds which, upon planting permits germination to proceed rapidly and to compete under wide range of environmental conditions (Woodstock, 1969), as the ability of seeds to produce seedlings well capable of increasing in length and volume, while still dependent on their own reserves (Germ, 1960), as the efficiency of seeds (Lindner, 1967), as the physiological stamina of seeds (Delouche, 1968), and characterised by the weight of a seedling after a period of growth in a given environment the amount of reserves that are present, the rapidity with which they are metabolised, and the efficiency of their metabolism favour seedling growth and development (Qualls and Cooper,1969). Synoptically, vigour might be described by such terms as integrity, robustness, sturdiness, adaptability, flexibility, resilience-none of them scientifically acceptable. It has also been defined as "the potential for rapid uniform germination and fast seedling growth under general field conditions (Ching, 1973). Vigour is the "sum total of all seed attributes, which favour rapid and uniform stand establishment in the field (Agarwal, 1993). At the biochemical level, it involves energy and biosynthetic metabolism, co-ordination of cellular activities and transportation and utilization of reserve foods.

Further, during ISTA congress, 1977 looking to the complexity of situation, the definition on vigour was adopted as "the sum total of those properties of the seed which determine the level of activity and performance of the seed or seed lot during germination and seedling emergence.

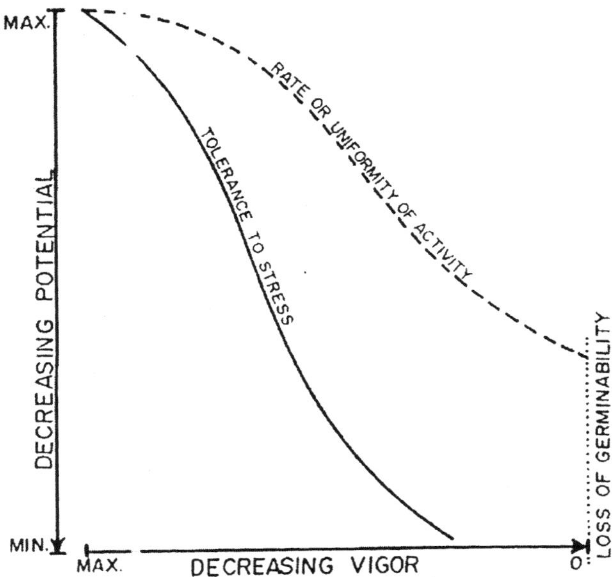

Fig.3: Relationships among the rate (intensity of responsibility) and tolerance (to environmental stresses) dimensions of seed potential and "vigour", their product

Woodstock (1969) also developed and presented a very interesting synthesis of the two primary manifestations of vigour; rate and /or intensity of seed performance or response and sensitivity (or tolerance) to environmental conditions (Fig.3).

Heydecker (1972) developed a diagrammatic scheme – based on a sequence of deteriorative changes postulated by Delouche (1969) and Roberts seed survival curves (1972) – which elegantly illustrated the rise (maturation) and fall (deterioration) in the vigour or performance potential (or capability) of a seed or seed lot. The scheme is reproduced in Figure 4.

Perry (1973) defined seed vigour " as a physiological property determined by the genotype and modified by the environment, which governs the ability of a seed to produce a seedling rapidly in soil, and the extent to which the seed tolerates a range of environmental factors.

The following concepts can be considered to clarify the meaning of vigour in terms of seed, seedling and plant performance (Copeland, 1976).

i. Speed of germination,
ii. Uniformity of germination and plant development under non-uniform conditions
iii. Ability to emerge through crusted soil
iv. Germination and seedling emergence from cold, wet and pathogen infested soil
v. Normal seedling morphological development
vi. Crop yield
vii. Storability under optimum or adverse conditions
viii. Initial higher capital (food reserves)

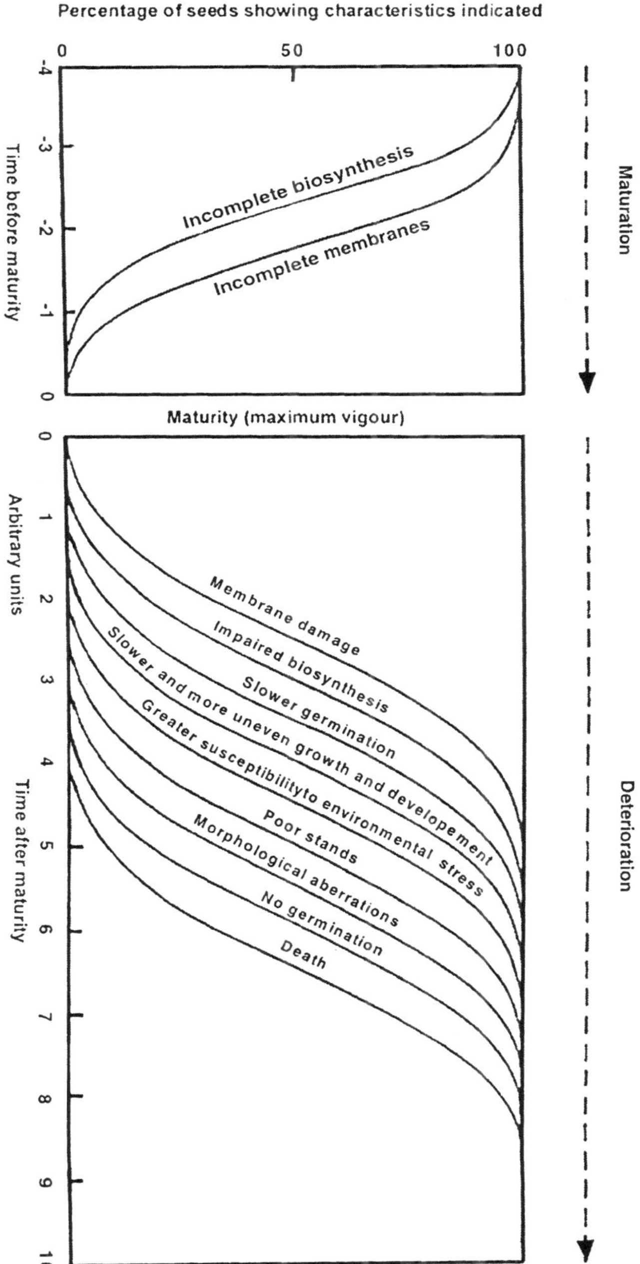

Fig.4: Diagram representing the typical time-course of events in the change in the quality of seed lot. The time scale is in arbitrary units but, very roughly, the units during maturation would be of the order of 1 to 2 days; whereas, during deterioration the units would be of the order of a day during typical accelerated aging treatments, or the order of a year for 'good' storage conditions. The curves are meant to represent cumulative normal distributions.

Looking to all these facts, seed vigour is a concept rather than specific property of a seed or seed lot. This important physiological quality parameter needs to be assessed to supplement germination and viability tests to gain insight into the performance of a seed lot in the field or in storage.

4. VIGOUR – GENETIC OR PHYSIOLOGICAL

There are two aspects of vigour – genetic and physiological. Genetic vigour can be seen in heterosis (hybrid vigour), that means the difference in vigour between two genetic lines or the difference in vigour between two genotypes of a crop. Physiological vigour can be seen in the difference in vigour between two seed lots from the same genetic line. Always physiological vigour has its basis in genetic vigour. It is very difficult to find out the cause of vigour particularly in genetically heterogeneous populations. The importance of seedling vigour at early stage has been realised in number of crops such as lettuce, chillies, tomatoes, cruciferae vegetables by nursery men and in tree species by the foresters at nursery (Heydecker, 1972).

5. FACTORS INFLUENCING SEED VIGOUR

Genetic constitution, environment during seed development and seed storage environment influence seed vigour.

5.1. Genetic constitution

Several factors *viz.*, hybrid vigour, susceptibility to seed damage, seed chemical composition and hard seededness are under genetic control and influence the expression of seed quality. Hybrid vigour is a component of heterosis and represents the measurable superiority of the hybrid progeny over its inbred parents. The superiority of the hybrid is greater under conditions of stress than under optimal conditions. The superiority of the hybrid progeny over its inbred parents has been attributed to more efficient mitochondria and extra enzyme systems for carbon assimilation (McDaniel and Sarkissian, 1968). Mechanical damage/seed coat cracking during harvesting and conditioning influences the seed vigour, which is also under genetic control. Different cultivars possess differing tolerances to mechanical abuse as reported in navy beans (Barriga, 1961) and snap bean (Wester, 1970). Breeding for improved nutritional quality at times results in poor seed quality. To overcome this problem, scientists are now attempting to identify the genes controlling nutritional quality without affecting seed vigour. Certain genes that are responsible for producing more vigourous seed have been identified (Nass and Crane, 1970). Hard seededness a genetically controlled trait though undesirable is known to protect the seed from ageing (Potts *et al.,* 1978). The plant breeders are considering all these genetically controlled seed vigour influencing factors in their breeding programmes for higher yields.

5.2. Environment during seed development

During seed development, the various factors that affect seed quality are relative humidity, temperature, frequent rains, soil moisture stress, insect and fungal invasion and soil nutrient deficiencies.

High humidity, warm temperatures and frequent rainfall during maturation, post maturation and preharvest stages adversely affect the seed quality (seed viability and vigour) and hence, is a serious seed production problem.

During the seed maturation stage, exposure of cotton crop to rains and soybeans to hot dry weather produces low quality seed (Simpson and Stone, 1935; Green *et al.,* 1965). Besides, warm and humid conditions predispose it seed to fungal invasions, which obviously influence seed quality (Nicholson *et al.,* 1972;Tekrony *et al.,* 1980 a).Delayed harvest of seed because of inclement weather results in loss of viability in a number of crops (Caldwell, 1972;Tekrony *et al.,* 1980b.). Edaphic environment especially soil moisture stress during the seed development results in shrivelled seed, which in turn results in poor vigour seed. Finally, the soil nutrient deficiencies also have an adverse affect on seed quality. For example, the groundnut crop harvested from boron and calcium deficient soils exhibited discoloration of the cotyledons due to boron deficiency and exhibited watery hypocotyls and physiological root break due to calcium deficiency (Cox and Reid, 1964; Sullivan, 1973).In other instances, a direct relationship between soil fertility and seed vigour has been reported. In wheat, nitrogen fertilization is known to increase protein levels in seed (McNeal *et al.*, 1971) and the seed with high protein levels, resultede in increased germination (Fox and Albrecht, 1957), seed vigour (Lowe and Ries, 1973) and subsequent crop yield (Ries *et al.,* 1970).

The production of good quality seed requires timely harvest at physiological maturity for maximal seed vigour. Seed development falls in three phases. The first phase begins with fertilization and rapid cell division. The second phase is characterized by cell expansion and accumulation of reserve materials by the embryo and endosperm. The third phase is characterized by water loss from the seed as seeds pass into a quiescent or dormant state before metabolic systems are re-activated upon imbibition (Bradford, 1986). In many orthodox seeds, such as cereals or other grain crops, desiccation coincides with the attainment of maximal dry weight and physiological maturity (Rosenberg and Rinne, 1986). Investigations relating maize seed quality to maturity has shown that maximal germination occur at high moisture levels early in development before the seed attains physiological maturity (TeKrony and Hunter, 1995). However, determining the optimum time of harvest for seeds that develop inside fleshy fruits is more complicated because seed maturation may precede fruit maturity (Coombe, 1976). In overripe tomato fruits, a high percentage of seeds may undergo precocious germination (vivipary) while in muskmelon fruits, seeds are either dead or had very low vigour if harvest was delayed until fruits start decomposing (Welbaum, 1993; Oluoch and Welbaum, 1996). When fruits are harvested prematurely, seeds require dry storage (after ripening), washing, or priming to improve vigour to the same level as mature seeds (Oluoch and Welbaum, 1996). Harvesting after fruit maturation and dry after ripening increased germination in various *Capsicum* species by overcoming seed dormancy. In fleshy fruits, the highest quality seed may be obtained from fruits harvested at edible maturity but before the onset of severe decomposition. Therefore, seeds must be harvested at the correct stage of development to obtain maximum viability and vigour.

Seed production in any one year is very dependent on weather conditions. For example, Table 1 shows the effect of the 1997-98 drought on the availability of seed in the local region. Much more seed was produced in the summer of 1996/97 than in the drought year 1997/98. As the best storage conditions for seed may be dependent on the length of storage required (Table 1) this has some practical implications. To ensure that, a seed supply is available every year it will be necessary to harvest more than is needed for any one year and store the surplus in cooler temperatures.

Table 1: Effect of the 1997-98 droughts on production of grass seed

Species	Localities	Flowers	Seed set
Austrodanthonia caespitosa	Yass	yes, but early	yes
Austrostipa scabra ssp. *Falcate*	Hume Barry Drive	yes	no
Bothriochloa macra	Paddys River area Canberra suburbs	yes	yes
Ehrharta stipoides	Yass "Tidbinbilla"	no	no
Joycea pallida	Aranda Black Mountain	about 1% plants flowered	v. low
Poa labillardierei	Smokers gap	yes	yes
Themeda triandra	Bruce "Tidbinbilla"	yes	no

Physiological causes for reduced storability may be ascribed to failure of accomplishing essential stages of late maturation events like incomplete embryo development , inadequate protection from desiccation or inadequate formation of storage proteins or chemical compounds necessary for storability (Hong and Ellis, 1990). The immature seeds to be more prone to processing damage (mechanical or heat) and may be more susceptible to infection (Schimdt, 2000). After physiological maturity, orthodox seeds can be harvested and after ripened under controlled conditions; while recalcitrant seeds tend to continue to accumulate drymatter upto full maturity and cannot be collected immature and after ripened (Schimdt, 2000).

Sasaki (1980) believed that immature dipterocarp seeds stored less than mature seeds and suggested wing browning should be used to assess readiness of seeds for harvest. In contrast, Panochit *et al.* (1986) showed that *Shorea roxburghii* seeds collected 2 weeks before seed fall stored much better than seeds collected at natural seed fall.

Sivakumar (2004) reported that in *Hopea parviflora*, immature seeds (3 weeks after anthesis) could be collected and stored to improve the storage life of these seeds. Mature seeds lose viability quicker than immature seeds. After 20 days of storage, mature seeds showed 32 % germination, while immature seeds showed 65% germination. In *Vateria indica*, there was no significant difference between mature and immature seeds on initial germination and storability. After 10 days of storage, mature and immature seeds showed 70 and 63 % germination, respectively (Table 2 and 3).

Table 2: Effect of seed maturity on germination characters of *Hopea parviflora* seeds stored at 10°C

Maturity status	Days of storage	Germination (%)	Germination energy (%)	Germination value
Matured (4 weeks after anthesis)	Initial	49.5 (44.73)	41.0	4.90
	20 DAS	32.0 (34.36)	23.5	3.10
Immature (3 weeks after anthesis)	Initial	25.5 (30.36)	21.5	1.17
	20 DAS	65.0 (53.75)	53.0	12.61
CD (P-0.05)	10.12			

Table 3: Effect of seed maturity on germination characters of *Vateria indica* seeds stored at 10°C

Maturity status	Days of storage	Germination (%)	Germination energy (%)	Germination value
Matured (10 weeks after anthesis)	Initial	88.0	72.0	10.05
	10 DAS	70.5	57.0	7.78
Immature (8 weeks after anthesis)	Initial	80.0	73.0	9.36
	10 DAS	63.0	48.0	6.45
CD (P-0.05)	NS			

Most of the recalcitrant seeds are big in size and do not have a hard seed coat. Hence, the seeds are damaged during fall. The damaged seeds are highly vulnerable for fungal infection and lose viability rapidly. The damaged seeds are highly prone to fungal infections. Minor damages to the seed coat serve as entry point for fungal attack which is a serious cause for seed deterioration under storage conditions where fungi are active i.e. at seed moisture content above 5.7% (Schimdt, 2000). Sivakumar (2004) reported that in *Vateria indica* there was no significant difference between mechanically damaged seed and control seeds on initial germination. But the storability of damaged seeds was poor. After 10 days of storage, the control gave 73 % germination while damaged seeds gave only 41% germination (Fig 5).

Irrigation should lead to more reliable seed supply. For example, the 100 seed weight for *Themeda triandra* collected on Mt Painter was 387 mg compared to 467 mg for seed collected from irrigated and fertilized plants at ANBG. However, some caution should be exercised, as there may be effects on the degree of dormancy in the irrigated seed lots that could be counter productive.

Environment during seed development including effect of soil moisture and fertility, and post-maturation/pre-harvest environment. Gusta *et al* (2003) found that Fall and April derived seed were considered 40% mature compared to 16% for May derived seed (Fig.6). and concluded that the earlier sown and thus earlier maturing crop avoided a critical period of heat stress later in the growing season.

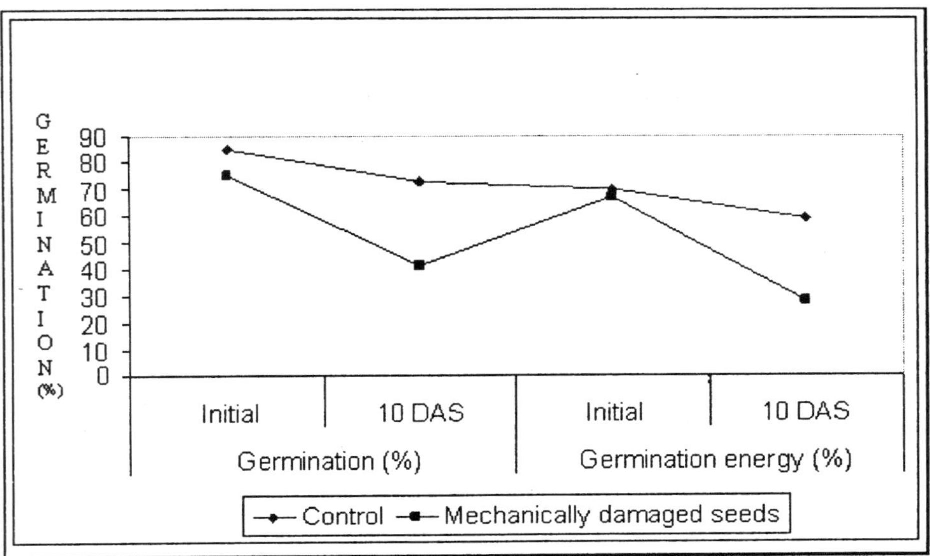

Fig.5: Effect of mechanical damage on germination characters of *Vateria indica* seeds stored at 20° C

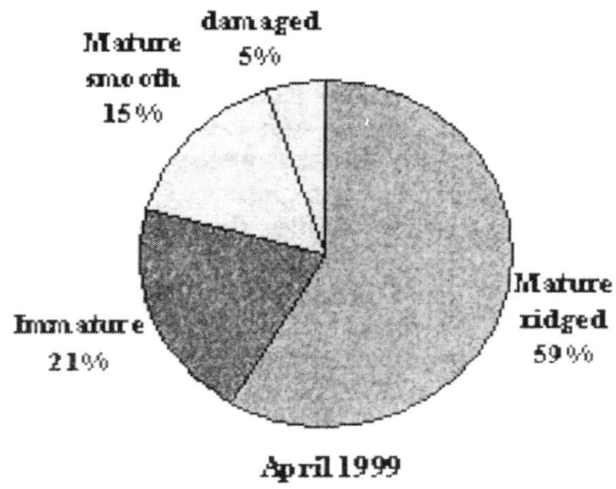

Fig.6: Effect of environment on seed quality on grass seed

Among various external and internal factors, seed development and seed position are the two intrinsic factors that determine seed performance (Rees, 1997). Initially seed performance is determined during seed development and maturation. Fruit development and seed maturation may occur independently. Thus seeds harvested at the same time can be at different developmental stages.

In chilli, seeds from basal region of the fruit exhibited higher germination and seedling vigour than those from tip and middle positions as reported by Doijode (1990).

Variation in performance of commercial lots of musk melon seed has been attributed to combining many individual fruit from different stages of development into a single lot (Oluoch and Welbaum, 1996). Development of seeds originated from early and late flowers on main raceme of *Brassica* was studied by Nanda *et al.* (1996) and reported that the rate of drymatter accumulation remained higher in late formed pods when compared to earlier seeds.

According to Sivasubramanian (1996), seeds collected from middle portion of the fruit recorded maximum vigour index than proximal and distal end in moringa. Murugesan (2003) concluded that the ash gourd fruits from the top nodes (> 46th node) could be utilized for seed purpose and from the bottom (27 – 30th node) and middle fruits (37th-40th node) could be utilized for vegetable purpose (Fig. 7; Plate 1).

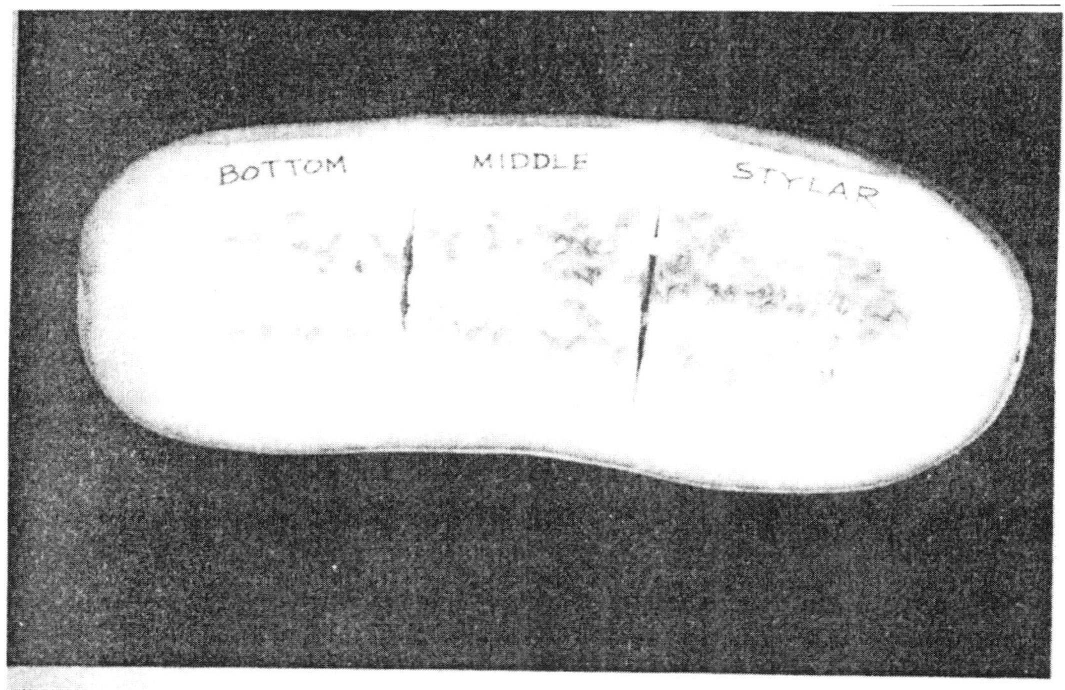

Plate 1: Peduncular, middle and stylar segments in the ashgourd cv. CO 2

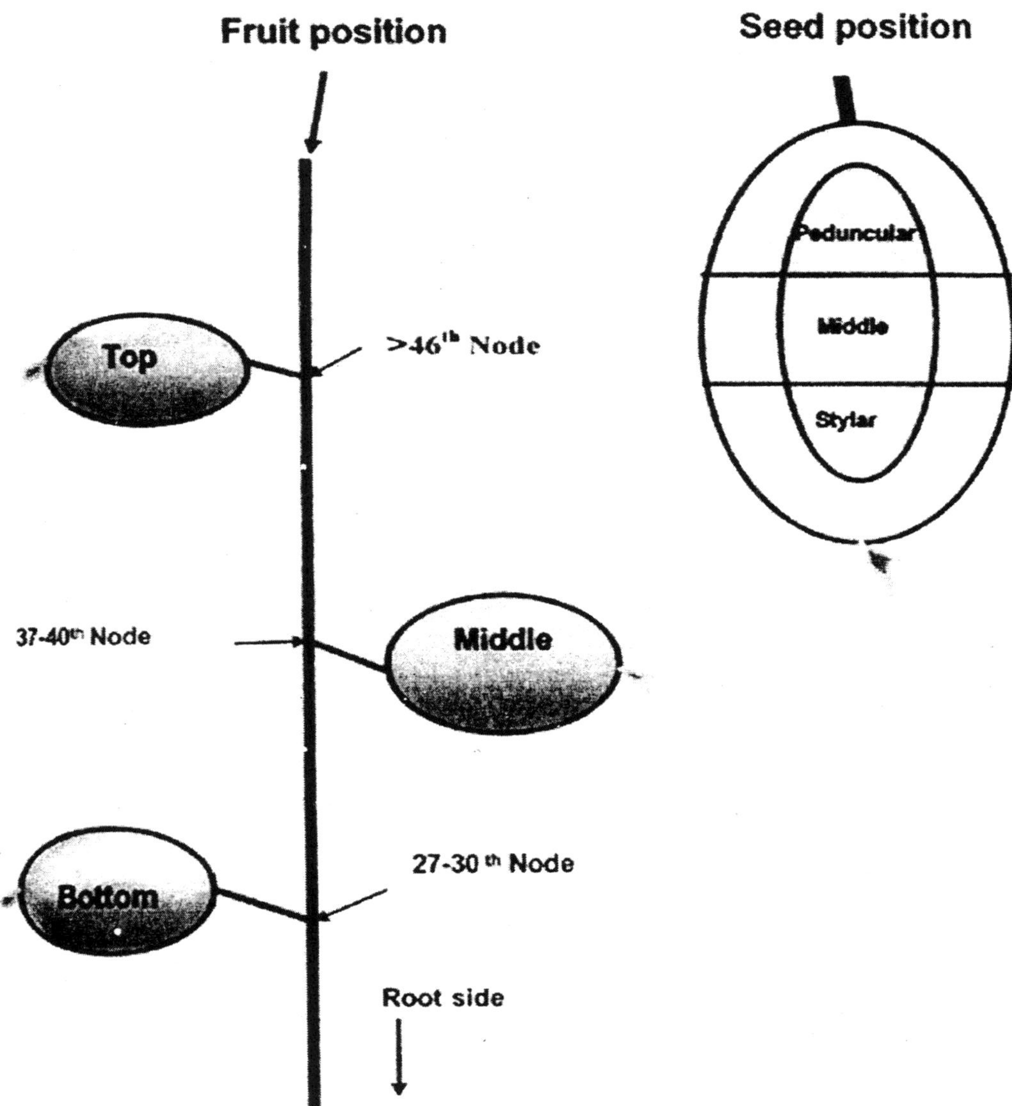

Fig. 7: Schematic diagram of ashgourd vine representing three fruit position viz., top middle and bottom and three seed position viz., peduncular, middle and stylar in the seed cavity

Ash gourd fruits can be harvested 60 DAA in summer season, when pedicel become dried or angular, and 70 DAA in rainy season, when fruits become thick with powdery ash and peduncular end of the fruit shoulder depressed fully. Leaving fruits beyond 80 DAA in rainy season invite fungal growth. During the final maturity stages, seed coat become dark brown and detached from the placenta, easily comes out from the pulp when cut open and scooped. At 70 DAA, almost all the fruit pedicel portion of summer crop became dried. Complete crop senescence occurred 170 days after sowing in summer, whereas in rainy season, vine and fruit attachment with plants existed beyond 180 days (Murugesan, 2003) (Fig.8)

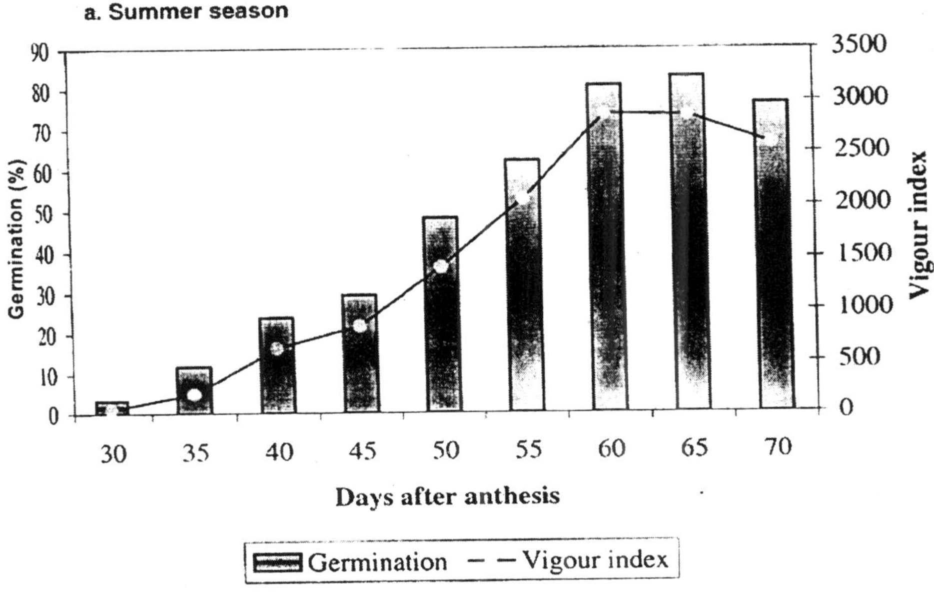

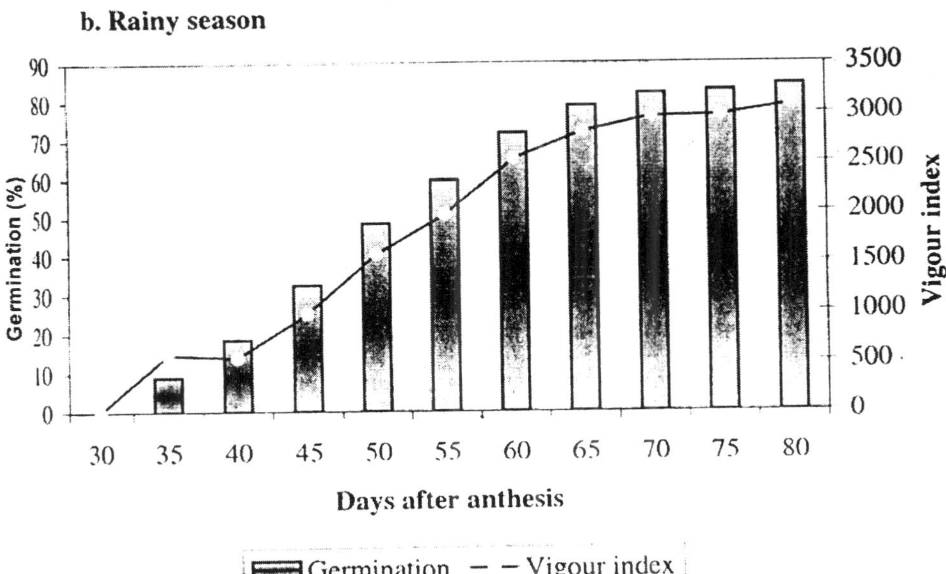

Fig.8: Germination and vigour index at different stages of seed development and maturation in ashgourd

Thomsen (2000) reported that ground should be cleared under the trees, covered by clean tarpaulins during collection to avoid inclusion of old seeds. Sivakumar (2004) revealed that in *Hopea parviflora*, collection of seeds from ground was found to affect the storability of seeds. Although initial germination was not affected, after 15 days of storage the crown collected seeds gave 21.5 % germination, while ground collected seeds gave nil germination (Table 4). In *Vateria indica*, the crown and ground collected seeds did not show difference in initial germination and germination after 10 days of storage.

Table 4: Effect of method of collection on germination characters of non desiccated *Hopea parviflora* seeds stored at 20°C

Method of seed collection	Days of storage	Germination (%)	Germination energy (%)
Ground collection	Initial	64.5	41
	15 DAS	0.0	-
Crown collection	Initial	53.0	34
	15 DAS	21.5	13
CD (P=0.05)		10.4	

5.3. Seed Storage

Seeds are usually stored at least for a brief period after harvest before planting; various factors like seed moisture, relative humidity of atmosphere, storage temperature, type of seed, physical condition of seed; microflora and insects influence seed viability and vigour.

Powell and Matthews (1978) suggested that the decline in vigour during periods of warehouse storage is an important factor responsible for the existence of some of the low vigour seed lots seen in peas. Evaluation of the storage potential of seed lots by use of a laboratory storage test such as conductivity measurements of leachates following 1 day in 94% RH at 45°C, would enable prediction of seed storage.

Seeds of lettuce (*Lactuca sativa* L.) and sunflower (*Helianthus annuus* L.) were stored hermetically at 35°C with 11 different moisture contents between 1·3 and 6·9%, and between 1·3 and 7·1% of fresh mass, respectively. Germination and vigour (mean germination time, root length, seedling dry weight) were determined after storage for 0, 8, or 16 weeks (sunflower) or 0, 8, 16, or 48 weeks (lettuce) in these environments followed by various humidification treatments (to avoid imbibition injury). The range of seed storage moisture contents over which deterioration was minimized depended upon the criterion of deterioration used, and varied somewhat between species. Comparison of these ranges for seeds stored for the longest durations showed that for some criteria seed performance was poorer ($P < 0{\cdot}05$) at both the lowest and highest moisture contents investigated than at certain of the intermediate storage moisture contents (e.g, most rapid germination occurred in sunflower following storage at 2·2-4·7% moisture content), whereas for other criteria all the drier storage moisture contents were superior to the more moist (e,g. greatest seedling growth occurred in sunflower following storage at 1·3-5·1% moisture content). But none of these results suggested that lettuce and sunflower seeds stored hermetically at 2·5-3·0% or 2·2-2·5% moisture content, respectively, were less vigourous than at any other moisture content tested. In both the species, these storage moisture contents are in equilibrium with about 8-10% relative humidity (R.H.) at 20°C, which is similar to and indeed marginally less than the 10-13% R.H. recommended following earlier studies on the longevity of seeds in hermetic storage at much warmer temperatures. Thus, these results show no evidence that the optimum seed moisture content for storage increases with decrease in temperature, at least over the range 35-65°C, as has been suggested elsewhere. We conclude that the international recommendation for the long-term seed storage for genetic conservation at 5 ± 1% moisture content should not be revised upwardly, and that in situations where refrigeration cannot be provided storage at even lower moisture contents is worthy of further investigation for those seeds in which desiccation at 20°C to equilibrium at 10% R.H. results in moisture contents well below 5% (Ellis *et al.,* 1995)

Veselova *et al.* (1998) suggesting that the loss of seed vigour during their ageing (and gradually accumulation of defects) seems to be a stepwise rather than a gradual process (viable seed with high vigour --> viable seed with low vigour --> nonviable seed).

Traditional assays (germination and vigour), as well as biochemical assays using isozyme loci, were performed to evaluate the effect of different seed storage conditions in *Coincya rupestris, Iberis pectinata, Moricandia moricandiodes* and *Vella pseudocytisus.* Seeds had been stored for 24–30 years under long-term (LT) and short-term (ST) conditions. Fresh seeds of the same species from natural populations (NP) were also collected in 1996 and used for comparison. Isozyme loci were used as markers of genetic deterioration and to determine the genetic diversity of samples. The germinability, vigour and enzyme activity of LT samples were comparable to values obtained using freshly collected NP samples, showing the effectiveness of the long-term storage conditions (Masselli *et al.,* 1999).

The influence of the storage environment on seed viability and antioxidant potential was examined for four species native to Western Australia: *Acacia bivenosa* DC., *Anigozanthos manglesii* D. Don, *Banksia ashbyi* E.G. Baker, and *Mesomelaena tetragona* (R. Br.) Benth. Seeds were stored at four water contents (at moisture content 5%, 11–15%, 20–23% and 50% relative humidity) at each of five temperatures (–196, –18, 5, 23 and 50°C), and seed germination and seedling vigour monitored over an 18-month period. Deterioration was apparent in all species (except *A. bivenosa*) stored at 50°C, with 11% RH maximizing longevity for *B. ashbyi* and *M. tetragona* seeds, and 5% or 11% RH preventing deterioration for *A. manglesii* seeds. Seed viability generally remained high for all species stored at 23°C or less. Notably, however, germination and seedling vigour of *A. manglesii* and *M. tetragona* seeds gradually declined when stored at –18°C, suggesting that storage at this temperature was detrimental. The antioxidant activity of lipid extracts of seeds after 18 months storage at 5, 23 and 50°C was also examined to determine whether the seed viability decline was associated with a loss of antioxidants. Antioxidant activity varied between storage treatments and was not related to seed viability (Merritt *et al.,* 2003).

In comparisons of six seed lots of different F_1 hybrid cultivars of cauliflower with similarly high laboratory germinations (above 90%) separation in germination was achieved after controlled deterioration (C.D.) at 24% moisture content (m.c.) and 45°C for 24 hours. This measure of vigour was related to the position of the lots on the seed survival curve and was highly predictive of the longevity of the lots when stored at 15% m.c. and 20°C for 12 and 16 weeks. When each seed lot was deteriorated at 24% m.c. for increasing times (from 0 to 36 hours) a reduction in the subsequent percentage germination was seen, which, using probit transformed percentages, was significantly and linearly related to the leakage of electrolytes into seeds soaked water over 24 hours. The case is made for an approach to the indirect evaluation of germination and vigour using C.D. followed by measurements of leakage that could be more discerning and rapid than the present laboratory germination test.

The progress of natural ageing with resultant loss of viability is not linear over time, but typically follows a sigmoid pattern. The loss of viability is initially slow, followed by a period of rapid decline. The higher the initial viability of the seed lots that enters into storage, the longer the seed viability under a given storage environment (Schimdt, 2000).

Sivakumar (2004) reported that seeds with low initial viability had short storability when compared to seeds with high initial viability. Moisture content of seed also varies from seed lot to seed lot. Seeds with high initial viability and low moisture content showed better storability. In *Vateria indica,* initial viability and moisture content affected the storability. Seeds with high initial viability and low moisture content were viable longer than seeds with low initial viability and high moisture content. Seed with 93 % initial viability and 45.42 % moisture content gave 70 % germination after 10 days of storage, while seeds with 62 % initial viability and 68.15% gave nil germination after 10 days.

Storage of *Hopea parviflora* seeds at 10° C was found to maintain the viability of seeds. Seeds stored at ambient temperature were viable only for less than 10 days. Seeds stored at 10° C gave 16.5 % germination after 60 days of storage. Storage of *Vateria indica* seeds at 20° C was found to improve the storability of seeds. After 10 days of storage, the ambient stored seeds gave 55 % germination, while seeds stored at 20° C gave 70 % germination.

6. EVOLUTION OF SEED VIGOUR TESTING

It was Fredrich Nobbe (Nobbe, 1876) during 1876 who first distinguished the concept of seed vigour from that of germination. He introduced the term *triekraft*, which meant shooting strength or driving force to convey the idea that, in addition to germination, speed and uniformity of emergence were important parameters of seed quality. Frank (1950) pointed out the differing concepts of germination testing between European and American laboratories. In Europe, germination tests were conducted under optimum and reproducible conditions, while seedling vigour was estimated by brick grit and soil tests. The American concept of germination was based on soil test results alone. Frank proposed that germination testing should be conducted under favourable conditions in order to obtain uniform test results. Further he defined his new term "vigour-as the plant producing ability of a seed lot in the field". Subsequently, many definitions have been proposed for seed vigour and the publication of two articles (Isely 1957 ; Delouche and Caldwell, 1960) directed to refocus on seed vigour testing. Isley (1957) defined seed vigour; as the sum of all the seed attributes, which favour rapid and uniform stand establishment, duly deleting the under favourable conditions' part of the Isley definition. The concept of seed vigour also states that the plant stand should succeed under a wide range of field conditions. However, the other issue of the definition i.e., sum of all seed attributes has been addressed (Woodstock, 1965) to achieve clarity, proposing seed vigour 'as the condition of seed health and natural robustness in seed, which, upon planting, permits germination to proceed rapidly under a wide range of environmental conditions'. Later, it was emphasized that the seed vigour was determined by both genetic and environmental components and defined it as the 'physiological property determined by the genotype and modified by the environment, which governs the ability of the seed to produce a seedling rapidly in soil and the extent to which the seed tolerates a range of environmental factors' (Perry, 1973). By this time, consensus was rapidly emerging on a definition of seed vigour.

The next major advance occurred in 1976, with development of standard seed vigour test procedures by Association of Official Seed Analysts (AOSA), which were resolved by vigour test committee of AOSA (1991). Subsequently, under the leadership of Dr. Woodstock, seed vigour testing progress report was published with details of developing, evaluating, codifying and standardizing vigour testing procedures. This document was a significant milestone and provided specific guidelines for the conduct of eight vigour tests. While, the International Seed Testing Association (ISTA) defined vigour as, the sum total of those properties of the seed which determine the potential level of activity and performance of seed or seed lot during germination and seedling emergence (Perry, 1978).). The AOSA vigour committee defined seed vigour as, "those seed properties which determine the potential for rapid, emergence and development of normal seedlings under a wide range of field conditions" (McDonald, 1980b). So by 1980, seed vigour definition has been approved by every major organization involved with the commerce and testing of seeds. AOSA committee also published AOSA vigour testing handbook in 1983 with the procedure of vigour tests (AOSA, 1983). The tests needed to stand the scrutiny of routine use before they could be declared as standardized. By 1987, accelerated ageing test was significantly revised and improved and became the first vigour test for soybeans. Utilization of vigour tests for seed quality assessment, however, experienced several difficulties: 1) mainly lack of standardization of the methodologies used, 2) the subjective interpretation of results and 3) discrepant results among the laboratories. In 1978, 52% of seed testing laboratories utilized vigour tests and by 1983, 60% and by 1990, 75% of the seed testing laboratories adopted one or more vigour tests (17). This indicates that the International Seed Testing Organization Committees, besides providing vigour testing protocols also educated the users of its value and limitations.

7. CHARACTERISTICS OF A VIGOUR TEST

The essential characteristics that a vigour test should possess have been described as

i) should be inexpensive

ii) should be able to conduct rapidly to minimise analyst time and germinator space

iii) should be simple and should be able to perform with out requiring additional staff with special background and training

iv) should be easily reproducible

v) should be able to predict field performance

8. METHODS OF MEASURING SEED VIGOUR

The challenge of vigour testing has been to identify a quantifiable parameter which is common to seed deterioration. The most neglected area of seed study remains in the physiological and biochemical characterization of seed deterioration. The principal reason for this lack of knowledge is that many events can create and contribute to loss in seed viability and it becomes difficult to identify the cause and effect of a specific deteriorative response. Although it is impossible to state with precision which events occur during seed deterioration, we can speculate on the probable sequence of changes. Such a hypothetical model (Fig. 9) has been developed by Delouche and Baskin (1973) and outlines some of the major parameters which can be utilized in measuring seed vigour.

Because a vigour test is a more sensitive index of seed quality than the germination test, any of the events which precede loss of germination could serve as vigour tests. The farther removed the parameter is from loss of germination, the more sensitive the measure of seed vigour. Thus, according to figure 5, the most sensitive vigour test might be one which monitors membrane degradation.

Membranes are essential for many metabolic events including respiration (cristae in mitochondria) which provides the seed with the energy molecules required for subsequent growth. The endoplasmic reticulum is also a membrane organelle on which many enzymes are formed as ribonucleic acid is translated. Thus, any impairment of membrane function can decrease the amount of ATP formed as an energy source, as well as retard the synthesis of specific enzymes essential for growth. Koostra and Harrington (1969) showed that membrane degradation occurred in deteriorated seeds, and this observation has been supported by others (Streeter, 1965; Villiers, 1973; Spencer *et al.,* 1973;). Beyond these studies ATP (Ching, 1973), RNA (Van Onckelen *et al.,* 1973) and respiration (Woodstock and Grabe, 1967) levels have been shown to decrease following accelerated ageing or natural ageing conditions. Subsequent to the loss in respiration and biosynthetic capacity, the rate of germination declines culminating in a loss of seed lot uniformity. Other associated events which occur during deterioration are the loss in storage capacity and the ability to resist to disease.

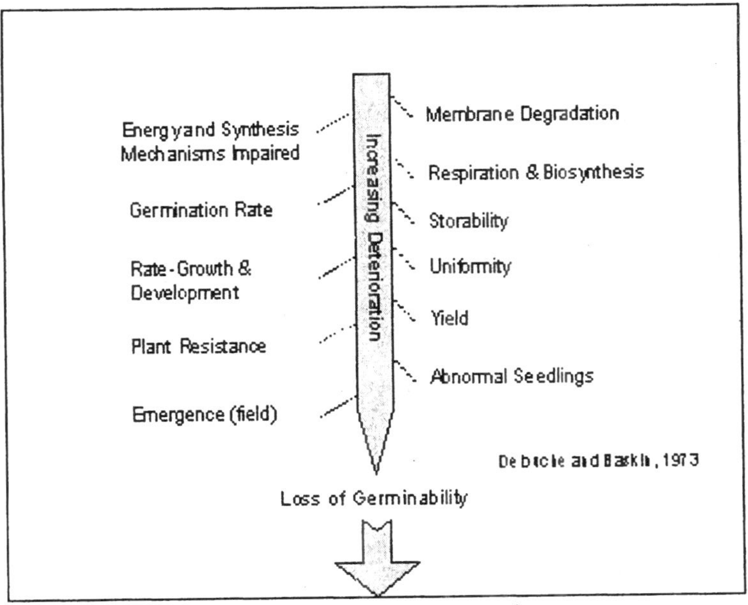

Fig. 9: Sequence of changes in seed during deterioration

Deteriorated seeds which are subjected to biological and environmental stresses also exhibit reduced field emergence which may or may not (Tekrony and Egli, 1977) be related to final yields of certain crops. Eventually, these subtle manifestations of loss in seed quality are expressed by an increasing incidence of abnormal seedlings – a component of the germination test. The final parameter and the most often employed – is seed germination underscoring the need for seed vigour tests to supplement present seed quality analysis.

As a result of these documented biochemical and physiological changes, which occur during seed deterioration, seed vigour tests have focused on measuring one or more of these parameters. Various vigour test classification systems therefore have incorporated a strategy, which separates seed vigour tests into groupings based on the parameters monitored. For example, Isley (1957) divided vigour test into direct and indirect tests.

9. KINDS OF SEED VIGOUR TESTS

The standard germination tests conducted under optimum conditions for seed germination correlate with field emergence, if the conditions at planting are near optimum (Perry, 1977: Egli and TeKrony, 1979; Luedders and Burris, 1979). However, under suboptimal field conditions, standard germination results usually over estimate field emergence (Delouche and Caldwell, 1960: TeKrony and Egli, 1977: Johnson and Wax 1978: Yaklich et al., 1979). Where as vigour tests predict seedling emergence under a wide range of field conditions.

Though many vigour tests have been suggested only a few have attained acceptance by seed analysts and seed testing organization.

9.1. Vigour tests based on physiological effects of ageing

9.1.1. Seedling growth rate test

Vigourous seeds efficiently synthesize new materials and rapidly transfer to the emerging embryonic axis, resulting in increased dry weight accumulation. The seedling growth rate test is based on this concept. Since seedlings growth rate is well correlated with vegetative development in field (Burris, 1976 and Pinthus and Kimel, 1979) this test offers substantial promise. However, small differences in seed moisture and light intensity significantly affect the seedling growth rate. This test has to be standardized for different cultivars of the crop as the seedling growth is genetically controlled (Burris, 1975).

9.1.2. Speed of germination

Speed of germination is one of the oldest seed vigour concepts. Seed lots with similar total germination can be further categorized into different vigour groups based on the rate of germination and growth. The rate of germination can be estimated by different methods. The simplest assessment of the rate of germination can be made from the first count or preliminary count in the germination test. First count was found to provide a good estimate of seedling vigour in Soybean (Burris *et al.,* 1969) and has been used as a vigour test (Perry, 1977). Number of days a lot requires to reach 90% germination (Belcher and. Miller, 1974) for high vigour seed and 50% germination for low vigour seed lots were also used as an index of vigourous seed. Maguire (1962) suggested the following formula for vigour rating.

$$\text{Vigour rating (X)} = \frac{\text{No. of normal seedlings}}{\text{Days of first count}} + \frac{\text{No. of normal seedlings}}{\text{Days of final count}}$$

Many more methods were used to calculate speed of germination, of all Maguire method was the most accepted. The advantage of this test is that it requires only little additional work compared to standard germination test. However, variation in temperature and moisture in the substrate and the test chamber should be avoided for good results.

9.1.3. Seedling vigour classification test

This test is in continuation to the standard germination test. The normal seedlings are further classified into strong and weak seedlings. Though the test is simple and easy it represents additional sources of variability. For this reason seed testing organization maintain active referee programme to help laboratories to identify their interpretation difficulties. However, this procedure needs standardization.

9.1.4. First count

The number of normal seedlings recorded in the first count represent the population of fast germinating seeds and thus functions as a vigour classification test with the exception that the strong normal seedlings in the second count are excluded. In soybean, the first count (four days) was found to provide an estimate of seedling vigour (Burris *et al.,* 1969) and has been used as a component of vigour index. It has been reported that foliage development, dry matter accumulation and seed yield from rapidly germinating soybean seeds exceeded those from slowly germinating soybean seeds from green house and field conditions. The advantage of the test is that very little work is required when compared to the standard germination test. Variation in temperature and moisture in the test chamber and substrata, however, may affect test results.

9.1.5. Mobilization efficiency (ME)

Important biochemical processes that are associated with deterioration occur in seed before the decrease in germinability. The formula suggested for the estimation of mobilization efficiency is as follows:

$$M.E(\%) = \frac{\text{Increase in dry weight of embryonic axis (Dry weight of seedlings)}}{\text{Decrease in dry weight of two cotyledons}} \times 100$$

Where decrease in weight of cotyledons =Actual dry weight of seeds in the beginning – dry weight of two cotyledons + dry weight of seed coat at the end of the experiment and the assumption in this is that the initial weight of embryonic axis is negligible. Mobilization efficiency test must be performed under complete darkness. It involves direct interpretation of overall biochemical and physiological processes and seems to be the most accurate index for detecting deterioration (Ramamoorthy and Karivaratharaju, 1989) thus the loss of viability and vigour in seeds.

9.1. 6. Stress tests

Stress tests subject seeds to the type of stresses they encounter in the soil during imbibition and germination. The ability of seeds to germinate under such conditions is used to assess seed vigour.

a) Cold test

The cold test is one of the oldest methods of stressing seeds and is most often employed for evaluating seed vigour in corn, cotton, sorghum and soybeans (Fiala, 1981). The test results of a cold test are usually expressed as the percentage of normal seedlings produced. Besides some difficulties *viz.*, lack of uniformity in field soil, moisture, pH and pathogen levels, the results of cold test remain consistent with in laboratories.

b) Brick gravel test (Hiltner test)

Brick gravel test provides mechanical stress to the germinating seeds and has been applied to a wide range of crops viz., cereals, large legumes and vegetables (Fuchs, 1981). Actually the test was developed for detecting seed borne (*Fusarium*) infection in some areas (Hiltner and Ihssen, 1911). Results of these studies indicated that this test was able to detect seed weakness other than those caused by fungi such as injury caused by frost, preharvest sprouting

(Schoorel, 1960) and hot-water treatment (Tempe,1963) which makes it useful as a vigour test.

c) Cool germination test

This test is to be conducted under standard laboratory condition at low temperature (18°C). This test is currently limited to use in warm season crops like cotton whose growth rate decrease with low temperature. The advantage of this test is that is similar to standard germination test and the same criteria for interpretation of normal seedlings is to be employed.

d) Osmotic stress

In osmotic stress test, the seeds are exposed to simulated drought stress conditions. Though many solutions are used for this stimulation, high molecular weight PEG (4000 or more) was considered satisfactory for simulating true drought (Manohar, 1966 and Parmer and R.P. Moore, 1968). The rate of germination under such conditions is markedly reduced and only the vigourous seeds can tolerate such great osmotic stress. This test does not require any equipment of training. However, this test did not stand as there was significant interaction between osmotic stress and temperature of germination (El-Sharkawi and I. Springuel, 1977).

9.2. Vigour tests based on biochemical changes during ageing

9.2.1. Conductivity test

This test is based on the leakage of solutes such as sugars, aminoacids and electrolytes from seeds into water. These leachates are detected by measuring the electrical conductivity of the steep water. This test is rapid, precise, inexpensive and simple. However, initial seed moisture (Simon and Wiebe, 1975) and seed size (Tao, 1978) can affect the electrical conductivity of leachates from seeds. Seed treatment chemicals, fungicides and antibiotics had to be removed before conducting this test.

In the routine EC test the value is the average for 50 seeds. Such expression presumes that all seeds are equally deteriorated and at times may not represent the seed lot. A seed lot, however, is composed of population of individual seed each with its own unique potential to perform in the field. A commercial instrument, automatic seed analyzer, is now available to monitor the electrolytic leakage of individual seeds. This instrument indicates whether high bulk leachate conductivities are due to the presence of a few badly damaged seeds or leakage from a whole population of deteriorated seeds. Recent studies suggest that it provides a more accurate appraisal of seed vigour in soybean (Mc Donald and Wilson, 1979 and Miles and Copeland, 1980), cotton (Hopper and Hinton, 1980), cowpea (Beighley and Hopper, 1981), navy bean (Suryatmana *et al.*, 1980) and corn (Joo *et al.*, 1980).

9.2.2. Tetrazolium (TZ) test

The TZ test is one of the most valuable techniques for analyzing seed quality and potential viability. It was developed by Lakon (1942) relies on the action of the TZ molecule to react with hydrogen atoms released as a result of dehydrogenase enzyme activity in living tissue. This results in the formation of a water insoluble red pigment called formazan. Living tissues therefore, appears red whereas dead cells remains colourless. Moore (1973) described the use of Tetrazolium test, the trained seed analyst evaluates for staining pattern and colour intensity. The analyst then subjectively places the seeds into vigour categories ranging from strong to weak. Though this test correlates well with seed vigour when interpreted by a qualified analyst, it is still subject to certain standardization difficulties viz., the ability of the analyst to ascertain whether a seed is vigourous and the failure of the test to detect seed treatment phytotoxicity and reveal seed dormancy.

9.2.3. Respiration

Seed germination and seedling growth require use of metabolic energy acquired from respiration. A decrease in the rate of respiration of germinating seeds has been showed to precede a decrease in the rate of seedling growth. Respiration rate measured during early hours of germination can detect injury caused by Gamma radiation. Significant positive correlation were observed between the rate of O_2 uptake during imbibition and seedling growth. However, other studies indicate that respiration does not always correlate with vigour (Byrd and Delouche, 1971; Bonner, 1974). Respiration tests are rapid and quantitative but require a respirometer and trained personnel. The mechanical injury, which lowers seed vigour, may increase rather than decrease respiration rates.

9.2.4. Glutamic Acid Decarboxylase Activity (GADA)

During seed germination, proteins are hydrolysed into amino acids by proteolytic enzymes. Amino acid decarboxylases are widely distributed in seeds. Glumatic acid decarboxylase activity is particularly important in seed germination as glutamic acid comprises a high percentage of total amino acids in seeds. Burris *et al.* (1969), reported little relationship between GADA and seedling vigour in soybean. In contrast, GADA was found to remain high when germination decreased in soybean seeds (James, 1968). A seed size effect in peanut with reference to GADA, was also observed (Baskin and Delouche, 1971). GADA is normally measured by the amount of CO_2 produced in the presence of glutamic acid.

Other enzymes like catalase (Leggatt, 1929-30) and peroxidase (McHargue, 1920) have been demonstrated to be related to seed germination, seedling growth rate, storability and yield. These are very sensitive indicators of physiological deterioration in storage.

9.2.5. Adenosine triphoshpate [ATP] content

ATP is the biological energy needed for every biosynthetic pathway as well as biological work. During germination, the embryo of a seed not only acts as the source for enzyme, the substrates, and co-factors for de novo synthesis of ATP (Anderson, 1970; Ching, 1972), but also has the salvage enzymes to convert. During seed formation, ATP content and Energy charge of seed increase, in concern with the biosynthetic needs for structural growth, cellular enzymes and organelles in the early stages and for food reserves at the latter stages (Ching *et al.,* 1977).

Cellular energy status expressed as ATP pool size and energy charge (EC) will be

$$Ec = \frac{ATP + 1/2\ ADP}{ATP + ADP + AMP}$$

The higher the energy status (for biosynthesis and biological work for the normal growing tissue) is associated with larger pool size with a high electrical conductivity.The high vigour wheat exhibited significantly higher activity of alpha amylase glutamin synthetase and acid phosphotase and had larger pool of ATP. When seed vigour was decreased by *Aspergillus rubra* infection, the ATP content of infected pea seeds was lower than that of non infected pea seeds.

9.2.6. Mitochondrial efficiency

Studies of the ultra structure of seeds have shown that mitochondria becomes organised and apparently active during germination (Bain and Mercer,1966; Ovcharov *et al.*,1970). In the early phases of germination there appears to be mitochondrial protein synthesis, and the ratio of mitochondrial RNA to mitochondrial protein increases (Mc Daniel, 1969). These increases may be accompanied by the increase in the number of mitochondria. Abu-Shakra and Ching (1967) studied mitochondrial activity in old and new soybean seeds, and found that isolated mitochondria of old soybean exhibited reduced phosphorylative efficiency as compared to vigourous (new) seed.

9.2.7. Glucose utilization

The utilization (in contrast to absorption) of externally supplied glucose by wheat and barley grains (Abdul - Baki, 1969) and groundnut (Ramamoorthy and Karivaratharaju, 1985) of different physiological ages were traced. The authors suggested that it provided and index of deterioration well ahead of measurement of seedling shoot growth, respiration and germination processes which lag behind insensitivity in increasing order. This decline in synthetic ability is therefore, an early danger signal which indicates that seeds are beginning to go downhill.

9.3. Vigour tests based on whole ageing process

9.3.1. Accelerated ageing test

Accelerated ageing test was developed by Delouche and Baskin (1973). This test was based on the fact that temperature and moisture content are two factors that determine seed ageing. If either of these factors is raised, the rate of ageing increases.

This test is rapid, inexpensive, simple, and useful to all species and requires no additional training for correct evaluation. However, the initial seed moisture should be considered when interpreting this test (McDonald Jr., 1977).

9.3.2. Controlled deterioration test

This test was proposed by Matthews (1980) and was however, one of the most promising vigour tests (Hampton and Coolbear, 1990). This test like accelerated ageing test increases the rate at which seeds age. The moisture content of the seed will be raised (around 20%) and then will be exposed to high temperature (45°C) for 24 hrs. After which normal germination test will be conducted. The results of the controlled deterioration test correlate well with both the field emergence and the storage potential of many vegetable seeds (Matthews, 1980).

9.4. Vigour test based on physical tests

9.4.1. Seed size test

Numerous studies have shown that seed size can affect yield (Kneebone and Cramer, 1955). Measurements of seed size can readily be adapted to normal seed testing laboratory routine and personnel. Seed size may be measured in a number of ways. One is to count the number of seeds per gram. A second is to determine the weight of 1000 seeds. Electronic counting equipment facilitates, both of these measurements. A third method is to determine the relative size of seed. This, may also be done in a number of ways. One is to determine the number of seeds in a certain volume or conversely, determine the volume occupied by 1000 seeds. Test weight is also related to these types of measurements.

Additional quality information may be obtained through the use of a set of hand screens. By making a nest of several screens, a seed sample may be divided into several size categories based on the size of the screens. The seed size distribution within the lot is then determined. Seed lots with the larger proportion of seeds in the large seed categories (seed recovery) are frequently superior to lots with the largest proportion of seeds in the small size categories. Information obtained from seed size determinations are useful in making planting rate recommendations, processing recommendations, and indicating potential yielding ability of seed lots.

Variation in seed size can affect vigour test evaluations. In large seeded kinds like beans, the smaller seeds in a lot may germinate more rapidly than the larger seeds, making them appear to be more vigourous in tests where length of root and / or hypocotyl is used to determine seed vigour. Greater emergence of small seeds of barley, black gram, greengram, cluster bean, maize, pigeon pea, soybean and triticale have been reported (Haskins and Gorz, 1975). McDonald (1975) added one additional grouping, a physical category which included seed size, shape, and density factors long associated with seed vigour because of their indication of seed maturity. Large seeds, on the other hand, usually produce heavier seedling making them appear more vigourous in tests where seedling weight is used as a measure of vigour.

9.4.2. Radiographic analysis of seeds

The X-Ray photographic method for analysing seed quality developed by Simak and Gustafsson (1953) is based on the principle that the different parts of a seed, such as testa, endosperm and embryo absorb soft X-rays to different extent and can consequently be differentiated on X-ray photograph. A clue to the vigour of a particular seed lot could be obtained from the impregnation pattern (both intensity and extent of necrotic areas). Since the extent of impregnation of contrast agent is directly proportional to the vigour of the seed tissues, it is possible to standardize seed-vigour attributes. A large number of seeds could be correlated with the extent of seedling vigour (Ramamoorthy and Karivaratharaju, 1991).

10. STANDARDIZATION OF SEED VIGOUR TESTS

Vigour testing of seed is surrounded by much confusion and controversy. The confusion is mostly associated with the multitude of vigour tests developed and promoted during the past 25 years, the lack of workable definitions of seed vigour and vigour tests, and excessive variability in vigour test results and their interpretation. The controversy in seed vigour testing derives from the fear of some and the hope of others that vigour tests will become more widely recognized as a major means of establishing the planting value of seed, and permitted or required in seed labelling.

10.1. Aspects of Standardization

Standardization of the two paramount aspects of seed vigour tests, viz., methodology and interpretation, is crucial for their routine and meaningful use. Although these two aspects are treated separately here, it is for convenience only. Test methodology, test results and their interpretation are so inextricably linked that a vigour test really has no meaning unless the results obtained can be interpreted in terms of a probable level of "seed performance" in storage or in the field. At least part of the confusion surrounding vigour testing arises from a conscious or unconscious disassociation of vigour test results and economically important-in terms of crop production-aspects of seed performance. Therefore, in considering the standarization to vigour test methodology and interpretation separately assume that the vigour tests considered are atleast generally relevant to important attributes of performance, and that the major concerns are uniformity and reproducibility of results, and their "practical" interpretation.

10.2. Levels of standardization

Vigour testing is or can be used for a variety of purposes ranging from educational campaigns for contract growers to identification of superior quality grades in seed marketing. The level or degree of standardization needed in vigour testing and the ease with which it can be achieved is largely determined by testing objectives or purposes.

The technology of vigour testing has been effectively used for over 25 years in the quality control laboratories of the larger seed companies. Obviously, these laboratories have been able to achieve an in-house level of standardization in vigour testing adequate to obtain the desired information. Many other seed companies with limited or no research and development capabilities have taken advantage of the considerable body of information on vigour testing published in the last 10 years and have established vigour testing in their quality control programs.

Standardization of vigour tests, or rather the lack thereof, is not a major constraint to their use in in-house quality control programs (James C. Delouche, 1976). Inadequate training, facilities and staff, poorly informed management, and improper use of the information derived from vigour tests are much more serious problems. In-house application of vigour testing technology and routine vigour testing, however, are quite different matters, especially when objectives of the latter are quite different maters, especially when objectives of the latter are to establish and promote a quality claim, or to verify such a claim. A much higher level of standardization, hence, uniformity in tests results and their interpretation, than presently realized must be achieved before vigour tests can be used in a routine and orderly manner in seed labelling and inspection. The challenge before us, therefore, is to accomplish the things that need to be done to extend the uses of vigour testing from the quality control laboratory into the traditional seed testing arena. These tasks will be easier if we recognize and take maximum advantage of the vast experience gained in in-house application of vigour testing technology.

11. CONTROL SAMPLES IN VIGOUR TESTING

An essential component of any laboratory testing program is the use of control or check samples. The primary purpose of control samples is to provide internal quality control to enable seed technologists to detect variations in test procedures. Seed vigour tests require precise time, temperature, and moisture that must be monitored and controlled during testing. Even small variations in these components can produce dramatic fluctuations in test results. The selection, maintenance, and use of control samples for each species tested is essential for laboratories engaged in vigour testing.

11.1. Control seed lot selection

Seed lots of each crop species tested should be selected annually by screening several samples for both germination and vigour. The sample selected should be relatively free of physical injury and disease, high in germination and moderate in vigour. Seed lots with extremely high vigour will not allow detection of minor differences in methodology and, therefore, would not make good control samples. For example, an acceptable control sample for corn or soybean may have a germination exceeding 95 or 90%, respectively, and vigour level of 70 to 80% and 80 to 90% for the accelerated ageing test for soybeans and the cold test for corn, respectively.

11.2. Seed storage and maintenance

Store enough seed of the control sample to provide an adequate supply for an entire year or testing season. Store the control samples in conditions that will preserve both germination and vigour. Prepackaging the control sample into sub-samples needed each day may be advantageous, removing from storage the number of samples needed for a week of testing. Two alternatives should be considered for storage and use of control (in order of preference):

- Divide the control seed lot into many sub-samples of ~500 seeds. Each sub-sample should be packaged in moisture-proof foil packets (heat sealed) or stored in a moisture-proof container at low temperatures (0 to -20 C) prior to testing. Control sub-samples must be allowed to equilibrate to room temperature (4 to 6 hours) prior to use.
- If it is not possible to handle the small control samples described above, subdivide the control sample into larger sub-samples. Place these samples into moisture-proof containers and store at low temperatures (0 to 10 C) prior to testing. Remove small sub-samples from the larger samples as needed for testing, place in sealed containers and allow to equilibrate to room temperature before use.

11.3. Seed moisture of control samples

The initial seed moisture of the control seed lot must be an acceptable seed moisture for the seed being tested. The moisture of the seed lot must be measured periodically to be certain that the seed has not decreased or increased in moisture during storage. Adjust the moisture to the initial level if necesary, as extremely dry seed can cause imbibitional injury and inaccurate vigour readings during testing.

11.4. Use of control samples

Control samples should be included as routine samples in every vigour test run and should not be identified as control or check seed until the test is completed. At the conclusion of the test run, the seed technologist or laboratory manager should determine if the vigour test results of the check are within the set parameters. If the sample falls outside the parameters, the problem should be identified and corrected and the test repeated. At frequent intervals, the vigour test results of the control samples should be summarized to evaluate uniformity of the samples and testing procedures.

Clearly, vigour testing for all species and regions has not yet achieved the same level of standardization possessed by the standard germination test. However, both seed testing organizations (AOSA and ISTA) and the seed trade are focusing on improving the reproducibility of vigour tests. Considering the important role that vigour tests have in plant breeding, seed production, quality control, and marketing programs, standardization is badly needed. Although future research and testing are required before vigour testing becomes a routine phase of seed testing, the promise of vigour testing in the future is bright.

12. VIGOUR AND CROP PRODUCTIVITY

12.1. Plant establishment

Field performance of a seed or seed lot is controlled by the environmental factors such as physical condition of the soil, depth of the seed planted, soil fertility and soil moisture but still it is not possible to obtain the desired result. Plant establishment can succeed only if the seedlings have sufficient vigour to maintain rapid growth and contact with water supply which rapidly recedes during growing season of arid region. In such natural ecological situation, vigour is also necessary if the seedling is to compete with neighbouring plants for light, water and mineral nutrients. Examination of a field of any crop will indicate that a number of the plants are non-productive. Such non-productive plants may be the result of abnormally low vigour seeds. Increasing seedling vigour to eliminate these non-productive plants could permit higher yields at a lower cost per acre (Pollock and Roos, 1972). The effect of vigour on stand density can be separated from the effect of vigour on productivity of individual plants by green house planting and vigour classification, followed by replanting to standard spacing in the field. This was the technique used by Scheer and Ellison (1960) who found that low vigour in asparagus seedlings was correlated with survival and yield in the field. However, such "thinning to stand" or "planting to stand" techniques are commonly used by Perry (Perry, 1969) in peas and Clark and Kline (1965) in tomatoes. Low vigour seeds emerged more slowly and took longer period. Final emergence was less than with high-vigour seeds (Powell *et al.,* 1991) in brassica vegetables. Seedlings produced from the seeds of low vigour were also less vigourous and more variable in height at the first leaf stage, produced less fresh and dry weight of seedlings.

The competitive interaction between plants of same genotype is one of the major considerations of vigour. Walley *et al.* (1966) showed for a number of species that seedling length was correlated with seed size. Black (1958) separated the clover seeds by size, assuming that large seeds would produce seedlings of high vigour than small seeds. When uniform size seeds were planted at equal plant densities, the number of plants surviving decreased throughout the growing season as competition between plants eliminated a percentage of the population. In these uniform plantings, there was no difference between large and small seeds in the percentage of seedling which died. However, when large and small seeds were planted together, the number of plants from the large seeds remained approximately constant, whereas two third of the plants from small seeds were eliminated by competition. The plants from small seeds received only 2 per cent of the sunlight at the time of foliage canopy development and concluded that the elimination of plants was caused by shading (Black, 1958).

In the field, large seeds frequently produce larger seedlings than small seeds, but size advantage is often lost during crop growth when interplant competition becomes the main limiting factor to crop yield in wheat and barley. In contrast, large seeds of legumes of several genera associated with improvements in crop yield.

Competition between species has been suggested as a means of weed control. Pavlycheko and Harrington (1934) found that cereal seeds competed well with weeds under adverse moisture conditions, under adequate moisture the weed seeds competed much more effectively. Guneyli *et al.* (1969) found that the competitive advantage of sorghum over weeds was related to rapid germination and early seedling vigour. This physiological seedling vigour has the potential role in weed control. This approach offers a potential method of non-chemical weed control.

The use of differing seed size physical parameters as discrimination criteria for seed among varieties and different species has been previcusly reported (Dehghan-Shoar *et al.* 1998; Illipronti *et al.* 1999). Nerson (2002) showed that small muskmelon seeds had the lowest percentage germination, emergence, and the lowest seedling growth demonstrating that there is an association between seed physical parameters and seed quality.

Seed colour was another physical trait that differed among the lettuce varieties permitting them to be classified as white- and dark-seeded genotypes. Similar responses have been described in watermelon (Duval and NeSmith, 2001) and for other species (Karivarhadaraaju *et al.* 2001).

The lack of relationship between seed size and germination could be explained by the more important influence of embryo size or weight; in general, larger seeds have larger embryos, which is associated with increased germination. Further, Mian and Nafziger (1994) showed that different seed sizes influence the water potential of winter wheat seeds and, therefore, the speed of germination. That difference can also affect the uniformity of germination and subsequent seedling development that causes a higher germination percentage in seeds with greater size, even though this may not necessarily occur with smaller seeds (Duval and NeSmith, 2001).

Patricia Peñaloza *et al.* (2005) indicated that large-seeded lettuce varieties had higher percentage germination, higher values, saturated salt accelerated ageing higher seed vigour image system index and more rapid and uniform greenhouse emergence. Black-seeded lettuce varieties possessed higher seed quality and less fungal invasion when evaluated by the SSAA test. The SVIS index significantly correlated with saturated salt accelerated ageing values and seedling emergence under greenhouse conditions suggesting it can be used as a measure of seed vigour. It is concluded that the SSAA and SVIS tests are practical and accurate determinants of lettuce seed quality and distinguish between high and poor quality lettuce seed lots (Table 5).

Poor field emergence of temperate (Pea, faba bean, field bean) and tropical (bengalgram, cowpea, long bean, soybean) grain legumes has been associated with rapid water uptake leading to imbibition damage reflected in the death of cells of the cotyledons and high solute leakage. Rapid water uptake by seed lots of pea and soybean results from a large proportion of seeds having cracks in the testa (Powell *et al.,* 1993). High vigour okra seeds emerged in the field more synchronously, grown quickly and evenly as compared to low vigour seeds (Tomer *et al.,* 1998).

12.2. Crop productivity

In the commercial production, yield differences are observed between two plants of same genotype but of different seed lots, comparing hybrid corn seed lots produced by different growers. Flaming (1966) found economically significant differences in yield between lots with the same level of germination. Similar results have been found in oats (Grabe and Frey, 1966), peas (Perry, 1969) tomatoes (Clark, and Kline, 1965), sorghum (Abdullahi and Vanderlip, 1972), cotton (Turner and Ferguson, 1972) and wheat seeds (Ries *et al.,* 1976).

Table 5: Percentage normal, abnormal, and dead seeds following a standard germination test, SSAA test after 72 hrs, and the SVIS vigour index for six lettuce varieties each containing four seed lots produced under different field conditions

Variety	Lot	Per cent germination						SVIS Vigour Index
		Standard			SSAA 72 h			
		Normal	Abnormal	Dead	Normal	Abnormal	Dead	
A	1	93	6	1	94	4	2	827
	2	91	7	2	94	4	2	808
	3	99	1	0	99	1	0	783
	4	98	1	1	86	11	3	862
	Variety	**95 a**	**4 b**	**1 c**	**93 a**	**5 c**	**2 b**	**820 c**
B	5	97	3	0	85	14	1	801
	6	94	4	2	90	6	4	727
	7	90	6	4	65	25	10	766
	8	98	2	0	98	2	0	861
	Variety	**95 a**	**4 b**	**1 c**	**84 a**	**12 b**	**4 b**	**789 c**
C	9	95	4	1	81	15	4	926
	10	96	3	1	93	1	6	935
	11	96	2	2	92	1	7	971
	12	100	0	0	90	10	0	916
	Variety	**97 a**	**2 b**	**1 c**	**89 a**	**7 bc**	**4 b**	**937 a**
D	13	81	13	6	0	11	89	605
	14	91	7	2	71	18	11	872
	15	90	8	2	65	28	7	928
	16	91	4	5	84	6	10	958
	Variety	**88 b**	**8 a**	**4 b**	**55 b**	**16 b**	**29 a**	**841 bc**
E	17	90	6	4	51	40	9	822
	18	88	8	4	73	20	7	969
	19	76	11	13	68	24	8	818
	20	78	8	14	54	25	21	796
	Variety	**83 c**	**8 a**	**9 a**	**61 b**	**27 a**	**12 b**	**851 b**
F	21	98	1	1	95	4	1	964
	22	93	4	3	86	10	4	803
	23	94	4	2	84	13	3	952
	24	94	5	1	79	18	3	942
	Variety	**95 a**	**4 b**	**1 c**	**86 a**	**11 b**	**3 b**	**915 ab**

Evaluation of vigour by measuring crop yield is complicated one. In fact, that the seed or seedling vigour can produce major differences in stand density. These differences by themselves can produce yield differences in stand as shown by Austin (1963) in onions, Austin and Longdon (1967) in carrots, Nobbe (1876) in peas and Kittock and Law (1968) in wheat. The seed size had no effect on percentage emergence, but large seeds can produce taller plants and greater yield (Smith and Camper, 1975) in soybean. Whereas greater yield, per plot and per plant due to heavier grains and number of pods per plant obtained from the plants raised from larger seeds of soybean (EL-Zahab and Zahran, 1976). They noted that although plants from larger seeds were taller and had a greater leaf area index compared with small seeds, the relative growth rate and net assimilation rates were similar. In cotton, high density seeds contained more lipids, carbohydrates, total nitrogen and mineral elements per seed and affected a more rapid transfer of reserves from the cotyledons into the seedling axes (Bartee. and Krieg, 1974). However, Hess (1977) and Krieg and Carrol (1978) reported that selection for high-density character tended to reduce seed size in the progeny. Similar advantage from large seeds of sorghum was observed (Maran Ville and Clegg, 1977) but there was no improvement in crop yield.

The relationship between seed vigour and the ultimate yield of the plants has not been clearly indicated. There have been reports of a positive association between seed quality, vigour and yield for barley (Perry and Harrison, 1977), cotton (Bishnoi, 1971), field bean (Vieira, 1966), oat (Grabe, and Frey, 1966), groundnut (Basrin, 1970), rice (Sittis rung, 1970), sorghum (Chamargo and Vaughn, 1973), and maize (Dungan and Koehler, 1944). Johnson and Wax (1978) planted a number of seed lots having variations in seed quality at a uniform seeding rate and reported a positive relationship among seed vigour, field emergence and yield. But they concluded that differences in yield were primarily related to differences in stand.

Seed vigour has got direct influence on the plant growth processes involved in the production of yield. The yield of any crop is determined by the solar radiation intercepted by the crop community, the efficiency with which intercepted solar radiation is converted to dry matter, and the proportion of the biomass to economic yield (Charles Edwards, 1982). Yield is a function of the characteristics of the individual plant (eg. photosynthetic efficiency), the plant community (e. g. LAI, plant density, leaf angle), the environment and the integration of growth processes over time in a changing environment.

Effects of seed vigour on total emergence, rate of emergence, the uniformity of emergence and stand establishment are well documented (Roberts, 1972 and Heydecker, 1972) in many crops. All of these factors can potentially influence dry matter accumulation by the plant or plant community and thus potentially affect yield. Total emergence determines plant density and there is strong relationship between plant density and yield (Willey, and Heath, 1969). Most of the plant tissues involved in the production of dry matter and yield are formed after seedling emergence. It seems unlikely that seed vigour would influence their ability to carry out physiological processes and accumulate dry matter. Seedling vigour did not affect the relative growth rate of soybean seedlings, provided they were free of physiological injury or necrotic lesions (Egli *et al.,* 1990). Soybean seedlings with necrotic lesions or physiological injury on cotyledon show greatly reduced growth rates. A greater proportion of such seedlings were produced from low-vigour seed lots, thus, the average size of seedlings from low-vigour seedlots may be smaller at any time after emergence. In this case, plant tissue developing after emergence could be free from physiological defects and capable of expressing maximum growth rate, however, the time required to reach maximum growth rates may be longer than for injured seedlings.

In rice, a number of agronomic traits including yield is reported to be significantly affected by seed vigour. Thus, to realise the full genetic potential of a variety, both germination and vigour should be maintained at high level at the time of sowing (Seshu. and Dadlani, 1993; Pandey *et al.,* 1989; Seshu *et al.,* 1988).

Siddique (1986) classified the seeds of five rice cutlivars as high, medium and low vigour based on the conductance of individual seed leachate and grew the plants from the respective seeds to maturity under field conditions. Reductions of 7.89 and 14.14 per cent grain yield was noted from medium and low vigour seeds, respectively, compared to high vigour seeds. This was mainly attributed to the higher number of panicles per plant and number of filled grains per panicle in the plants raised from high vigour seeds. The production of vigourous seed is generally associated with ideal growth conditions of the mother plants. Conditions that disfavour complete seed development, therefore, result in lack of uniform germination, low seedling vigour and decreased seedling performance under field conditions. Increased assimilate transport towards the developing reproductive structures is required for production of high quality seed

Kanzikwera *et al.* (2000) reported that application of N and K rates of 120 and 132.8 kg ha^{-1}, respectively, were best for production of TPS with high seedling emergence and vigour (Table 6).

Table 6: Effect of mother plant-applied N and K on performance of Kisoro x Rutuku hybrid TPS in a screen house at Namulonge (seed from 1999 long rains)

Treatment		SDM (%)	SDW (g)	CVE	Seedling emergence (%)		
					10 DAS	15 DAS	21 DAS
N (kg ha^{-1})	**K (kg ha^{-1})**						
0	0	19.0a	3.2	7.36a	38.3b	43.3bc	48.3b
0	132.8	7.1b	3.4	9.28a	56.7a	60.0a	60.0a
0	265.6	15.7a	2.4	5.76b	34.7b	33.3c	38.3b
120	0	17.8a	2.7	10.24a	56.7a	61.7a	63.3a
120	132.8	10.5b	2.3	9.28a	51.7a	56.7a	60.0a
120	265.6	6.3b	4.0	6.08b	31.7b	36.7bc	40.0b
240	0	13.3a	2.0	9.92a	58.3a	61.7a	63.3a
240	132.8	22.4a	3.8	5.76b	30.0b	33.3c	41.7b
240	265.6	14.7a	3.7	6.08b	31.7b	38.3bc	40.0a
CV (%)	24.2	26.9	19.8	24.3	20.2	19.3	

Values followed by the same letter or not followed by a letter in a column are not significantly different at P≤0.05; CVE= Coefficient of Velocity of Emergence, SDM = seedling dry matter, SDW = seedling dry weight

The time from sowing to seedling establishment is thus a crucial phase in crop production and can have a major impact on final yield and postharvest seed quality. Seed germination is a triphasic process which begins with the rapid initial water uptake (phase I), a plateau phase with little change in water content (phase II), and an increase in water content that coincides with radicle growth (phase III) (Bradford, 1986). The length of phase three is important since germination is considered complete when embryo growth is initiated. The soil environment on the other hand is never conducive for rapid germination and seedling growth. Adverse temperatures, soil crusting, limited or excess water, salinity, and disease pathogens all combine to create a stressful environment for germination and seedling emergence. Water relations are central to understanding the adaptive strategy that allows the seed to germinate and emerge under adverse conditions.

Water relations: The expression and quantification of biophysical mechanisms closely related to seed germination processes has generated fundamental parameters for understanding the influence of physiological processes on seed vigour. Using the moisture release curves approach, psychrometric measurements have demonstrated in lettuce (Bradford, 1986) and muskmelon (Oluoch, 1996) that primed seeds have a lower water potential (ψ) and a higher turgor potential (ψ_p) than nonprimed seeds. The effect is correlated to the improvement in germination rates after priming. It has been suggested that during priming, seeds may accumulate solutes that are retained through drying and rehydration. The hydrostatic turgor pressure of the embryonic axes may increase through the accumulation of osmotic solutes allowing the radicle to exceed the yield threshold of the tissues surrounding the radicle.

Some species of grass produce many sterile flowers, which produce no seed. The seeds on native grass heads ripen gradually so that many under-ripe seed units are collected at harvest and ripe seeds may have already been shed and lost. The result is a lower seed yield than would be expected in some species. In addition, insects can damage seed. Hand collected seed units for all species were dissected to estimate the approximate percentage of intact, undamaged and fully-grown seed in the collections. The results are shown in Table 7.

Table 7: Seed quality of hand collected seed

Species	% intact seed
Austrodanthonia caespitosa	10%
Austrostipa scabra	50%
Bothriochloa macra	100%
Ehrharta stipoides	30%
Joycea pallida	10%
Poa labillardierei	26%
Themeda triandra	10%

Similarly, a sample of *Austrodanthonia caespitosa* 'seed' collected with a mechanical harvester contained only 10% seed by weight. Only 25% of the florets contained seed. In practical terms 1 kg of raw 'seed' contained about 120,000 germinable seed (100g).

Later assessment of seed quality for all species except *Themeda triandra* and *Austrostipa scabra* was done by soaking the seed in water for a few hours then examining under a microscope. This makes the seed coat transparent so that the presence and size of the seed can be easily observed. It also facilitates the removal of the seed coat for germination tests where this is desirable. In the case of *Ehrharta stipoides* seed maturity is indicated by the seed coat changing colour from green to green/brown. Germination of green seed is about onc †hird less than mature seed.

Themeda triandra is commonly collected with a large amount of hay and the seed units may represent only about 1% by weight of the collection. This may be true for other species too. Recently commercial seed suppliers in the local region have been developing more efficient seed harvesting and cleaning machinery. We therefore investigated the efficiency of various harvesting methods, including:

- hand collection
- collection with a hand held harvester based on a modified grass trimmer (Grass Seeds Australia)
- collection using larger brush or suction type harvesters drawn behind 4-wheel motor bikes or tractors

The relative performance of each is very dependent on the site and the purity of the sward and it is therefore not possible to make fair comparisons between them. However when seed harvested by any of these methods has been examined under the microscope prior to germination tests indicated that there is only minimal damage from harvesting.

For example, *Ehrharta stipoides*, harvested with a suction machine and then cleaned by a seed merchant of all extraneous plant material other than the seed coat, was compared with uncleaned material form the same collection. The seed coat was then manually removed prior to germination testing. Germination was the same for both treatments (99% + after 7 days). There was no physical damage to the seeds from the cleaning process.

Similar results were obtained for *Themeda triandra* collected with a cutting harvester developed by John Betts and Tony Wilson and cleaned using a machine constructed by Tony Wilson. Five batches of 100 seeds were randomly selected and assessed for damage (Table 8).

Table 8: Effect of mechanical harvesting on seed quality

Seed condition	%
Undamaged full seed	88
Callus broken off	1
Top end of seedcoat chipped, inner seed intact	6
Cracked or abraded seedcoat	1
Dent or crack in coat and seed	1
Total full (hard) seed	96
Empty seed	4

Machine harvesting therefore causes no problems with seed quality. Removal of seed coats may however be more difficult.

Plots planted with low vigour seed yield 5 -15 % less than those planted with vigourous seed even though the number of plants per unit area is the same. This yield loss is the summation of reduced levels of the various components of yield. In corn, for example, lower vigour seed produce a higher percentage of barren plants, slightly fewer ears per plant, and slightly smaller ears with slightly reduced shelling percentages as compared to vigourous seed. These "slight" reductions add upto a 10 – 15 % loss in yield.

The influence of seed vigour on plant performance is most dramatically manifested in vegetable crops, especially those produced for their fleshy roots such as radish and turnips. Root development in plants from low vigour seed is slow and many of the roots do not reach marketable size by the time when the crop is "normally" harvested. In other vegetable crops, low vigour seed contributes substantially to non uniformity of maturity as well to lower yields.

Crop production is limited by the vigour of the seed planted just as it is by the quantity and distribution of precipitation, rate and timing of fertilization, effectiveness of weed control, variety planted and so on. This constraint on productivity will be eliminated or at least minimized only when farmers begin to demand higher quality seed and seedsmen can consistently supply it.

13. CURRENT STATUS OF SEED VIGOUR TESTS

The cold test has been most commonly employed for evaluating seed vigour in corn and soybeans. This test is difficult to standardize because it exposes seeds to soil microorganisms to measure their ability to resist attack under cold and wet conditions. It is difficult to standardize the substrate (soil), microflora to maintain a constant level of pathogeneicity. A recent study of Byrum and Copeland (1995) who has conducted cold test on four corn seed lots with varying levels of quality tested at 10 different laboratories in the U.S. Corn Belt representing official, commercial and crop improvement associations indicated that the cold test for corn is as repeatable as the standard germination test. This study illustrates the present state of the art of cold test in the United States and demonstrates the potential for the eventual standardization of all vigour tests.

Among vigour tests, the electrical conductivity (EC) of seed soaked in water has been satisfactorily used to evaluate seed vigour of pea (Matthews and Powell 1981), soybean (Tao, 1978 and Hampton and Coolbear, 1990) and maize (Yaklich *et al.,* 1979 and Tao, 1978). This test has been considered by many researches as an adequate index of seedling emergence in the field (Marcos Filho *et al.,* 1985 and Matthews and Carver, 1971) and the results can be reproducible in other laboratories. Detailed studies were conducted by Loeffler *et al.* (1988) and Oliveria *et al.* (1984) with soybean and they considered EC test as an efficient index of seedling emergence in the field. Extensive work done with soybean seed harvested at different time and stored in distinct environments have revealed that the information provided by EC test was able to differentiate levels of physiological quality of seeds and to estimate the potential of field emergence of seedling more precisely (Marcos Filho *et al.,* 1985). More recently scientists (Prete *et al.,* 1994 and Vieira *et al.,* 1999) have reported a highly significant negative correlation between the electrical conductivity and the field emergence of soybean seedlings. The test was also sufficient to classify seed lots and cultivars into vigour levels.

In lentil and soybean standard germination test and accelerated ageing are the most important laboratory tests that correlate well with field parameters and may be used to predict field performance (Marcos Filho *et al.,* 1985 and Tomer *et al.,* 1993). While in sorghum a number of vigour tests *viz.*, accelerated ageing test, electrical conductivity test, catalase activity in four day old seedling and Hiltner's test were found to be suitable to assess seed vigour (Dighe, 1993).

In rice, standard germination, rate of germination, root length, seedling dry weight, germination after accelerated ageing and dehydrogenase activity have positive significant correlation with field emergence (Seshu and Dadlani, 1993). Conductance of individual seed leachates can also be used to classify seed vigour groups in rice (Siddique, 1986). Besides germination rate index, germination over 25%, PEG germination at 15°C and amalyse activity also showed significant positive correlation with field emergence in rice (Shenoy *et al.,* 1990).

Cold test, controlled deterioration test and NaCl stress test were found to be good indicators of seed vigour for maize (Padma and Reddy, 1998).

Though good work was done at individual laboratories of National and International level with good expertise, but the work on large scale was not done like the study conducted by Byrum and Copeland (1995) for cold test and Marcos Filho *et al.* (1985) for conductivity test.

A vigour test suitable for one crop may not be suitable indicator for another crop. In a number of studies, higher prediction accuracy using vigour index was attained when standard germination test was included. It is always wise to develop vigour tests based on such variable which are least affected by the genotypic difference with in crop species. Therefore comparisons of seed vigour should be made between seed lots of a particular variety or between closely related varieties of any crop species as they are grown under different agroclimatic and soil conditions.

14. PRECAUTIONS TO BE TAKEN WHILE CONDUCTING VIGOUR TESTS

The vigour tests should be conducted in precise conditions. More care should be taken while conducting vigour tests compared to standard germination test. Variation in temperature, moisture or RH results in variation in rate of growth or rate of biochemical process based on which vigour is estimated. For example 1°C difference in temperature may have considerable effect on seedling growth test or seed deterioration in accelerated ageing test. Similar is the situation with relative humidity and seed moisture. So equipment with great precision should be designed and used for vigour test.

Dormancy also interferes with the interpretation of vigour test results. If dormancy is suspected, the seed should be checked by tetrazolium test.

Under favourable environmental and soil conditions, the variation in field emergence of high germinating lots is less while under unfavourable conditions it become significant. In adverse soil conditions or in adverse field environment planting high vigour seed will result in high field emergence compared to low vigour seed. By doing so, one can prevent replanting and associated delayed maturity and yield reductions due to poor seed standards.

Vigour tests designed based on the stress conditions in that locality may give good results under field conditions. For example, in sub tropics and tropics where humidity and temperature plays an important role on seed ageing, exposing the seed to high humidity and high temperature like accelerated ageing or CD tests may give good results. Similarly cold test for crops of temperature origin, PEG test for crops grown in drought prone areas and NaCl stress test for crops grown in saline soils may be adopted. To sum up vigour tests developed considering the agroclimatic conditions, soil condition and seed aging process, might give better results.

15. LIMITATIONS OF SEED VIGOUR TESTS

High vigour seed performs well in adverse field and soil environment up to a limited range. If the conditions prevailing after plating are too severe even the most vigourous seeds cannot perform satisfactorily. While standardizing the vigour tests acceptable vigour level for different production seasons, crops and genotypes has to be evaluated. Certain crops and genotypes have the ability to compensate for minor stand difference following emergence and this may result in little difference in final yield. Many previous studies concerning seed vigour have focused on the relationship of laboratory germination and vigour to field emergence, where as there is less published information relating seed vigour to other aspects of crop performance.

At the present state of art on seed vigour, sufficient evidence is not available to show a strong relationship between yield and vigour in the absence of stand differences. However, the compensated seed rates used for low vigour seed may not be useful for direct sown crops for maintaining uniform plant stand on par with high vigour seed. For vegetable and rice crops grown in nurseries compensated seed rates can be adopted. While standardizing the vigour tests all these factors have to be considered to identify a viable vigour test for a crop species.

16. APPLICATIONS OF VIGOUR TEST RESULTS

The aim of seed vigour testing is to maintain the seed standard. The environmental factors that influence the seed quality are already identified. Many surveys were carried out to identify the ideal seed production and seed storage areas, where high vigour seed can be produced and maintained respectively.

State Seed Certification Agency personnel should educate the farmers to follow good agronomic practices, plant production and protection technology and to maintain certification standards. The seed produced in ideal seed production areas with good management should be tested for vigour before marketing it as 'vigour tested seed'. This gives assurance to the farmer about the standard of the seed. It also helps farmers in decision making. So in adverse soil conditions and environmental conditions high vigour seed can be used to avoid replanting. Besides vigour testing for all species and regions has not yet achieved the same level of standardization possessed by the standard germination test. However, both seed testing organisation and the seed trade are focusing on improving the reproducibility of vigour tests. Considering the important role that vigour tests can have in plant breeding, seed production, quality control and marketing programmes, standardization is badly needed. Although future research and testing are required before vigour testing becomes a routine phase of seed testing, the promise of vigour testing in the future is bright.

CONCLUSION

A vigour test is an important part of a quality assurance program but should only be used in conjunction with all other tests performed on a seed lot. The vigour test(s) used must be based on the type of seed and the conditions that the seed may be exposed to. Finally, careful interpretation of the result is required and must be based on experience in vigour testing.

Although a high vigour seed lot is preferable, the producer should always ensure that management practices are such to provide the most favourable conditions for the germinating seed. This will allow the seed to perform as close to the maximum potential as indicated by the germination value.

REFERENCES

Abdul-Baki, A.A.1969. Relationship of glucose metabolism to germinability and vigour in barley and wheat seeds. Crop Science,9: 732-37.

Abdullahi, A. and R.L. Vanderlip. 1972. Relationship of vigour tests and seed source and size to sorghum seedling establishment. Agron. J., 64: 143-144.

Abu-Shakra, S.S. and Ching T.M. 1967. Mitochondrial activity in germinating new and old soybean seed. Crop Science, 7 : 115-118.

Ader, F. 1965. Zur Definition eines einheitlich anwendbaren Begriffs der Triebkraft. Proc. Int. Seed Test. Assoc., 30: 1005-1012.

Agrawal, R.L. 1993. Seed Technology, Oxford and IBH publication Ltd., New Delhi, p.583.

Anderson, J.D. 1970. Physiological and biochemical differences in deteriorating barley seed. Crop Science, 10 : 36-39.

Association of Official Seed Analyst (AOSA) 1991. Rules for testing seeds. Journal of Seed Technology, 12: 1-109.

Association of Official Seed Analysts (AOSA). 1983. Seed vigour testing hand book. East Lansing, MI, USA, pp. 93.

Association of Official Seed Analysts (AOSA).1976. News Letter of Association of Official Seed Analysts, 50(2).

Austin, R.B. 1963. Yield of onion from seed as affected by place and method of seed production. J. Hort. Sci., 38: 277.

Austin, R.B. and P.C., Longdon. 1967. Some effects of seed size and maturity on the yield of carrot crops. J. Hort Sci., 42: 339-353.

Bain, J.M. and F.V. Mercer. 1966. Subcellular organisation of the cotyledons in germinating seeds and seedling of Pisum sativum L. Australian Journal of Biological Science, 19: 69-84.

Barriga, C. 1951. Effects of mechanical abuse of navy bean seed at various moisture levels. Agronomy Journal, 53: 250-251.

Bartee, S.N. and D.R., Krieg, 1974. Cotton seed density: Associated physical and chemical properties of 10 cultivars. Agron. J., 66(1): 433-435.

Baskin, C.C. and J.C. Delouche. 1971. Differences in metabolic activity in peanut seed of different size classes. Proceedings of Association of Official Seed Analysts, 61:73-77.

Basrin, C.C. 1970. Relation of certain physiological properties of peanut seed to field performance and storability. Ph.D. Thesis, Mississippi state University, USA.

Beighley, D.H. and N.W. Hopper. 1981. The relationship of chemical composition and electrical conductivity of cowpea seed to field performance. Agronomy Abstracts, : 117.

Belcher, E.W. and L. Miller 1974. Influence of substrate moisture level on the germination of sweet gum and sand pine seed. Proceedings of the Association of Official Seed Analysts, 65: 88-89.

Bishnoi, U.R. 1971. Deterioration of cotton seeds under warm moist storage conditions and its consequences in terms of seed and plant responses. Ph.D, Thesis, Mississippi state University, USA.

Black, J.N. 1958. Competition between plants of different initial seed sizes in swards of subterranean clover with particular reference to leaf area and the light microclimate. Aust. J. Agr. Res., 9: 299.

Bonner, F.T. 1974. Tests for vigour in cherrybark oak acorns. Proceedings of Association of Official Seed Analysts , 64: 109-144.

Bradford, K.J. 1986. Manipulation of seed water relations via osmotic priming to improve germination under stress conditions. Hort Science, 21 : 1105-1112.

Burris, J.S. O.T. Edje and A.H. Wahab. 1969. Evaluation of various indices of seed and seedling vigour in soybean, Glycine max (L.) Merr. Proceedings of the Association of Official Seed Analysts, 59: 73-81.

Burris, J.S. 1975. Seedling vigour and its effect on field production of corn. Proceedings of the 30th Annual corn and Sorghum Research Conference, : 185-193.

Burris, J.S. 1976. Seed / seedling vigour and field performance. Journal of Seed Technology, 1: 58-74.

Byrd, H.W. and J.C. Delouche. 1971. Deterioration of soybean seed in storage. Proceedings Association Official Seed Analysts, 61:41-57.

Byrum, J.R. and L.O. Copeland. 1995. Variability in vigour testing of maize (Zea may L.) seed. Seed Sci &Technol., 23: 543-549.

Caldwell, W.P. 1972. Relationship of preharvest environmental factors to seed deterioration in cotton. Ph.D. Dissertation, Mississippi State University.

Chamargo, C.P. and L.E. Vaughn. 1973 .Effect of seed vigour on field performance and yield of grain sorghum. Proc. Assoc. Off. Seed Anal., 63: 135-147.

Charles Edwards, D.A. 1982. Physiological determinats of crop growth, Academic Press, Austrailia North Ryde NSW.

Ching, T.M. 1972. Metabolism of germinating seeds. In : T.T. Kozlowski (ed.) Seed Biology, Academic Press, New York / London, 2 : 103 - 218.

Ching, T.M., 1973. Asdenosine triphosphate content and seed vigour. Pl. Physiol., 51(2): 400-402.

Ching, T.M., S. Hedtake, M.C. Bougler and W.C. Kronstand. 1977. Correlation of field emergence rate and seed vigour criteria in barley cultivars. Crop Science, 17:312-314.

Clark, B.E. and D.B. Kline. 1965.Relationship between the vitality and the performance of Fireball tomato seed. Ann. Rep Agric. Exp. Stn., :1.

Coombe, B.G. 1976. The development of fleshy fruits. Annual Review of Plant Physiology, 27 : 507-528.

Copeland, O. 1976. Principles of Seed Science and Technology. Burgers Publication Co. Ltd. Minnesota, USA.

Cox, F.R. and P.H. Reid 1964. Calcium – boron nutrition as related to concealed damage in peanuts. Agronomy Journal, 56: 173-176.

De Tempe, J. 1961. Seed weakness. Proc., Inter. Seed. Test. Ass., 22: 3-11.

Dehghan-Shoar, M.; Hampton, J. and Haslett, Steve. 1998. Identification of, and discrimination among, lucerne (Medicago sativa L.) varieties using seed image analysis. Plant Varieties and Seeds, 11 (2) : 107-127.

Delouche, J.C. 1968. Physiology of seed storage, 23rd Corn and Sorghum Res Conf. Amer. Seed Trade Ass., :83-90.

Delouche, J.C. and C.C. Baskin. 1973. Accelerated aging techniques for predicting the relative storability of seed lots. Seed Sci & Technol., 1: 427-452.

Delouche, J.C. and W.P. Caldwell.' 1960. Seed vigour and vigour tests, Proc. Assoc. Off. Seed Anal., 50: 124-129.

Dighe, R.S. 1993. Evaluation of seed vigour tests in sorghum (Sorghum bicolor). Seed Res., Special Vol. 2: 801-805.

Doijode, S.D. 1990. Influence of seed position in fuit on seed viability and vigour during ambient storage of chilli fruit. Capsicum News letter, 10 : 62 -63.

Dungan, G.H. and B. Koehler. 1944. Age of seedcorn in relation to seed infection and yielding capacity. J. Ain Soc Agron., 36: 436-443.

Duval, J. and D.Nesmith. 2001. Difference in seed coat thickness between the cultivars. Resistance to gas diffusion through the seedcoat and hilum of watermelon (Citrullus lanatus Matsum & Nakai). Proceedings of the Florida State Horticultural Society, 114 : 323-326.

Egli, D.B and D.M. TeKrony. 1979. Relationship between soybean seed vigour and yield. Agronomy Journal, 71: 755-759.

Egli, D.B., D.M Tekrony, and R.A. Wiralaga. 1990. Effect of soybean seed vigour and size on seedling growth J. Seed Technol., 14: 1-12.

Ellis, R. H., T. D. Hong and E. H. Roberts. 1995. Survival and Vigour of Lettuce (Lactuca sativa L.) and Sunflower (Helianthus annuus L.) Seeds Stored at Low and Very-low Moisture Contents. Annals of Botany, 76: 521-534, 1995

El-Sharkawi, H.M. and I. Springuel. 1977. Germination of some crop plant seeds under reduced water potential. Seed Sci & Technol., 5: 677-688.

EL-Zahab, A and M. Zahran. 1976. Effect of seed size on early growth, yield contributing variables of soybean. Zeitschrift for Acker und Pflanzenbau, 143(2): 148-165.

Ferguson, J.M. 1993. AOSA perspective of seed vigour testing. Journal of Seed Technology, 17(2): 101-104.

Fiala, F. 1981. Cold test. In: Hand book of vigour test methods (Ed. D.A. Perry), ISTA, Zurich, pp. 28-36.

Flaming, A.A. 1966. Effects of seed age, producer and storage on corn production. Agron. J., 58: 227.

Fox, R.L. and W.A. Albrecht. 1957. Soil fertility and the quality of seeds. Missouri Agricultural Experiment Station Research Bulletin, 619.

Frank, W.J. 1950. Introductory remarks concerning a modified working of the international rules for seed testing on the basis of experience gained after the world war. Proceedings of the International Seed Testing Agency, 16: 405-430.

Fuchs, H. 1981. Hiltner Test. (In) Handbook of vigour test methods (Ed. D.A. Perry). International Seed Testing Association, Zurich, pp. 28-36.

Germ, H., 1960. Methodology of the vigour test for wheat, rye and barely in rolled filter paper. Proc. Int. Seed Test Assoc., 25: 515-518.

Grabe, D.F. and K.F. Frey. 1966. Seed quality and planting rates affect oat yields, Lowa Farm Sci., 20(8): 7-9.

Green, D.E., E. L. Pinnell, L. E. Cavanah and L. F. Williams. 1965. Effect of planting date and maturity date on soybean seed quality. Agronomy Journal, 57: 165-168.

Guneyli, E., O.C. Burnsde, and P.T. Nordquist. 1969. Influence of seedling characteristics on weed competitive ability of sorghum hybrids and inbred lines. Crop Sci., 9: 713.

Gusta, L.V., E.N. Johnson, N.T. Nesbitt, and K.J. Kirkland. 2003. Effect of Seeding Date on Canola Seed Vigour. Can. J. Plant Science. In press

Hampton, J.G. and P. Coolbear. 1990. Potential versus actual seed performance can vigour testing provide an answer? .Seed Sci & Technol., 18 : 215-225.

Haskins, F.A. and H.J. Gorz. 1975. Influence of seed size, planting depth and companion crop on emergence and vigour of seedlings in sweet clover. Agronomy Journal, 67 : 652-66.

Hess, D.C. 1977. Selecting of increased seed density in cotton. Proceedings of the Beltwide Cotton Production Research Conf. 1977, Atlanta, Georgia p.84-86.

Heydecker, W. 1972. Vigour, In: Viability of seeds (ed) Roberts, E.H., Academic Press Ltd. New York, P.209-252.

Hiltner, L. and G. Ihssen. 1911. Uber das schlechte Auflaufen und die Auswinterung des Getreides infolge Befalls des Saatgutes durch Fusarium. Landw. Jahrubuch fur Bayern., 1: 20-60, 315-362.

Hong, T.D. and R.H. Ellis. 1990. a comparison of maturation drying, germination and desiccation tolerance between developing seeds of Acer pseudoplatanus and Acer platanoides. New Phytoloist, 116 : 589-596.

Hopper, N.W. and H.R. Hinton. 1980. The use of electrical conductivity as a measure of cotton seed quality. Agronomy Abstracts, : 109.

Illipronti, R., C.Langerak and W. Lommen. 1999. Variation in physical seed attributes relates to variation in growth of soybean seedlings within a seed lot. Seed Science and Technology, 27,(1) : 339-357.

Isely, D. 1957. Vigour tests. Proceedings of the Association of Official Seed Analysts, 47: 176-182.

ISTA (1995). Hand book of vigour test methods. 3rd edition (eds. Hampton & D.M. Tekrony), ISTA, Zurich, Switzerland, p. 117.

James C. Delouche. 1976. Standardization of vigour tests. J. Seed Tech., 1 (2) : 75 – 85.

James, E.1968. Limitations of glutamic acid decarboxylase activity for estimating viability in beans (Phaseolus vulgaris L.). Crop Science, 8 : 403-404.

Johnson, R.R, and L.M. Wax. 1978. Relationship of soybean germination and vigour tests to field performance, Agron J., 70(2): 273-278.

Joo, P.K., B. A. Orman, A. M. Moustafa and M. Hafdahl. 1980. Can leachate electro-conductivtiy be a useful tool for corn seed emergence potential evaluation? Agronomy Abstracts, : 109.

Kanzikwera, C.R., D.S.O. Osiru, J.S. Tenywa, E. Adipala and A.S. Bhagsari. 2000. Effect of mother plant-applied nitrogen and potassium on seed weight and seedling emergence of true potato seed. African Crop Science Journal, 8 (4) :. 375-385

Karivarhadaraaju, T., P.Srimathi and K..Malarkodi,. 2001. Seed coat colour as a parameter for seedling quality of amla (Embrica officinalis L. Gaertn.). Advances in Plant Sciences, 1 (1) : 271-273.

Kittock, D.L. and A.G. Law. 1968. Relationship of seedling vigour to respiration and tetrazolium chloride reduction by germinating wheat seeds. Agron. J., 60: 268-288.

Kneebone, W.R. and C. L. Cremer. 1955. The relationship of seed size to seedling vigour in some native grass seed species. Agronomy Journal, 47 : 472-477.

Koostra, P.T. and J.F. Harrington. 1969. Biochemical effects of age on membranal lipids of cucumis sativus L. Seed. Proc.Int. Seed Test. Assoc., 34 : 329 – 340.

Krieg, D.R. and J.D. Carroll. 1978. Cotton seedling metabolism as influenced by germination temperature, cultivar and seed physical properties. Agron. J., 70(1):21-25.

Leggatt, C.W. (1929-1930). Catalase activity as a measure of seed viability. Scientific Agriculture, (Ott.)10: 73-110.

Lindner, H.1967.Moglichkeiten and Grenzen der 'Trieb kraft' von getreide mitdem keimrollen tests. In: Physiologie, Okalogie and Biochemie der Keimung, Vol 2 (ed), H.Borriss, 965-967 Ernst-Morti 2-Arndt Universitat, Greifswald.

Loeffler, T.M., D.M. Tekrony and D.B. Egli. 1988. The bulk conductivity test as an indicator of soybean seed quality. Journal of Seed Technology, 12 : 37-53.

Lowe, L.P. and S.K. Ries. 1973. Endosperm protein of wheat seed as a determinant of seedling growth. Plant Physiology, 51: 57-60.

Luedders, V.D. and J.S. Burris. 1979. Effect of broken seed coats on field emergence of soybeans. Agronomy Journal, 71: 877-879.

Maguire, J.D. 1962. Speed of germination – aid in selection and evaluation for seedling emergence and vigour. Crop Science, 2 : 176-177.

Manohar, M.S. 1966. Effect of osmotic systems on germination of peas (Pisum sativum L.). Plant, 71: 81-88.

Maran Ville, J.W. and Clegg,M.D. 1977. Influence Of seed size and density on germination, seedling emergence and yield of grain sorghum. Agron. J., 69(2): 329-330.

Marcos Filho, J., R.V. Carvalho, S.M. Cicero and C.G.B. Demetrio. 1985. Soybean seed physiological quality and performance in storage and field. Anais da Escola Superior de Agriculture "Luiz de Queiroz" 42, p. 195-249.

Maselli, S., F. Pérez-García and I. Aguinagalde. 1999. Evaluation of Seed Storage Conditions and Genetic Diversity of Four Crucifers Endemic to Spain Annals of Botany, 84: 207-212.

Matthews, S. 1980. Controlled deterioration: a new vigour test. In: Seed Production (ed.) P.D. Hebblethwaite. Butterworths Co. Ltd., London, pp.647-660.

Matthews, S. 1993. .Seed vigour and crop performance. Seed Res., Special. 1: 21-27.

Matthews, S. and A.A. Powell. 1981. Electrical conductivity test. (In) Hand book of vigour test methods (Ed D.A. Perry), ISTA, Zurich, p. 37-42.

Matthews, S. and M.F.F. Carver. 1971. Further studies on rapid exudates tests indicative of potential field emergence. Proceedings of Association of Official Seed Analysts, 36: 307-312.

Mc Donald M.B. Jr. and D.O. Wilson. 1979. An assessment of the standardization and ability of the ASA-610 to rapidly predict soybean germination. Journal of Seed Technology, 4(2): 1-12.

McDaniel, R.G. 1969. Relationships of seed weight, seedling vigour and mitochondrial metabolism in barley. Crop Science, 9 : 823-27.

McDaniel, R.G. and I.V. Sarkissian. 1968. Mitochondrial heterosis in maize. Genetics, 59: 465-475.

McDonald Jr. M.B. 1977. The influence of seed moisture on the accelerated aging vigour test. Journal of Seed Technology, 2 (1): 19-28.

McDonald, M.B. 1980a. Assessment of seed quality. Horticultural Science, 15(6) : 784-788.

McDonald, M.B. 1980b. ASA-610 ability to detect changes in soybean seed quality. Journal of Seed Technology, 5(1): 56-66.

McDonald, M.B. 1994. The history of seed vigour testing. Journal of Seed Technology.

McDonald, M.B. Jr. 1975. A review and evaluation of seed vigour tests. Proceedings of Association of Official Seed Analysts, 65:109-139.

McHargue, J.S. 1920. The significance of the peroxidase reaction with reference to the viability of seeds. Journal of American Chemical Society, 42 : 612-615.

McNeal, F.H., M.A. Berg, P.L. Brauig and C.F. McGuire. 1971. Productivity and quality response of five spring wheat genotypes (T. aestivum) to nitrogen fertilizer. Agronomy Journal, 63: 908-910.

Merritt D.J. Senaratna T. Touchell D.H.Dixon K.W. Sivasithamparam K. 2003. Seed ageing of four Western Australian species in relation to storage environment and seed antioxidant activity . Seed Science Research, 13(2) 155-165

Mian, M.A. and E.D.Nafziger, 1994. Seed size and water potential effects on germination and seedling growth of winter wheat. Crop Science, 34 (1) :169-171.

Miles, D.F., Jr. and L.O. Copeland. 1980. The relationship of vigour tests and field performance in soybeans (Glycine max (L.) Merr.), Agronomy Abstracts, : 111.

Murugesan, P. (2003). Harvesting, post harvest handling and conditioning of fruit and seed in ashgourd. Ph.D. Thesis, Tamil Nadu Agricultural University, Coimbatore -3.

Nanda, R., S.C. Bhargava and M.Gautam. 1996. Effect of flowering, time on rate of drymatter accumulation, duration of filling period and total oil content in brassica seeds. Indian J. Pl. Physiol., 1 (2) : 88-92.

Nass, H.G. and P.L. Crane. 1970. Effect of endosperm mutants on germination and early seedling growth rate in maize (Zea mays L.). Crop Science, 10 : 139-140.

Neeb, O. 1970. Keimfahigkeit, Triebkraft und Feldaufgang bei Zuckerriibensaatgut, In: Viability of seeds (ed.) E.H. Roberts, Chapan and Hall Co. Ltd., p.212.

Nerson, H. (2002). Relationship between plant density and fruit and seed production in muskmelon. Journal American Society for Horticultural Science, 127 (5) : 855-859.

Nicholson, J.F., O.D. Dhingra and J.B. Sinclair 1972. Internal seed - borne natue of Sclerotinia sclerotiorum and Phomopsis spp. and their effects on soybean seed quality. Phytopathology, 61: 1261-1263.

Nobbe, F. 1876. Handbuch der Samen Kunde, Wiegandt-Hempel-Parey, Berlin.

Nutile, G.E. 1964. Effect of desiccation on viability of seeds. Crop Sci., 4: 325-328.

Oliveira, M.A., S. Matthews and A.A. Powell. 1984. The role of split seed coats in determining seed vigour in commercial seed lots of soybean as measured by the electrical conductivity test. Seed Sci & Technol., 12: 659-668.

Oluoch, M.O. and G.E.Welbaum, 1996. Effect of postharvest washing and post-storage priming on viability and vigour of six year old muskmelon (Cucumis melo L.) seeds from eight stages of development. Seed Sci. & Technol., 24 : 195-209.

Oluoch, M.O.1996. Effects of priming and development on vigour and longevity of muskmelon (Cucumis melo L.) seeds. Blacksburg, 1996. Thesis (Ph.D.)- Virginia Polytechnic Institute and State University.

Ovcharov, K.E., Doman, N.N. and B.A. Popov 1970. Correlation between mitochondrial ultra structure and germinating capacity of maize seeds. Fiz. Rast.,17: 329-36.

Padma, V. and B.M.M. Reddy. 1998. Standardization of laboratory vigour test for maize. Seed Res., 26(2): 134-137.

Pandey, M.P., D.V. Seshu, and M. Akber. 1989. Genetic control of embryo size and implications of that trait on seed vigour in rice, IRRI, Seminar, June 17, 1989. IRRI, Los Banos, Philippines.

Panochit, J., P. Wasuwanich and A.K.Hellum. 1986. collection and storage of seeds of Shorea roxburghii. The Embryon, 2 : 62-67.

Parmer, M.T. and R. P. Moore. 1968. Carbowax 6000, mannitol and sodium chloride for simulating drought conditions in germination studies of corn (Zea mays L.) of strong and weak vigour. Agronomy Journal, 60: 192-195.

Patricia Peñaloza, Gerardo Ramirez-Rosales, Miller B. McDonald and Mark A. Bennett. 2005. Lettuce (Lactuca sativa L.) seed quality evaluation using seed physical attributes, saturated salt accelerated aging and the seed vigour imaging system. Plant Biotechnology, 8(3):

Pavlychenko, T.K. and J.B. Harrington. 1934. Competitive efficiency of weeds and cereal crops. Can. J. Res., 10: 77.

Perry, D. A. 1969. A vigour test for peas raised on seedling evaluation. Proc. Pat. Seed Test. Ass., 34: 265-270.

Perry, D. A. 1978. Report of the vigour test committee 1974-1977. Seed Sci & Technol., 5: 159-181.

Perry, D.A. (1973). Seed vigour and stand establishment. Horticultural Abstracts, 42: 334-342.

Perry, D.A. 1970. Seed vigour and field establishment. Hort. Abstr., 2: 334-342.

Perry, D.A. 1977. A vigour test for seeds of barley (Hordeum vulgare) based on measurement of plumule growth. Seed Sci &Technol., 5: 709-719.

Perry, D.A. and J.G. Harrison. 1977. Effects of seed deterioration and bed environment on emergence and yield of spring-sown barley. Ann. Appl. Bio., 86(2): 291-300.

Pinthus, M. J. and U. Kimel 1979. Speed of germination as a criterion of seed vigour in soybeans. Crop Science, 19: 291-292.

Pollock, B. M. and E.E. Roos. 1972. Seed and seedling vigour. In: Seed Biology Vol III (ed.) T.T. Kozlowaskl, Academic Press Ltd., New York, pp.313-387.

Potts, H.C., J.D. Duangpatra, W.G. Hairston and D.C. Delouche. 1978. Some influences of hard seededness on soybean seed quality. Crop Science, 18: 221-224.

Powell, A.A., Thornton, J.M. and Mitchell, J.A. 1991. Vigour differences in brassica seed and their significance to emergence and seedling variability. J. Agri. Sci. (Cambridge), 116: 369-373.

Powell, A.A., W. D. Abdullah, N. Legesse, M.D.A. Oliveira and S. Matthews. 1993. Selection of grain legume genotypes for improved field emergence. Special Vol., 2: 691-699.

Powell, A.l.i.s.o.n. .A. and S. Matthews. 1978. Rapid evaluation of the storage potential of seed peas. Acta Hort., 83:133-140

Prete, C. E. C., S. M. Cicero and M.V. Folegatti. 1994. Soybean seedling field emergence and relationship of imbibition and electrical conductivity of the seeds. Semina, 15: 32-37.

Qualls, M. and C.S. Cooper. 1969. Germination, growth and respiration rates of birdsfoot trefoil at three temperatures during the early non-photosynthetic stage of development. Crop Sci., 9: 758-760.

Ramamoorthy, K. and T.V. Karivaratharaju. 1985. Mobilization efficiency - A direct test to assess seed viability in groundnut (Arachis hypogaea L.). Seeds and farms, XV (1) : 28-30.

Ramamoorthy, K. and T.V. Karivaratharaju. 1991. Elite Tea Seed Production. TNAU, Newsletter, 20 (11): 2.

Rees, M. 1997. Evolutionary ecology of seed dormancy and seed size. In : Silvertown, J.W. Franco, M.Harper (eds.) Plan life histories, ecology, phylogeny, and evolution, Cambridge, Cambridge University Press, pp 121- 142.

Ries, S. K., O. Moreno, W. F. Meggitt, C. J. Schweizer and S. A. Ashkar. 1970. Wheat seed protein – chemical influence on and relationship to subsequent growth and yield in Michigan and Mexico. Agronomy Journal, 62: 746-748.

Ries, S.K., G. Ayers, V. Wert and E.H. Everson. 1976. Variation in protein, size and seedling vigour with position of seed in heads of winter wheat cultivars. Can. J. Pl. Sci., 56(4): 823-827.

Roberts, E. H, 1972. Viability of Seeds. Chapman and Hall Ltd, London.

Roberts, E. H. 1986. Quantifying seed deterioration. (In) Physiological of seed deterioration (eds. MB. McDoland, Jr and C.T. Nelson), CCSA Special Publication Number 11: 101-123, Crop Sciecne Society of America, Madison, Wisconsin, USA.

Rosenburg, L.A. and R.W.Rinne. 1986. Moisture loss as a prerequisite for seedling growth in soybean seeds (Glycine max (L.) Merr.). Journal of Experimental Botany,.43 :.307-317.

Sasaki, S. 1980. Storage and germination of dipterocarp seeds. Malaysian Forester, 43 : 290-308.

Scheer, D.F. and J.H. Ellison. 1960. Asparagus performance and related to seedling vigour. Proc. Amer. Soc. Hort. Sci., 76: 370.

Schimdt, L. 2000. Guide to handling of Tropical and subtropical forest seed. Danida Forest Seed Centre, Humlebaek, Denmark.

Schoorel, A.F. 1960. Report on the activities of the vigour test committee. Proceedings of the International Seed Testing Association, 25: 519-525.

Seshu, D.V and M. Dadlani. 1993. Seed vigour in rice (Oryza sativa) its assessment and impact on crop performance. Seed Res., Special, Vol.2: 700-706.

Seshu, D.V., V. Krishnasamy, and S.B. Siddique. 1988. Seed vigour in rice, In: Rice Seed Health, IRRI, Manila, Philippines.

Shenoy, V. V., M. Dadlani and D. V. Seshu. 1990. Association of laboratory assessed parameters with field emergence in rice. The monanoic acid stress as a seed vigour test. Seed Res., 18: 60-66.

Siddique, S.B. 1986. Variability in seed dormancy and tolerance to accelerated ageing of seed in rice (Oryza sativa L.) Ph.D. thesis, Cornell University, USA pp. 146.

Simak. M and A. Gurtafson. 1953. X-ray photography and sensitivity in forest tree species, Hereditas, 89 : 458-68.

Simon, E. W. and H. H. Wiebe. 1975. Leakge during imbibition, resistance to damage at low temperature, and the water content of peas. New Phytology, 74: 407-411.

Simpson, D. M. and B. M. Stone. 1935. Viability of cottonseed as affected by field conditions. Journal of Agricultural Research, 50: 435-447.

Sittisrung, P. 1970. Deterioration of rice seed in storage and its influence on field performance. Ph.D. Thesis, Mississippi State University, USA.

Sivakumar, V. 2004. Studies on factors affecting storage life of Vateria indica and Hopea parviflora seeds. Ph.D. Thesis, Institute of Forest Genetics and Tree Breeding, Coimbatore -2.

Sivasubramanian, K. 1996. Studies on certain aspects of seed quality in moringa cv. PKM 1. Ph.D. Thesis, Tamil Nadu Agrcultural University, Coimbatore – 3.

Smith, T.J. and H.M. Camper. 1975. Effects of seed size on soybean performance. Agron. J., 67(5): 681-684.

Spencer,G.F., F.R.Earle, I.A.Wolff and W.H.Tallent. 1973. Oxygenation of unsaturated fatty acids in seeds during storage. Chem. Phys. Lipids, 10 : 191-202.

Streeter, J.G. 1965. Possible mechanisms in the loss of seed viability. Assoc. Off. Seed Anal. Newslett., 39 (3) : 27-35.

Sullivan, G.A. 1973. Effects of dolomitic limestone, gypsum and potassium on planting seed quality, yield and grade of peanuts (Arachis hypogea L.) Ph.D. Dissertation, North Carolina State University, Raleigh.

Suryatmana, G., L. O. Copeland and D. F. Miles. 1980. Comparison of laboratory indices of seed vigour with field performance of Navy bean (Phaseolus vulgaris L.). Agronomy Abstracts, : 113.

Tao, J.K. 1978. Factors causing variations in the conductivity test for soybean seeds. Journal of Seed Technology, 3: 10-18.

TeKrony, A., D. Phillips and D. B. Egli. 1980b. The effect of field weathering on soybean seed viability and vigour. Crop Science, 72: 749-753.

TeKrony, D. M and D.B. Egli. 1977. Relationship between laboratory indices of soybean seed vigour and field emergence. Crop Science, 17: 573-577.

TeKrony, D. M., D.B. Egli and J. Balles. 1980a. The effect of the field production environment on soybean seed quality. (In) Seed Production (ed. P.D. Hebblethwaite), Butterworth and Co. Ltd., London.

Tekrony, D.M.and J.L.Hunter. 1995. Effect of seed maturation and genotype on seed vigour in maize. Crop Science, 35 : 857-862, 1995.

Tempe, J. D. 1963. The use of correlation coefficients in comparing methods for seed vigour testing. Proceedings of the International Seed Testing Association, 28: 167-172.

Thomsen, K.2000. Handling of desiccation and temperature sensitive tree seeds. DFSC Technical Note, 56 : 6.

Tomer, R. P. S. and A. K Narwal and U. Verma. 1998. Seed vigour studies in okra (Bhindi). Seed Tech. News, 28(4): 44.

Tomer, R. P., P. S. Kumari, T. P. Yadava and D. P. Deswal. 1993. Seed vigour studies in Lentil (Lens culinaris). Seed Res., Special Vol. 2: 763-768.

Turner, J.H. and D. Ferguson. 1972. Field performance of cotton grown from field and partially filled seeds. Crop Sci., 12: 868-871.

Van Onckelen, H.A., R.Verbeek, M.A.Stone and A.A.Khan. 1973. The metabolism of barley embryo under accelerated aging condition and its relation to alpha amylase synthesis and RNA metabolism of aleurone . Plant Physiol., 51 : 1-39.

Veselova., T., V. Veselovsky and E. Leonova. 1998. Vigour loss in pea seeds. a population approach. 11th Congress of the Federation of European Societies of Plant Physiology which will be held in Varna, Bulgaria from 7-11 September, 1998.

Vieira, C. 1966. Effect of seed age on germination and yield of field bean. Turrialba 16: 396-398.

Vieira, R. D., J A. Paiva – Aguero, D. Perecin and S. R. M. Bitten Court (1999). Correlation of Electrical conductivity and other vigour tests with field emergence of soybean seedlings. Seed Sci & Technol., 27: 67-75.

Villiers, T.A. 1973. Aging and the longevity of seeds in field conditions p.265-288. In: W.H.Heydecker (ed.), Seed Ecology. The Pennsylvania State University Press, University Par, PA.

Walley, D. B., C. M. McKell, and L.R. Green. 1966. Seedling vigour and the early non-photosynthetic stage of seedling growth in grass. Crop Sci., 6: 147-150.

Welbaum, G.E. 1993. Water relations of seed development and germination in muskmelon (Cucumis melo L.): VIII. Development of osmotically distended seeds. Journal of Experimental Botany, 44 :1245-1252, 1993.

Wester, R. E. 1970. Nonbleached, a genetic marker for quality seed in lima beans. Seed World, 160(7): 24.

Willey, R. W. and S.B. Heath. 1969. The quantitative relationship between plant population and crop yield. Adv Agron., 21: 281-321.

Woodstock, L.W. 1965. Seed vigour. Seed World, 97(5):6.

Woodstock, L.W. 1969. Seedling growth as a measure of seed -vigour. Proc. Int. Seed Test. Assoc., 34: 273-280.

Woodstock, L.W. and D.F.Grabe. 1967. Relationship between seed respiration during imbibition and subsequent seedling growth in Zea mays L. Plant Physiol., 42 : 1071-1076.

Yaklich, R. W., M. M. Kulik and J. D. Anderson. 1979. Evaluation of vigour tests in soybean seeds: Relationship of ATP, conductivity and radio active tracer multiple criteria laboratory tests to field performance. Crop Science, 19: 806-810.

12

Tree Seed Physiology

M. Djanaguiraman[1], Mallika Vanangamudi[1] and R. Arun Kumar[2]
[1]*Department of Crop Physiology, Tamil Nadu Agricultural University Coimbatore- 641 003*
[2]*Department of Horticulture, Adhiparasakthi Agricultural College, G. B. Nagar Kalavai- 632 506*

1. INTRODUCTION

Silviculture, a branch of forestry is gaining momentum today as it is involved in reviving the past nature. Silviculturist, must maintain the present extent of forest cover and must attempt, wherever possible, to make its composition, at least some what like that of the natural forest. At the same time they have to ensure a yield of timber or forest products such as fodder, food, fuel to meet ever-increasing demand, and last, but not the least one must keep in mind that the forest is not just a wood producing factory, but plays a vital role in the water balance of the area and in preventing the soil erosion, thus following 'eco' preservation and development.

The forest in the country occupies technically about 22% of geographical area and recent aerial survey has indicated only about 12% as functional forest area. The rest can be assumed as non-productive, degraded forest land. The remote sensing data exhibits fast depleting forest cover and the real picture makes the environmentalists and scientists to worry more. Hence, afforestation must be taken as people's movement and should be done on war-footing basis. Although afforestation of the species was carried out through various Governmental and Non-Governmental bodies like Social Forestry, Farm Forestry, SIDA, World Bank, CIDA, USAID and DANIDA, the total area under forest cover kept depleting still. The sole reason is not the deforestation alone. Failure of the afforestation programme also plays major role in the reduction of forest cover.

A major limiting factor in the afforestation programme on wastelands is drought stress after transplanting. The change of success can be increased with proper utilization of growth stimulation chemicals and growth regulators by promoting root growth and developing suitable microbial inoculants for tree species used in afforestation programme.

Availability of water is an important limiting factor for the establishment and subsequent growth and development of a seedling. Because, almost every physiological and biochemical process occurring in seedlings as well as in grown up trees is some way affected or limited by the availability of water during the initial phase of plant growth. The inconsistent precipitation pattern must then be managed to improve the establishment of seedlings by increasing the water availability for long duration. There are many ways to sustain the water availability to the seedlings after transplanting. Plastic mulching is one such technique used to collect the rain water around the seedlings and conserve it for long period of time. Hira *et. al.* (1990) have reported that this technique assists in seedling establishment and growth by increasing salt free soil water zone around plant roots and conserving harvested rain water.

Root pruning is normal method of conditioning carried out in the nursery to increase this ratio by the formation of fibrous root system and reduced top growth. These changes increase water absorption and improve the internal water balance (Rook, 1971). Root pruned seedlings develop a hardened woody stem and show increased height increment and field survival rates after out planting (Tanaka *et al.,* 1976).

However in many cases, alternative methods of conditioning seedlings are used *e.g.* restricted watering or drought hardening. The transpiration rate of the moisture stress conditioned seedlings was decreased by 30 per cent and water use efficiency increased by 67 per cent as a result in case of *Pinus taeda* (Seiler and Johnson, 1985). By the economic use of water, this method decreases the transplanting shock and increases the survival of seedlings.

Primary internal factor which causes the drought stress in the seedlings is the excessive transpiration. Of the total water absorbed by the plants, 99 per cent is lost by way of transpiration. In such a situation, the help of some external material may be sought out to curtail the transpirational loss. Any material applied on transpiring plant surface with an aim to reduce the water loss from the plant is known as anti-transpirant. Zelitch (1969) made elaborate review of anti-transpirants. They grouped them into three categories namely stomata closing, file forming and reflective types based on their mechanism of action.

Another approach to increase the seedling establishment is through biofertilizers treatment. The role of biofertilizers has already been proved extensively in the field of Agriculture and Horticulture, but information regarding its applicability in forest species is scanty in India. However, in forestry, few research reports are available to demonstrate that bio-fertilizers accelerate the growth (Mukerji *et. al.,* 1996); biomass (Reddy *et. al.,* 1996) and enhance uptake of N (Mathur and Vyas, 1996); P (Dela Cruz *et. al.,* 1988); K (Merina Prem Kumari, 1991) and other micro nutrients and thereby the survival rate of planted seedling is enhanced.

Unlike bio-fertilizers, the growth regulators and growth stimulants have been extensively used in the forestry for enhancing the growth and development of seedlings under nursery conditions (Kumaran *et. al.,* 1994) because of their major role in enhancing the root growth (Palani *et al.,* 1996; Kumaran *et al.,* 1996). Hence, it is clear that plant growth regulators, chemicals, biofertilizers, root pruning, can positively influence the seedling establishment, and whereas, stresses like drought, salinity, high temperature etc. are not conducive for seedling establishment.

2. BIOFERTILIZER INOCULATION ON SEEDLING GROWTH AND DEVELOPMENT

Biofertilizers role in annual crops has been proved extensively, but its impact on forest species is sparse. Development of suitable biofertilizer for tree species is one of the important management technologies for higher tree seedling establishment in field. The biofertilizer containing live cells of nitrogen fixing micro-organisms or phosphorus solubilizers have helped in minimizing the dependence on inorganic fertilizers. It also helps to reduce environmental pollution. Biofertilizers can stimulate the growth (Sudhir *et al.*, 1994; Mukerji *et. al.*, 1996), biomass (Sekar *et al.*, 1995; Reddy *et al.*, 1996), biochemical parameters (Niranjan *et al.*, 1990) and uptake of N (Mathur and Vyas, 1996), P (Dela Cruz *et al.*, 1988), K (Merina Prem kumari, 1991) and other micronutrients. Ultimately, the survival rate of planted seedling is enhanced.

The synergistic effect of different biofertilizer inoculants such as *Azospirillum*, Phosphobacteria and AM fungi on the seedling growth of *Syzygium montanum*, *S. arnottianum* and *Elaeocarpus oblongus* was also reported by Sekar *et al.* (1996). The combined inoculation of the above said biofertilizer significantly increased the shoot length of *S. montanum*, *S. arnottianum* and *E .oblongus*. Likewise, the combined inoculation also enhanced shoot collar diameter in the above species over control. This enhanced growth might be due to the increased uptake of nutrients, synergistic effect with other inoculants, ability to produce more growth regulators, resist to root pathogens and survival under adverse soil condition and temperature (Srinivas, 1987; Harley and Smity, 1988).

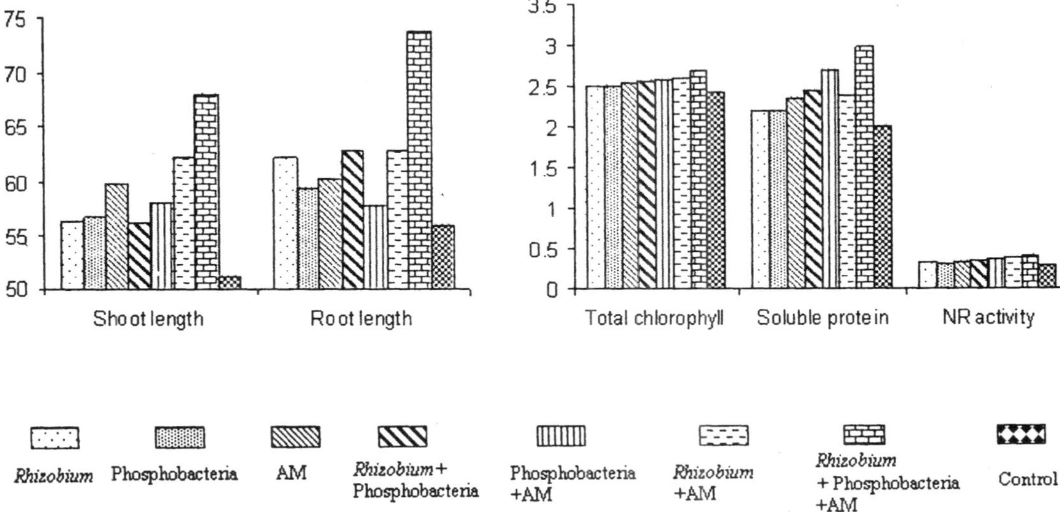

Fig.1: Effect of biofertilizer on shoot length (cm), root length (cm), total chlorophyll(mg-1g), soluble protein (mg-1g) NRase activity (µg-1ghr-1) of pungam seedling

Enhanced shoot and root length (Fig. 1), total dry matter and nutrient uptake due to triple inoculation with *rhizobium*, phosphobacteria and AM were observed in pungam seedlings (Venkatesh *et al.*, 1998). Dual inoculation (*Rhizobium* +

phosphobacteria or phosphobacteria + AM or *Rhizobium* + AM) proved as good as the triple inoculation with regard to P and K yield. However, the biochemical and physiological indices like total chlorophyll, soluble protein and NR activity (Fig.1) revealed that triple inoculation was the best. Thus, triple inoculation could enhance the pungam seedling survival (Venkatesh *et al.,* 1998).

The impact of arbuscular mycorrhizae (AM), *rhizobium* and phosphobacteria on elite tree seedlings production was found to be good. In, *Tamarindus indica* the combined inoculation of *Rhizobium* + phosphobacteria + AM improved the shoot length, root length and total leaf area compared to the uninoculated control and other treatments (Ilango, 2001). The increase in plant growth attributes might be due to increased uptake of nutrients in mycorrhizal associated plants and their synergistic effect with other inoculants. Apart from this, the rhizosphere effect through microbial activity modifies the plant itself by providing plant growth substances and increasing the availability of nutrients at the root zone (Jakobsen *et al.,* 1994).

3. GROWTH STIMULANTS ON GROWTH AND ESTABLISHMENT

Seed germination and seedling growth are known to be regulated by exogenous hormones in many tree species (Kumaran *et al.,* 1994). Similarly, treatments with inorganic ions like KCl (Palani *et. al.,* 1995) and KNO_3 (Maideen *et al.,* 1990) have also been reported to enhance germination in tree species. Vanagamudi *et al.* (1999) reported that, compared to water soaking all the pre-sowing chemical treatments (IAA, IBA, GA_3, NaH_2PO_4 and KCl) recorded enhanced germination but the magnitude of increase was maximum in IBA followed by NaH_2PO_4 in *Syzygium cuminii* (Table 1). The enhanced germination due to IBA may be ascribed either to diffusion of the growth regulator (Mathur *et al.,* 1971) or its antagonistic effect on the inhibitor present in dormant seed (Khan, 1968).

Table 1: Effect of pre-sowing growth stimulants on germination and seedling growth in *Syzygium cuminii*

Treatment	Germination (%)	Shoot length (cm)	TDMP (mg seedling^{-1})	Vigour index
Water soak	56	9.77	911	1108
IAA 100 ppm	61	13.43	1297	1466
IBA 100 ppm	65	12.25	1299	1475
GA_3 100 ppm	64	12.15	1188	1473
NaH_2PO_4 1%	66	12.98	1381	1464
KCl 1%	52	9.01	1139	967
KNO_3 1%	65	10.06	1098	1329
Dry seed	46	9.49	922	857
CD (P=0.05)	0.63	2.97	NS	89

Seed germination and seedling vigour studies in *Pongamia pinnata* with growth regulators and chemicals *viz.*, IAA, IBA, GA_3 (100 ppm), $ZnSO_4$, KH_2PO_4, $FeSO_4$ and $MgSO_4$ at 1% (W/V) and water showed that seeds soaked in IBA 100 ppm for 12 hr. recorded the maximum germination percent which was 48%

higher than the untreated dry seeds. However, the shoot length and vigour index was improved by GA_3 100 ppm. Among the salt solution, none of them were at par with the growth regulators, however, treatment with KH_2PO_4 was found to be the best in salt (Venkatesh *et al.*, 1997).

The increase in germination and growth by plant growth regulator may be due to enhanced rate of metabolism (Verma and Tandon, 1988). The increased growth by GA_3 might be due to mobilization of starch which acts as a respiratory substance, as a raw material for the synthesis of other PGR and also by increase in the somatic pressure of cell leading to immediate enhancement in cell elongation (Nanda and Dhindsa, 1967) and promotes the synthesis of RNA which has the property of making the cell response to auxin (Yangishima and Shimoda, 1973). Vanagamudi *et al.* (2000) reported that soaking neem seed in 2.5% PEG for a period of 24 hr. was found to be the best for viability and vigour, whereas, the higher concentrations (10 and 12.5%) proved ineffective and were comparable to control. The beneficial effect of the lower concentrations may be ascribed to the progressive rupture of the endosperm (Guedes *et al.*, 1981).

Investigation on the effect of growth stimulants like IAA (100 ppm), IBA (100 ppm), GA_3 (100 ppm), Kinetin (100 ppm), NAA (100 ppm), Ethrel (100 ppm), KCl (1%), KNO_3 (1%), $ZnSO_4$ (1%), KH_2PO_4 (1%), $FeSO_4$ (1%) and $MgSO_4$ (1%) on germination and morpho-physiological attributes of pungam seedlings revealed that seeds soaked in IBA (100 ppm) gave high germination and root length (Table 2). GA_3 (100 ppm) enhanced the shoot length, total leaf area, total dry weight and total chlorophyll content more than other treatments. The values for most of the above parameters recorded for IBA (100 ppm) were comparable with those of GA_3 (Table 2). KH_2PO_4 (1%) induced a greater soluble protein content than the water soaked seeds and other growth promoting chemicals (Venkatesh *et al.*, 2000).

Table 2: Influence of pre-sowing treatments on germination, shoot length, root length and vigour index.

Treatment	Germination (%)	Shoot length (cm)	Root length (cm)	Vigour index
IAA	51.6	21.5	36.2	2987
IBA	66.7	17.5	34.5	3473
GA_3	61.1	35.0	29.5	3764
Kinetin	61.1	19.7	29.0	2521
NAA	61.0	20.0	15.0	1873
Ethrel	35.0	20.0	32.5	1838
KCl	58.3	25.0	21.0	2073
KNO_3	36.9	25.5	27.0	1937
$ZnSO_4$	38.3	20.0	29.0	1882
KH_2PO_4	63.3	22.6	26.0	3083
$FeSO_4$	29.2	17.5	30.5	1401
Water soaking	53.3	16.5	26.5	2417
Dry seed	45.0	14.0	19.0	2017
CD (P=0.05)	4.8	4.8	NS	519

The enhanced germination by auxin (IBA) may be attributed to their indirect effect through change in the membrane permeability, solubilisation of carbohydrates through synthesis of different enzymes responsible for promotive effects and production of some precursors needed for germination. The auxin may exert their primary effect on the cell wall and change subcellular proton concentration (Jann and Aman, 1977). The increase in chlorophyll content by GA_3 may be through enhanced synthesis of phytochrome and synthesis of chloroplasts (Prasad and Mohammad, 1987).

The positive effect of foliar spray of growth hormones on seedling growth attributes in *Albizia lebbeck* was studied by Ilango (2001). Growth hormones *viz.*, IBA (300 ppm), GA_3 (300 ppm), ABA (300 ppm), kinetin (300 ppm), maleic hydrazide (300 ppm), cycocel (3%), ethrel (3%) and alar (300 ppm) were used in foliar application along with water spray. Among the growth hormones, GA_3 maximised the shoot length (Table 3), which indicates that growth regulator was limiting factor for plant growth (Bhatnagar and Talwar, 1978), and also it enhanced endogenous auxin level through either biogenesis or destruction of auxin. The growth retardants namely alar produced the longest root by reducing the vegetative growth (Table 3). This might be due to the translocation of more photosynthates to the roots (Bhattacharjee, 1984). Although growth retardants such as CCC, ethrel and maleic hydrazide have an inhibitory effect on the growth and development of plants, they help to accumulate a considerable amount of photosynthates resulting in a higher total dry weight. Apart from this, the increased chlorophyll content by this group of chemical may probably due to subdued chlorophyllase enzyme activity (Paricha *et al.*, 1977).

Table 3: Influence of growth regulators on shoot length, root length, DMP and total chlorophyll in *Albizia lebbeck*.

Growth stimulant	Shoot length (cm)	Root length (cm)	DMP (g plant^{-1})	Total chlorophyll (mg g^{-1})
IBA	43.2	47.7	4.58	4.06
GA_3	56.3	46.8	4.97	4.72
ABA	36.2	48.4	5.92	3.52
Kinetin	38.5	45.2	7.61	3.14
Maleic hydrazide	34.8	52.3	4.31	3.80
CCC	35.8	51.1	10.70	3.98
Ethrel	32.2	41.4	9.66	3.82
Alar	29.3	62.1	5.42	4.29
Control	28.3	38.3	4.21	3.66
CD (P=0.05)	0.97	1.22	0.71	0.40

4. SEEDLING GRADE AND ESTABLISHMENT OF TREE SPECIES ON FIELD

Afforestation programme is being carried out through various governmental and non-governmental bodies like World Bank, SIDA, CIDA, USAID, DANIDA *etc.* The total area under forest cover is still declining. This is not only due to deforestation, but also due to the failure of afforestation programmes. The main reason for the area reduction is due to non availability of good quality planting stock material. Seedling morpho-metric attributes like root collar diameter, seedling height and terminal bud condition have long been used to grade forest tree seedlings into different classes of implied survivability and growth after transplanting in the field. However, reports on morphological grades have been contradictory. Many papers report on optimum performance of tall seedlings (Lauer 1987; Singh *et al.*, 1988; Shiver *et al.*, 1990). However, in drought years medium–sized seedlings survive better than small or large seedlings (Venator, 1983). Venkatesh *et al.* (2003) reported that tall seedlings for field planting will produce trees capable of optimum survival; these seedlings will also have best resistance to death of the terminal bud.

The work on pungam seedlings by Venkatesh *et al.* (2003) revealed that during initial evaluation, the shoot length increased with the increase in seedling height class from 25.2 cm (20-30 cm height class) to 56.9 cm (51-60 cm height class). Higher seedling height class (51-60 cm) recorded more total leaf area and total dry weight than lower height classes (20-30, 31-40 and 41-50 cm height). The performance of height classes after planting showed that the tallest class was superior to others. The tallest seedlings (51-60 cm) survived better (94.2%) than seedlings of the other height classes *viz.*, 41-50 cm (84.5%), 31-40 cm (75.2%) and 20-30 cm (41.2%) (Table 4). The magnitude of increment in respect to the growth parameters after three months of planting showed a close parallel with the seedling height (Table 5). The dominance of taller seedlings over smaller ones was also reported by Shiver *e. al.* (1990) for Pinus spp and Singh *et al.* (1988) for spruce. Grading of nursery stock will be especially worthwhile where early height advantages reduces competition for light and space or protect seedlings from animals (Miuko, 1974).

Table 4: Effect of seedling height class on initial seedling parameters of pungam seedling

Seedling height class (cm)	Shoot length (cm)	Shoot collar diameter (mm)	Total leaf area (cm^2 plant^{-1})	Root length (cm)	Total dry weight (g plant^{-1})
20-30	25.2	4.4	51.8	34.9	3.81
31-40	34.2	5.3	80.0	33.5	5.98
41-50	46.2	5.5	111.9	36.7	8.45
51-60	56.9	5.7	123.5	35.3	9.13
CD (P= 0.05)	2.4	NS	30.4	NS	1.51

Table 5: Effect of seedling height class on survival, growth and development of pungam seedlings after field planting.

Seedling height class (cm)	Survival (%)	Shoot length (cm)	Shoot collar diameter (mm)	Total leaf area (cm^2 plant^{-1})	Root length (cm)	Total dry weight (g plant^{-1})
20-30	41.7	30.9	6.8	100.1	40.7	6.98
31-40	75.2	38.6	10.0	159.4	38.9	9.33
41-50	84.5	49.0	11.3	189.2	45.2	11.20
51-60	94.2	61.8	12.5	204.5	51.0	12.55
CD (P= 0.05)	5.6	2.9	1.3	18.4	4.1	0.55

5. WATER USE EFFICIENCY AND SEEDLING ESTABLISHMENT

There is a considerable interest in estimating water use efficiency for forests to interpret the selection of species appropriate for sites with limited water supply. Seasonal variations in the estimate of water use efficiency were attributed to the effects of soil water and air saturation deficits on transpiration. Honeysett *et al.* (1992) measured changes in root-zone water storage for contrasting *Eucalyptus nitens* and *E. delegatensis* in years when water was limiting and non-limiting. The value of leaf area index for *E nitens* was approximately doubles that for *E. delegatensis*, the value for water-use efficiency much higher for *E. nitens* than that for *E. delegatensis*. In view of the preceding discussions of the regulation of photosynthesis and transpiration by soil and atmospheric water deficits, it is anticipated that estimates of water-use efficiency will vary widely between species and growing conditions. Farquhar *et al.* (1982) showed that the trade-off between photosynthesis and transpiration in leaves of C_3 plants, regulated by stomatal conductance. Olbrich *et al.* (1993) showed a two-fold variation in water-use efficiency, calculated from measurements of stem wood production and tree transpiration over a period of 4 months for four year old clones of *E. grandis*. There was a clear difference in fractionation in all annual rings between trees indicating that the carbon isotope signature is a useful index of susceptibility of trees to drought. Differences in the degree of fractionation across growth rings were closely related to the seasonal development of soil and atmospheric water deficits. The relationship between isotope fractionation and basic wood density was weak, but more-careful measurements of the isotopic signature across individual growth rings provided evidence that the decrease in density at the drought-prone site was attributable to reduced cambial activity occurring earlier in the season, resulting in less latewood production. Further progress in the application of measurements of carbon isotope fractionation to phloem samples in *E. globules* in relation to fractionation in leaves and across annual rings is anticipated to be useful for forest managers to define the current nutrient and water status of trees (Pate and Arthur, 1998). Higher values of mass-to area ratio and nitrogen content for populations from arid sites suggested that higher water-use efficiency is attributable mainly to higher photosynthetic capacity.

Mulches have a definite controlling effect on the evaporation for soil moisture and hence play an importance role in establishment of tree seedling. Under field conditions, surface soil in just stirred and soil becomes air-dry. By adopting this practice, some amount of moisture is lost from the upper layer. After the preliminary loss of moisture, the soil mulch effectively controls further evaporation from deeper layers by breaking the capillary continuity. This system is highly suitable for medium to heavy textured soils.

5.1. Plastic as mulch

Locally available material like black coloured low density polyethylene (LDPE) sheet can be very well used as mulch material. Growth and establishment of tree seedling has been increased in the mulched field due to better water use and reduced evaporation. It acts as a barrier between the incident radiation and soil surface leading to reduction in evaporation losses.

Sivakumar (1995) showed that, when the *Albizia lebbeck*, *Azadirachta indica*, *Eucalyptus tereticornis* and *Tamarindus indica* seedlings were mulched with black coloured (50 cm x 50 cm size, LDPE sheet of 200 guage thickness) increased the soil moisture content by collecting the rain water and conserving the moisture up to 10 days. Increased plant height, diameter, dry matter and leaf area growth were observed in *A. lebbeck*, *A. indica* and *E. tereticornis* seedlings, with pronounced effect in *Eucalyptus tereticornis* but no transplanting shock was observed in any species (Table 6). Plastic mulching significantly increased the transpiration rate of *A. lebbeck*, *A. indica* and *E. tereticornis* seedlings in all the days of observation but, reduced in *T. indica*. A reverse effect was observed in case of diffusive resistance which was reduced in *A. lebbeck*, *A. indica* and *E. tereticornis* seedlings but increased in *T. indica*. Plastic mulching significantly improved the RWC and total chlorophyll content and soil temperature by 2.94°C (Table 7). The increase in temperature was observed under both 5 and 25 cm depths (Sivakumar, 1995).

Table 6: Effect of plastic mulching on height increment (cm), diameter increment (mm), dry matter increment (g/plant) and leaf area increment in *Albizia lebbeck*, *Tamarindus indica*, *Azadirachta indica* and *Eucalyptus tereticornis* seedling.

Tree species	Height increment		Collar diameter increment		Dry matter increment		Leaf area increment	
	Control	PM	Control	PM	Control	PM	Control	PM
T. indica	3.67	4.24	0.95	0.93	2.27	2.39	61.5	60.6
A. lebbeck	4.66	7.51*	0.96	1.30*	2.27	3.63*	64.1	128.9*
A. indica	5.68	8.46*	0.85	1.32*	3.78	5.65*	71.3	110.6*
E. tereticornis	8.02	12.09*	1.30	1.68*	4.97	6.80*	165.8	216.1*

PM – Plastic Mulching * Significant at 5% level.

Table 7: Effect of plastic mulching on transpiration rate ($\mu g\ cm^{-2}\ s^{-1}$), diffusive resistance ($S\ cm^{-1}$), RWC (%) and total chlorophyll content ($mg^{-1}\ g$) of *Tamarindus indica*, *Albizia lebbeck*, *Azadirachta indica* and *Eucalyptus tereticornis* seedling.

Tree species	Transpiration rate		Diffusive resistance		RWC		Total chlorophyll content	
	Control	PM	Control	PM	Control	PM	Control	PM
T. indica	3.72	3.43*	2.81	3.67*	79.7	84.4*	1.62	1.66*
A. lebbeck	5.67	6.00*	2.83	2.32*	77.1	79.9*	3.87	4.42*
A. indica	2.99	3.29*	3.68	3.31*	86.4	87.6*	2.03	2.14*
E. tereticornis	3.84	4.17*	3.44	3.33*	82.2	85.4*	1.59	1.68*

PM – Plastic Mulching * Significant at 5% level.

5.2. Use of antitranspirants

The possibilities of reducing transpiration are of greatest importance during initial tree species establishment. Any material that is applied to transpiring plant surfaces for reducing water loss from the plant is called antitranspirant. Many herbicides, fungicides and growth hormones possess antitranspirant characteristics. Antitranspirants may be of a) stomatal closing type (b) file forming types (c) reflectant type or (d) growth retardant.

Some of the fungicides like Phenyl Mercuric Acetate (PMA) and herbicides like atrazine in low concentrations serve as antitranspirants by inducing stomatal closing. The plastic and waxy materials which form a thin film on the leaf surface retard the escape of water due to formation of a physical barrier, *e.g.*, mobileaf, hexadeconol, silicone. The success of these chemicals is limited since they also reduce photosynthesis. Some of the white materials form a coating on the leaves and increase the leaf reflectance (albedo). Transpiration loss is reduced by reflecting the radiation, reducing leaf temperature and vapour pressure gradient. Spraying of kaolin (5%) forms thin layer on the leaf surface and reduces transpiration. Some of the growth retardant (cycocel) induce stomatal conductance and thus reduce transpiration losses.

Table 8: Effect of antitranspirants spray on height increment (cm), collar diameter increment (mm), dry matter increment (mg plant^{-1}) and leaf area increment (cm^2 plant^{-1}) in *A .lebbeck* and *E. tereticornis* seedlings.

Treatments	Height increment		Collar diameter increment		Dry matter increment		Leaf area increment	
	A. lebbeck	*E. tereticornis*	*A. lebbeck*	*E. tereticornis*	*A. lebbeck*	*E. tereticornis*	*A. lebbeck*	*E. tereticornis*
Control	4.66	8.02	0.96	1.30	2.20	4.98	64.5	165.8
Rice gruel	5.07	7.88	0.95	1.36	2.29	4.22*	76.6	151.2
Cow dung	5.83	7.87	0.94	1.33	2.55	4.07*	77.4	161.0
Lime	6.91*	13.6*	1.12*	1.56*	2.79*	5.76*	92.6*	196.2*
Kaolin	6.4*	13.8*	1.20*	1.52*	2.88*	6.30*	94.0*	197.5*
Paraffin	5.3	7.73	0.97	1.00*	2.42	4.23*	83.5	166.1
Disprin	5.22	7.26	1.02	1.07*	2.37	4.20*	83.9	148.3
Cycocel	2.86*	5.41*	0.7*	1.16*	1.73*	3.95*	39.1*	112.9*
Paraquat	2.90*	4.83*	0.7	0.82*	1.51*	1.82*	25.5*	80.0*

* Significant from control at 5% level.

Spraying antitranspirant to reduce the water loss from the seedling was tried with the following antitranspirants *viz.*, rice gruel (v/v), cowdung slurry (1:4), lime solution (2%), kaolin (5%), liquid paraffin (0.5%), disprin (500 ppm), cycocel (500 ppm) and paraquat (50 ppm) were sprayed to 100 seedlings along with 0.5% Tween 20 solution (Sivakumar, 1995). The result revealed that none of the antitranspirant spray improved the height, diameter, dry matter and leaf area growth of *T. indica,* but lime and kaoline improved the above characters in *A. lebbeck*, *A. indica* and *E. tereticornis* seedlings (Table 8). Cycocel and paraquat sprays reduced the growth parameters in all the species. The transpiration rate was reduced in all the species by all the antitranspirants (Table 9). The tree species behaved differentially to each antitranspirant spray. However, the antitranspirant spray increased the diffusive resistance and RWC in all the species studied. Cycocel and paraquat increased the total chlorophyll content in all the four species studied (Table 9). The leaf water potential was increased in *T. indica* by liquid paraffin followed by paraquat and rice gruel, whereas, in *A. lebbeck*, cowdung slurry followed by kaolin spray increased the leaf water potential. The survival percentage was increased in *T. indica* by liquid paraffin followed by rice gruel and disprin; in *A. lebbeck* by plastic mulching followed by cowdung slurry and lime; in *A. indica* by plastic mulching followed by kaolin and disprin; and in *E. tereticornis* by plastic mulching and drought hardening.

Table 9: Effect of antitranspirants spray on transpiration rate (μg cm^{-2} s^{-1}), diffusive resistance (S cm^{-1}), RWC (%) and total chlorophyll content (mg g^{-1}) in *A. lebbeck* and E. *tereticornis* seedlings.

Treatments	Transpiration rate		Resistance		RWC		Total chlorophyll content	
	A. lebbek	*E. tereticornis*	*A. lebbek*	*E. tereticornis*	*A. lebbek*	*E. tereticornis*	*A. lebbek*	*E. tereticornis*
Control	5.67	3.84	2.83	3.44	77.1	80.2	3.87	1.6
Rice gruel	4.64*	2.84*	4.04*	5.24*	77.6*	84.6*	3.68*	1.31*
Cow dung	3.5*	3.28*	5.92*	4.79*	80.4*	85.8*	3.24*	1.47*
Lime	3.58*	2.60*	4.38*	5.10*	80.4*	85.3*	4.04*	1.69*
Kaolin	4.27*	2.97*	4.13*	5.48*	80.4*	86.9*	4.23*	1.66*
Paraffin	4.83*	3.39*	4.71*	5.18*	79.3*	83.9*	3.45*	1.64
Disprin	4.93*	3.27*	3.41*	4.41*	78.4*	84.9*	3.85	1.58
Cycocel	4.25*	3.0*	4.59*	5.60*	78.4*	84.3*	4.36*	2.04*
Paraquat	4.51*	2.76*	4.05*	6.24*	79.0*	83.3	3.97*	1.72*

* Significant from control at 5% level.

In an overall assessment of the water use efficiency (WUE) experiments, the establishment and the growth of the seedlings were improved by plastic mulching, drought hardening, root pruning and antitranspirant sprayings. Considering the cost and the manual work involved in other treatments, lime and kaolin sprayings can be recommended for large scale afforestation programmes.

6. EFFECT OF PRUNING ON GROWTH AND DEVELOPMENT

Pruning and thinning are silvicultural operations that are associated with changes in canopy size and condition. Green pruning, the removal of live branches from below, reduces canopy size of individual trees. Thinning, the selective removal of a proportion of the stand basal area, also reduces canopy size at the stand level and changes the distribution of solar radiation, and increases the supply of water and nutrients to the remaining tree canopy. A quantitative understanding of the physiological responses to both practices can be used as a basis for the management. The reduction in canopy size with green pruning triggers physiological responses similar to those observed following defoliation (Trumble *et al.*, 1993). In Eucalyptus, the responses occur at two levels. The first is an increase in leaf-level photosynthesis (*A*). In *E. nitens*, this can be as much as 175% of *A* measured in unpruned controls (Pinkard *et al.*, 1998). The second response is an increase in the rate of leaf development and delayed senescence of existing leaves (Pinkard and Beadle, 1998). The magnitude and duration of the photosynthetic response is related to the severity of pruning (Pinkard *et al.*, 1998) though these are ultimately constrained by the maximum rates of carboxylation. In practice, the physiological changes that lead to an increased capacity for biomass production at tree level are balanced by an increasing reduction in crown size. At a silvicultural level, the balance point is the pruning severity beyond which a significant reduction in growth is observed.

The enhanced levels of light-saturated leaf-level photosynthesis, A_{max} observed following thinning are related to the increase in the fraction of incident solar radiation penetrating to the middle and lower parts of the canopy compared to that in unpruned controls (Medhurst, 2000). The magnitude of this response increases with time from thinning and is associated with increased levels of leaf nitrogen concentration and changes in its distribution commensurate with its allocation according to the prevailing light environment and to maximise carbon gain (Field, 1983; Givnish, 1988). Parallel increases in leaf phosphorus concentration also occurred and the response for both nutrients was associated with decreases in specific leaf area (Medhurst, 2000). In the absence of water or nutrient stress, responses to thinning are optimised by retaining these improved physiological conditions within a particular silvicultural regime for the maximum possible period between thinning and harvest. Following pruning of *E. nitens*, there were no changes in leaf nitrogen or phosphorus concentrations, but pruning is associated with increases in nitrogen-use efficiency (Pinkard *et al.*, 1998). However, thinning *E. nitens* was not associated with any changes in nitrogen-use efficiency (Medhurst, 2000). Thinning is also associated with increases in tree and stand-growth efficiency, the amount of stem wood produced per unit leaf area. In a thinning fertilizer experiment with *E. marginata*, increased efficiency in response to thinning was limited mainly to individual tree basal area increment with faster-growing trees being more efficient (Stoneman and Whitford, 1995). However, application of fertilizer to unthinned stands was associated with increased water stress which limited rates of stand growth in spite of an increase in growth efficiency (Stoneman *et al.*, 1997).

Experiment was designed to investigate the effect of root pruning on growth and field establishment of *Albizia lebbeck*, *Azadirachta indica*, *Eucalyptus tereticornis* and *Tamarindus indica* was conducted by Sivakumar (1995). The seedlings were grown without polythene sheets below to allow the root to come out from the bags. After 60 days, the roots coming out from the bags were pruned with pruning shears at every 20 days interval. The result of the experiment showed that in all the species taken for study, root pruning increased height increment except *T. indica*. The increase in height over control in *A. lebbeck*, *A. indica* and *E. tereticornis* was 33, 15 and 26% respectively. Likewise, the collar diameter also increased in the above species, over control. Maximum increase was observed in *A. lebbeck* (33% over control) and *E. tereticornis* (33% over control). Dry matter and leaf area increment followed the same trend. Transpiration rate significantly increased over control in *A. lebbeck* (9%) and *E. tereticornis* (15%), but reduced in *T. indica* (14%) seedlings due to root pruning (Table 10). However, the diffusive resistance was increased by 14% in *T. indica* and decreased by 16 and 12% in *A. lebbeck* and *E. tereticornis*, respectively. Significant increase in RWC was observed in root pruned seedlings of *T. indica*, *A. lebbeck* and *E. tereticornis*.

However, an increase in total chlorophyll content and root volume was observed in root pruning treatment (Table 11).

Table 10: Effect of root pruning on height increment (cm), collar diameter (mm) dry matter increment (g plant^{-1}), leaf area increment (cm^2 plant^{-1}) and transpiration rate (μg cm^{-2} s^{-1}) in *T. indica, A. lebbeck, A .indica* and *E. tereticornis* seedling.

Tree species	Height increment		Collar diameter		Dry matter increment		Leaf area increment		Transpiration rate	
	C	RP	C	RP	C	RP	C	RP	C	RP
T. indica	3.67	3.50	0.95	0.95	2.27	2.00	61.5	73.4	3.72	3.19*
A. lebbeck	4.66	6.20*	0.96	1.28*	2.29	3.61*	64.1	142.4*	5.67	6.19*
A. indica	5.68	6.56*	0.85	1.0	3.78	4.10	71.3	76.1	3.00	2.94
E. tereticornis	8.02	10.12*	1.30	1.74*	4.97	6.77*	165	239.5*	3.84	4.43*

C - Control RP – Root pruned * Significance at 5% level

Table 11: Effect of root pruning on diffusive resistance (S cm^{-1}), RWC (%), total chlorophyll content (mg^{-1} g), and root volume (mm^3 plant^{-1}) in *T. indica, A. lebbeck, A. indica* and *E. tereticornis* seedling.

Tree species	Diffusive resistance		RWC		Total chlorophyll content		Root volume	
	Control	Root pruned	Control	Root pruned	Control	Root pruned	Control	Root pruned
T. indica	2.81	3.21*	79.7	80.2*	1.62	1.868	9.2	9.3
A. lebbeck	2.83	2.37*	77.1	79.6*	3.87	4.42*	7.9	12.3*
A. indica	3.68	3.89	86.4	86.4	2.03	2.01	10.1	10.3
E. tereticornis	3.44	3.01*	82.4	86.1*	1.59	1.69*	8.1	14.6*

C - Control RP – Root pruned * Significance at 5% level.

7. DROUGHT HARDENING AND FIELD ESTABLISHMENT

Woody plants have developed different mechanisms to cope with water stress. Some species tolerate drought (Kramer and Kozlowski 1979) by avoiding tissue desiccation through early and tight stomatal closure, low cuticular transpiration or foliage reduction. Other species tolerate drought by withstanding some degree of tissue dehydration without protoplasmic damage by enhancing the stability of cell membranes, osmotic adjustment and increasing cell wall elasticity (Gebre *et al.,* 1994). A variety of compounds can contribute to osmotic adjustment, but soluble carbohydrates, inorganic ions and amino acids predominate in most species (Dichio *et al.,* 2003). Osmotic adjustment and increased cell wall elasticity permit greater cell turgor at low tissue water potential, which may enable plants to maintain gas exchange and growth during drought. Holm oak (*Quercus ilex* L.) In natural populations enhance their dehydration tolerance by osmotic adjustment in response to summer drought (Kyriakopoulos and Richter 1991), and seedlings are vulnerable to transplanting stress and to summer drought, having high mortality and slow growth and high morality (Baeza *et al.,* 1991). Drought hardening applied in the nursery before planting may improve field establishment of seedlings by increasing their drought tolerance. In two conifer species, drought hardening increased seedling survival and was related to morphological changes that occurred during nursery cultivation (Van den Driessche, 1991). Restricted watering applied during seedling cultivation can induce osmotic adjustment, increase cell membrane stability to dehydration (Gebre *et al.,* 1994; Collet and Ghuel, 1997) and reduce both stomatal and residual transpiration (Edwards and Dixon, 1995; Villar-Salvador *et al.,* 1999).Drought hardening applied in the nursery increased drought tolerance of Holm oak seedlings primarily by affecting physiological traits rather than morphological attributes. Hardening promoted osmotic adjustment, reduced RT and stomatal conductance (*gs)* and enhanced plasmalemma stability. Seedlings subjected to drought-hardening attained the highest osmotic adjustment. Moderate and severe conditioning also reduced stomatal conductance and new root growth after drought stress recovery. Drought conditioning did not improve transplanting performance under either xeric or mesic conditions. The potential performance advantage in the field of drought-hardened seedlings is probably limited to those conditions where seedlings experience water stress immediately after planting.

Table 12: Effect of drought hardening on height increment (cm), collar diameter increment (mm) and dry matter increment (g plant^{-1}) in *Tamarindus indica, Albizia lebbeck, Azadirachta indica* and *Eucalyptus tereticornis.*

Species	Height increment		Collar diameter increment		Dry matter increment	
	C	DH	C	DH	C	DH
T .indica	3.67	3.44	0.95	0.82	2.27	2.07
A. lebbeck	4.66	5.82	0.96	0.91	2.27	2.02
A .indica	5.68	7.35*	0.85	1.07*	3.78	4.80*
E. tereticornis	8.02	15.94*	1.30	1.84*	4.97	7.29*

C – Control DH – Drought hardened plants

Hardening of tree species sapling is an important management practice for assured field establishment. To induce drought hardening, exposure to sublethal stress regime has proved to be an effective approach to withstand later lethal stress. This process is generally referred as acclimation. To prove this concept, 90 day old *Albizia lebbeck, Azadirachta indica, Eucalyptus tereticornis* and *Tamarindus indica* seedlings were watered to saturation, when the leaf water potential fell below -1.0 MPa up to 180 days. Thereafter, seedlings were saturated with water till the leaf water potential has reached -1.4 MPa. During this acclimation process, the associated physiological changes were studied by Sivakumar (1995). The result indicated that drought hardening increased the height by 98.8 and 29.4% in *E. tereticornis* and *A. indica*, respectively, over control. Significant increase in collar diameter was observed in drought hardened *A. indica* and *E. tereticornis* (Table 12).

The same trend was also observed for dry matter and leaf area increment. All the tree species showed remarkable reduction in transpiration rate, while the maximum reduction was observed in *E. tereticornis* (44%) followed by *A .lebbeck* (35%), *A. indica* (34%) and *T. indica* (21%). However, the drought hardening increased diffusive resistance in all the species, wherein, the highest percentage of increase was observed in *E. tereticornis* (150%) followed by *T. indica* (130%), *A. lebbeck* (124%) and *A. indica* (96%). Drought hardening significantly reduced RWC and total chlorophyll content in all the species and the reduction was subdued progressively as the days progressed in all the species. However, increased proline content was observed in all the species studied. Maximum accumulation of proline was observed in *A. indica* followed by *E. tereticornis* and *T .indica* (Table 13).

Table 13: Effect of drought hardening on leaf area increment (cm^2 $plant^{-1}$), transpiration rate ($\mu g\ cm^{-2}\ s^{-1}$), diffusive resistance ($S\ cm^{-1}$) and proline content ($mg\ g^{-1}$) in *Tamarindus indica, Albizia lebbeck, Azadirachta indica* and *Eucalyptus tereticornis.*

Species	Leaf area increment		Transpiration rate		Diffusive resistance		Proline content	
	C	DH	C	DH	C	DH	C	DH
T .indica	61.5	69.1	3.72	2.91*	2.81	6.48*	0.77	2.05*
A. lebbeck	64.1	67.8	5.67	3.66*	2.83	6.36*	0.51	1.11*
A .indica	71.3	121.4*	3.00	1.98*	3.68	7.24*	0.96	3.90*
E. tereticornis	165.8	224.4*	3.84	2.12*	3.44	8.62*	0.63	2.12*

C – Control DH – Drought hardened plants

8. NUTRIENT ENRICHMENT

Although there are many possible causes of plantation failure, initial nutrient deficiency is often a factor contributing to early planting check of white spruce seedlings (Sutton, 1995). Nursery grown seedlings often exhibit nutrient deficiency soon after planting because slow regeneration and extension of the root system restricts exploitation and absorption of soil nutrients for most of the growing season (Burdett, 1990). Hence, growth occurring after planting is highly dependent on internal reserves for nutrient requirements. The ability of plants to accumulate nutrient reserves for subsequent retranslocation and reutilization is the rationale for nutrient loading forest planting stock during nursery culture. The aim of nutrient loading is to counter initial planting check by building up the nutrient reserves of seedlings in the nursery by applying large amounts of fertilizer (Timmer and Munson, 1991). Loaded container-grown seedlings exhibit the capacity to retranslocate nutrients for current growth rather than conserving them in old tissue, which is characteristic of non loaded seedlings (Malik and Timmer, 1995). Moreover, the nutrient reserves that accumulate in response to exponential loading are more readily available for retranslocation than the nutrient reserves that accumulate in response to conventional loading. Nutrient loading of nursery seedlings to counter planting check offers several advantages over fertilization after planting in the field: (1) nutrient loading requires less fertilizer and is less costly to apply; (2) it avoids the problem that added nutrients may inadvertently enrich competing vegetation rather than target seedlings; (3) and enhanced growth of crop trees may suppress competing vegetation (Malik and Timmer, 1996). Exponentially increasing nutrient additions may be more effective and less toxic because enrichment is raised progressively to match exponential growth and nutrient demand. This schedule induces steady-state growth and nutrient accumulation in the plant that may favour field performance (Timmer, 1997).

McAlister and Timmer (1998) reported that high nutrient reserves in the seedlings at the end of the nursery rotation resulted in increased biomass production (40-190%) after planting, which was related to the ability of the seedlings to retranslocate internal reserves for new growth, despite little or no net uptake of nutrients during the first season after planting. Compared with conventionally loaded seedlings, retranslocation was greater in exponentially loaded seedlings, which had accumulated larger and more readily available nutrient reserves during the nursery phase. The shortness of this study limits its usefulness for predicting the persistence of the loading response after planting, but we postulate that the high nutrient status of loaded seedlings at the end of the nursery rotation will contribute to future growth through increased nutrient storage and retranslocation, thus prolonging the loading response.

9. PROPAGATION THROUGH TISSUE CULTURE

Tamarind is naturally propagated through seeds, though it can also be propagated rarely through cuttings to a limited extent. Genetic improvement of the species for higher biomass production is beset with problems due to its long generation time, heterozygous nature due to cross- pollination. Since adult trees are difficult to propagate through conventional techniques using cuttings, biotechnology can play an important role in the program of improvement. Therefore, micro propagation of elite genotypes occurring in natural stands is the most practical option to improve productivity in neglected tree species.

Micropropagation of *Tamarindus* species has been reported earlier from the adaxial surface of the cotyledons of 12 day old seedlings with BAP; shoot tip explants from *in vitro* grown seedlings with IAA and IBA (Knop and Nataraja, 1990) and hypocotyl segments (Sonia *et al.,* 1998).

In all these cases, the explants used were obtained from seedlings or from juvenile material. However, there are no reports yet on the regeneration from the explants obtained from the mature trees, tend to release polyphenols in the medium leading to poor establishment, stunted growth and slow multiplication. This study provides information on the *in vitro* regeneration of *Tamarindus indica* using the explants obtained from the young branches of adult trees and compares the explant response in terms of release of polyphenols, establishment of cultures and other micropropagation parameters.

The process of micropropagation of *Tamarindus indica* involves three stages *viz.* enlargement and sprouting of lateral buds, proliferation of shoots and rhizogenesis. The protocols were established for high frequency micropropagation of *Tamarindus indica* involving enhanced formation of adventitious shoots from the nodal segments of the branches from matured trees. Pre-existing lateral buds from nodal segments of both young and mature shoots, sprouted after 4-5 weeks after inoculation in the initiation medium. Release of polyphenols was noted in both types of explants initially which has greatly affected culture establishment. Exudation of polyphenols into the medium could be avoided to a large extent by washing the explants in running water and then drying them partially. This step was crucial for the initial establishment of the cultures. Addition of activated charcoal significantly reduced the polyphenol leaching. Sprouting of buds was far more frequent in explants from younger branches than in older branches. In *Tamarindus,* the apical portions of mature shoots and the zone near the floral and apical meristems were more suitable, since these regions are relatively more juvenile and have greater potential for regeneration. Nevertheless, the young shoots are understandably more juvenile than any part of living.

10. ABIOTIC STRESS

10.1. Elevated CO_2

The first physiologically meaningful contact between plant and atmosphere takes place at the leaf and most subsequent effects of increasing CO_2 concentration are linked to changes in CO_2 assimilation. The undisputed response to increasing CO_2 is an increase in photosynthesis, but a host of questions have arisen concerning longer-term effects after growth and development at a higher CO_2 concentration. The key question, relative to increasing CO_2 concentration on forests, is how much photosynthesis will increase as atmospheric CO_2 concentrations rise and what bearing this will have on higher-level processes. In trees growing outdoors, rooted in the ground, the rate of photosynthesis were almost always higher in elevated CO_2. Photosynthesis was stimulated 40–80%, the photosynthetic response might be reduced by nutrient deficiency or conversely, enhanced in combination with other stresses or unaffected by stress. Many of the differences in CO_2 effects can be explained by temperature differences. As discussed by Long (1991), the relative affinity of Rubisco for CO_2 decreases markedly with increasing temperature, but elevated CO_2 concentrations increase the competitive inhibition of oxygenation such that the relative stimulation of assimilation by elevated CO_2

increases with temperature and the temperature optimum for assimilation increases with increasing CO_2. Experiments with temperature manipulation treatments confirm this with higher CO_2 enhancement ratios for trees growing in temperatures raised 2–4 °C above the ambient chambers. Light attenuation within a mature forest canopy and the interactions between CO_2 and leaf acclimation to light environment are important factors in evaluating CO_2 responses at the canopy level.

Stomatal conductance (*gs*) may remain lower in foliage grown under elevated CO_2, because of reduced stomatal density. If lower *gs* reduce intercellular CO_2 concentrations (*Ci*) in elevated CO_2 grown foliage, then E/A ratios at a common *Ca* would not represent differences in biochemical capacity at equivalent conditions. From the range of responses obtained from *A*/*Ci* curves, it is apparent that prolonged growth at elevated CO_2 does not result in a consistent down-regulation of photosynthetic parameters. The pattern does suggest a potential decrease in both *Vc*max and *J*max, particularly late in the season, concurrent with decreases in measured Rubisco content (and thus activity per unit leaf area), although decreases in Rubisco activity, measured biochemically (Turnbull *et al.*, 1998).

In most cases, leaf mass per unit area (SLW) is higher with growth at elevated CO2 and in many cases, leaf nitrogen concentrations decrease while starch and soluble sugar concentrations increase (Korner and Miglietta, 1994). With respect to the N reallocation hypothesis, Drake *et al.* (1997) pointed out that at higher temperatures and increasing CO_2, a leaf can sustain a substantial loss in Rubisco content (which accounts for a significant fraction of foliar N) without an effect on assimilation rate. Nevertheless, it is important to emphasize that changes in leaf biochemistry, including seasonal declines in *Vc*max or Rubisco, do not eliminate a photosynthetic response to elevated CO_2. All of the evidence from field-grown trees suggests a continued, and surprisingly consistent, stimulation of photosynthesis, 60% for a 300 ppm increase in CO_2. There is, at present, little reason to expect a long-term loss of sensitivity to CO_2 as suggested by earlier pot studies of trees.

10.2. Nutrient Stress

The temporal distribution of soil nutrients is heterogeneous and thus the uptake, storage and later remobilization of brief nutrient pulses may be critical for growth in nutrient limited habitats. Warren *et al.* (2004) investigated the response of photosynthesis and the major nitrogen (N) fractions in needles of 2-year-old douglas-fir (*Pseudotsuga menziesii* (Mirb.) Franco) seedlings to a 15-day nutrient pulse (containing 250 ppm N). The nutrient pulse (N pulse) was imposed in late July, toward the end of the seedlings third growing season and subsequent changes in photosynthesis and needle N fractions were examined over the following 3 months. Needles are sites of photosynthesis and putative storage organs. Thus, they tested two hypotheses: (1) N from the N pulse is quickly synthesized from soluble non-protein N into soluble proteins, especially Rubisco, and (2) the N pulse increases photosynthetic rates and thus growth. Warren *et al.* (2004) also examined an alternative hypothesis that Rubisco functions also as a storage protein, in which case it would predict increases in amount of Rubisco in response to the N pulse without concomitant increases in photosynthesis. Soluble non-protein N was the most dynamic N pool and may have constituted a temporary storage reservoir; however, the quantitative significance of soluble non-protein N is questionable because this pool was at most only 7% of total N.

Concentrations of Rubisco were unaffected by the N-pulse treatment and there was little evidence that Rubisco served as a storage protein. Nutrient-pulse seedlings added twice as much dry mass as controls during the 3 months post-treatment (Warren *et al.,* 2003). Over the same period, the maximum rate of light-saturated photosynthesis (*A*max) declined to low rates in control seedlings, whereas *A*max increased in N-pulse seedlings. Nevertheless, treatment and temporal trends in N and Rubisco content per unit area were poorly related to *A*max, and it seems likely that photosynthesis was limited by additional factors, perhaps thylakoid proteins or an inadequate supply of other nutrients. The twofold greater growth in N-pulse seedlings compared with control seedlings may have been a consequence of modest increases in the amount of foliage and large treatment differences in *A*max.

Treatment and temporal trends in *A*max were poorly related to N and Rubisco contents per unit area, and it seems likely that photosynthesis was limited by thylakoid proteins. In Douglas-fir seedlings, concentrations of Rubisco were relatively unaffected by the N-pulse treatment, casting doubt on the universality of the hypothesis that Rubisco functions as a storage protein.

10.3. Drought Stress And Mycorrhizal Inoculation

The lack of mycorrhizal fungi on root systems is a leading cause of poor plant establishment and growth in a variety of forest, restoration, agricultural, suburban and urban landscapes. At afforestation sites devoid of mycorrhizal propagules (*e.g.*, cutover lands, treeless areas, mining spoils), non mycorrhizal seedlings never develop fully and often die. Moreover, it has been demonstrated that specific endomycorrhizal or ectomycorrhizal fungi are able to improve the survival and early growth of various tree species in the field. Mycorrhizal association is estimated to occur in 95% of native vegetation from undisturbed sites, whereas it occurs in less than 1% of vegetation from disturbed sites. However, mycorrhizal symbiosis can be reestablished and the benefits of mycorrhizal association can be achieved by inoculating seedlings before they are transplanted to disturbed sites.

Plant hydraulic conductance has been used to quantify tree behavior in the soil–plant–atmosphere continuum (Lamhamedi *et al.,* 1992) and to assess the functional significance of mycorrhizal fungi in water transport (Nelsen, 1987). Morte *et al.* (2000) suggested that mycorrhization could increase water uptake by increasing effective root hydraulic conductivity. The presence of abundant and well developed rhizomorphs on the root systems of trees inoculated with *Rhizopogon roseolus* and *Scleroderma citrinum*. The architecture of the extrametrical phase of mycorrhizas influences resistance to water flow through the soil–root interface and rhizomorphs have been shown to absorb water and facilitate its transport over ecologically significant distances (Cairney, 1992). Hubbard *et al.* (1999) compared old and young *Pinus ponderosa* var. *ponderosa* Dougl. trees to test the hypothesis that lower hydraulic conductance limits photosynthesis.

They suggested that water transport in trees is regulated by the hydraulic conductance of the soil–root–shoot–leaf pathway, whereas whole tree hydraulic conductance is a function of path lengths and permeability of these tissues. Because stomatal conductance and photosynthesis depend on the transport of water from soil to leaf and atmosphere, changes in whole tree hydraulic conductance may affect leaf gas exchange. Water relations, measured as gas exchange, were significantly affected by inoculation with *R. roseolus* and *S. citrinum*, mainly when soil water availability was low. Sands and Theodorou (1978) detected no significant difference in transpiration rates between drought treated non-inoculated radiata pine seedlings and drought treated *Rhizopogon* inoculated seedlings. In contrast, other authors have found positive effects of mycorrhization under drought conditions. Pallardy *et al.* (1983) studied tissue water relations of mycorrhizal and non-mycorrhizal *Pinus echinata* Mill. seedlings during drying cycles and found that trees inoculated with *Pisolithus tinctorius* showed significantly greater root and shoot dry mass than non-inoculated trees.

Chlorophyll fluorescence parameters indicated that trees mycorrhizal with *R. roseolus* or *S. citrinum* suffered less severe stress than control trees when soil water availability was low. The decrease in *Fv/Fm* is a good indicator of photoinhibitory damage caused by light when plants are subjected to environmental stresses (Ogren and Oquist, 1985). At low water availability, leaves exhibiting low assimilation rates may be exposed to high irradiance stress, especially at midday.

Under such conditions, the photosynthetic apparatus suffers photo damage as a result of over-excitation of photosystem II (Kyle, 1987). The decrease in *Fv/Fm* may because by an increase in *Fo*, as a result of the destruction of the photosystem II reaction centers or by a decrease in *Fv*, indicating an increase in non-photochemical quenching. *R. roseolus* and *S. citrinum* mycorrhizal trees showed a decrease in *Fv/Fm* with respect to control trees, suggesting a mitigation of the negative effects of water deficit. Under conditions of low water availability, inoculation of radiata pine seedlings with *R. roseolus* or *S. citrinum* resulted in increased water uptake that was sufficient to permit the stomata to remain open for most of the day. The ability to maintain open stomata and photosynthesis during the early stages of a drought could increase the supply of carbon for growth, particularly new root growth, which requires current photosynthates (Van den Driessche, 1987).

Moreover, an increased rate of photosynthesis also supplies additional carbon for osmotic adjustment, thereby promoting turgor maintenance (Tan *et al.*, 1992) and permitting higher water uptake from dryer soils. Therefore, inoculation of radiata pine with *R. roseolus* or *S. citrinum* provides the tree with a greater capacity to tolerate limited soil water availability.

10.4. Salinity Stress

A greenhouse study was undertaken to evaluate the response to sodium chloride (control, 68 mM and 137 mM) of different species and genotypes of genus *Populus* L. (*P.* × *euramericana* (Dode) Guinier clones: I-214, Flevo, I-MC and Triplo; *P. deltoides* × *P. alba* clones: M30 and M31; *P. alba* L. clones: Jalon-F-1–22, Jalon F-1–4, Guadalquivir F-21–38, Guadalquivir F-21–39, Guadalquivir F-21–40 and Montanana F-2–9; and *P. euphratica* Oliv.). Survival, visual symptomatology, height growth, dry weight both in shoots and roots, photosynthesis rate and leaf concentrations of Na^+ , K^+ and Ca^{2+} were evaluated by Sixto *et al.* (2005). There were differences in survival between clones (0 per cent for the most sensitive and 100 per cent for the most tolerant) but growth was significantly reduced in all cases except for dry weight of *P. euphratica* at 68 mM. The photosynthesis rate was significantly reduced in all clones at the end of the experiment (60 days) but not in the same way in the middle (30 days). The leaf levels of Na^+, K^+ and Ca^{2+} were also significantly different between clones. Intra and interspecific variability have been detected in the genus, especially in *P. alba.* Decrease of the net photosynthesis rate seems to be associated with the increase of salt concentration as well as the time exposure of the saline treatments. The changes produced in the photosynthetic system by the presence of salt may be a good indicator of the behavioural differences, though it needs to be borne in mind that other techniques, such as the emission of chlorophyll fluorescence, are faster and offer easier detection. The significantly higher concentrations of Na^+ shown in leaves of the sensitive clones in relation to the more tolerant might explain the differences found between the responses, whether due to the toxic effect that the ion can produce, as described by Maas (1993) or to its interference with other physiological processes such as photosynthesis, changes in water status, soluble protein, etc. It should be taken into account that difference in Na^+ concentrations may be a good indicator of response differences. The negative correlation found between concentration and acrial dry weight, in a similar way to that described by Graig *et al.* (1990), could confirm this.

CONCLUSION

Deforestation has already caused very serious problems leading to soil erosion, flooding and changes in local climate. The amount and distribution of the rainfall are the reasons for the success or failure of an afforestation programme. The inconsistency in the rainfall pattern makes afforestation programme difficult and phasing of out planting indecisive. Any delay in starting of the rain after transplanting, leaves the seedling prone to drought. The success of the seedling establishment can be increased with proper utilization of growth stimulating chemicals and growth regulators by promoting root growth and development. Apart from this, development of suitable microbial inoculants for tree species may also increase the survival percentage of the tree seedlings. Enhanced growth rate of the tree seedlings can be achieved by the foliar sprays of plant growth regulators. Root pruning is normal method of conditioning tree seedling in nursery. Root pruned plants had increased water absorption, hardened woody stem and good field survival rates. Reducing water loss through foliar spray of antitranspirants is also proved to be the best. Hence, to increase the field establishment, growth of seedlings the above physiological strategies has to be followed.

REFERENCES

Baeza, M.J., A. Pastor, J. Martin and M. Ibanez. 1991. Mortalidad post-implantacion en repoblaciones de *Pinus halepenesis*, *Quercus ilex*, *Ceratonia siliqu* y *Tetraclinis articulata* en la provincia de Alicante. *Stvdia Oecologia,* **8** : 139–146.

Bhatnagar H.P. and K.K. Talwars. 1978. Studies on the effect of growth regulators on growth and trachied characters of *Pinus caribaea* seedlings. *Indian Forester,* **104**: 333-334.

Bhattacharjee, S.K. 1984. Effect of growth regulating chemicals on growth, flowering and tuberous root formation of *Dalulia variables. Punjab Horticulture Journal,* **24** ; 138-143.

Burdett, A.N. 1990. Physiological processes in plantation establishment and the development of specifications for forest planting stock. *Canadian Journal of Forest Research,* **20**: 415--427.

Cairney, J.W.G. 1992. Translocation of solutes in ectomycorrhizal and saprotrophic rhizomorphs. *Mycological Research,* **96**: 135–141.

Collet, C. and J.M. Ghuel 1997. Osmotic adjustment in sessile oak seedlings in response to drought. *Annals of Forestry,* **54**: 389–394.

Dela Cruz R.E., M.Q. Manalo, N.S. Aggangan. and J.D. Tambalo. 1988. Growth of three legume trees inoculated with VAM fungi and *Rhizobium. Plant and Soil,* **108**: 111-115.

Dichio, B., C. Xiloyannis, K. Angelopoulos, V. Nuzzo, S.A. Bufo. and G. Celano. (2003). Drought-induced variations of water relations parameters in *Olea europaea. Plant and Soil,* **257**: 381–389.

Drake, B.G., M.A. Gonzalez-Meler, and S. P. Long. 1997. More efficient plants : A consequence of rising atmospheric CO_2? *Annual Review of Plant Physiology and Plant Molecular Biology,* **48**: 609–639.

Edwards, D.R. and M.A. Dixon. 1995. Mechanisms of drought response in *Thuja occidentalis* L. I. Water stress conditioning and osmotic adjustment. *Tree Physiology,* **15**: 121–127.

Farquhar, G.D., M.H., O'Leary and J.A. Berry. 1982. On the relationship between carbon isotope discrimination and the intercellular carbon dioxide concentration in leaves. *Australian Journal of Plant Physiology,* **9**: 121–137.

Field, C. 1983. Allocating leaf nitrogen for the maximization of carbon gain: leaf age as a control on the allocation program. *Oecologia,* **56**: 341–347.

Gebre, G.M., M.R. Kuhns and J.R. Brandle. 1994. Organic solute accumulation and dehydration tolerance in three water-stressed *Populus deltoides* clones. *Tree Physiology,* **14**: 575–587.

Givnish, T.J. 1988. Adaption to sun and shade: a whole-plant perspective. *Australian Journal of Plant Physiology,* **15**: 63–92.

Graig, G.F., D.T. Bell. and C.A. Atkins. 1990. Response to salt and waterlogging stress of ten taxa of *Acacia* selected from naturally saline areas of western Australia. *Australian Journal of Botany,* **38**: 619 –630.

Guedes A.C., D.J. Cantliffe. and T.A. Neli. 1981. Morphological changes during lettuce seed priming and subsequent radicle development. *Journal of American Society of Horticultural Science,* **106**: 121-126.

Harley J.L. and S.E. Smith. 1988. Mycorrhizal symbiosis Academic Press. London, pp.483.

Honeysett, J.L., C.L.Beadle. and C.R.A. Turnbull. 1992. Evapotranspiration and growth of two contrasting species of eucalyptus under non-limiting and limiting water availability. *Forest Ecology Management,* **50**: 203–216.

Hubbard, R.M., B.J. Bond and M.G. Ryan. 1999. Evidence that hydraulic conductance limits photosynthesis in old *Pinus ponderosa* trees. *Tree Physiology,* **19**: 165–172.

Hira ,G.S. ,J.S. Kalkai and S.K.Shakya 1990.Use of plastic sheet for plantation in problem soils.In proceedings of the XI International Congress on the use of plastics in agriculture.26th Feb to 2nd Mar.1990., New Delhi. India. Rotterdam. Netherlands ,A.A.Balkema,161-165. ISBN 90-6191-998-3

Ilango K. 2001. Effect of some microbial inoculants on growth, biomass and nutrient uptake of tamarind (*Tamarindus indica*) seedlings. Ph.D Thesis submitted to Forest College and Research Institute, Tamil Nadu Agricultural University, Mettupalayam.

Jakobsen, I., E.J. Joner. and J. Larsen. 1994. Hyphal phosphorus transport, a keystones to mycorrhizal enhancement of plant growth pp. 133-146. *In* S. Gianinazzi and H. Schueep (Eds). Impact of Arbuscular Mycorrhizae on Sustainable and Natural Ecosystemss. Birkhauser Verlag, Swizerland.

Jann R.C. and R.D. Aman. 1977. What is germination? pp.77-109. *In* A.A. Khan, Eds. The physiology and biochemistry of seed dormancy and germination. North - Holland publishing company, Ansterdam.

Khan, A.A. 1968. Inhibition of gibberellic acid induced germination by abscisic acid removal by cytokinins. *Plant Physiology,* **43**: 1463-1465.

Knop, M.S. and K. Nataraja. 1990. *In vitro* plantlet regeneration from shoot tip cultures of *Tamarindus indica* (L). *Indian Journal of Forestry. 13*, 30-33.

Korner C.H. and F. Miglietta. 1994. Long term effects of naturally elevated CO_2 on mediterranean grassland and forest trees. *Oecologia,* **99**: 343–351.

Kramer, P.J. and T.T. Kozlowski. 1979. Physiology of woody plants. Academic Press, San Diego, p.811.

Kumaran, K., M. Palani, R. Jerlin. and C. Surendran. 1994. Effect of growth regulators on seed germination and seedling growth of neem (*Azadirachta indica*). *Journal of Tropical Forest Science,* **6**: 529-532.

Kumaran, K., C.Surendran and M.Palani. 1996. Effect of presowing chemical treatment on germination and seedling growth in neem (*Azadirachta indica* A. Juss.). *Indian Journal of Forestry,* **19**: 87-88.

Kyle, D.J. 1987. The biochemical basis for photoinhibition of photosystem II. *In* Photoinhibition. D.J. Kyle, C.B. Osmond and C.J. Arntzen (Eds.). Elsevier Science, Amsterdam, pp 197–226.

Kyriakopoulos, E. and H. Richter. 1991. Dessication tolerance and osmotic parameters in detached leaves of *Quercus ilex* L. *Acta Oecologia,* **12**: 357–367.

Lamhamedi, M.S., P.Y. Bernier. and J.A. Fortin. 1992. Hydraulic conductance and soil water potential at the soil–root interface of *Pinus pinaster* seedlings inoculated with different dikaryons of *Pisolithus* sp. *Tree Physiology,* **10**: 231–244.

Lauer. D.K. 1987. Seedling size influences early growth of long leaf pine. *Tree Pianter's Notes,* **38**: 16-17.

Long, S.P. 1991. Modification of the response of photosynthetic productivity to rising temperature by atmospheric CO_2 concentrations: Has its importance been underestimated? *Plant Cell and Environment,* **14**: 729–739

Maas, E. 1993. Salinity and citriculture. *Tree Physiology,* **12**: 195 –216.

Majada, J. 2004. Effectiveness of mycorrhizal inoculation in the nursery on growth and water relations of *Pinus radiata* in different water regimes. *Tree Physiology,* **24**: 65–73

Maideen, S.K., A.S. Jacqueline. and R.S.Vinaya Rai. 1990. Presowing chemical treatment to hastens germination of *Cauarina equiseifolia. International Tree Crops Journal,* **6**: 173-181.

Malik, V. and V.R. Timmer. 1995. Interaction of nutrient loaded black spruce seedlings with neighbouring vegetation in greenhouse environments. *Canadian Journal of Forest Research,* **25**: 1017--1023.

Malik, V. and V.R. Timmer. 1996. Growth, nutrient dynamics and inter-specific competition of nutrient loaded black spruce seedlings on a boreal mixed wood site. *Canadian Journal of Forest Research,* **26**: 1651--1659.

Mathur D.D., G.A. Couvillon, H.M. Vines. and C.H. Hendrshoot. 1971. Stratification effects on endogenous gibberelllic acid in peach seed. *Horticulture Science* ,**6**: 538-539.

Mathur, N. and A. Vyas. 1996. Physiological change in *Ziziphus mauritiana* by different VAM fungi. *Indian Forestry,* **122**: 501-505.

McAlister, J.A. and V.R. Timmer. 1998. Nutrient enrichment of white spruce seedlings during nursery culture and initial plantation establishment. *Tree Physiology,* **18**: 195-202.

Medhurst, J.L. 2000. Physiological responses of *Eucalyptus nitens* to thinning. Ph.D. thesis, submitted to University of Tasmania, Hobart, Australia.

Merina Prem Kumari, S. 1991. Response of certain horticultural crops to inoculation of VAM fungi *Azospirillium* and Phosphobacteria. M.Sc., (Ag.) Thesis, submitted to Tamil Nadu Agricultural University, Coimbatore, Tamil Nadu, India.

Miuko, G. 1974. Effect of seedling size on growth of field planted *Pinces radiates. Forestry Technical papers, Forestry Commission, Victorias,* **21**: 58-68.

Morte, A., C. Lovisolo. and A. Schubert. 2000. Effect of drought stress on growth and water relations of the mycorrhizal association *Helianthemum almeriense–Terfezia claveryi. Mycorrhiza,* **10**: 115–119.

Mukerji, K.G., B.P.Chamola, A.Kaushik and N.Sarwar. 1996. Vesicular arbuscular – mycorrhiza potential biofertilizer for nursery raised multipurpose tree species in tropical soils. *Annals of Forestry,* **4**: 12-20.

Nanda K.K. and R.S. Dhindsa. 1967. Effect of gibberellic acid (GA_3) on nitrogen content of soybean (*Glycine max* L.) and mechanism of gibberellin action. Proceedings International Symposium on Plant Growth Substances Calcutta University, Calcutta, January 23-28.

Nelsen, C.E. 1987. The water relations of vesicular–arbuscular mycorrhizal systems. *In* Ecophysiological of VA Mycorrhizal Plants. G.R. Safir. (Eds.) CRC Press, Boca Raton, FL, pp 71–91.

Niranjan R, S. Banwarilal. and M. Rao. 1990. Studies on the effect of *Rhizobium* and endomycorrhizal interaction in *Dalbergia sissoo. In*: Proceedings of a National Conference on Mycorrhiza. Hisar Agricultural University, Hissar, India. Feb 14-16, pp.205-27.

Ogren, M. and G. Oquist. 1985. Effect of drought on photosynthesis, chlorophyll fluorescence and photoinhibition susceptibility in intact willow leaves. *Planta,* **166**: 380–388.

Olbrich, B.W., D. Le Roux, A.G. Poulter, W.J. Bond. and W.D. Stock. 1993. Variation in water use efficiency $\delta^{13}C$ and levels in Eucalyptus grandis clones. *Journal of Hydrology,* **150**: 615–633.

Palani M., M.G. Dasthagir. and K. Kumaran. 1995. Effect of presowing chemical treatment on germination and seedling growth in *Acacia nilotica. International Tree Crop Journal,* **8**: 189-192.

Palani, M., M.G.Dasthagir, K.Kumaran and R.Jerlin. 1996. Effect of presowing treatment on growth attributes of *Albizia lebbek* (L.) Benth. *Annals of Forestry,* **4**: 85-88.

Pallardy, S.G., W.C. Parker, R.K. Dixon, H.E. Garrett and B.A. Thielges. 1983. Tissue water relations of roots and shoots of droughted ectomycorrhizal shortleaf pine seedlings. *In* Physiology and Genetics of Intensive Culture. Proceedings of Seventh North American Forest Biology Workshop, University of Kentucky, Lexington, KY, pp 368–373.

Paricha. P.C., B.K. Ghosh. and N.C. Saroo. 1977. Further studies on the significances of cycocel is enhancing drought resistance in rice. *Science and Cultures,* **43**: 230-231.

Pate, J. and D. Arthur. 1998. $\delta^{13}C$ analysis of phloem sap carbon: novel means of evaluating seasonal water stress and interpreting carbon isotope signatures of foliage and trunk wood of *Eucalyptus globulus. Oecologia,* **117**: 301–311.

Pinkard, E.A. and C.L.Beadle. 1998. Aboveground biomass partitioning and crown architecture of *Eucalyptus nitens* following green pruning. *Canadian Journal of Forest Research,* **28**: 1419–1428.

Pinkard, E.A., C.L.Beadle, N.J.Davidson. and M.Battaglia. 1998. Photosynthetic responses of *Eucalyptus nitens* (Deane and Maiden) Maiden to green pruning. Trees ,**12**: 119–129.

Prasad, R. and G. Mohammad. 1987. Effect of gibberellic acid, nutrients and bavistin on chlorosis and growth of *Grevillia* species. *Journal of Tropical Forestry*,**3**: 156-159.

Reddy. B., D.J. Bagyarak and B.C. Mallesha. 1996. Selection of efficient VA-mycorrhizal fungi for acid time. *Indian Journal Microbiology,* **36**: 13-16.

Rook,D.A. 1971 .Effect of undercutting and wrenching on growth of *pinus radiate* D.Dan seedlings. *Journal of Applied Ecology,*: 477-490

Sands, R. and C. Theodorou. 1978. Water uptake by mycorrhizal roots of radiata pine seedlings. *Australian Journal of Plant Physiology,* **5**: 301–309.

Seiler,J.R. and J.D.Johnson.1985. Photosynthesis and transpiration of loblolly pine seedlings as influenced by moisture stress conditioning. *Forest Science*,**31**: 742-749

Sekar I., K. Vanangamudi. and K.K. Suresh. 1995. Effect of biofertilizers on the seeding biomass, VAM colonization, enzyme activity and phosphorus uptake in the Shola tree species. *My Forest,* **31**: 21-26.

Sekar, I., K. Vanangamudi. and K.K. Suresh. 1996. Influence of bio-fertilizers on the seedling growth of shola tree species. *Journal of Tree Science,* **15**: 54-57.

Shiver, B.D., B.E. Borders, H.H. Page. and S.M. Raper. 1990. Effect of some seedling morphological and planting quality variables on seedling survival in the Georgias Piedmont. *Journal of Applied Forestry,* **14**: 109-114.

Singh, O., S. Rhagat. and S.P. Chaukiyal. 1988. Effect of seedling size on the growth of transplanted spruce seedlings in the nursery. *Indian Forestry,* **11**(*14*): 470-473.

Sivakumar, V. 1995. Studies on silvicultural and physiological strategies to improve field establishment and growth of tree seedlings. M.Sc Thesis submitted to Forest College and Research Institute, TNAU, Mettupalayam.

Sixto, M., J.M. Grau, N. Alba. and R. Alia. 2005. Response to sodium chloride in different species and clones of genes *Populus* L. *Forestry,* **78**: 93-104.

Sonia J. P.K., A. Gulati and S. Dahiya. 1998. Direct organogenesis in hypocotyls cultures of *Tamarindus indica. Biologia Plantarum,* **41**: 331-337.

Srinivas K. 1987. Studies on the occurrence of vesicular – arbuscular mycorrhizal fungi and their interaction with some forest tree species. M.Sc (Ag.) Thesis, submitted to Tamil Nadu Agricultural University, Coimbatore – 641 003.

Stoneman, G.L. and K.Whitford. 1995. Analysis of the concept of growth efficiency in *Eucalyptus marginata* (Jarrah) in relation to thinning, fertilising and tree characteristics. *Forest Ecology Management,* **76**: 47–53.

Stoneman, G.L., D.S. Crombie, K. Whitford, F.J. Hingston, R. Giles, C.C. Portlock, J.H. Galbraith. and G.M. Dimmock. 1997. Growth and water relations of *Eucalyptus marginata* (Jarrah) stands in response to thinning and fertilisation. *Tree Physiology,* **17**: 267–274.

Sudhir, M.U., N.J. Sonali, S.S. Paude, S.J. Gaikwad. and A.S. Jawarkar. 1994. Microbial Technology for raising seedlings of fast growing tress. *Indian Journal of Forestry,* **17**: 243-248.

Sutton, R.F. 1995. White spruce establishment: initial fertilization, weed control, and irrigation evaluated after three decades. *New Forestry,* **9**: 123--133.

Tan, W., T.J. Blake and T.J.B. Boyle. 1992. Drought tolerance in faster- and slower-growing black spruce (*Picea mariana*) progenies: II. Osmotic adjustment and changes of soluble carbohydrates and amino acids under osmotic stress. *Physiology Plantarum,* **85**: 645–651.

Tanaka,Y., J.D.Walstad and J.E. Borrecco.1976.The effect of wrenching on morphology and field performance of Douglas fir and loblolly pine seedlings. *Canadian Journal of Forest Research,* **6**: 453-458.

Timmer, V.R. 1997. Exponential nutrient loading: a new fertilization technique to improve seedling outplanting performance on competitive sites. *New Forestry,* **13**: 279--299.

Timmer, V.R. and A.D. Munson. 1991. Site-specific growth and nutrient uptake of planted *Picea mariana* in the Ontario Clay Belt. IV. Nitrogen loading response. *Canadian Journal of Forest Research,* **21**: 1058--1065.

Trumble, J.T., D.M. Kolodney-Hirsch. and I.P. Ting. 1993. Plant compensation for arthropod herbivory. *Annual Review of Entomology*, **38**: 93–119.

Turnbull, M. H., D.T. Tissue, K.L Griffinm, G.N.D. Rogers. and D. Whitehead. 1998. Photosynthetic acclimation to long-term exposure to elevated CO_2 concentration in *Pinus radiata* Don. Is related to age of needles. *Plant Cell and Environment*, **21**: 1019–1028.

Van den Driessche, R. 1987. Importance of current photosynthate to new root growth in planted conifer seedlings. *Canadian Journal of Forest Research*, **17**: 776–782.

Van den Driessche, R. 1991. Influence of container nursery regimes on drought resistance of seedlings following planting. I. Survival and growth. *Canadian Journal of Forest Research*, **21**: 555–565.

Vanangamudi K., A. Venkatesh, Jayaprakash, Mallika Vanagamudi, J.R. Umarani. and R.S. Vinaya Rai. 1999. Effect of presowing treatments of germination seedling growth the *syzygium cumini*. *Journal of Tropical Forest Science*, **11**: 846-848.

Vanangamudi K., Mallika Vanangamudi A. Venkatesh, R.S. Vinya Rai, R. Umarani and S. Balaji. 2000. Effect of osmotic priming on seed germination and vigour of neem (*Azadirachta indica*). *Journal of Tropical Forest Science*, **12**: 153-155.

Venator. C.R. 1983. First years survival of morphologically graded loblolly pine seedlings in central Louisianas. *Tree planter's Notes*, **34**: 34-36.

Venkatesh A., Malika Vanangamudi, K. Vanangamudi, R. Umarani. and K.T. Parthiban. 1997. Improving seed germination and seedling growth of *Pongamina pinnata* (L.) Pierre through pre-sowing seed treatment. *Journal Tree Science*, **16**: 75-80.

Venkatesh A., Mallika Vanangamudi, K. Vanangamudi, K.T. Parthiabn, V. Ravichandran. and R.S. Vinaya Rai. 2000. Effect of growth stimulants on seed germination and morpho-physiological attributes in pungam (*Pongamia pinnata*). *Journal of Tropical Forest Science*, **12**: 643 – 649.

Venkatesh, A., Mallika Vanangamudi, K. Vanangamudi, V. Ravichandran and R.S. Vinaya Rai. 1998. Impact of biofertilizers on morpho-physiological attributes in pungam (*Pongamia pinnata* (L.) Pierre) seedlings. *Tropical Agricultural Research and Extension*, **1**: 7-11.

Venkatesh, K. Mallika Vanangamudi. and K. Vanangamudi.2003. Effect of seedling grade on growth and survival of pungam. *Journal of Tropical Forest Science*, **15**: 28-30.

Verma A.N. and P.T. Tandon. 1988. Effect of growth regulators on germination and seedling growth of *Pinua kesiya* (Royle ex Gord) and *Schima khasiana* Dyer. *Indian Journal of Forestry*, **11**: 32-36.

Villar-Salvador, P., L. Ocana, J. Penuelas. and I. Carrasco. 1999. Effect of water stress conditioning on the water relations, root growth capacity, and the nitrogen and nonstructural carbohydrate concentration of *Pinus halepensis* Mill. seedlings. *Annals of Forestry*, **56**: 459–465.

Warren, C. R., N. J. Livingston. and D. H. Turpin. 2004. Photosynthetic responses and N allocation in Douglas-fir needles following a brief pulse of nutrients. *Tree Physiology*, **24**: 601–608.

Warren, C.R., N.J. Livingston. and D.H. Turpin. 2003. Response of Douglas-fir seedlings to a brief pulse of ^{15}N labeled nutrients. *Tree Physiology*, **23**: 1193–1200.

Yanagishima, N and C. Shimoda. 1973. Auxin and Yeast. *Botanical Review*, **39**: 11.

Zelitch.I 1969. Stomatal control. *Annual Review of plant Physiology*, **20** : 329-350.

13

Seed Source on Tree Seed Quality

K. Rajal[1], A. Bharathi[2], S. Marimuthu[3], Mallika Vanangamudi[4] and S. Kauitha[1]

[1] *Central Institute for Cotton Research, Coimbatore 641 003*

[2] *Department of Seed Science and Technology, Tamil Nadu Agricultural University Coimbatore 641 003*

3. Department of Plant Breeding and Genetics, Adhiparasakthi Agricultural College, G. B. Nagar, Kalavai 632 506

[4] *Department of Crop Physiology, Tamil Nadu Agricultural University Coimbatore 641 003*

1. INTRODUCTION

A characteristic feature of all living organisms is the immense natural variability they exhibit for various characters. Broadly three types of variations are recognized at the intraspecific level *viz.*, tree-to-tree variation on the same site, site-to-site variation and geographical variation from different parts of the species range. Natural selection usually leads to the establishment of geographical races or provenances within species which have widespread ranges. Success in the establishment and productivity of forest tree plantation is governed largely by the species used and the source of seed within species (Lacaze, 1978). The largest, cheapest and fastest gains in most forest tree improvement programmes will accrue if use of suitable species and seed sources within species is assured (Zobal and Talbert, 1984). Provenance research is therefore of paramount importance. Provenance denotes the original geographic area from which seed or other propagules are obtained, whereas seed source is the place where seeds are collected for large scale planting purpose. The objectives of provenance testing are to select best performing seed source for large scale plantations and for use of best adopted and most productive base population for future genetic improvement work. The thumb rule in forestry is "use seed from a local source for commercial planting until some other source is proven superior to the local source by appropriate testing.

2. HISTORICAL BACKDROP IN PROVENANCE TESTING

De Vilmorin a French nurseryman is credited with having established the first recorded well documented provenance test in 1821 on *Pinus sylvestris* (Scotch pine). He used seed origin in a series of unreplicated planting and reported his findings in 1857, which shown much geographic variation, which had hitherto been attributed to environmental influences, was genetic in nature (Wright, 1976) In 1908, the International Union of Forest Research Organization (IUFRO) undertook a series of replicated cooperative tasks on scotch pine with 13 seed lots, and planting them simultaneously in 11 locations spread over different countries. During 1929-36, Langlet, a forest geneticist of Sweden collected seeds from 582 stands from all parts of Sweden. In 1938, International of forest research organization laid a second series of cooperative test in Scotch pine using replicated plots and the coverage was half of the species range.

In India provenance trial on teak was initiated by Professor M. V. Laurie during 1928-30 and in chir pine by Sir Harry champion. International provenance test were also conducted in *Tectona grandis* and *Gmelina arborea* by DANIDA. The present interest in provenance testing is being expanded to many tropical and temperature tree species.

3. SEED SOURCE VARIATION

Agro ecological conditions largely comprising of edaphic and environmental factors have more than one effect on the performance of seeds. The possibility is that seeds are also by their place of birth (Heydecker, 1972), altitudinal variation, environmental variations (Mathur *et al.*, 1985). The use of sound seeds from stands of high quality is widely recognized as the best means of ensuring fast growth and healthy plantations capable of yielding high quality wood (Aldhous, 1972).Better growth, quality and adoptability can be achieved through careful selection of the best seed sources and trees when raising seedlings for a planting programmes (Kjaer and Foster, 1996). Genetic variation is important for long term adaptations of species (Falk and Holsinger, 1991), populations under stress may respond through natural selection. At the individual tree level, low genetic diversity can lead to inbreeding depression affecting growth, survival and adoption. Seed is a biological product, for any species and for a particular collection year, the behavior of seed cannot be predicted, unless it is examined in a competent laboratory for various characteristics, which need to be made at several stages during its progress from the parent tree to the nursery bed. Efficient seed management results in uniform germination better survival and improved seedling vigour and quality. Hence, supply of quality seeds with good germination and high vigour is paramount important for elite seedling production and establishment of viable tree plantation.

3.1. Seed Source on Physical Quality of Seed

Differences in seed source due to environment variation affect the physical seed quality in terms of seed size and its weight. The physical characteristics of seed are strongly related to the physiological potential of the seed. Several scientists have made affirmative statement on the positive influence of seed size and weight over germinability (Manonmani *et al.*, 1996; Manonmani and Vanangamudi, 1997; Venkatesh, 2000). According to Bharathi (1999) neem seeds collected from Western zone (Coimbatore) recorded higher 100 seed and kernel weight than those collected from other sources/places (Table. 1).

Table 1: Effect of seed sources on physical seed quality of neem

Seed sources	100 seed weight (g)	100 kernel weight (g)
Ambasamudram	20.6	13.9
Bhavanisagar	23.4	13.7
Coimbatore	25.8	12.2
Dharmapuri	24.2	13.7
Kovilpatti	22.8	11.5
Mettupalayam	21.3	11.8
Vamban	221.6	15.9
Salem	21.4	11.4
Theni	25.4	12.9
Thiruchirappalli	21.9	12.6
Vridhachalam	18.7	10.4
CD (P = 0.05)	3.749	2.027

In neem, seeds collected from high rainfall zone recorded maximum seed and kernel weight (Fig.1) compared to other agro-ecological zones (Sivagnanam, 1995). Evaluation of seed source for high quality seed production in *Acacia nilotica* ssp. *indica* (Venkatesh, 2000) exhibited that the seeds collected from Kovilpatti expressed more seed length of 8.3 mm which was on par with seeds collected from Vellore, Tirupattur, Dharmapuri, Coimbatore, Virudhunagar and Madurai. Other sources recorded less seed length ranging from 7.5 to 7.7 mm. Seeds collected from Virudhunagar expressed a wider breadth (7.2 mm) followed by Madurai (7.1 mm), Coimbatore (6.9 mm) and Kovilpatti (6.9) which were on par with each other (Table 2). According to Mani (2000), the neem seeds collected from south zone (E. Kumaralingapuram) registered the highest 100 seed weight (Fig.2), while seeds collected from Western zone (Sadhyamangalam) recorded the least.

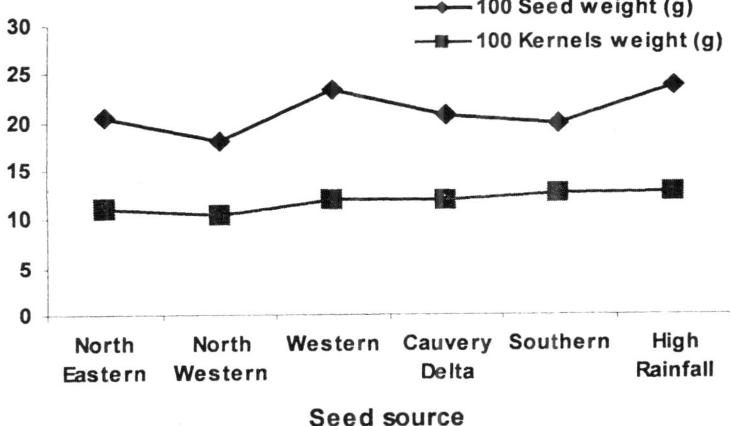

Fig.1: Effect of seed source on physically seed quality of neem

Table 2: Seed physical characteristics, sample size and purity on *A . nilotica*

Sources	Seed length (mm)	Seed breadth (mm)	100 seed weight (g)	Working sample size (g)	Purity analysis	
					Pure seed (%)	Inert matter (%)
Vellore	8.2	6.7	13.48	337.00	99.19	0.81
Tirupattur	8.1	6.7	14.36	357.33	97.21	2.79
Tiruvannamalai	7.6	6.4	12.72	322.33	98.95	1.05
Salem	7.5	6.3	12.39	355.33	97:9	2.61
Dharmapuri	8.2	6.3	14.38	351.67	98.21	1.79
Erode	7.7	6.4	14.52	365.00	99.67	0.35
Coimbatore	8.2	6.9	15.21	384.33	97.17	2.83
Aliyar	7.6	6.6	13.17	333.00	98.53	1.47
Virudhunagar	7.8	7.2	13.81	345.67	97.80	2.20
Kovilpatti	8.3	6.9	11.36	286.33	96.49	3.51
Madurai	8.2	7.1	11.84	332.00	97.95	2.05
Pararamakudi	7.7	6.7	12.00	304.67	98.15	1.85
Cumbum	7.6	6.5	11.93	300.33	95.15	4.85
Periyakulam	7.6	5.6	12.45	295.00	97.64	2.36
Mean	7.9	6.6	13.11	333.57	97.82	2.18
SEd	0.30	0.23	3.61	16.034	0.842	0.813
CD (P=0.05)	0.59	0.45	0.473	31.820	1.635	1.613

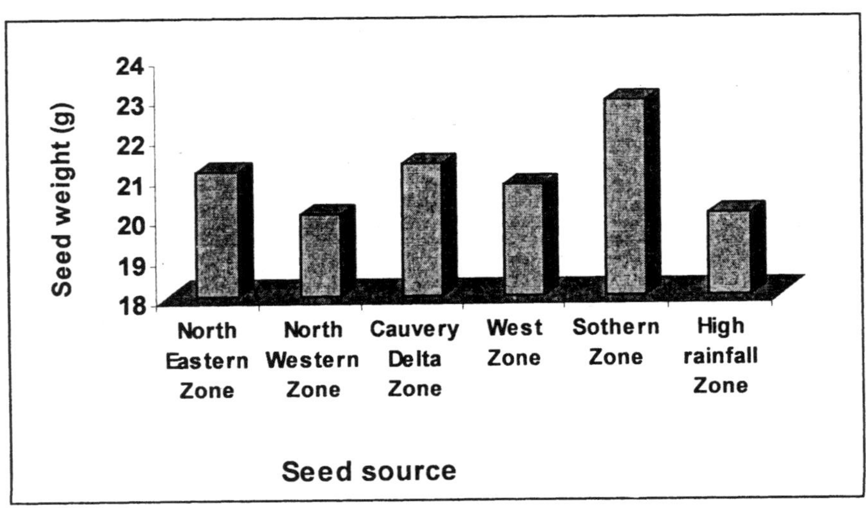

Fig. 2: Seed source on physical seed quality in neem

In Sandal (Manonmani and Vanagamudi, 1997), the higher and lower seed length was recorded in Coimbatore (8.0 mm) and Mettupalayam (6.3 mm), respectively. The thickness of seed recorded by Mettupalayam was more (7.6

mm) followed by Coimbatore (7.2 mm), while less thickness was recorded in Vellore (6.0 mm). (Table 3)

Table 3: Seed physical characteristics and sample size in *S.album*

Sources	Seed length (mm)	Seed thickness (mm)	100 seed weight (g)	Working sample size (g)
Coimbatore	8.0	7.2	13.454	335.94
Tirupattur	6.9	6.0	13 .122	326.58
Vellore	7.4 '	6.0	14.955	354.06
Mettupalayam	6.4	7.6	15.922	442.38
Mean	7.20	6.70	14.363	364.74
CD(0.05)	0.65	0.33	7.8	11.32

3.2. Germination and Vigour

Variation in germination and vigour due to different seed sources is attributed to variations in fertility status of soil, soil types, climate factors and rainfall in the place of production conducive climate of production places during the seed development and maturation results in high photosynthetic efficiency with considerable filling of the seeds, accumulation of food reserves and source to sink ratio.

Many scientists have made affirmative statement on the impact of seed source on germination of tree crops. According to Sivagnanam (1995), the neem seeds collected from southern zone outperformed for germination (Fig.3) than these seeds collected from other zones. Bharathi (1999) reported that the seeds collected from western zone (Coimbatore) followed by the source Bhavanisagar (Western zone) recorded higher germination and vigour in neem (Table 3).

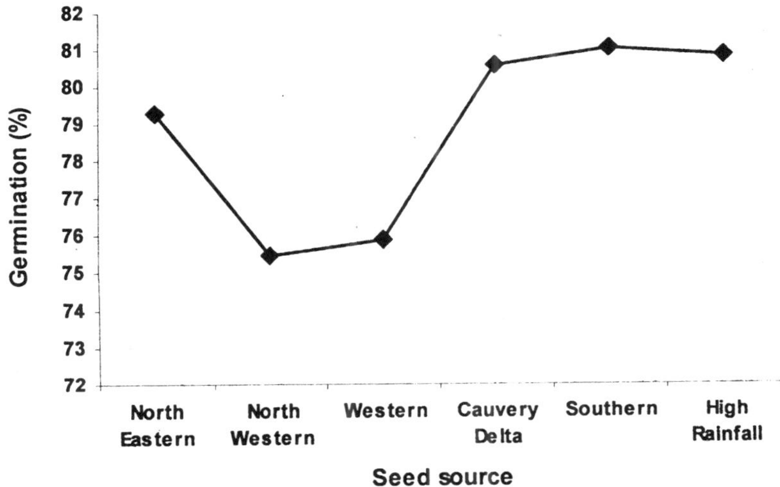

Fig. 3: Seed source on germination of neem

Table 4: Seed source on physical seed quality of tree crops

S. No.	Crop	Reference	Findings
1.	Teak	Murthy (1973)	The seed weight and diameter were independent of the size and there were large variation in seed weight and diameter both between and within the provenance
2.	Teak	Manonmani and Vanangamudi (1997)	The drupes collected from Topslip source out performed others. This source contained bigger drupes and more one seeded drupes indicating their close relationship which resulted in more weight
3.	Teak	Masilamani *et al.* (1999)	Drupes collected from five sources involving two forest types and five states of India *viz.,* Topslip (Tamil Nadu), Pecchi (Kerala), Adilabad (Andhra Pradesh), Dharwad (Karnataka) and Akola (Maharastra) exhibited variation in seed weight
4.	*Acacia*	Suresh (1994)	Significant variation for 100 seed weight was observed in seeds collected from different source (provenance)
5.	*Acacia*	Vanangamudi *et al.* (1998)	Seeds collected from seven seed sources exhibited significant variations in their physical character (seed size)
6.	Sandal	Bagchi and Sharma (1989)	Seeds obtained from various locations showed significant variation in seed weight
7.	Sandal	Kumaran (1991)	Seeds collected from nine seed sources showed variation in seed length and width and seed weight
8.	Neem	Surendran *et al.* (1993)	Seeds collected from different seed sources exhibited variation in seed weight
9.	Neem	Sindhu (1995)	Seeds of neem collected from different places in Karnataka and Andhrapradesh recorded high genetic variability for seed length, width and weight
10.	Neem	Mishra *et al.* (1997)	Seeds collected from five different states exhibited wide variation for seed morphological parameters. Seeds from Kanpur recorded maximum seed breadth, while seeds from Karnataka state excelled other parameters
11.	*Albizia*	Natarajan (1999)	Among twelve seed sources of *Albizia,* Coimbatore source recorded more seed breadth and weight, while seeds from Tirunelveli and Anamalai WLS registered the least
12.	*Albizia*	Vasudeva *et al* (1999)	Among the seeds collected from 24 trees in different parts of Karnataka, variation in seed characteristics was more in the southern part of the state than in the northern part
13.	*Cassia siamea*	Masilamani (1992)	Seeds collected from Coimbatore recorded maximum 1000 seed weight than other zones
14.	*Prosopis*	Masilamani (1992)	Seeds collected from Tiruppur possessed maximum 1000 seed weight than the seeds collected from other provenance
15.	Ber	Srimathi (1997)	Physical variations were observed in seeds collected from different seed source

Table 5: Impact of seed source on germination and vigour of neem

Seed sources	Germination (%)	Vigour index
Ambasamudram	90.0	39.0
Bhavanisagar	96.0	42.4
Coimbatore	98.0	33.2
Dharmapuri	76.0	21.9
Kovilpatti	82.5	36.5
Mettupalayam	76.0	34.3
Vamban	75.0	27.9
Salem	70.0	27.7
Theni	84.0	36.7
Thiruchirappalli	67.5	30.6
Vridhachalam	57.5	22.9
CD (P = 0.05)	14.040	7.418

Similarly, the seeds obtained from Chennai (North Eastern Zone) and Pondicherry (North Eastern Zone) recorded the highest germination of 100 and 98 per cent (Table 4), respectively in neem (Mani, 2000).

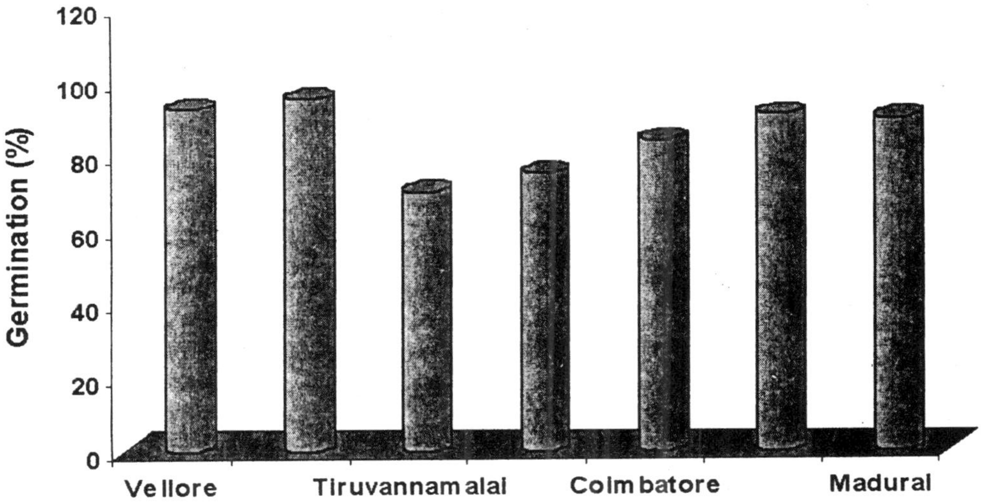

Fig. 4: Seed source on germination of *Acacia*

Venkatesh (2000) observed significant variations in germination among the seed sources of *Acacia nilotica* sub sp. *indica* (Fig. 4.). Higher germination was noticed in the seeds obtained from Tirupattur (95%), Vellore (92%) (North east zone), Cumbum (91%) and Madurai (90%) (Southern zone).

2.3. Performance of Seed Sources under Nursery Conditions

Venkatesh (2000) reported that Vellore excelled other sources recording the highest values for root length, and dry weight at 6th month of sowing under nursery conditions. The next best seed source was Tirupattur. The poor performing seed sources were Kovilpatti for shoot length and Tiruvanamalai for shoot length and dry weight (Table 6).

Table 6: Effect of seed source on germination of neem

Seed source	Germination (%)	Vigour index
Chennai (NEZ)	100	2745
Pondicherry (NEZ)	98	3340
Krishnagiri (NWZ)	65	1972
Vazhapady (NWZ)	97	2915
Aduthurai (CDZ)	84	2359
Trichy (CDZ)	88	3130
Andipatty (WZ)	73	1893
Palladam (WZ)	90	2839
Madurai (SZ)	81	2177
Shengottai (SZ)	58	1049
CD (P - 0.05)	7.299	11.420

NEZ - North Eastern Zone
CDZ - Cauvery Delta Zone
SZ - Southern Zone
NWZ - North Western Zone
WZ - Western Zone

In case of *A. lebbeck* (Natarajan, 1999) at four months after sowing, the seedlings raised from Coimbatore registered the longest root length of 49.0 cm followed by Madurai (44.7 cm), Arupukottai (43.3 cm), Tirunelveli (42.7 cm) and Aliyar (42.7 cm) which were on par with each other. Root length of seedlings from Trichy (29.2 cm) and Chinglepet (21.2 cm) were shorter. Seedlings from Arupukottai recorded longer shoot (27.8 cm) which was followed by Madurai (2.6.4 cm), whereas Ramanathapuram (18.5 cm) recorded shorter shoot length. Shoot collar diameter was more in Aliyar and Tirunelveli (0.9 cm) followed by Arupukottai (0.8 cm), Trichy (0.7 cm) and Ramanathapuram (0.7 cm) and they were on par with each other. The seedlings obtained from Aruppukottai registered more total dry weight (11.34 g /seedling) followed by Coimbatore (10.86 g /seedling). A minimum of 5.62 g /seedling was recorded in Chinglepet (Table 7).

Table 7: Seed source on germination of tree crops

S. No.	Crop	Reference	Findings
1.	Teak	Gupta and Pattanath (1975)	Greater variations were observed in germination of teak seeds collected from 18 different sources in India
2.	Teak	Dasappa (1990)	Seeds collected from different regions exhibited wide variation in germination
3.	Teak	Kandya (1997)	Seeds collected from distant sources of Madhya Pradesh showed variation in germination
4.	Teak	Manonmani and Vanangamudi (1997)	Teak drupes collected from Topslip of Tamil Nadu registered higher germination and seedling vigour than other seed sources
5.	Teak	Masilamani *et al.* (1999)	*Tectona grandis* drupes collected from five sources involving two forest types and five states of India *viz.*, Topslip (Tamil Nadu) Peechi (Kerala), Adilabad (Andhra Pradesh) (moist deciduous forest), Dharwad (Karnataka) and Akola (Maharashtra) (dry deciduous forest) exhibited variations in germination and seedling vigour
6.	*Acacia*	Joshi *et al.* (1983)	The germination percentage varied widely among the locations. Seeds from Midnapur of West Bengal recorded maximum germination of 96.6 per cent
7.	*Acacia nilotica*	Suresh (1994)	Seeds from different seed sources reported significant variation for germination and vigour
8.	*Acacia nilotica*	Ginwal *et al.* (1995)	Variation in germination was observed in five subspecies of *Acacia nilotica* belonging to 30 provenance
9.	*Acacia nilotica*	Vanangamudi *et al.* (1998)	Seeds collected from seven seed sources exhibited significant variations in germination and vigour
10.	Sandal	Bagchi and Kulkarni (1985)	Seeds from 8 selected sandal trees recorded germination percentage ranging from 25-61 per cent and showed significant differences among the trees
11.	Sandal	Veerendra and Sarma (1990)	Seeds collected from one of the 11 Karnataka provenance showed more seedling vigour than other seed sources
12.	Neem	Dwivedi (1993)	Wide variation in germination was noted in seeds collected from different provenance in India
13.	Neem	Mishra *et al.* (1997)	Neem seeds collected from five different states showed wide variation for germination. Seeds from Karnataka was superior in recording germination to other seed sources
14.	*Albizia*	Bahuguna *et al.* (1989)	Seeds of *A. falcatoria* from Philippines provenance showed significantly better germination and seedling vigour than the Malaysian provenance
15.	*Albizia*	Natarajan (1999)	Among 12 seed sources, seeds collected from Tirunelveli and Annamalai WLS registered the highest germination of 94 and 91 per cent, respectively
16.	*Casuarina*	Jambulingam (1990)	Seeds collected from riverine source was superior in germination and seedling vigour to inland and coastal seed sources
17.	*Casuarina*	Misra and Banerji (1995)	Among nineteen provenances of *casuarina* species collected from Australia, Hawaii, Egypt, Israel and India *C. cunninghamiana* and *C. equisetifolia* collected from Australia, Egypt and Israel were found to be the best interms of germination and seedling vigour
18.	*Cassia siamea*	Masilamani (1992)	Seeds collected from Coimbatore recorded the maximum germination and seedling vigour than other seed source
19.	*Dalbergia latifolia*	Bangarwa *et al.* (1995)	Significant differences in germination were observed between the progenies of different locations, whereas progenies from the same location exhibited no difference
20.	*Dalbergia sissoo*	Gera *et al.* (1999)	Variation in germination was noted among ten seed sources

Table 8: Performance of seed sources of *A. nilotica* under nursery condition at six months after sowing

Sources	Root length (cm)	Shoot length (cm)	Shoot collar diameter (mm)	Total dry weight (g/seedling)
Vellore	32.4	83.0	5.2	9.046
Tirupattur	31.0	82.5	5.2	7.900
Tiruvanamalai	23.0	57.8	5.0	4.852
Salem	21.5	81.3	5.2	6.774
Dharmapuri	23.0	69.5	5.8	5.067
Erode	22.5	68.1	5.1	5.146
Coimbatore	.26.5	75.5	5.4	8.931
Aliyar	22.3	69.0	4.8	7.834
Virudhunagar	23.5	69.0	5.3	6.816
Kovilpatty	20.8	72.0	4.8	5.157
Madurai	29.0	70.0	4.9	5.485
Paramakudi	23.5	65.5	4.9	5.614
Cumbum	29.5	64.0	5.0	6.146
Periyakulam	26.0	70.5	4.5	6.043
SEd	2.10	2.38	NS	0.497
CD (P=0.05)	4.51	5.10		1.562

According to Manonmani and Vanagamudi (1997) teak seeds collected from Mudumalai (37.9 cm) followed by Topslip (33.7 cm) expressed significantly higher values for shoot length than Siruvani and Kallar (21.0 cm). For root length, Mudumalai and Kallar had performed better. In case of shoot collar diameter, Mudumalai recorded significantly higher value compared to other sources. It was less in Siruvani. Seeds collected from Mudumalai recorded significantly higher total dry matter (11.3 g) followed by Topslip (9.92 g) (Table 8). This suggests that environmental and physiographic conditions plays an important role in deciding survival and establishment of seeds.

Mani (2000) found that neem seeds collected from Chennai (North Eastern zone) and Uthupatty (Southern zone) performed better than other seed sources in recording seedling length (Fig.5), shoot diameter, number of leaves and dry weight.

Table 7a: Performance of *A.lebbeck* seed sources at four months after sowing under nursery conditions

Sources	Root length (cm)	Shoot length (cm)	Shoot collar diameter (cm)	Total dry weight (g / seedling)
Tirunelveli	42.7	20.8	0.9	10.12
Ramanathapuram	34.2	18.5	0.7	7.05
Madurai	44.6	26.4	0.6	8.19
Arupukottai	43.3	27.8	0.8	11.34
Anamalai WLS	36.3	21.8	0.5	8.72
Aliyar	42.7	23.7	0.9	9.65
Coimbatore	49.0	25.2	0.7	10.86
Mettupalayam	36.7	23.3	0.7	7.74
Erode	33.0	23.7	0.6	6.20
Salem	31.2	26.0	0.5	5.65
Chinglepet	29.2	19.2	0.7	5.62
Trichy	21.2	19.8	0.7	6.08
SEd	1.94	1.55	0.09	0.173
CD (P=0.05)	3.99	3.20	0.20	0.358

Table 8a: Performance of seed sources of *Tectona grandis* at three months after sowing under nursery conditions

Sources	Shoot length (cm)	Root length (cm)	Shoot collar diameter(cm)	Total dry weight (g / seedling)
Siruvani	21.0	19.4	5.4	6.18
Mudumalai	37.9	23.8	9.4	11.93
Topslip	33.7	20.7	7.0	9.92
Kallar	31.4	23.3	6.9	6.91
CD (P=0.05)	5.86	1.79	1.92	1.146

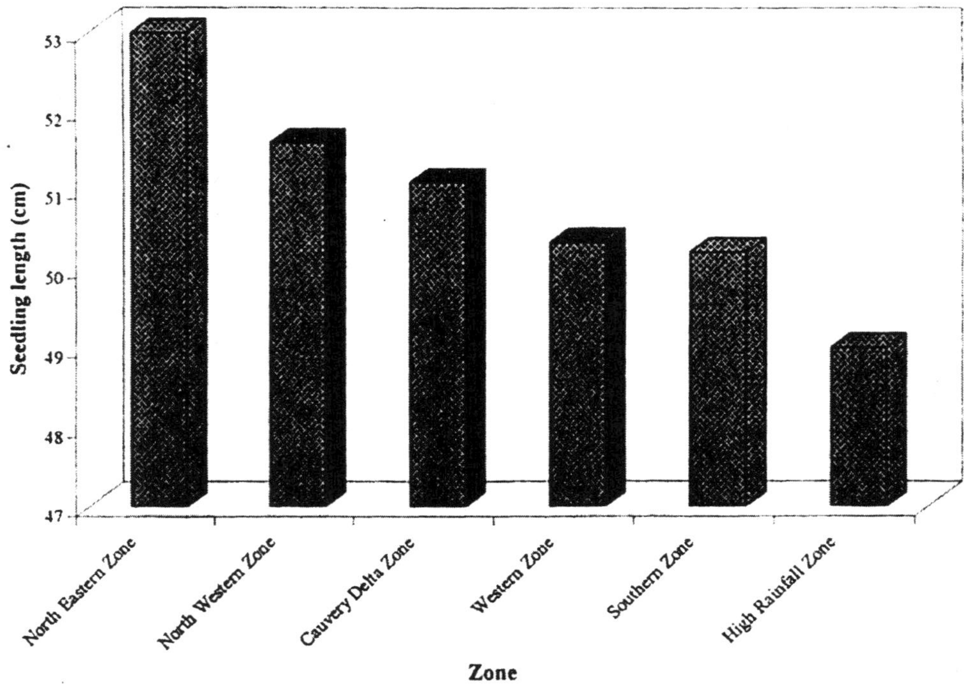

Fig. 5: Seed source on seedling characters of neem under nursery

3.4. Performance of Seed Sources under Field Conditions

Venkatesh (2000) reported that *Acacia* seeds collected from Tirupattur, Dharmapuri and Aliyar registered maximum survival (73%) which was on par with Kovilpatti, Paramakudi and Periyakulam. Tiruvanamalai and Virudhunagar (60%). In Madurai recorded the minimum survival. The tallest plant (107.0cm) was observed in Tirupattur and the shortest (60.0 cm) in Vellore

In case of *A. lebbeck* (Nataraja, 1999), seedling survival was higher in Coimbatore (93.0%), followed by Erode (90.0 %). Ramanathapuram recorded lower survival of 83 per cent which was on par with Arupukottai (84.0 %) and Trichy (85.0 %). The tallest plant of 48.0 cm was recorded in Mettupalayam, which was on par with Aliyar 47.6. The shortest plant was recorded in Salem (29.5), which was on par with Trichy (29.6 cm) and Chinglepet (31.3 cm). Shoot collar diameter was high in Erode (0.86 cm) while it was less in Arupukottai (0.52 cm) (Table 9).

Performance of neem seedlings in the field at twelve months after planting showed that E.Kumaralingapuram of southern zone out performed other seed sources in terms of yield and other growth parameters, while the performance of seed sources Sathankulam (Southern zone) and Krishnagiri North Western zone was poor in the field (Mani, 2000)

Table 9: Performance of *A. lebbeck* seedlings in the field at three months after planting

Seed source	Survival Plant (%)	Height diameter (cm)	Shoot collar (cm)
Tirunelveli	89	45.2	0.63
Ramanathapuram	83	39.6	0.57
Madurai	86	44.2	0.57
Arupukottai	84	4; 1.6	0.52
Anamalai WLS	86	35.6	0.71
Aliyar	88	47.6	0.57
Coimbatore	93	37.5	0.57
Erode	86	48.0	0.53
Mettupalayam	90	46.9	0.86
Salem	86	29.5	0.56
Chinglepet	87	31.3	0.60
Trichy	85	29.7	0.54
Cd (P = 0.05)	2.9	1.87	0.025

REFERENCE

Aldhous, J.R 1972. Nursery practices. Forestry commission Bulletin No.43, London

Bagchi, S.K. and H.D. Kulkarni. 1985. Germination of open pollinated seeds and survival of seedlings from the selected trees of *Santalum album. My Forest,* **21(3):** 221-224.

Bagchi, S.K. and V.P. Sharma. 1989. Biometrical studies on seed characters of *Santalum album L.* Silvae *Gen.,* **38 (3-4):** 152-153.

Bahuguna, V.K., K.P. Unnikrishnan and U.C. Dhaundiyal. 1989. Studies on the performance of Philippines and Malaysian provenance of *Albizia falcatoria (L.)* Febrez at nursery stage under north Indian tropical conditions. *Indian For.,* **115(4):** 209-215.

Bangarwa, K.S., V.P. Singh and R.P.S. Tomer. 1995. Progeny testing for seed quality parameters in *Dalbergia sissoo. Seed Sci. & Technol.,* **23(1):** 253-257.

Bharathi,A. 1999. Studies on handling management and storage of neem seed (*Azadirachta indica* A. Juss). Ph.D.,thesis, TamilNadu Agricultural University, Coimbatore -3.

Dasappa. 1990. Nursery techniques in teak. (*Tectona grandis* Linn f.), for afforestation. *My Forest,* **26(1):** 23-31.

Dwivedi, A.P. 1993. National level neem provenance trail at Jodhpur. In: Proc. of the Intl. Consultation of Neem Importance. Kasetsart University, Thailand. pp. 18-28.

Falk, D. A., Hoisnger. 1991. Genetics and conservation of rare plants. Oxford University Press, London pp 150-160.

Gera,M., Gera, N and H.S. Ginwal.1999. Seed source variation in germination and early growth among ten indigenous populations of Dalbergia sisso Roxb. Indian For., 125 (12): 1190 – 1197

Ginwal, H.S., P.S. Rawat, Mohit Gera, Neelu Gera and R.L. Srivastava. 1995. Study on the pattern of seed germination of various sub species cum. provenances of *Acacia nilotica* wild. ex. Del. under nursery conditions. *Indian For.,* **121(1):** 29-38.

Gupta, T.N. and P.G. Pattanath. 1975. Factors affecting germination behaviour of teak seeds of eighteen Indian origins. *Indian For.,* **10(10):** 584-588.

Heydecker, W. 1972. Vigour In: "Viability of Seeds", (Ed. E.H. Roberts). Chapman Hall. Londen pp. 209-252.

Jambulingam,R.1990. Recent development in research on *Casuarina* Tamil Nadu. Variation in population advances in *Casuarina* research and utilization. In : Proc. Of the Second International Workshop, Cairo, egypt, January,pp. 15 – 20

Joshi, H.B., R.N. Loganey, L.K. Patnaick, D.C. Choudhari and R.S. Chauhan. 1983. Troup's the silviculture of Indian trees, Vol IV. Revised and enlarged by the Editorial Board, F.R.I. and Colleges. Dehra Dun.

Kandya, A.K., 1997. Quality characteristics of *Tectona grandis* seeds from eight sources of Madhya Pradesh: A study. IUFRO symposium on "Innovation in Forest Tree Seed Science and Nursery Technology", November 22-25, Pt. Ravishankar Shukla University, Raipur, India. p. 165.

Kjaer, E.D., and Foster, 1996. The Economics of tree improvement of Technical Notes 43. The world Bank and Danida Forest seed center. Denmark.

Kumaran, K., M. Palani, M. Jerlin and C. Surendren. 1991. Effect of growth regulators on seed germination and seedling growth of neem (*Azadirachta indica). J. Trop. For. Sci.,* **6(4):** 529-532.

Lacaze. J.F., 1978. Advances in species and provenance selection. *Unasylva,* **30**. 119-120: 17-20.

Laurie, M.Y. 1974. Tree planting practices in African Savannas. .FAO *For. Dev. Paper* **No. 19**. FAO, Rome.

Mani, G. (2000). Studies on seed source variation, standardization of seed testing procedures and nursery technology in neem (*Azadirachta indica* A. Juss). Ph.D. Thesis, Tamil Nadu Agricultural University, Coimbatore – 3.

Manonmani, V. and K. Vanangamudi. 1997. Investigations on seed source variation in teak (*Tectona grandis*). National symposium on Natural resources management system, St. Joseph's College, Trichy, February 26-28.

Manonmani, V., K. Vanangamudi and R.S.V. Raj. 1996. Effect of seed size on seed germination and vigour in *Pongamia pinnata. J. Trop. For. Soi.,* **9(1):** 1-15.

Manonmani, V., Vanangamudi, K and Vinaya Rai, R.S. 1996. Effect of seed size on germination and vigour in *Pongamia pinnata* L. pierre. *J. For. Trop. Sci.,* **9(1):** 5.

Manonmani, Y., and Vanangamudi, K.1997. Investigations on seed source variation in teak *(Tectona grandis).* National symposium on *"Natural resources management*

Masilamani, P.. 1992. Production, processing and storage technology of seeds of *Cassia siamea* Lamb, *Hardwickia binata* Roxb and *Prosopis juliflora* Swartz. Dc. M.Sc. (Ag.) Thesis, Tamil Nadu Agricultural University, Coimbatore.

Masilamani, P., C.Dharmalingam and K.Annadurai. 1999. Seed source variation in seeds and seedling attributes of teak (*Tectona grandis* L.). In : National symposium on forestry Towards 21st century, Forest College and research Institute, Mattyupalayam , Sep. 27 – 28.p. 25

Mathur, R. S., K. K. Sharma and M. M. S. Rawat. 1985. Germination behaviour of various provenances of *Acacia nilotica* spp. *Ind. For.,* **111**: 435-449.

Mishra, C.M., Prabhakar Dubey, B.S. Burfal, A.C. Banerjee and S.L. Singh. 1997. Performance of *Azadirachta indica* provenance at seed and seedling stage, IUFRO symposium on "Innovations in Forest Tree Seed Science and Nursery Technology", November 22-25, Pt. Ravishankar Shukla University, Raipur, India. p. 144.

Misra, C.M and A.C. Banerji. 1995. Provenance variation in *Casuarina* species with reference to germination and growth . J. Trop. For., 11(3) : 209 – 211

Murthy,K.A.V.R.G. 1973. Problems of teak seeds . Germination studies and seed problems . In: Intl. Symposium on seed processing – IUFRO Working Party 5.2.01.06. Vol. II, p. 20 Baegen, Norway, pp, 21 – 24

Natarajan,S. 1999. Investigations on seed source variations, standardization of seed testing procedure and nursery techniques in *Albizia lebbeck* L. Benth, Ph.D., thesis , TamilNadu Agricultural University, Coimbatore -3.

Sindhu, V.H.C. 1995. Variation studies in provenances of *Azadirachta indica. Indian For.*, **11:** 1053-1056.

Sivagnanam, K. 1995. Growth biology, reproductive strategy and seed production spectrum in neem (*Azadirachta indica* A. Juss.). Ph.D. Thesis, Tamil Nadu Agricultural University, Coimbatore.

Srimathi, P. 1997. Research focus on seed collection, processing and storage of amla (*Emblica officinalis*, Gaertn.) jamun (*Syzygium cuminii* skeels.) and ber (*Zizyphus mauritiana* Lamk). Ph.D. Thesis, Tamil Nadu Agricultural University, Coimbatore.

Surendran, C., R.S.V. Rai, K. Sivagnanam, M. Shanmugam, K. Kumaran, A. Ragupathi, K. Vanangamudi and I. Vimala. 1993. Tree improvement and seed management in Indian neem. In: Genetic improvement of neem : strategic for the future. Proc. Intl. Consultation on neem improvement held at Kasetswart University, Bankok, Thailand, pp.29-39.

Suresh, K.K. 1994. Growth performance of *Acacia nilotica* under diverse ecosystems. Ph.D. Thesis, Tamil Nadu Agricultural University, Coimbatore.

Vanangamudi, K., R. Umarani, A. Bharathi and A. Venkatesh. 1998. Effect of seed source on physical and physiological qualities of *Acacia nilotica* seeds. *Seed Res.*, **26(2):** 114-116.

Vasudeva, R., G.B. Shashidhara and M.A. Haseeb.1999. Identification of plus trees of *Albizia lebbeck* L. Benth in Karnataka. In: National symposium on forestry Towards 21st century, Forest College and research Institute, Mattyupalayam, Sep. 27 – 28.p. 25

Veerendra, S.C.H. and C.R. Sarma. 1990. Variation studies in sandal (*Santalum album* L.). 1. Time of emergence and seedling vigour. *Indian For.*, **116(7):** 568-571.

Vekatesh, A. 2000. studies on seed source variations, standardization of seed testing procedure and nursery techniques in Acacia nilotica (L) Wild. Sub. spp indica Benth. Ph.D.,thesis, TamilNadu Agricultural University, Coimbatore -3.

Wright, J.W. 1976. Introduction to Forest genetics. Academic press, New york, 463 pp.

Yadav, V.K., P.K. Khare and G.P. Mishra. 1986. Quick test for viability and vigour index

Zobel, B.J and J.Talbert. 1984. Applied Tree Improvement. Johnwiley and Co P 505.

14

Seed Testing for Tree Crops

K. Raja[1], A. Bharathi[2], K. Vanangamudi[3], Mallika Vanangamudi[4] and S. Kavitha[1]

[1] *Central Institute for Cotton Research, Coimbatore – 641 003*

[2] *Department of Seed Science and Technology, Tamil Nadu Agricultural University Coimbatore – 641 003*

[3]*Department of Plant Breeding and Genetics, Adhiparasakthi Agricultural College G.B.Nagar, Kalavai – 632 506*

[4] *Department of Crop Physiology, Tamil Nadu Agricultural University Coimbatore – 641 003*

1. INTRODUCTION

The physical, physiological and biochemical qualities of the seed need to be tested at all the stages of seed multiplication system i.e. seed to seed bed. Justice (1972) has defined the following objectives for developing rules for seed testing (a) to provide methods by which the quality of seed samples can be determined accurately, (b) to prescribe methods by which seed analysts working in different laboratories in various countries throughout the world can obtain uniform results, (c) to relate the laboratory results as far as possible, to planting value, (d) to complete the tests within the shortest period of time possible, commensurate with the above mentioned objectives, and (e) to perform the test in the most economical manner.

The aim of the seed testing in forestry is to assess the value of a seed lot with regard to number of germinable seeds in the lot. This helps to determine the quantity of seeds required per unit area of the nursery bed. Although tree species and genera from the tropics and southern hemisphere (notably *Eucalyptus*) have began to figure in the ISTA lists, the balance is still overwhelmingly weighted towards north temperate species. Many tropical species are not included at all. Their omission simply reflects the lack of reliable information from research on the best methods of testing these species and genera (ISTA, 1993). Hence, standardization of seed testing is paramount important for assessing the quality of tree crops.

2. STANDARDIZATION OF SUBMITTED AND WORKING SAMPLE FOR IMPORTANT TREE CROPS

In seed testing, standardization of submitted and working sample size is important to have a true representation of the bulk seed lot and maintain uniformity for submission of seeds. The working sample weight depends on the seed size of the species in question. The ISTA (1993) rules are intended to provide a minimum of 2500 seeds for all except the very large seeded species, for which a minimum of 500 seeds is prescribed. These quantities are considered sufficient for most of the usual tests (purity, authenticity, seed weight and germination), but an additional 10g of seeds of most species is needed, if moisture content is to be determined. The weight of submitted sample should be double to that of the working sample (ISTA, 1993). Among the tropical species listed by ISTA, the prescribed minimum size of working sample range from 2g for *Eucalyptus deglupta* to 1000g for *Tectona grandis.* Based on research carried out by different scientists following the ISTA rules, the working and submitted sample size for different tree crops are presented hereunder

S. NO.	Tree crop	Source	Remarks
1	*Acacia nilotica*	Venkatesh (2000)	The mean weight of working sample was 333.57g (weight of 2500 seeds). The weight of submitted sample arrived at was 3335.7g or 3.336 kg
2	*Albizia lebbeck*	Natarajan (1999)	The mean weight of working sample was 276 g (weight of 2500 seeds). The weight of submitted sample arrived at was 2760 g or 2.760 kg.
3	*Santalum album*	Manonmani (1997)	The mean weight of working sample was 365 g (weight of 2500 seeds). The weight of submitted sample arrived at was 3650 g or 3.650 kg
4	*Tectona grandis*	Manonmani (1997)	The mean weight of working sample was 1863.4g (weight of 2500 drupes). The weight of submitted sample arrived at was 18634.5 g or 18.6345 kg
5	*Azadirachta indica*	Mani (2000)	The mean weight of working sample was 584g. The weight of submitted sample arrived at was 5836 g or 5.836 kg.

3. STANDARDIZATION OF PURITY ANALYSIS TEST FOR IMPORTANT TREE CROPS

The seed sample can contain impurities such as weed seeds, seeds of other crops, detached seed structures, leaf particles and other inert materials. The purity test is significant as this factor affects the number per unit weight of genuine seeds and the germination per cent of the seed lot (ISTA, 1993). When purity analysis is done, it is the first test to be carried out because subsequent tests are made only on the pure seed component. The result of the purity analysis for important tree crops is as follow

Acacia nilotica (Venkatesh, 2000): The mean pure seed was 97.82%. Among the different seed sources, higher pure seed of 99.67% was observed in Erode. It was lower in Cumbum (95.15%). The inert matter was more in Cumbum (4.85%) and less in Erode (0.35%).

Albizia lebbeck (Natarajan, 1999): The mean pure seed was 99.11%. Higher pure seed of 99.70% was recorded in Anamalai WLS and lower in Trichy (98.3%), among the different seeds sources. The inert matter was more (1.8%) in Trichy and less in Anamalai WLS (0.3%).

Azadirachta. indica (Mani, 2000): The pure seed fraction varied bfrom 98.90 per cent to 99.90 per cent with an average of 94.4 per cent. Hence, it is concluded that the seed lots of neem should posses a pure seed fraction of 99.4 per cent

4. STANDARDIZATION OF 100 SEED WEIGHT FOR IMPORTANT TREE CROPS

S. NO.	Tree crop	Source	Remarks
1	*Acacia nilotica*	Venkatesh (2000)	The mean 100 seed weight was 13.110 g. Seeds collected from Coimbatore recorded the highest 100 seed weight of 15.210 than other seed sources. The lowest of 11.360 g was recorded in Kovil patti.
2	*Albizia lebbeck*	Natarajan (1999)	The mean 100 seed weight was 10.776 g. It was higher (13.500 g) in Coimbatore and lesser in Anamalai WLS (8.011g), among the different seed sources
3	*Santalum album*	Manonmani (1997)	The mean 100 seed weight was 14.363g. Seeds obtained from Mettupalayam recorded the maximum 100 seed weight (15.922 g) and Tirupattur recorded the minimum (13.122 g).
4	*Tectona grandis*	Manonmani (1997)	A higher weight of 100 drupes was noticed for Tiruvarur (115.840 g) followed by Top slip. The mean 100 drupe weight was 74.580 g. The Standard deviation (SD) calculated was below 4, which confirms ISTA (1993) standards
5	*Azadirachta indica*	Mani (2000)	Hundred seed weight was determined by observing in forty six seed lots. The weight ranged from 18.173g to 28.830g, with an average of 21.500g. The standard deviation (SD) calculated was below 4 (0.280), which confirms ISTA (1993) standards. Therefore, the hundred seed weight reported as 21.50g

5. STANDARDIZATION OF GERMINATION MEDIA FOR IMPORTANT TREE CROPS

Medium plays an important role in germination tests because seeds have characteristics requirements for moisture and oxygen. Thus small seeds are best geminated on top of the paper rather than in sand. Seeds which require light for germination also need top of the paper medium. Type of medium best suited depends upon the physical condition of the seeds. Generally speaking, sand as medium, is best suited for large seeds, and paper for small seeds. However, large seeds may also be germinated on paper (Gupta and Pattanath., 1975). Standardization of germination medium is to be done before recommending for adoption. The medium used must be non–toxic to the germinating seedlings and must provide adequate aeration and moisture. The best medium for germination test for different tree crops is given below

Acacia nilotica (Venkatesh, 2000): The acid scarified seeds (4 x 1 00 seeds) were sown in red soil, red soil + sand (50: 50), sand, vermiculite, between paper and top of the paper. More germination (97%) and vigour index (914) were noticed in sand medium followed by vermiculite. While top of the paper recorded minimum values (Table 1).

Table 1: Standardization of medium for germination test of *A. nilotica*

Sources (S)	Germination (%)				Vigour index			
	Coim Batore	**Tiru pathur**	**Cumb um**	**Dharma puri**	**Coim batore**	**Tiru-pathur**	**Cumb um**	**Dharma puri**
Red soil	93	91	92	90	677	667	687	626
Red soil +sand	97	96	94	88	792	785	704	696
Sand	98	98	97	95	923	933	943	859
Vermiculite	96	97	95	94	871	855	752	726
Between paper	84	78	76	76	802	748	729	716
Top of the paper	81	76	72	71	405	339	333	331
	M	S	M x S		M	S	M x S	
CD (P=0.05)	1.36	1.11	2.71		23.7	19.4	47.4	

A. lebbeck (Natarajan, 1999): The acid scarified seeds (4 x 100 seeds) were sown in top of paper, between paper and sand. The germination and vigour index were worked out. A maximum of 77 per cent germination was recorded in sand followed by between paper medium (74.9 per cent), whereas top of the paper recorded 72.1 per cent. Maximum vigour index of 1836 was recorded in sand medium (Table 2).

According to Mani (2000), high germination of 76 per cent in neem was observed in sand medium followed by sand + soil medium (60 per cent). Germination was only 15 per cent (Plate 1a,b,c,d), when seeds were sown on top of the paper sand as a germination substrate is preferred for tree species of larger seeds. ISTA (1993) recommended sand medium for *T. grandis*. Damage of seedlings due to fungi is always less in sand than paper.

a) Germination test in step in germinator

b) Roll towel medium

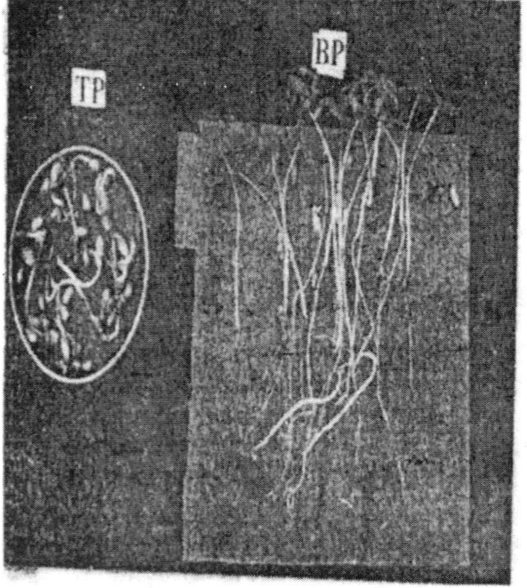

) Top of the paper and roll towel media

d) Sand + soil, sand & soil media

Plate 1: Standardization of germination media for neem

Table 2: Standardization of medium for germination of *A. lebbeck*

Sources (S)	Germination (%)				Vigour index		
	Sand	Soil	Top of paper	Roll towel	Sand	Soil	Roll towel
Tirunelveli	78	77	74	85	1938	1758	1544
Ramanathapuram	81	79	78	73	2127	1996	1961
Madurai	76	71	71	76	1851	1550	1491
Arupukottai	68	69	68	72	1630	1781	1510
Anamalai WLS	72	69	81	81	1613	1762	2052
Aliyar	89	87	76	77	2059	2132	1865
Coimbatore	92	85	88	82	2744	2502	2423
Mettupalayam	81	81	73	77	1994	2029	1972
Erode	76	71	66	69	1692	1655	1562
Salem	72	66	64	65	1497	1393	1589
Chengalpet	69	68	60	66	1498	1678	1483
Trichy	66	60	61	72	1391	1336	1386
	L	M	L x M		L	M	L x M
CD (P=0.05)	1.16	0.67	2.33		54.3	27.1	94.0

6. STANDARDIZATION OF VIGOUR TESTS FOR IMPORTANT TREE CROPS

Heydeckar (1972) fittingly described vigour as the condition of a seed which was at the highest of its potential power when all the factor that may detract its quality were absent and those that made good seed were present in the right proportion, promoting a satisfactory performance over a maximum range of environmental conditions

Performance of seed sources under stress conditions *viz.,* high acidity, alkalinity, salinity and drought is an important test on the physiological potential of the sources to screen their superiority for specific condition. Moreover, when the seed lots are exposed to lower and higher limits of stress conditions, it helps to find out the tolerance limit of the species. Accelerated ageing is highly related to the vigour of seed lot and hence to the capacity of the lots to perform well under field conditions (ISTA, 1993). Testing the seeds under lower or higher pH has also been recommended as stress test. The vigour tests also help to differentiate the seed lots that have same germination percentage but with differential responses in field emergence and seedling vigour.

6.1. NaCl test

Bangash (1977) reported that germination decreased with increasing salinity when tested in six tree species including *Albizia lebbeck*. Seedling survival under field conditions was also affected by salinity. *Acacia arabica, Albizia lebbeck* and *Zizyphus jujubha* were tolerant to saline conditions.

Germination, shoot and root length of *Acacia nilotica, Pongamia pinnata* and *Prsopis juliflora* were reduced and mortality increased with increase in the electrical conductivity, sodium absorption ratios and residual sodium carbonate when compared with control (Tomer and Yadav, 1980). According to Misra and Singh (1981), germination of *Acacia arabica, Acacia catechu* and *Albizia lebbeck* in solutions of NaCl + Na_2SO_4 + $CaCl_2$ was little affected at electrical conductivities upto 12, 8 and 12 m.mhos, respectively. Gupta and Pattanath (1987) reported that under sandy soils safe and toxic limit of salinity of *Albizia lebbeck* was 2.5-11.0 dSm^{-1}. A significant decline in germination, seedling vigour and biomass was noticed in various tree species as the salinity level increased (Kathiresan *et al.*, 1996; Sharma *et al.*, 1998). This could be attributed to osmotic effect of NaCl on seed hydration and toxic effect of NaCl on seed embryo or endosperm cell membranes. Hayward and Wadleigh (1949) reported that saline substrate may affect the germination of the seeds in two ways (i) there may be enough soluble salts in the seed bed to build up osmotic pressure of the soil to point which will retard or prevent intake of water, and (ii) certain salt constituents may be toxic to the embryo and seedling.

According to Natarajan (1999), the maximum reduction in germination was obtained for all seed late with 1.0% NaCl concentration. A similar results was obtained with 0.75% concentration for five seed lots *viz.,* Coimbatore, Ramanathapuram, Madurai, Chenglepet and Trichy out of the nine seed lots (Table 3; 4). The trend of result did not match with 0.25 and 0.50% concentrations. Hence, 0.75% concentration of NaCl is recommended for testing the salinity or salt tolerance of the seeds.

Table 3: Standardization of NaCl salt stress vigour test in *A. lebbeck* Germination (%)

Seed lot (L)	Control	NaCl concentrations (C)				Mean
		0.25%	0.5%	0.75%	1.0%	
Tirunelveli	85.0	69.0	59.0	49.0	38.5	60.1
Ramanathapuram	73.0	62.0	57.0	43.5	35.5	53.0
Madurai	82.5	67.0	57.5	46.0	35.0	57.6
Aliyar	74.0	61.0	50.0	42.0	39.5	53.3
Coimbatore	70.0	62.5	52.0	45.0	38.0	53.5
Erode	64.5	59.0	49.0	38.5	33.5	48.9
Salem	62.0	51.0	43.0	35.0	30.5	44.3
Chengalpet	56.0	48.5	40.5	31.0	25.0	40.2
Trichy	54.0	46.5	38.0	28.5	23.5	38.1
Mean	69.0	58.5	48.9	38.8	33.2	
CD(P=0.05)		L	C	L x C		
		0.60	0.45	1.34		

Table 4: Standardization of NaCl salt stress vigour test in *A. lebbek* – Vigour index

Seed lot (L)	Control	NaCl concentrations (C)				Mean
		0.25%	0.5%	0.75%	1.0%	
Tirunelveli	1904	1434	1054	790	553	1147
Ramanathapuram	1491	1167	828	615	451	910
Madurai	1708	1306	969	629	414	1005
Aliyar	1497	1141	821	572	473	901
Coimbatore	1304	1095	807	570	433	842
Erode	1146	871	720	473	365	751
Salem	1081	859	614	418	329	660
Chengalpet	974	788	569	354	262	589
Trichy	950	757	535	328	250	564
Mean	1340	1046	769	528	392	
CD(P=0.05)		L	C	L x C		
		32.0	23/0	71.6		

Venkatesh (2000) infered that 1.0% concentration of NaCl could be recommended for salt stress test in *Acacia nilotica* and identified Tirupathur and Paramakudi seed sources as the highly tolerant sources towards salinity. Mani (2000) observed higher reduction in germination of neem for all the seed lots with 1.0% concentration of NaCl (Fig. 1). The ranking of six seed lots was not found to be similar with any other concentration *viz.*, 0.25, 0.50 and 0.75 per cent. Hence, 1.00% concentration of NaCl is recommended for testing salinity or salt tolerance of the *Azadirachta indica* seed lots.

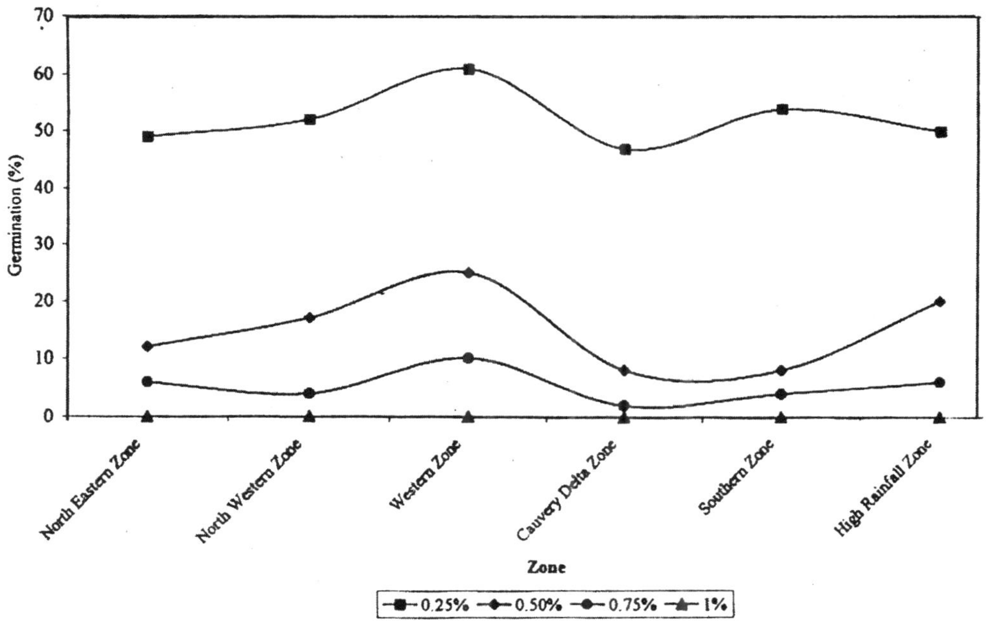

Fig. 1: Standardization of NaCl test for neem

6.2. pH test

Albizia lebbeck seeds recorded 40-60 per cent germination under high range of 9-11 pH and some fodder *Albizia* sp. recorded maximum germination of 78.6 per cent under pH 8 (Roy, 1990). The rooting potential was the highest in acidic pH of 6 and low in 3 and 4 and death of seedlings was noticed at pH 2 in *Rhizophora mucronata* (Kathiresan *et al.*, 1996) and in *R. capiculata*. According to Natarajan (1999), pH of 8 was suitable to test the vigour of *A. lebbeck* towards alkalinity tolerance (Table 5 and 6).

Table 5: Standardization of pH vigour test in *A. lebbeck* – Germination (%)

Seed lot (L)	Control	pH value (P)							Mean
		4	5	6	7	8	9	10	
Tirunelveli	72.3	20.7	26.3	32.7	49.0	51.7	61.7	56.7	46.4
Aliyar	69.3	27.3	34.0	38.0	60.3	63.0	56.7	37.3	48.3
Coimbatore	80.6	33.7	39.0	44.7	58.3	74.3	68.3	60.0	57.4
Chengalpet	85.3	14.3	23.7	40.7	50.3	57.0	59.3	51.7	47.8
Trichy	64.6	20.0	28.3	33.7	55.7	62.3	52.7	46.7	45.5
Mean	74.5	23.2	30.3	37.9	54.7	61.7	59.7	50.5	
CD(P=0.05)		L	P	L x P					
		0.58	0.73	1.64					

Table 6: Standardization of pH vigour test in *A. lebbeck* – Vigour index

Seed lot (L)	Control	pH value (P)							Mean
		4	5	6	7	8	9	10	
Tirunelveli	2146	344	539	679	1115	1393	1801	1445	1183
Aliyar	1597	494	723	987	1835	1768	1885	969	1277
Coimbatore	2079	546	706	939	1423	1949	2002	1257	1362
Chengalpet	2207	194	401	746	1258	1594	1633	1036	1378
Trichy	1679	335	501	671	1259	1585	1432	985	1056
Mean	1942	383	574	804	1378	1056	1742	1138	
CD(P=0.05)		L	P	L x P					
		23.2	29.3	65.6					

Studies by Venkatesh (2000) on the response of provenance to pH revealed that Tirupathur (North east zone) recorded the lowest percentage of reduction in germination and vigour at pH 12, over neutral pH 7. The most affected source was Dharmapuri. Mani (2000) found that the germination of neem was gradually decreased form pH 4 (79 per cent) to pH 8 (53 per cent). Thereafter, a sharp decline was noticed. The germination was only 3% with pH 10 and nil at pH 11 (Table 7). Among the different sources, the high rainfall zone sources recorded the lowest percentage of reduction in germination even at pH of 10 (Plate 2a to i) over the neutral pH (pH 7). In pH 11 and 12, similar trend in germination and vigour was observed for the seed lots. Therefore, it is recommended that a pH of 10 is suitable to test the vigour as the neem seeds for alkalinity tolerance.

Plate 2: Standardization of pH test for neem

Table 7: Standardization of pH vigour test in *Azadirachta indica* - Germination (%)

Seed lot (L)	pH value (P)								
	4	5	6	7	8	9	10	11	12
North Eastern Zone	76	74	70	68	57	36	12	0	0
North Western Zone	75	70	67	62	48	37	15	0	0
Western Zone	82	79	76	73	54	46	11	0	0
Cauvery delta Zone	77	71	69	61	49	30	4	0	0
Southern Zone	84	82	78	70	56	42	8	0	0
High rainfall Zone	82	78	73	71	52	31	19	0	0
Mean	79.33	75.6	72.2	67.5	52.66	37.0	11.5	0	0
	CD(P=0.05)		L		P		L X P		
			2.254		2.760		NS		

6.3. PEG test

Osmotic priming with PEG is known to reduce germination time, synchronize germination, and improve the percentage germination and seedling stand (Guedas *et al.*, 1981). Umarani *et al.* (1997) reported that soaking in 0.25 per cent PEG for one day was advocated to maximize viability and vigour in neem. According to Edgar Piedrahita (1997), osmoconditioning of *Pinus patula* seeds with PEG 6000 at -0.6, -1.0 and -1.6 MPa resulted in reducing the time of initial germination, increasing the vigour of the seeds and synchronizing germination time. Vanangamudi *et al.* (1998) advocated that soaking of neem seeds in 2.5 per cent PEG for duration of one day for maximizing germination (Fig. 2). Evaluation of seedling quality in neem through osmotic priming with PEG showed that the seeds primed in 0.25% PEG 6000 at 15 $\pm$ 0.5°C with 1000 lux illumination for 1 hr was the best in improving germination (Fig. 3) and the seed and osmotic solution should be 1:1 (Bharathi, 1999).

According to Natarajan (1999), for neem the PEG concentration ranging from 0.25 per cent to 1.25 per cent was evaluated for their effect on the germination. Up to 1.0% concentration, there was an increase in germination. Only in 1.25% concentration, there was a decrease in germination (Table 8) however, the rate of decrease varied with the seed lots. Seed lots from Coimbatore recorded the lowest decrease in germination, while Chengalpet recorded the highest. It is hence, conceivable that seed vigour in the drought resistance point of view can be evaluated by subjecting the seeds to 1.25% concentration of PEG.

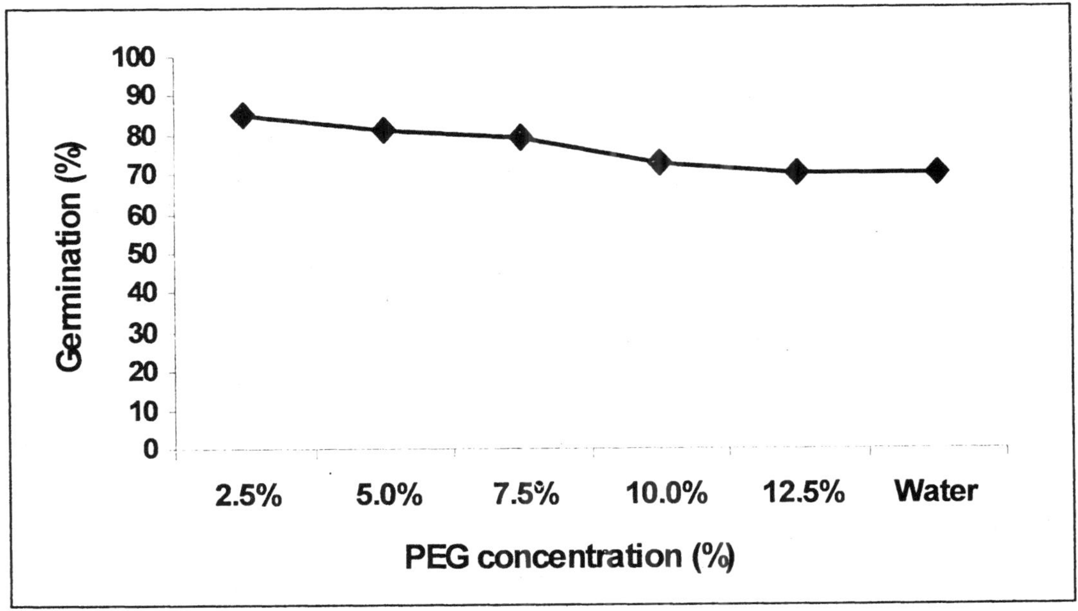

Fig.2: Standardization of PEG test for neem

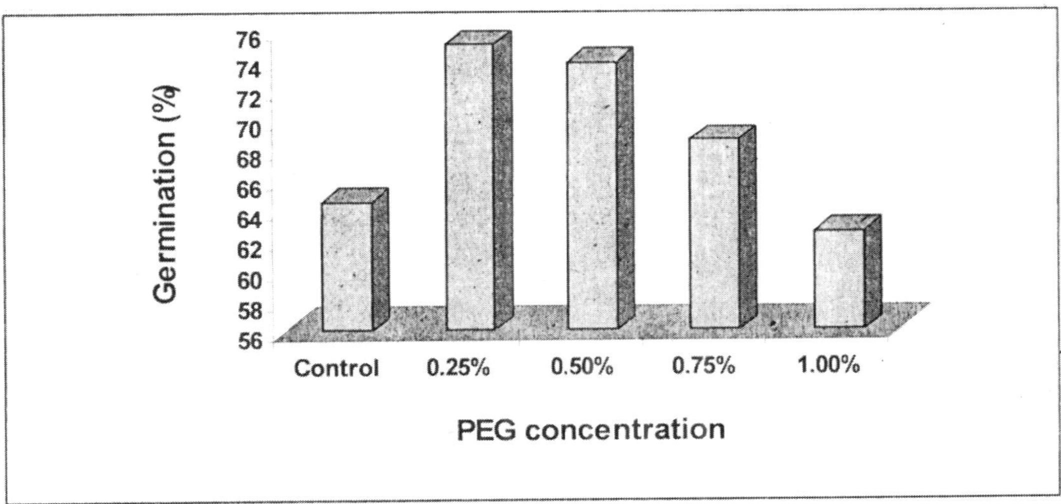

Fig. 3: Influence of PEG test on germination of neem

Table 8: Standardization of Poly Ethylene Glycol (PEG) vigour test in *A. lebbeck* – Germination (%)

Seed lot (L)	Control	PEG concentrations (C)					Mean
		0.25%	0.5%	0.75%	1.0%	1.25%	
Tirunelveli	75.0	72.5	81.0	84.0	74.0	69.5	76.0
Aliyar	71.0	66.0	71.5	77.5	68.5	64.5	69.8
Coimbatore	84.0	80.5	86.0	89.0	84.0	80.0	83.9
Chengalpet	78.5	83.0	88.5	79.5	73.5	68.5	78.6
Trichy	68.5	73.5	79.5	83.0	82.0	77.0	77.3
Mean	75.4	75.1	81.3	82.6	76.4	71.9	
CD(P=0.05)		L	C	L x C			
		0.97	1.06	2.37			

Venkatesh (2000) observed that *Acacia nilotica* seeds performed well, if they were soaked in PEG solution for 24 hr at -15 bars concentration. In a study, the PEG at different concentrations *viz.*, -5, -10 and -15 bars were evaluated for their effect on the germination and vigour in neem (Mani, 2000). The results exhibited that in -5 and -10 bars, there was an increase in germination. But in -15 bars, there was a decrease in germination (Fig. 4). However, the rate of decrease varied with seed lots. Cauvery delta zone recorded the lowest percentage reduction (20.8) in germination, while North Eastern Zone recorded the highest (28.9). It is hence, conceivable that neem seed vigour in the drought resistance point of view could be evaluated by subjecting the seeds to PEG at -15 bars.

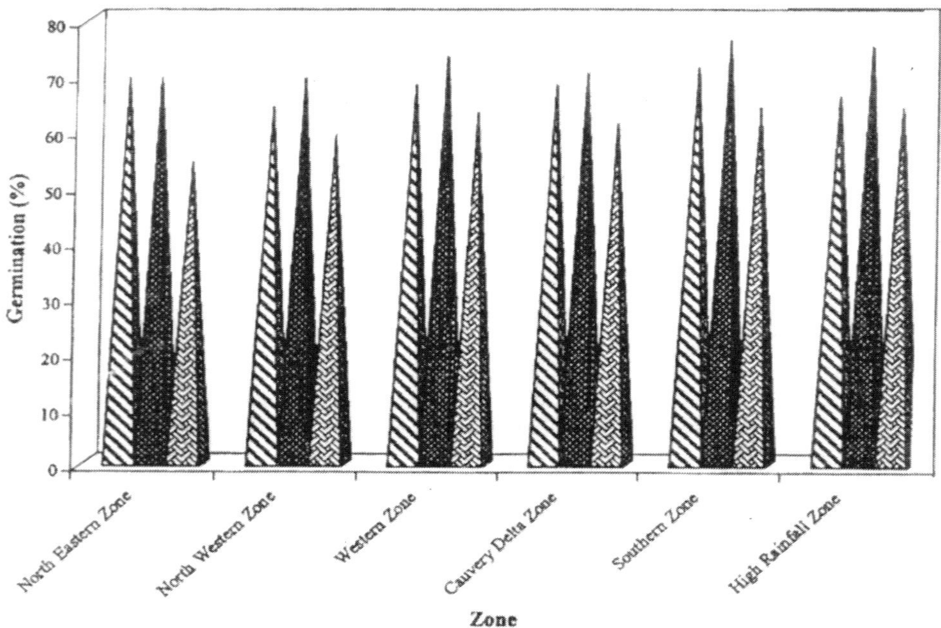

Fig. 4: Influence of PEG test on germination of *Acacia nilotica*

6.4. Accelerated ageing test

Scarified seeds of *Acacia auriculiformis, Acacia nilotica* and *Leucaena leucocephala* subjected to short term accelerated ageing at 100% RH and 40°C temperature showed loss of vigour and viability, compared to control (Sur *et al.,* 1987). Vanangamudi *et al.* (1998) subjected the sees of *Pinus roxburghii* to different artificial ageing at 41°C and 100% RH for different periods. The germination of the seeds after ageing varied from 15 to 69 per cent. The teak drupes aged for 15 days registered higher germination (53.2%) than non-aged seeds (21.4 per cent) and it recorded more seedling vigour (Manonmani, 1997). Neem and jamun seeds collected from Coimbatore and Mettupalayam when subjected to ageing test behaved differently for viability and seedling vigour. Germination of jamun dropped to less than 50 per cent on 4th day after ageing, whereas it took 6 days for neem (Vanangamudi *et al.,* 1997). According to Natarajan (1999), the accelerated ageing for 8 days is recommended for estimating the vigour status of seed lot of *Albizia.* Maximum reduction in germination percentage was obtained for all seed lots after 10 days of accelerated ageing (Table 9). A similar result was obtained after 8 days of accelerated ageing for eight out of 12 seed lots viz., Tirunelveli, Ramanathapuram, Aruppukottai, Annamalai WLS, Aliyar, Salem, Chengalpet and Trichy. However, after 6 days of accelerated ageing, results of only six seed lots matched with the 8 days and 10 days of accelerated ageing. No such similarity was observed in the 2 and 4 days of accelerated ageing. Therefore, an accelerated ageing test for 8 days is required to test the vigour of the *Albizia* seed lot.

Table 9: Standardization of accelerated vigour test for *Albizia lebbeck* - Germination (%)

Seed lot (L)	0 DAA	2 DAA	4 DAA	6 DAA	8 DAA	10 DAA	Mean
Tirunelveli	93	88	81	74	61	48	74
Ramanathapuram	89	84	76	71	66	61	74
Madurai	86	77	67	59	56	26	62
Arupukottai	88	79	61	50	31	16	54
Anamalai WLS	90	84	78	69	64	50	72
Aliyar	79	69	60	57	42	17	53
Coimbatore	83	79	67	58	50	39	62
Mettupalayam	93	82	78	67	58	49	71
Erode	86	80	70	62	55	25	63
Salem	69	60	53	39	17	6	41
Chinglepet	67	61	56	43	27	12	44
Trichy	61	57	50	32	21	6	38
Mean	82	74	66	56	46	29	59
CD (P= 0.05)		L	A	LxA			
		1.300	0.919	3.18			

A study to reveal the possibility of utilizing different vigour tests for differentiating the high and low vigour seed lots of neem by Bharathi (1999) exhibited that the highest positive direct effect by tetrazolium chloride test and paper pierceing test against field emergence percentage (Table 10). The germination test and brick grit test had low negative direct effects. Accelerated ageing test also had very low negative indirect effects. Hence, it is concluded that the tetrazolium chloride test and paper piercing test can be very well employed as vigour tests for neem to differentiate low and high vigour seed lots.

Table 10: Standardization of accelerated vigour test for neem

Seed lot (L)	Standard germination test	Accelerated ageing test	Paper piercing test	Tetrazolium chloride test
Ambasamudram	75	34	64	76
Kovilpatti	80	36	65	83
Theni	74	35	62	75
Salem	84	39	69	83
Dharmapuri	87	43	72	88
Coimbatore	85	41	71	85
Mettupalayam	81	35	67	82
Bhavanisagar	85	42	70	86
Vridhachalam	84	38	68	84
Tricy and Vamban	78	37	65	79
Mean	81.3	38.1	67.1	82.2
CD (P= 0.05)	3.668	2.301	2.816	3.154

Table 11: Standardization of accelerated ageing vigour test in neem - Germination (%)

Days after ageing (D)	North Eastern Zone	North Western Zone	Western Zone	Cauvery Delta Zone	Southern Zone	High Rainfall Zone	Mean
0 DAA	76	75	82	77	89	82	79
1 DAA	64	69	78	67	78	75	71
2 DAA	52	49	55	50	61	58	54
3 DAA	49	46	53	49	55	52	50
4 DAA	42	42	48	47	49	47	45
5 DAA	36	38	43	41	44	41	40
6 DAA	35	37	40	38	40	39	38
7 DAA	26	24	26	29	27	22	25
8 DAA	18	17	22	20	13	15	17
9 DAA	11	10	16	14	8	9	11
10 DAA	2	1	7	5	4	1	30
11 DAA	0	0	0	0	0	0	0
Mean	37	37	42	51	42	40	
	CD (P=0.05)		L		D		L x D
			2.869		3.885		9.518

DAA – Days after Accelerated Ageing

Accelerated ageing of seeds of *Dalbergia sissoo* was effective in a vigour test only at temperatures in excess of 43°C for 72 hr. Venkatesh (2000) recommended that accelerated ageing of *Acacia nilotica* seeds for 10 days at 40% and 98+2% RH was optimum to test the vigour of seed lots. In accelerated ageing test of neem (Mani, 2000), significant reduction in germination was evident on the third day of ageing and a reduction below 50 per cent was noticeable on six day after ageing and a drastic reduction (below 10 per cent) in germination percentage was obtained for all seed lots after 10 days of accelerated ageing and it dropped to zero on 11th day. A similar result was obtained after nine days of accelerated ageing for all the six seed lots Table 11). Therefore, accelerated ageing test for 9 days is employed to test the vigour of neem seed lots.

7. VIABILITY TEST

The tetrazolium test is useful for quick testing of seed viability (Lakon, 1949). Tetrazolium staining is a dependable and accepted biochemical method for testing the inherent germination of a seed lot (ISTA, 1993). This is especially important in forest seeds which take much longer time to germinate or may exhibit varying types of dormancy.

Table 12: Standardization of tetrazolium test for neem seed viability (%)

Tetrazolium concentration and duration	Water soaking (W)		
	24h	12 h	Mean
0.1% TTZ @ 40°C for 1 hr	91.5	82.5	87.0
0.2% TTZ @ 40°C for 1 hr	93.0	92.5	92.8
0.1% TTZ @ ambient for 1 hr	2.0	2.0	2.0
0.2% TTZ @40°C ambient for 1 hr	2.5	3.0	2.8
0.1% TTZ @40°C for 1/2 hr	25.5	6.5	16.0
0.2% TTZ @40°C for 1/2 hr	27.0	35.5	31.3
0.1% TTZ @ 40°C and ambient for 1/2 hr	0.0	0	0
0.2% TTZ @ 40°C and ambient for 1/2 hr	2.0	2.5	2.3
CD (P=0.05)	T	W	T x W
	1.072	0.536	1.517

Bharathi (1999) proved that for quick viability test of neem, the endocarp should be removed very carefully using fine scissors without any damage to the kernel and they should be preconditioned in cold water for 24 hr. for easy removal of the kernel coat. The preconditioned kernels should be soaked in 0.2% tetrazolium solution with pH range of 6.5 - 7.5 (Fig. 5) and kept in dark (hot air oven) maintained at 40°C for 1 hr (Table 12) for proper staining to facilitate effective separation of viable and non viable seeds (Plate 3).

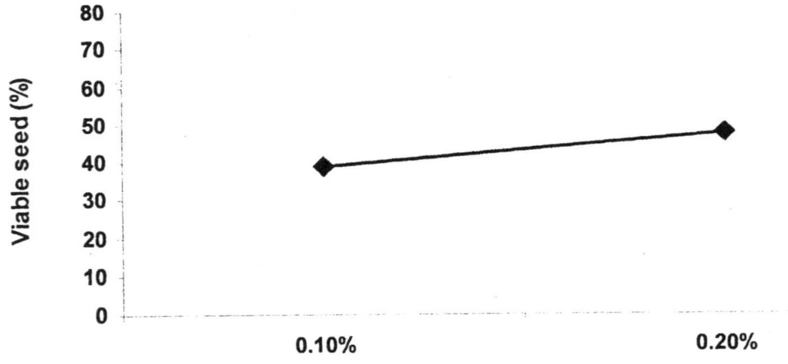

Fig.5: Effect of tetrazolium solution concentration on viable seed of neem

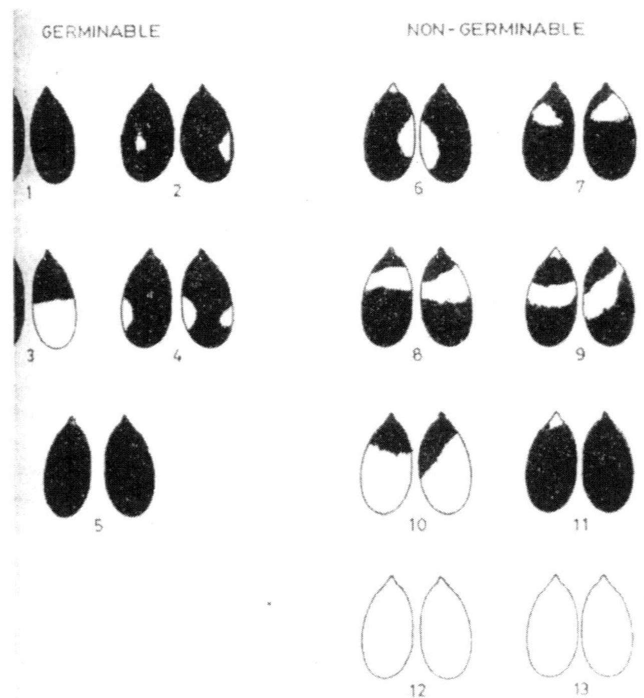

Plate 3: Viability evaluation of neem

1. Seed completely stained
2-4 Minor unstained areas on cotyledons
5. Extreme tip of embryo unstained
6. More than tip of embryo unstained, minor spots on cotyledon
7. Unstained area at juncture of cotyledons and embryonic axis

8. Embryonic axis bisected from cotyledons by unstained area
9. Unstained areas on embryonic axis and at point of attachment of cotyledons to axis
10. More than one half cotyledons node unstained
11. Embryonic axis completely unstained
12. Seeds stained off-colour, greyish red,(or) glassy (or)transparent
13. Seed completely unstained

According Natarajan (1999), for conducting quick viability (tetrazolium) test of *Albizia lebbeck* the seeds should be preconditioned by nipping off the seed coat followed by 24hr. water soaking. The best suited preparation method was longitudinal splitting of cotyledons (Plate 4). A concentration of one percent tetrazolium solution was found to be optimum for proper staining to facilitate effective separation of viable and non viable seeds (Fig.6). For the viability test of *Acacia nilotica* subsp. *indica* benth, the seeds should be preconditioned by acid scarification for 60 min followed by 12 hr water soaking. According to Venkatesh (2000), the best suited preparation method was longitudinal splitting of the cotyledons. A concentration of 0.75% tetrazolium solution with pH range of 6.5 - 7.5 and kept in hot air oven maintained at 35 ± 2°C 2 hr. was optimum for proper staining for effective separation of viable and non viable seeds.

Plate 4: Quick viability test – preparation methods for *Albizia lebbeck*

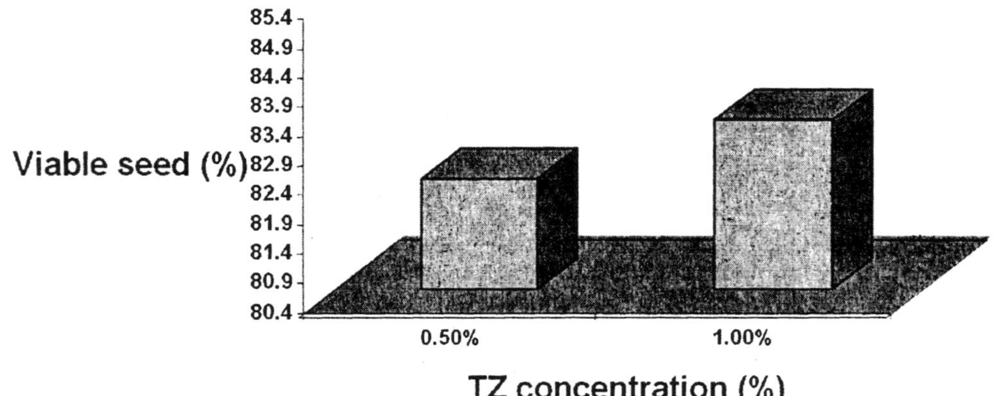

Fig.6: Effect of tetrazolium solution concentration on viable seed of *Albizia lebbeck*

Results of three seed lots of *Lagerstroemia parviflora* bisected longitudinally and stained with 0.1% triphenyl tetrazolium chloride for 5 hr were in agreement with germination tests (Babeley and Kandya, 1986). *Chloroxylon sweitinia* seeds cut and soaked in 0.1% tetrazolium solution were used to differenciate the viable and non viable seeds (Yadav *et al.,* 1986).

According to Karivaratharaju *et al.* (1987), the seeds of *Dendrocalamus strictus* with longitudinal splitting stained in 0.1 per cent tetrazolium chloride for six hours showed several abnormalities of embryo. In *Acacia nilotica,* chipping of the seed coat and soaking in distilled water for 24 hr followed by soaking in 1% tetrazolium chloride and incubated at 30°C in dark for 24 hr proved to be the best staining method (Neelugeera *et al.,* 1998). Presoaking of the seeds for 4 hr followed by removal of the seed coat without damaging the embryo or splitting of cotyledon with the retention of embryo in one of the cotyledons and soaking of the above preconditioned seed in 1% tetrazolium solution for 5 hr at 40°C was ideal for perfect staining of *Pongamia pinnata* (Jerlin, 1998). The staining of *Casuarina equisetifolia* seeds with 0.25, 0.50, 0.75 and 1.00% TZ solutions revealed that the intensity of staining increased as the concentration of solution increased. The intensity was very dark with 1.00 per cent, while 0.75 per cent resulted in optimum staining of the seeds (Umarani, 1999).

REFERENCES

Babeley, G.S. and A.K. Kandya. 1986. Use of TTC for rapid testing of the viability of *Lagerstroemia parviflora* seeds. *J. Trop. For.,* **2**(3): 226-231.

Bangash, S.H. 1977. Salt tolerance of forest trees species as determined by germination of seeds at different salinity. *Pak. J. For.,* **27**(2): 93 - 97

Bharathi, A. 1999. Studies on handling management and storage of neem seed (*Azadirachta indica* A. Juss). Ph.D.,thesis, Tamil Nadu Agricultural University, Coimbatore - 3.

Edgar Piedrahita 1997. Effect of osmotic priming with PEG 6000, KNO_3, NaCl and mannitol on germination of *Pinus patula* seeds. IUFRO symposium on "Innovations in Forest Tree Seed Science and Nursery Technology", November 22-25, Pt. Ravishankar Shukla University, Raipur, India. p. 58.

Guedas, A.C., D.J. Cautliffe and T.A. Nell. 1981. Morphological changes during lettuce seed priming and subsequent radicle development. *J. Am. Soc. Hort. Sci.,* **106**: 121-126.

Gupta, T.N. and P.G. Pattanath. 1975. Factors affecting germination behaviour of teak seeds of eighteen Indian origins. *Indian For.,* **10**(10): 584-588.

Hayward, H.E. and C.H. Wadleigh. 1949. Plant growth on saline and alkaline soils. *Adv.Agron.,* **1**: 1 - 38.

Heydecker, W. 1972. Vigour In: Viability of Seeds, (Ed.) E.H. Roberts. Chapman Hall. London, pp. 209-252.

ISTA. 1993. International rules for seed testing. *Seed Sci. & Technol.,* **21**: 1-288.

Jerlin, 1998. Seed technological studies in Pungam (*Pongamia pinnata* L. Pierre). Ph.D. Thesis, Tamil Nadu Agricultural University, Coimbatore – 641 003.

Justice, O.L. 1972. Essentials of seed testing. In : Seed biology. Vol. 3 (ed. T.T. Kozlowski). Academic Press, New York and London. p. 301

Karivaratharaju, T.V., N. Natarajan and R.S.V. Rai. 1987. Studies on the seed of *Dendrocalamus strictus. Van Vig.,* **25**(1-2): 29-31.

Kathiresan, K., P. Moorthy and S. Ravikumar. 1996. A note on the influence of salinity and pH on rooting of *Rhizophora mucronata* Lamk. seedlings. *Indian For.,* **122**(8): 763-764.

Lakon, g. 1949. Topographischer Nachwies der akeimfahigkeit der Gettreidefruchte durch Tetrazolium-Salze. *Ber Deutsch. Bot. Ges.,* **60**: 299-305.

Mani, G. (2000). Studies on seed source variation, standardization of seed testing procedures and nursery technology in neem (*Azadirachta indica* A. Juss). Ph.D. Thesis, Tamil Nadu Agricultural University, Coimbatore – 3.

Manonmani, Y. 1997. Investigations on seed collection, storage and germination of teak. Ph.D., thesis , Tamil Nadu Agricultural University, Coimbatore -3.

Misra, L.M and S.L. Singh. 1981. Seed germination studies on three prominent tree species of southern Uttar Pradesh. *Ann. Arid zone,* **20** (3): 193 – 198

Natarajan, S. 1999. Investigations on seed source variations, standardization of seed testing procedure and nursery techniques in *Albizia lebbeck* L. Benth, Ph.D., thesis , Tamil Nadu Agricultural University, Coimbatore -3.

Neelu geera , Mohitgera and Mamta Purohit.1998. Tetrazolium test for the seeds of *Acacia nilotica* Wild.ex. Del. *Indian For.,* **124**(12): 1039 - 1043

Roy, M.M. 1990. Effect of pH on germination of some fodder *Albizia. J. Tree Sci.,* **9**(!): 47 - 49

Sharma, A., M.U. Kukkadia, D.B. Jadeja and B.G. Vashi. 1998. Effect of different salinity levels on germination and initial growth parameters of different afforestation tree species in nursery stage. *Indian. J. For.,* **20**(2): 156-159.

Sur, K., A.K. Lahiri and R.N. Basu.1987. Improvement of germinability of some forest tree seeds by acid scarification and hydration – dehydration treatments. *Indian Agriculturist,* 31(2): 115 – 122.

Tomer,O.S. and J.S.P. Yadav, 1980. Effect of saline irrigation water of varying EC, SAR and RSC levels on germiantiona nd seedling growth of some forest species. *Indian J. For.,* **3**(4): 306-314.

Umarani,R. 1999. Studies on the physiological and biochemical basis of seed germination and deterioration in *Casuarina equisetifolia.* Ph.D. Thesis, Tamil Nadu Agricultural University, Coimbatore – 3.

Umarani, R., K.Vanangamudi, Mallika Vanangamudi and R.S. Vinaya Rai, 1997. Effect of osmostic priming on germination and vigour of neem seed. In: IUFRO. Symposium on "Innovations in Forest Tree Seed Science and Nursery Technology". November 22 – 25. Pt. Ravishankar Shukla University, Raipur, India. P. 156

Vanangamudi, K., R. Umarani, A. Bharathi and A. Venkatesh. 1998. Effect of seed source on physical and physiological qualities of *Acacia nilotica* seeds. *Seed Res.,* **26**(2): 114-116.

Vekatesh, A. 2000. studies on seed source variations, standardization of seed testing procedure and nursery techniques in *Acacia nilotica* (L) Wild. Sub. spp *indica* Benth. Ph.D.,thesis, TamilNadu Agricultural University, Coimbatore -3.

Yadav, V.K., P.K. Khare and G.P. Mishra. 1986. Quick test for viability and vigour of *Chloroxylon sweitinia. J.Trop., For.,* **2**(3): 213 – 216.